Secure and Sustainable Energy System

Secure and Sustainable Energy System

Editors

Farhad Taghizadeh-Hesary
Han Phoumin

MDPI • Basel • Beijing • Wuhan • Barcelona • Belgrade • Manchester • Tokyo • Cluj • Tianjin

Editors
Farhad Taghizadeh-Hesary
Tokai University
Japan

Han Phoumin
Economic Research Institute for ASEAN and East Asia (ERIA)
Indonesia

Editorial Office
MDPI
St. Alban-Anlage 66
4052 Basel, Switzerland

This is a reprint of articles from the Special Issue published online in the open access journal *Sustainability* (ISSN 2071-1050) (available at: https://www.mdpi.com/journal/sustainability/topical_collections/Secure_and_Sustainable_Energy).

For citation purposes, cite each article independently as indicated on the article page online and as indicated below:

LastName, A.A.; LastName, B.B.; LastName, C.C. Article Title. *Journal Name* **Year**, *Volume Number*, Page Range.

ISBN 978-3-0365-5713-7 (Hbk)
ISBN 978-3-0365-5714-4 (PDF)

Contents

About the Editors

Farhad Taghizadeh-Hesary

Farhad Taghizadeh-Hesary is an Associate Professor of Economics at Tokai University in Japan. In addition, he is a vice president and co-founder of the International Society for Energy Transition Studies (ISETS). He is a recipient of the Excellent Young Researcher status from the Ministry of Education of Japan (MEXT). He is also a visiting professor at Keio University (Japan), the Technology Studies Institute (Iran), Chiang Mai University (Thailand), and a distinguished research fellow at the University of Economics Ho Chi Minh City (Vietnam). Currently, he serves as the editor-in-chief of the Journal of Environmental Assessment Policy and Management and as an associate editor/board member of several scholarly journals. He has guest-edited special issues for several journals, including Energy Economics, Energy Policy, Resources Policy, and Finance Research Letters. His research credits include authoring more than 250 academic journal papers and book chapters and editing 17 books published by Springer Nature, Routledge, World Scientific, and Asian Development Bank Institute. He was nominated as a top global scholar in green finance based on a recent journal paper published in Renewable Energy (Elsevier) in 2022. In addition, In 2022, he was listed as of the world's top scientists in economics and energy fields based on the Stanford-Elsevier ranking. He holds a Ph.D. in economics from Keio University with a scholarship from the government of Japan (MEXT).

Han Phoumin

Han Phoumin has more than 20 years of professional experience working at various international and inter-governmental organisations and multi-disciplinary research consortiums related to energy market and technologies, environment, integrated water resource management, governance, and economic development in the region of ASEAN and East Asia. He specialised in economic development and policy and applied econometrics. Much of his career in the past 12 years revolved around power sectors, especially with sustainable hydropower development, renewable energy policy research (i.e., biomass power generation competitiveness studies, solar and wind), energy efficiency and conservation, clean coal technology, energy security, and energy demand and supply forecasting. He has been serving as an expert for APEC on energy security for oil and gas emergency responses since 2013 until now. He is also one of the peer review experts for Peru's energy subsidy removal commissioned by USAID in 2015. He also has led a number of projects in the EAS region related to energy policy and planning, and also contributed articles and special issues for academic journals. In 2022, he joined Elsevier as a member of the International Advisory Board for Energy Policy.

Preface to "Secure and Sustainable Energy System"

Alarming reports from the Intergovernmental Panel on Climate Change (IPCC) have shown that climate change urgently needs to be addressed, and in 2015, United Nations (UN) members agreed to keep the global temperature increase below 2°C through their Nationally Determined Contribution. The UN also added 'Climate Action' to their Sustainable Development Goals (SDGs). However, IPCC and the UN Environment Programme (UNEP) reports highlight that further actions must be taken to fulfil these SDGs. Increasing the share of sustainable energy resources in the energy baskets would not only reduce Green House Gas (GHG) emissions—in line with the SDGs and the Paris Agreement—but would also increase energy security.

Several developed and developing economies are still adhering to pro-fossil fuel energy policies. The extra GHG generated by new coal-fired power plants could more than wipe out any reductions in emissions made by other nations. One of the most significant barriers to developing a sustainable energy system is the low level of investment. The lack of long-term financing, the low rate of return, the existence of various risks, and the lack of capacity of market players are major challenges to developing sustainable energy systems.

With this background, this Special Issue aims to contribute to the climate actions which called for the need to address GHG Emissions, keeping global warming to well below 2°C through various means, including accelerating renewables, clean fuels, and clean technologies into the entire energy system. As long as fossil fuels (coal, gas and oil) are still used in the foreseeable future, it is vital to ensure that they are used cleanly through abated technologies. Financing clean and energy transition technologies is vital to ensure the smooth transition towards net zero emission by 2050. This Special Issue collected 17 high-quality empirical studies that assess the challenges of developing secure and sustainable energy systems and provide practical policy recommendations. The editors wish to thank the Economic Research Institute for ASEAN and East Asia (ERIA) for funding several papers that were published in this Special Issue.

Farhad Taghizadeh-Hesary and Han Phoumin
Editors

sustainability

MDPI

Article

Multi-Criteria Decision-Making System for Wind Farm Site-Selection Using Geographic Information System (GIS): Case Study of Semnan Province, Iran

Hossein Yousefi *, Saheb Ghanbari Motlagh and Mohammad Montazeri

Renewable Energies and Environmental Department, Faculty of New Sciences and Technologies, University of Tehran, Tehran 14395-1561, Iran; saheb.ghanbari@ut.ac.ir (S.G.M.); montazery.mohamad@ut.ac.ir (M.M.)
* Correspondence: hosseinyousefi@ut.ac.ir

Abstract: Selecting the best place for constructing a renewable power plant is a vital issue that can be considered a site-selection problem. Various factors are involved in selecting the best location for a renewable power plant. Therefore, it categorizes as a multi-criteria decision-making (MCDM) problem. In this study, the site selection of a wind power plant is investigated in a central province of Iran, Semnan. The main criteria for classifying various parts of the province were selected and pairwise compared using experts' opinions in this field. Furthermore, multiple restrictions were applied according to local and constitutional rules and regulations. The Analytic Hierarchy Process (AHP) was used to weigh the criteria, and according to obtained weights, wind speed, and slope were the essential criteria. Moreover, a geographic information system (GIS) is used to apply the weighted criteria and restrictions. The province's area is classified into nine classes according to the results. Based on the restrictions, 36.2% of the total area was unsuitable, mainly located in the north part of the province. Furthermore, 2.68% (2618 km^2) and 4.98% (4857 km^2) of the total area are the ninth and eightieth classes, respectively, which are the best locations for constructing a wind farm. The results show that, although the wind speed and slope are the most essential criteria, the distance from power facilities and communication routes has an extreme impact on the initial costs and final results. The results of this study are reliable and can help to develop the wind farm industry in the central part of Iran.

Keywords: wind farm site selection; multi-criteria decision-making system; Analytic Hierarchy Process; Semnan province; ArcGIS

Citation: Yousefi, H.; Motlagh, S.G.; Montazeri, M. Multi-Criteria Decision-Making System for Wind Farm Site-Selection Using Geographic Information System (GIS): Case Study of Semnan Province, Iran. *Sustainability* **2022**, *14*, 7640. https://doi.org/10.3390/su14137640

Academic Editors: Farhad Taghizadeh–Hesary and Han Phoumin

Received: 8 May 2022
Accepted: 20 June 2022
Published: 23 June 2022

1. Introduction

In recent years, environmental problems, such as global warming, climate change, pollution, and problems with traditional fossil resources (such as increased extraction costs and non-renewability) that have been a source of human energy for many years, have caused doubt about the use of these resources and increased the tendency to employ more renewable resources [1]. Countries ratified the Paris Climate Agreement in 2015 to control this critical situation. Under the agreement, Iran voluntarily pledged to reduce greenhouse gas emissions by 4% and 8% by 2030 and 2050, respectively [2]. Using renewable energy resources, such as solar, wind, waves, and biofuels, play a crucial role in reaching this goal.

Utilizing wind energy can play an important role in enabling Iran to meet the standards set for this country in the Paris Agreement. Studies have also shown that hybridizing wind turbines with other energy sources reduces carbon emissions [3]. Applying wind energy creates employment and helps reduce CO_2 [4]. Research shows that wind energy use in Canada, Sweden, China, and Germany will increase significantly by 2025 [5]. Meanwhile, it has a more extended history than other renewable sources in Iran. It is also low-cost and attractive to investors and can reduce dependence on fossil fuels [6].

Wind energy also has adverse effects, such as noise, unpleasant visual impact (tourism industry), habitat occupation, and the extinction of some bird species [7]. Therefore, it is necessary to minimize these adverse effects according to the existing criteria for locating and constructing wind farms.

MCDM is a method to evaluate multiple conflicting criteria in decision-making problems. MCDM could be applied for various applications, some of which are mentioned in references [8–10]. In addition, it is a suitable solution to deal with different and sometimes contradictory criteria for choosing the best places to use renewable resources [11]. There are several methods for weighing criteria in MCDM. Some of these methods include AHP [12], analytic network process (ANP) [13], fuzzy measures [14], Entropy [15], Swara [16], Dematel [17], Standard deviation [18], etc.

The ANP is a generalization of AHP. ANP is used to solve more complex decision-making problems, which AHP is not suitable for solving [19]. According to some scientists, the use of an exact number to compare the alternatives, unbalanced scale of judgment, inability to manage uncertainty, and inaccuracy in pairwise comparisons have caused some doubts about AHP [20,21]. Nevertheless, it is no secret that AHP is one of the most essential and widely used methods in MCDM. In some papers, due to interval judgment instead of fixed judgment [22], the Fuzzy AHP method is used to weigh the criteria. In 2021, Nguyen et al. used a hybrid fuzzy AHP MCDM to organize the priorities of the medical community and government during the COVID-19 crisis in Vietnam [23]. In another study by Nguyen, fuzzy AHP and machine learning approaches were used to predict vaccination intention against COVID-19 [24]. Entropy is another weighing method that has been widely used in recent years. The main difference between Entropy and other methods is to remove the human factor from the decision-making process [25], which enhances this method's accuracy. In 2015, Zhao and Gou developed a hybrid MCDM system based on Entropy to evaluate China's economic, social, and environmental benefits of the renewable energy sector [26]. In other studies, MCDM systems based on Entropy were used to investigate the sustainable development factors worldwide [27–30]. There are two general views on the Entropy method. According to some literature, Entropy is reliable and effective [31]. However, from the other point of view, Entropy results do not always consider the importance of the indexes [32]. The Swara method is similar to AHP in that the expert's opinion specifies the importance and prioritization of the alternatives. In the end, the weight of the attributes calculates by considering two main features. According to this method, all the attributes are compensatory and independent [16]. The Dematel is similar to Swara, except that Dematel is used to solve very complex subjects. In the decision-making process of Dematel, the expert's opinion uses to develop the pairwise comparison matric, and it has three main features. The attributes are compensatory and independent from each other. The qualitative attributes convert to quantitative attributes [16]. Moreover, the Swara and Dematel methods have been extensively used in MCDM problems, especially in the renewable energy sector [33–35]. In this study, the AHP method was used to solve the site-selection problems due to following reasons:

1. It is widely used because it is easy to understand and apply.
2. It is very compatible with GIS which is extensively used for land analysis and site-selection problems.
3. Possibility of hierarchical modeling, adoption with verbal judgments, and consistency verification [36].
4. AHP can be combined with other methods, including mathematical programming, fuzzy sets, genetic algorithms, neural networks, etc. [36].
5. It considers both quantitative and qualitative criteria to interpret the problem [37].
6. AHP can apply various sensitivity analyses to criteria [38].
7. AHP facilitates the decision-making process, using the pairwise comparison among the criteria [38].
8. AHP can consider the consistency and inconsistency of the alternatives, which is one of the essential benefits of this method [38].

9. In site-selection problems, where the main goal is to select the best places, simple methods such as AHP are sufficient, and more complicated methods such as fuzzy AHP do not necessarily lead to different results [39].

In this study, GIS is employed to create and use map layers. GIS is a powerful tool for MCDM, and it can utilize topological, structural, and ecological information to perform calculations based on criteria and sub-criteria. Topological, structural, and ecological information can be displayed as layers in GIS, and by overlapping these layers, criteria, and restrictions, the final suitable and inappropriate locations can be determined. Afterward, a suitable area can be weighed according to various criteria. These criteria are determined by the type of problem and the area. After classifying the suitable areas according to these criteria, the importance and value of each area were determined, and decision-making was carried out according to the categorized final map. Figure 1 illustrates how the GIS tool works in integrating information layers.

Figure 1. Integrating information layers using GIS [40].

2. Literature Review

GIS is widely applied in site selection issues. Xu et al. [41] used GIS, Interval Analytic Hierarchy Process (IAHP), and stochastic VIKOR to find the best area for wind farms in the Wafangdian region, China. Colak et al. [42] also employed GIS and AHP to find an optimal area for a photovoltaic farm in Malatya, Turkey. Castro-Santos et al. [43] studied Galicia coastal area in Spain for floating offshore wind farm site selection via GIS. In addition to its application in selecting optimal areas for renewable energy farms, GIS is used for rainwater harvesting [44], power plants [45], landfills [46], pressurized irrigation [47], and electric vehicle charging stations [48] site selection.

Many articles have been published related to wind farm site selection using GIS in Iran. Moradi et al. [38] measured wind energy potential in Alborz province (central regions of Iran) through MCDM and GIS. They used AHP to weigh the criteria, and according to their results, 20% of the area was suitable for wind farms. Noorollahi et al. [6] conducted the same work for Markazi province (west of Iran), and based on the results, 28% of the area was suitable for wind farms. In 2020, in a study by Ahmadi et al. [49], different parts of Iran were reviewed to build a wind-powered pump storage plant, and according to the results, the Gilane-Gharb dam was the best area and the capacity of which was estimated at 31 MW.

Table 1 represents several other studies on wind farm site selection in Iran and other countries.

Table 1. Other renewable resources site selection research in Iran and other countries.

Ref	Location	Type of Site Selection	Applied Method
[50]	Shahrood, Khorramdareh, Zabol, and Abadeh In Iran	Wind farm	TOPSIS
[51]	Izmir, Turkey	Wind farm	MCDM-(best-worst method) (BWM)
[52]	Northeast of Iran	Wind farm	Equal importance criteria
[53]	China	Offshore wind farm	MCDM-intuitionistic linguistic aggregation operators
[54]	China	Wind farm	MCDM-Fuzzy
[55]	India	Wind farm	MCDM-Fuzzy AHP
[56]	Sudan	Wind farm	MCDM-Fuzzy AHP
[57]	Mauritius	Wind farm	MCDM-AHP

One of the main practical uses of the decision-making systems is to improve risk assessment ability and overcome the adverse effects [58] of the site-selection projects. Multi-dimensional assessment and making accurate decisions are vital prerequisites to starting a business. Moreover, location selection is one of the essential parts of the business due to the long-term impacts on risks and costs of the projects [59]. Furthermore, as another practical benefit of decision-making systems, the simultaneous use of GIS and MCDM reduce the cost and time of the site-selection problems and increases the accuracy. At the same time, GIS-MCDM-based decision-making systems take multiple environmental, social, economic, and sustainability parameters to account to make the best decision among the various alternatives [60].

This research aimed to determine the suitable area for wind farms in Semnan province, Iran. Semnan has the lowest population density among the provinces of Iran and is also the seventh province in terms of area. Moreover, due to unfavorable climatic conditions, the possibility of agriculture is less in many parts of Semnan province, such as the southern areas. Therefore, there is much usable land in many parts of the province. Semnan province is also one of the central provinces of Iran, located near the capital (Tehran) and other large provinces, such as Khorasan and Isfahan. The proximity of the province to energy highways and energy consumption centers increases the importance of this strategic province for energy production. The construction of fossil power plants, such as steam, combined cycle power, and plants, seems very irrational due to high water consumption, pollution, and contradiction with the hot and dry climate of the province. For the above reasons, renewable sources, namely wind energy, seem a very reasonable and justifiable option. Criteria and sub-criteria were specified to evaluate the wind farm potential in the province. To solve this decision-making problem, AHP will be applied to weigh and compare the criteria. The main selected criteria for this study are wind speed, slope, power lines, power stations, urban areas, highways, and roads. These criteria were chosen according to similar previous studies and the opinion of the experts. Finally, areas with the most potential for the wind farm will be specified separately. In Section 3, the study area, electricity consumption, and social information of the Semnan are described. In Section 4, the AHP method is presented, and the weights of the criteria will be calculated. In Section 5, the weighted criteria and multiple restrictions will be applied. In Section 5, the final categorized map will be presented using the data of previous sections, and the results will be discussed and compared with other papers.

3. Study Area

Semnan is one of the central provinces of Iran, located in the east of Tehran province, south of the Alborz mountains, and north of Dashte-e kavir. This province is centered in Semnan city, and Shahrood, Garmsaar, and Damqaan are the other important cities of Semnan province (Figure 2). The province covers an area of 97.491 square kilometers, which is 5.9% of the country's total area. This province is the seventh province in Iran in terms of area. In the last official census in 2016, the province's population was 702,000, and the relative population density was 2.7 people per km^2. This vast province is home to less than one percent of the country's population [61].

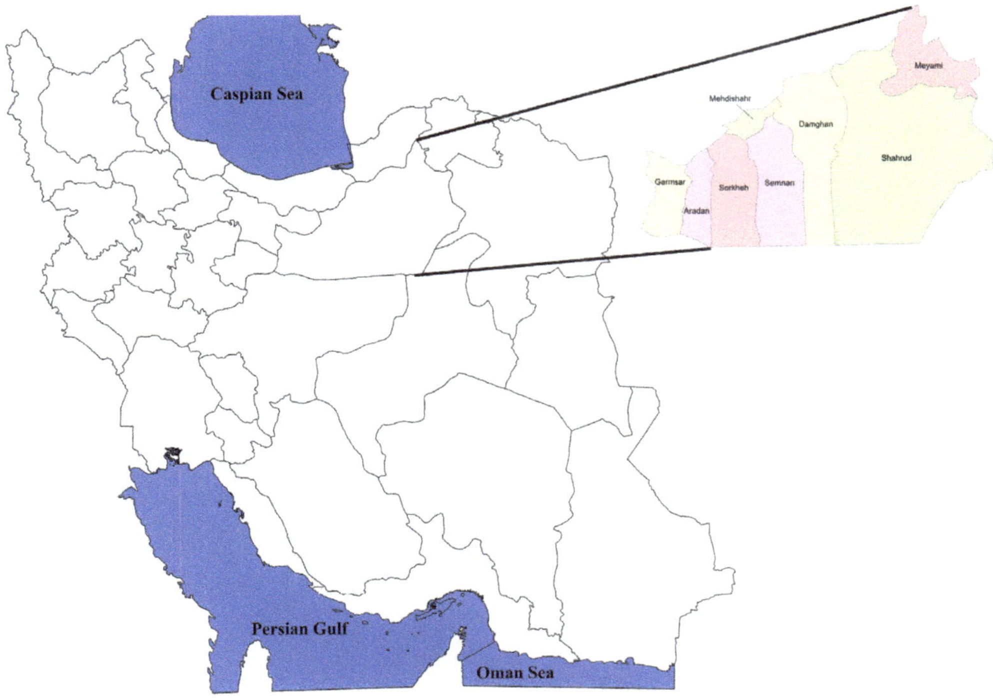

Figure 2. Location of Semnan in Iran map.

With an average daily temperature of 24 °C and a maximum temperature of 39 °C, Semnan is considered a warm province. The best months to travel to Semnan are June to September, and the worst months are November to March [62].

Iran's electricity grid has expanded considerably in recent decades, and most of the electricity generation in this grid is provided by fossil thermal power plants. According to the pattern obtained in the last decade, electricity consumption in Iran is increasing by 6% annually. Figure 3 depicts the trend of the increase in electricity production and consumption in Iran's electricity grid from 1980 to the last decade. According to this figure, the electricity network will face several problems in supplying electricity in the near future.

Semnan has two large power plants named Shahid Bakeri and Shahid Bastani, whose net production in 2018 was equal to 2,433,779 MWh. In 2018, out of 377,050 electricity subscribers in Semnan province, 76.88% were household, 13.69% commercial, 1.62% agricultural, 5.88% general, 1.32% industrial, and 0.58% street lighting. Moreover, 20.19% of the total electricity sold was for household consumption, 8.69% for general consumption, 23.28% for agricultural consumption, 40.3% for industrial consumption, 5.21% for commercial consumption, and 2.33% was allocated to street lighting. Figure 4 demonstrates the amount of electricity sold to various subscribers in 2018 [64,65].

Based on the annual consumption pattern, electricity consumption in Semnan is growing every year, and wind resources can be a key to meeting this demand.

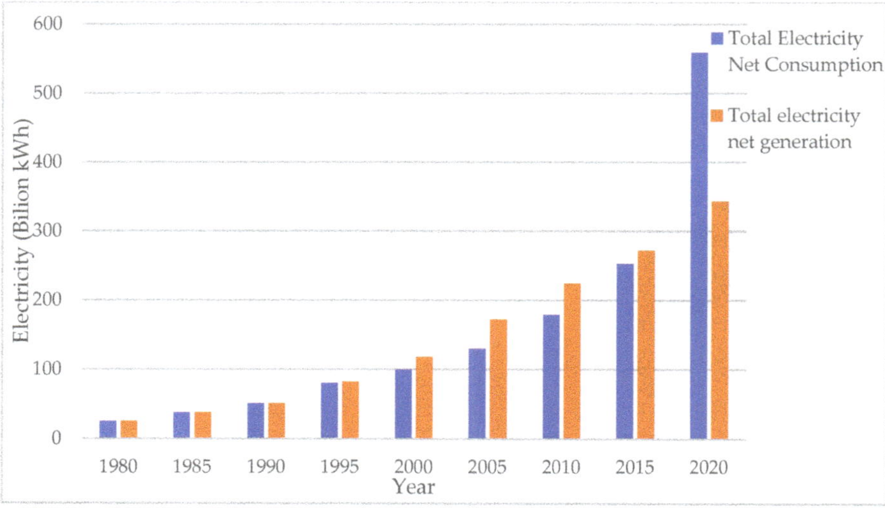

Figure 3. Electricity production and consumption growth in Iran (based on the data from [38,63]).

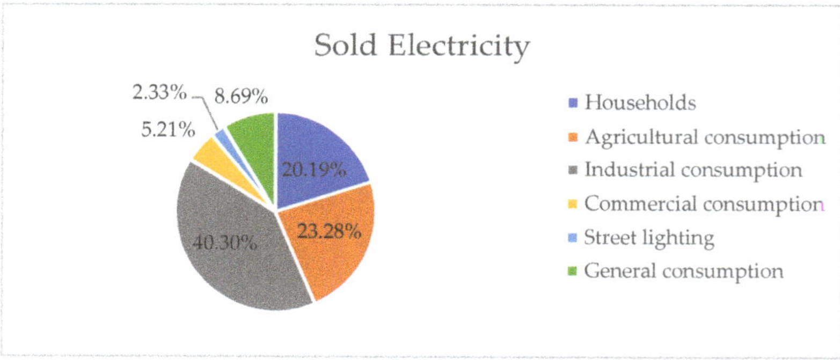

Figure 4. Electricity consumption of different groups of subscribers in Semnan [64,65].

4. Analytical Hierarchy Process

MCDM techniques are applied to solve site selection problems [66]. AHP is one of the most widely used methods in MCDM, introduced by Saaty in 1980 [12]. One of the advantages of AHP is the pairwise comparison among the criteria of the problem. Through this method, sensitivity analysis can be carried out on the criteria and sub-criteria by offering several choices. AHP can minimize the impact of taste decisions and orientations in problem-solving, which is another important advantage of this method [38].

In the AHP method, the problem becomes hierarchical, consisting of four levels. These levels are the problem goal, the criteria, the sub-criteria, and the final choices [67,68].

In the AHP method, the criteria are compared two by two. To compare the criteria, they are given points ranging from one to nine, which can be seen in Table 2 [12,69,70].

Thus, the adjustment ratio (CR), indicating the degree of coherence of decision makers' opinions, is calculated according to Equation (2). The appropriate value for CR is below 0.1, and if it exceeds this value, decision-makers should reconsider their views in pairwise comparison. If the pairwise comparison does not involve inconsistencies, the principal eigenvalue (λ_{max}) is at least the same as the number of columns or rows ($\lambda_{max} = n$) [38].

Table 2. Pairwise comparison scales in AHP [69].

Scale	Numerical Rating	Reciprocal
Extreme importance	9	1/9
Very to extremely strong importance	8	1/8
Very strong importance	7	1/7
Strong to very strong importance	6	1/6
Strong importance	5	1/5
Moderate to strong importance	4	1/4
Moderate importance	3	1/3
Equal to moderate importance	2	1/2
Equal importance	1	1

Consistency index (*CI*) could be calculated from Equation (1) [12]:

$$CI = \frac{\lambda_{max} - n}{n - 1} \tag{1}$$

The consistency ratio (*CR*) is calculated from Equation (2):

$$CR = \frac{CI}{RI} \tag{2}$$

Table 3 represents the Random Index (*RI*) values used to calculate *CR*. As mentioned, the *CR* value should be less than 0.1. Otherwise, the decisions made in the pairwise comparison should be reconsidered [71].

Table 3. Random index values according to Saaty and Tran [70].

N	1	2	3	4	5	6	7	8	9	10
RI	0	0	0.52	0.89	1.11	1.25	1.35	1.4	1.45	1.49

The suggested criteria for this decision-making problem are wind speed, slope, distance from power lines, distance from substations, distance from highways and roads, and urban areas. Figure 5 portrays the decision process hierarchy for the wind farm.

All of the criteria were compared, the weight of each was determined, and incompatibility rate calculations were performed for the criteria. Table 4 shows a pairwise comparison of the criteria for selecting wind farm locations. Some university experts made this pairwise comparison. Appendix A (Table A1) shows the academic information of these experts.

As can be seen in Table 4, wind speed is an essential criterion, and investigating the wind speed of the study area could help to estimate the wind power potential, and larger wind turbines can be installed to generate more power in areas with higher wind energy potential. Lands with lower slopes are usually prioritized. Because increasing the slope can increase the initial cost of construction and the maintenance cost. The distances from power lines and power stations also have a direct impact on the project cost. In remote areas that do not have access to the facilities of the electricity network, the construction of power stations and lines can incur huge initial and maintenance costs to project investors. Moreover, the remoteness of urban areas, roads, and highways could cause higher investment costs, such as constructing new access roads. Remoteness from urban areas, where most of the electricity consumption occurs, can also be technically problematic. Because increasing the distance between power the producer and consumer and lengthening the power transmission lines will cause more voltage drop and power loss, and maintenance of these long power transmission lines can be tedious and costly.

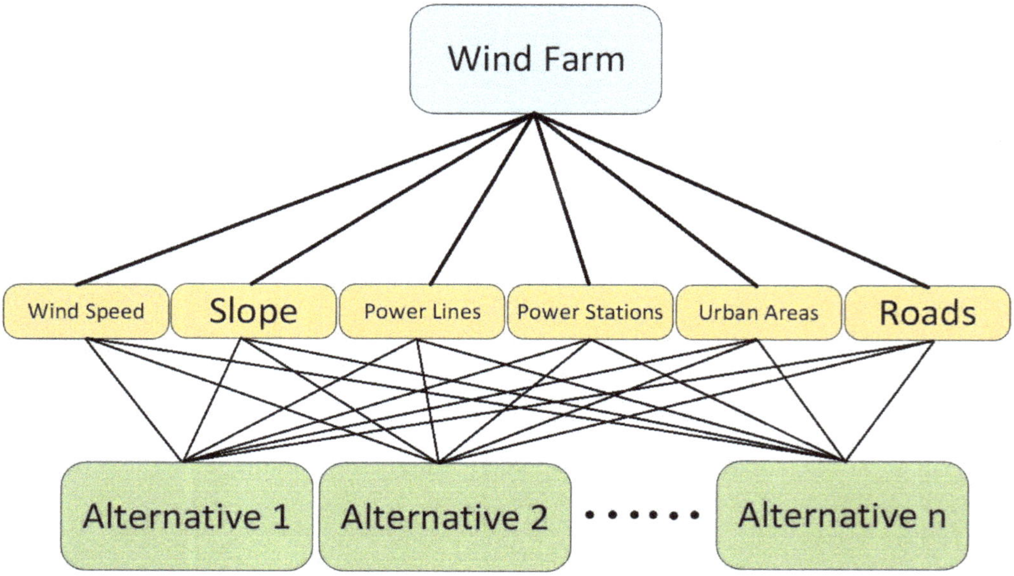

Figure 5. Decision process hierarchy for wind farm.

Table 4. Pairwise comparison for wind farm.

Criteria	Wind Speed	Slope	Power Lines	Power Stations	Urban Areas	Highways	Roads
Wind Speed	1	2	3	3	2	3	3
Slope	0.5	1	2	2	1	2	2
Power Lines	0.33	0.5	1	0.5	2	2	2
Power Stations	0.33	0.5	2	1	2	2	2
Urban Areas	0.5	1	0.5	0.5	1	1	1
Highways	0.33	0.5	0.5	0.5	1	1	1
Roads	0.33	0.5	0.5	0.5	1	1	1

Table 5 represents the pairwise comparison matrix that is calculated using the pairwise comparison of the criteria shown in Table 4. The last row of this matrix shows the sum of each column. Table 6 represents the normalized pairwise matrix calculated by dividing each element of Table 5 by its last row number. The last column of Table 6 shows each criterion weight from the average of each row of the normalized matrix. In the next step, the consistency of the AHP results investigates using the CR value. Table 7 represents the consistency matrix. The last row of this matrix represents the weight of each criterion. The elements of this matrix calculate by multiplying the elements of each column of Table 5 by that column's weight. Furthermore, one of the columns of Table 7 shows the weighted sum value, which shows the sum of each row of consistency matrix elements. The last column of Table 7 shows the ratio of weighted sum value to weights in each row. λ_{max} calculates by averaging the numbers of the last column of Table 7, which is equal to 7.29. Finally, CI and CR calculate using Equations (1) and (2) [24].

Table 5. Pairwise comparison matrix of the MCDM problem.

Criteria	Wind Speed	Slope	Power Lines	Power Stations	Urban Areas	Highways	Roads
Wind Speed	1	2	3	3	2	3	3
Slope	0.5	1	2	2	1	2	2
Power Lines	0.33	0.5	1	0.5	2	2	2
Power Stations	0.33	0.5	2	1	2	2	2
Urban Areas	0.5	1	0.5	0.5	1	1	1
Highways	0.33	0.5	0.5	0.5	1	1	1
Roads	0.33	0.5	0.5	0.5	1	1	1
Sum	3.32	6	9.5	8	10	12	12

Table 6. Normalized pairwise comparison matrix of the MCDM problem with the weights of the criteria.

Criteria	Wind Speed	Slope	Power Lines	Power Stations	Urban Areas	Highways	Roads	Weights
Wind Speed	0.3012	0.3333	0.3157	0.375	0.2	0.25	0.25	0.2893
Slope	0.1506	0.1666	0.2105	0.25	0.1	0.1666	0.1666	0.1730
Power Lines	0.0993	0.0833	0.1052	0.0625	0.2	0.1666	0.1666	0.1262
Power Stations	0.0993	0.0833	0.2105	0.125	0.2	0.1666	0.1666	0.1502
Urban Areas	0.1506	0.1666	0.0526	0.0625	0.1	0.0833	0.0833	0.0998
Highways	0.0993	0.0833	0.0526	0.0625	0.1	0.0833	0.0833	0.0806
Roads	0.0993	0.0833	0.0526	0.0625	0.1	0.0833	0.0833	0.0806

Table 7. Consistency matrix of the criteria with the weighted sum value.

Criteria	Wind Speed	Slope	Power Lines	Power Stations	Urban Areas	Highways	Roads	Weighted Sum Value	Weighted Sum Value/Weights
Wind Speed	0.2893	0.346	0.3786	0.4506	0.1996	0.2418	0.2418	2.1477	7.423
Slope	0.1446	0.173	0.2524	0.3004	0.0998	0.1612	0.1612	1.29265	7.471
Power Lines	0.0954	0.0865	0.1262	0.0751	0.1996	0.1612	0.1612	0.905269	7.172
Power Stations	0.0954	0.0865	0.2524	0.1502	0.1996	0.1612	0.1612	1.106569	7.3668
Urban Areas	0.1446	0.173	0.0631	0.0751	0.0998	0.0806	0.0806	0.71685	7.1823
Highways	0.0954	0.0865	0.0631	0.0751	0.0998	0.0806	0.0806	0.581169	7.209
Roads	0.0954	0.0865	0.0631	0.0751	0.0998	0.0806	0.0806	0.581169	7.209
Weights	0.2893	0.173	0.1262	0.1502	0.0998	0.0806	0.0806		$\lambda_{max} = 7.29$

After performing the calculations, each criterion's final weights were obtained in the AHP method, as shown in Figure 6. The CR factor for this weighted criterion was 3.58%, implying that the pairwise comparison matrix is suitable and does not require change.

According to Figure 6, the most significant criterion is wind speed, which was predictable. After wind speed, the slope of the terrain, distance from power stations, distance from power lines, distance from urban areas, and distance from highways and roads are respectively important.

After calculating the weights, the buffer areas will be applied considering multiple restrictions. These restrictions include the distance from communication routes and railways, urban areas, and environmentally restricted areas. Moreover, according to the references, high-altitude lands specify as a buffer area due to the high cost of the construction process. These restrictions were taken into account by considering international and national standards and regulated technical, electrical, environmental, and economic principles. Figure 7 exhibits the methodology of the present paper. The various section of this study, including categorizing the study area using AHP weights and finding restricted areas, are shown in this figure. Finally, the final map can be obtained by overlying the categorized and restricted maps of the study area.

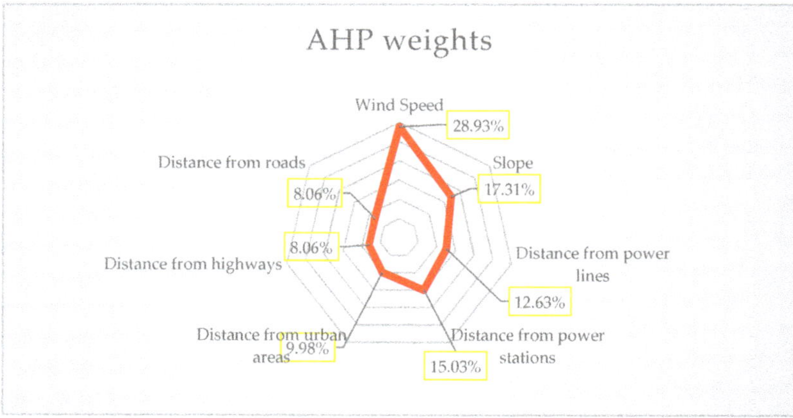

Figure 6. Final criteria weights according to the AHP method (CR = 3.7%).

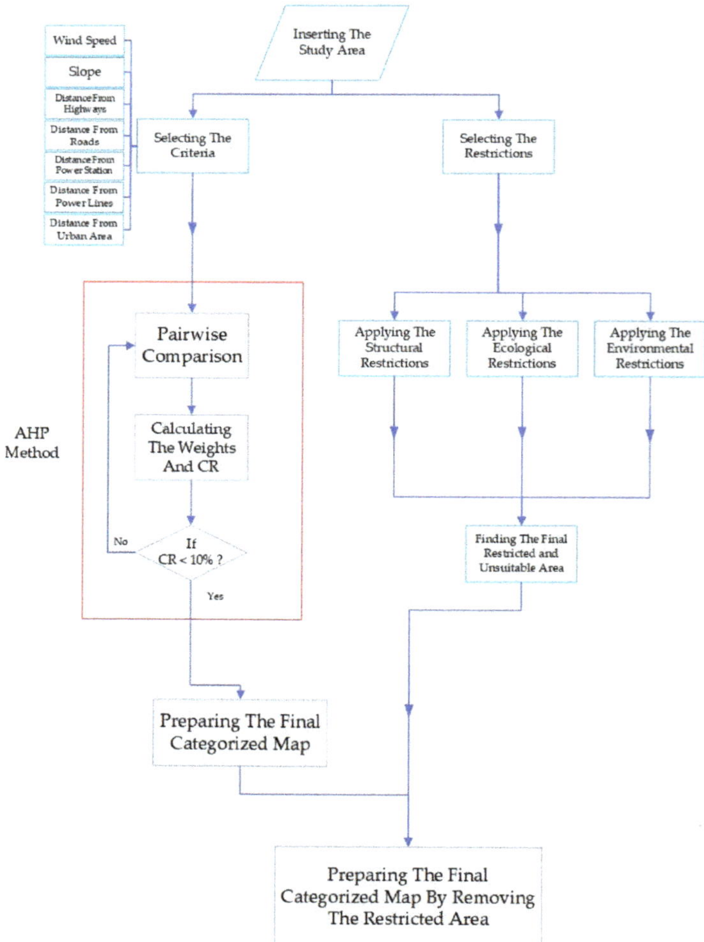

Figure 7. Methodology of study.

5. Materials and Methods

Selecting appropriate points for wind farm site selection is a complex process. The methodology for selecting suitable areas is determined by the calculated weights and the restrictions specified for ecological, structural, and topological criteria and sub-criteria.

A two-step methodology has been used to select the best areas for wind farms:

1. The province has been divided into two areas according to restrictions, suitable and unsuitable.
2. The best areas have been chosen according to the weighted criteria among suitable regions.

In the first step, some restrictions are applied to divide the province into two suitable and unsuitable parts. The restrictions are set for the amounts and distances of each topological, ecological, and structural features.

A conceptual model will be developed after dividing the province into suitable and unsuitable areas. Figure 7 depicts the conceptual model for this study. Specific criteria and restrictions are defined in this model. The required data, assessments, and the characterizations of the study area are collected. According to the defined criteria and collected data, the map layer will identify, layers will integrate based on the conceptual model, and the final suitable and the unsuitable area will be represented.

Finally, the main weighed criteria with the AHP method will be used to classify the appropriate area and determine the best area for the wind farm.

5.1. Restrictions

Topological, ecological, and structural restrictions were specified based on the local and constitutional rules and regulations and recent studies. Sloping and high lands were less considered due to problems in installing turbines, complex and costly repairs, and the maintenance of turbines. Proximity to faults was also avoided due to the dangers it may cause the structures. To ensure security and minimize the problems of turbines for the general public, the permitted distance of turbines from communication routes, fuel, and energy transmission lines, and airports must also be observed. One of the disadvantages of wind turbines is the environmental hazards in the habitats of various animals, particularly birds. Therefore, environmental considerations keep the turbines far from the protected areas and rivers. By applying these restrictions, unsuitable areas for wind farm construction will be identified. These areas will be shown in red on the map and will be removed from the final desired map, which will be investigated by applying the criteria.

5.1.1. Topological Restrictions

The topological restrictions are represented in Table 8. The digital elevation model was used to create a slope map of the province. According to the references mentioned in Table 8, unsuitable areas with an altitude of more than 2000 m and slopes of more than 30% were considered. A distance of fewer than 500 m from the faults was also considered unsuitable. Figure 8 shows the suitable and unsuitable areas based on the topological restrictions mentioned in Table 8. The red areas indicate inappropriate locations, and the green areas indicate appropriate areas.

Table 8. Topological restrictions.

Sub-Criteria	Buffer Zones	References
Elevation (m)	>2000	[6,72]
Slope (percent)	>30	[73]
Faults (m)	<500	[6]

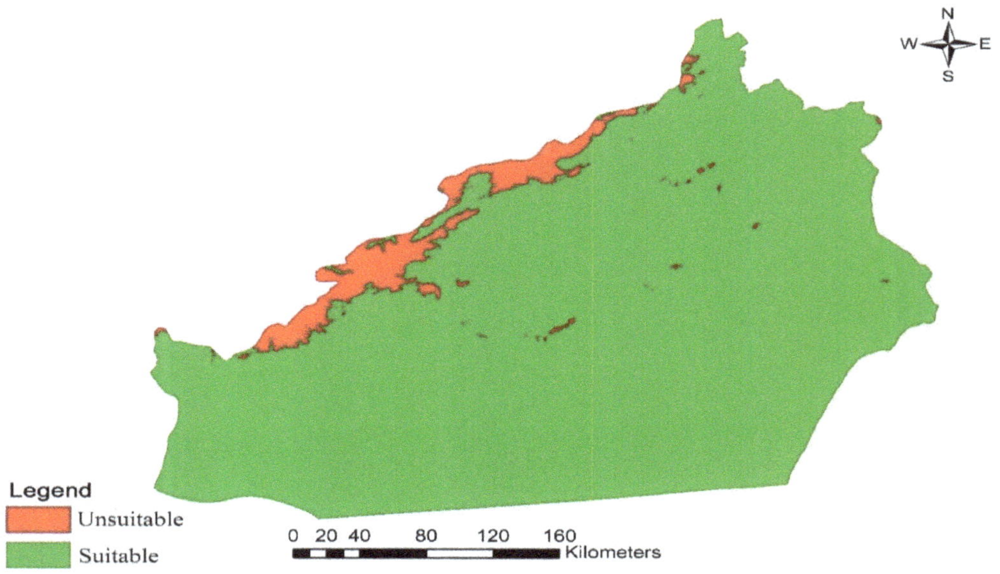

(a)

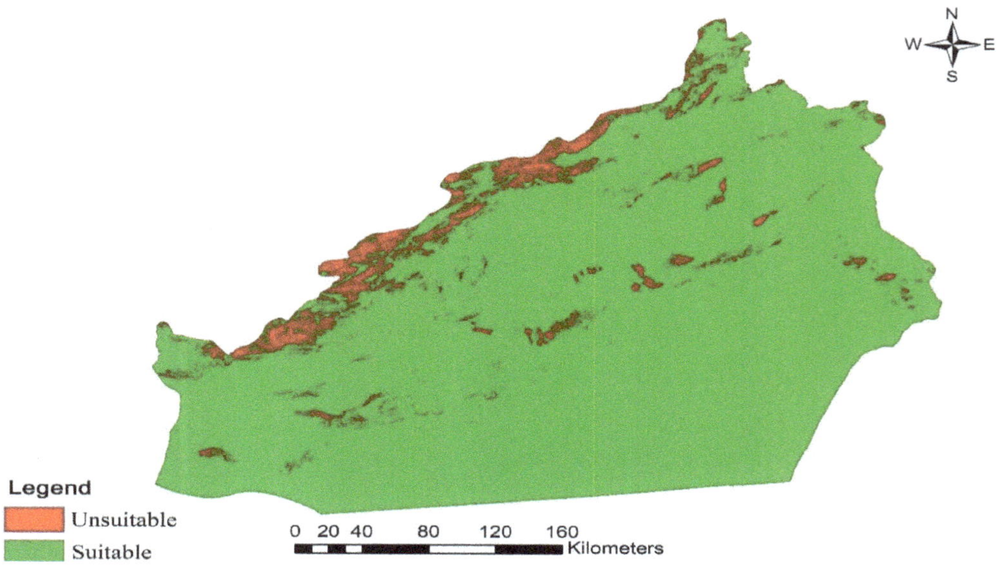

(b)

Figure 8. *Cont.*

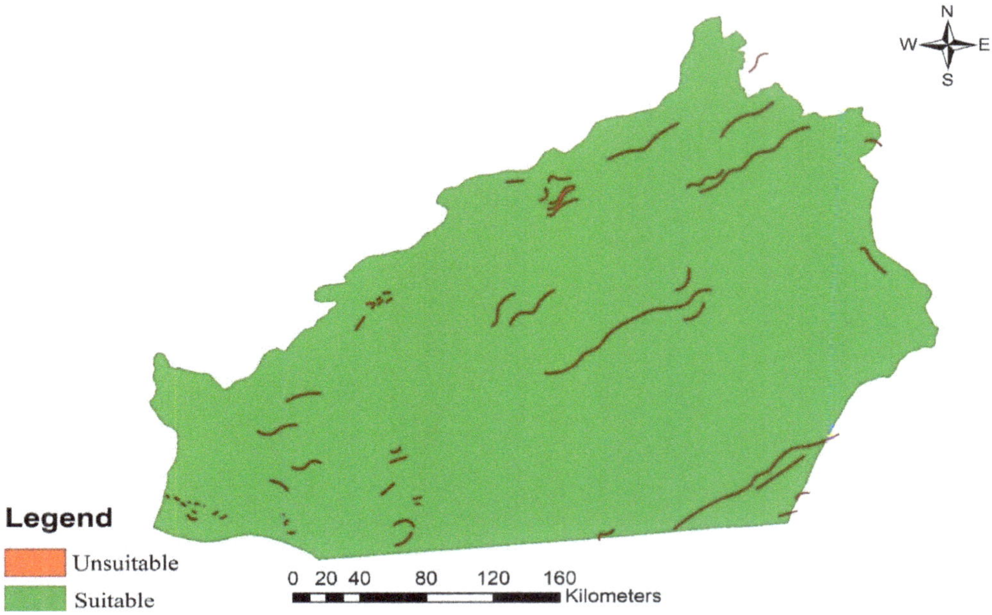

(c)

Figure 8. Topological restrictions (**a**) elevation map. (**b**) slope map. (**c**) faults buffer zone.

5.1.2. Structural Restrictions

The structural restrictions are given in Table 9. According to the latest maps prepared in the planning studies of the Semnan province, carried out by the Planning and Budget Organization of the Ministry of Interior in 2016, the location of roads, oil and gas transmission lines, high voltage power lines, substations, railways, and the airports were identified [74]. The buffer areas are specified based on the mentioned references in Table 9.

Table 9. Structural restrictions.

Sub-Criteria	Buffer Zones	References
Highways and roads (m)	<500	[6]
Oil and gas transmission lines (m)	<500	[6,72]
High voltage power lines (m)	<250	[6,72,75]
Substations (m)	<250	[72]
Railways (m)	<300	[6]
Airports (m)	<2500	[6]

Figure 9 illustrates the suitable and unsuitable structural areas according to the buffer zones of Table 9. The red areas are unsuitable buffer zones, and the green areas are suitable according to structural restrictions. The area of Semnan province is equal to 97,491 km^2. Based on the calculations made according to Table 9, 3864.2 km^2 (3.96%) of the province's total area is among the unsuitable areas.

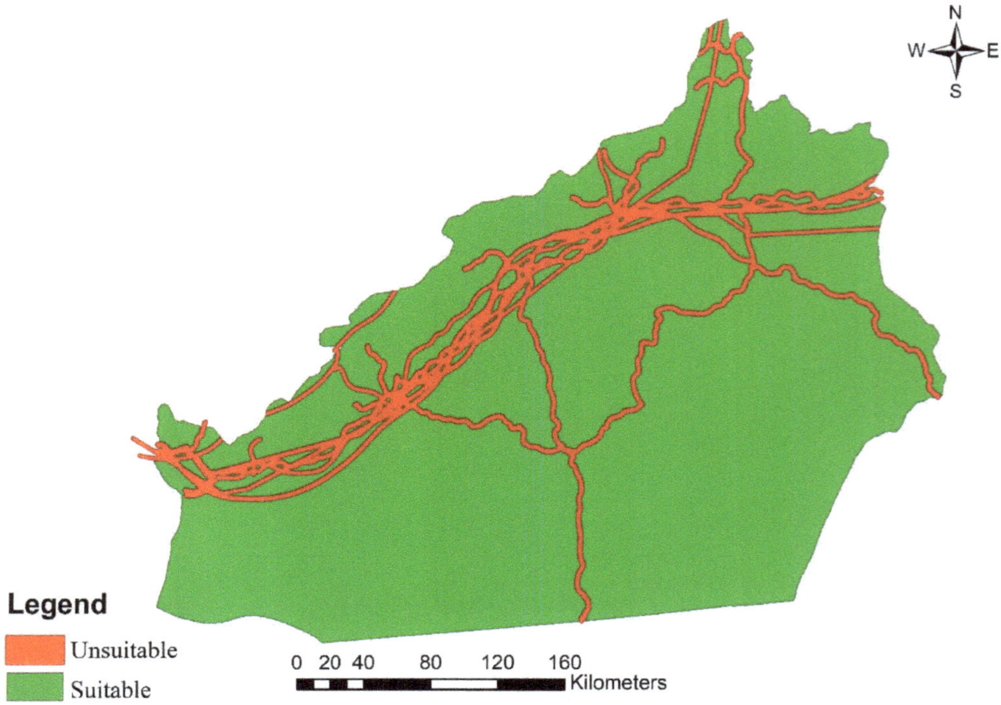

Figure 9. Suitable and unsuitable areas according to structural restrictions.

5.1.3. Ecological Restrictions

Table 10 shows the ecological restrictions. Given the planning studies of the Semnan province [74], the location of environmental protected areas, urban and rural areas, and water bodies and rivers were identified, and buffer zones were specified according to the mentioned references in Table 10.

Table 10. Ecological restrictions.

Sub-Criteria	Buffer Zones	References
Environmental protected areas (m)	<2000	[6]
Urban areas (m)	<2500	[76]
Water bodies (m)	<1000	[6]
Rivers (m)	<500	[6]

Figure 10 exhibits the suitable and unsuitable ecological areas according to the buffer zones of Table 10. According to the ecological restrictions, the red and green areas are unsuitable and suitable zones. Based on the calculations made according to Table 10, 18,052.24 km^2 (18.51%) of the province's total area is among the unsuitable areas.

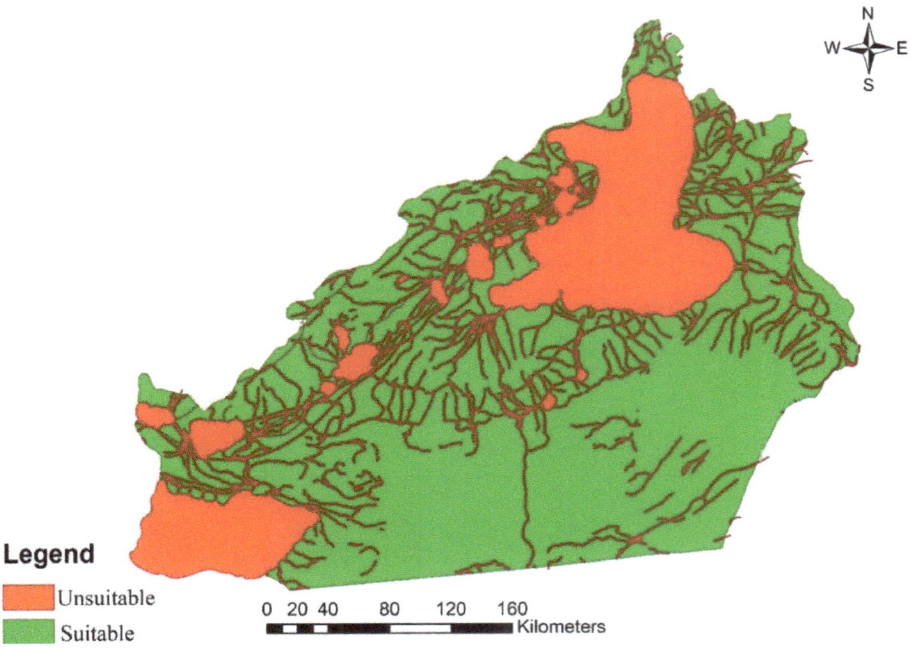

Figure 10. Suitable and unsuitable areas according to ecological restrictions.

Figure 11 shows Semnan province by applying all the restrictions and buffer areas. The areas marked in red and green are unsuitable and suitable, respectively. The total unsuitable areas are equal to 35,094.042329 km^2, which is 36.2% of the total province area.

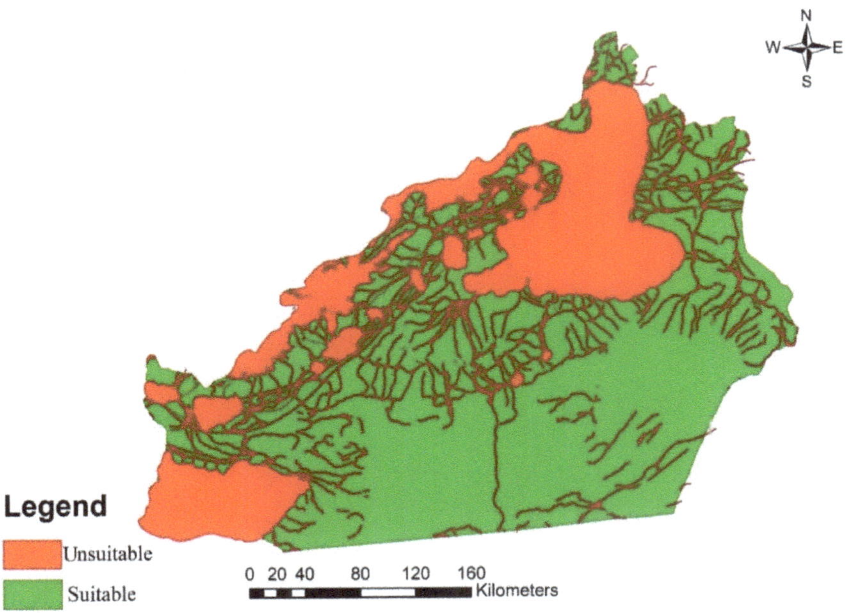

Figure 11. All buffer areas of Semnan province.

5.2. Executing AHP Weights

In almost all studies, the most crucial criterion for solving wind farms' site selection problems is always the region's wind potential [76–80], which also can be seen in calculating the weights of the criteria in the current work. Higher wind speeds in the area mean that it is possible to use larger turbines.

Data from the Renewable Energy and Energy Efficiency Organization (SATBA) were used to prepare the wind map of the province. For this purpose, the wind data of 13 stations in and around the province were employed. Appendix A (Table A2) shows the entire data of SATBA for the 13 stations. Moreover, Table 11 reveals the speed data for the 13 stations. A wind map of the whole province was prepared through interpolation techniques, portrayed in Figure 12. The wind map was reclassified into nine classes. The areas with the highest wind speed were given the highest score, and those with the lowest wind speed were given the lowest score. The minimum and maximum wind speeds were between 3.6 and 5.3 m/s. Moreover, the central areas of the province, which are generally uninhabited and include desert and flat lands, have the best wind speeds.

Table 11. Wind speed data of SATBA stations.

Row	Station	Latitude (Deg)	Longitude (Deg)	Average Wind Speed (m/s)
1	Qom	50.88	34.64	4.92
2	Vesf	50.94	34.19	4.92
3	Friruzhuh	52.77	35.75	4.33
4	Aqqala	54.45	37.01	3.6
5	Marave tappe	55.95	37.9	3.83
6	Bojnurd	57.33	37.47	5.25
7	Davaran	56.88	36.44	3.71
8	Rudab	57.31	36.03	5.21
9	Afriz	59	33.45	4.7
10	Kahak	53.32	35.14	4.21
11	Moalleman	54.56	35.21	5.3
12	Hadadeh	54.73	36.26	4.95
13	Semnan	53.39	35.58	3.64

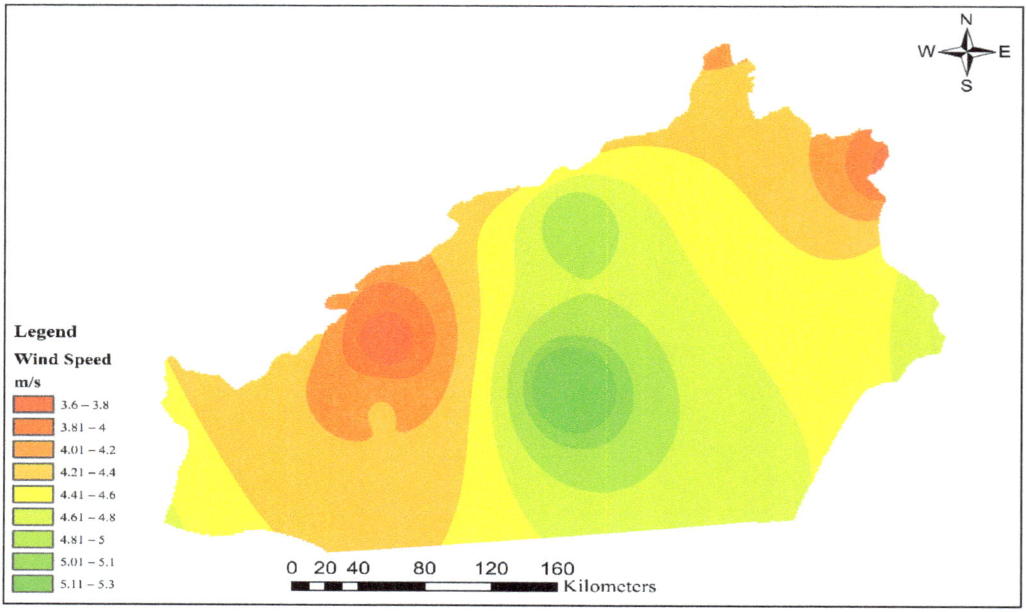

Figure 12. Wind map of Semnan province (m/s).

The rest of the criteria were classified according to the pairwise comparison, and the weights were specified with the AHP method. The classified maps are represented in Figure 13. Each map is categorized into nine classes.

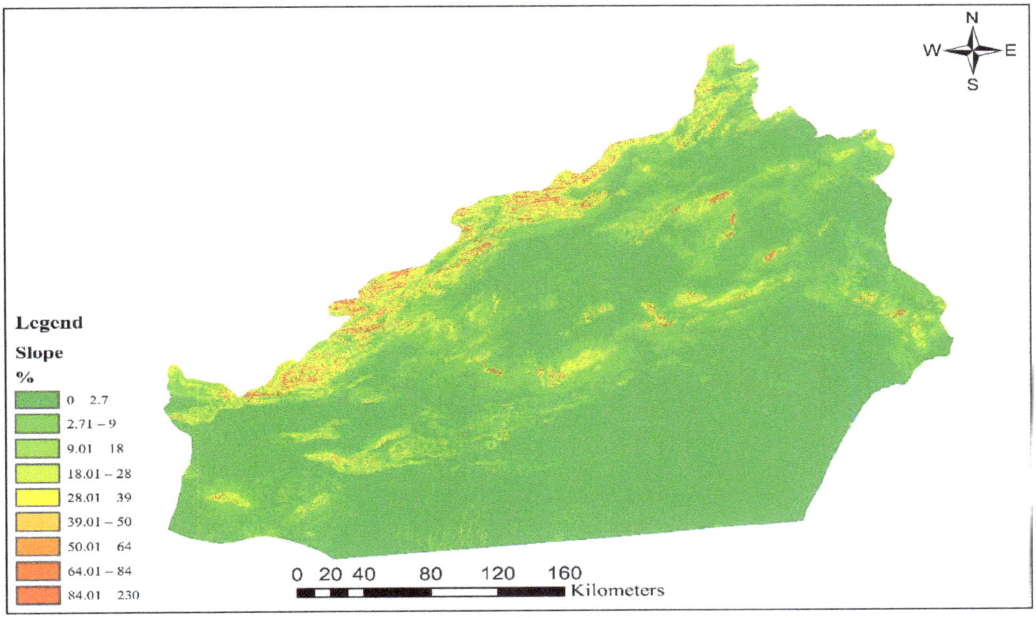

(**a**)

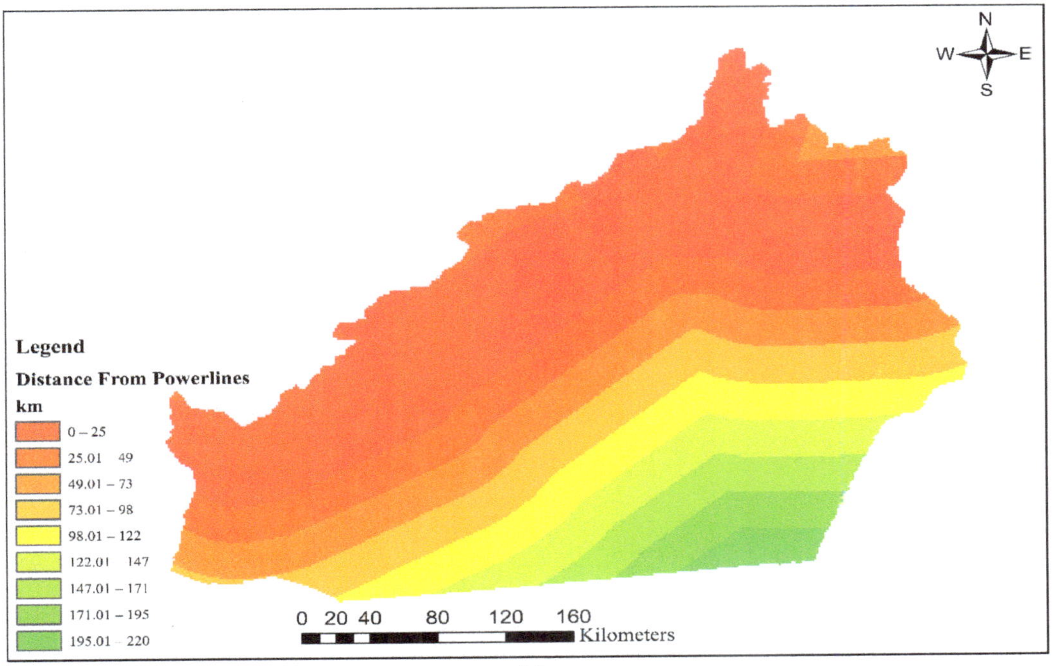

(**b**)

Figure 13. *Cont.*

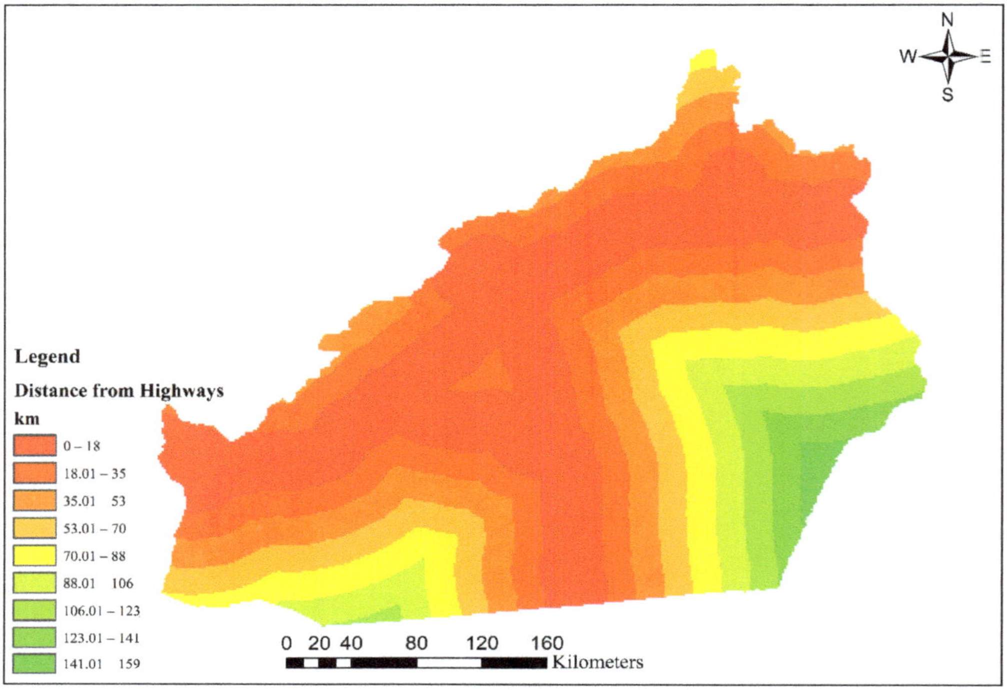

(c)

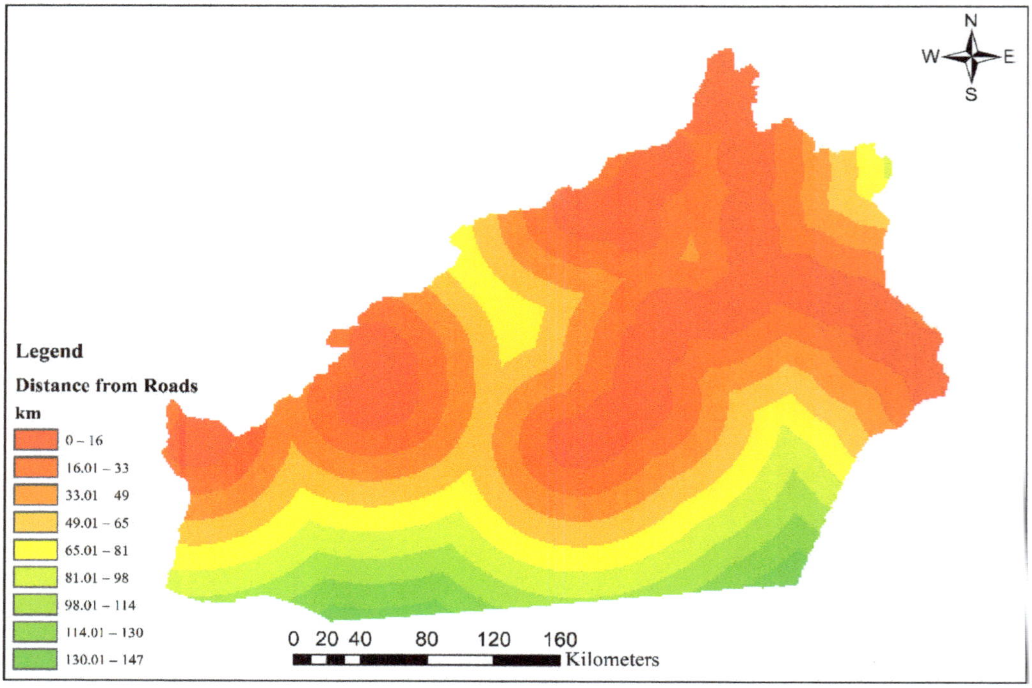

(d)

Figure 13. *Cont.*

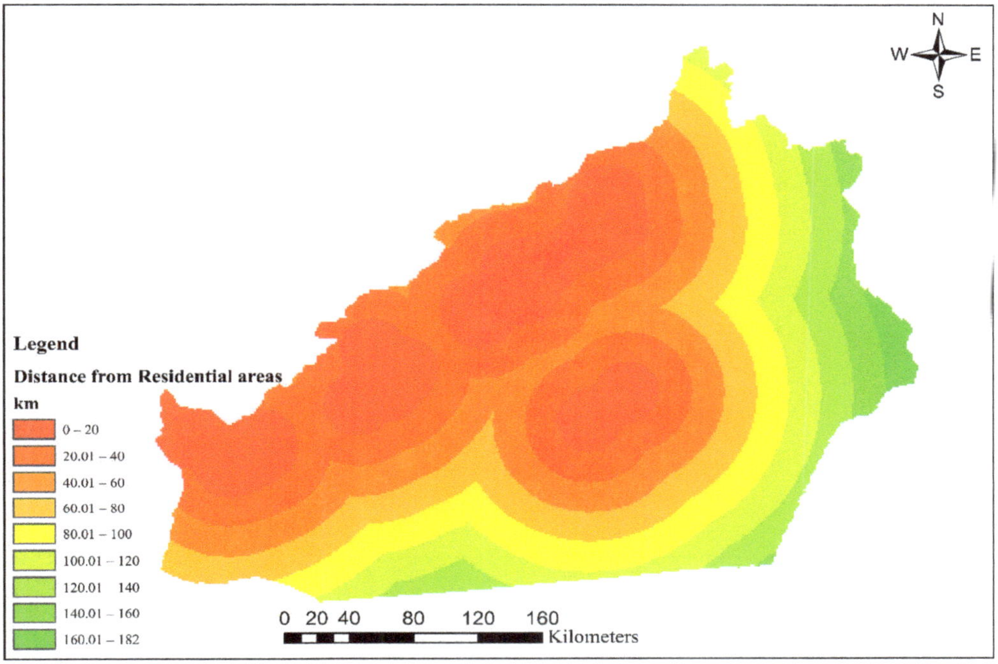

(**e**)

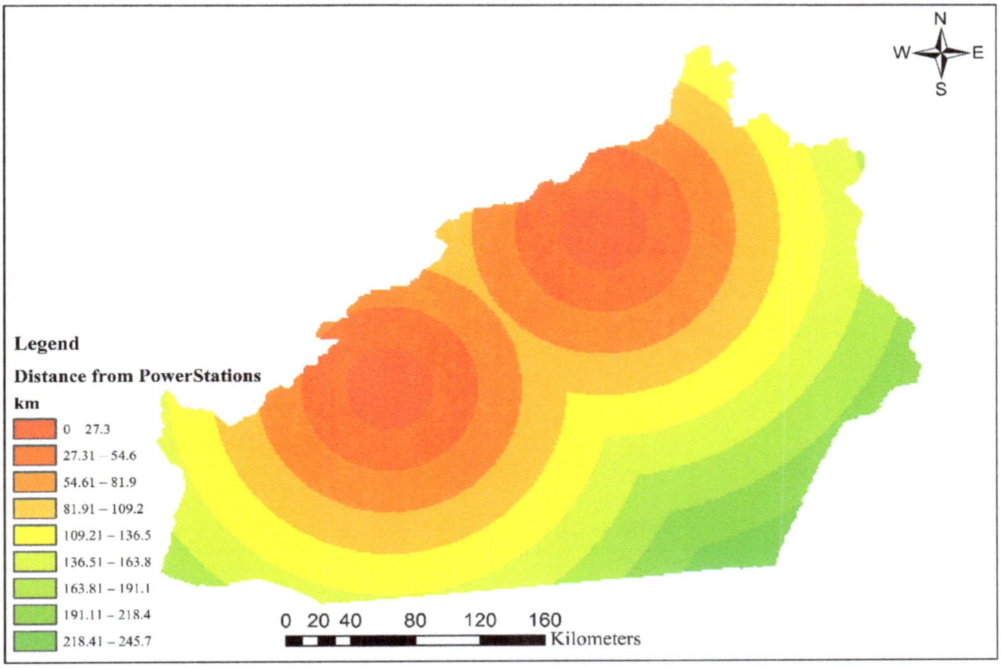

(**f**)

Figure 13. Classified map of the province based on different criteria (**a**) Slope map. (**b**) Distance from powerlines. Distance from (**c**) highways and (**d**) roads. (**e**) Distance from the residential area. (**f**) Distance from power stations).

A slope map was prepared via a digital elevation model (DEM). It was then reclassified, and each class was rated. On account of the importance of the low slope area, these areas were given a higher score.

The distance from other features, such as residential areas, power lines, power stations, and highways and roads, was also classified into nine categories. The areas closer to communication roads and stations and power lines are more important. Therefore, the shortest distance was given the highest score and the farthest distance the lowest score.

6. Results and Discussions

In this section, the results will be displayed. The regional data and the method shown in Figure 7 were applied to show the map's layers, constraints, and classifications. Furthermore, the results of similar studies are mentioned and compared, and the wind energy in the province is compared with other provinces.

After executing the AHP weights for the criteria, the classified map is illustrated in Figure 14. The province was categorized into nine classes, and the green and red areas, respectively, are the best and worst areas for this province. No restrictions were applied in this figure. Looking more carefully at restrictions and comparing them with the slope and elevation map, it can be seen that many of the green areas that are prone to wind farm construction are located in places with high altitudes and high slopes.

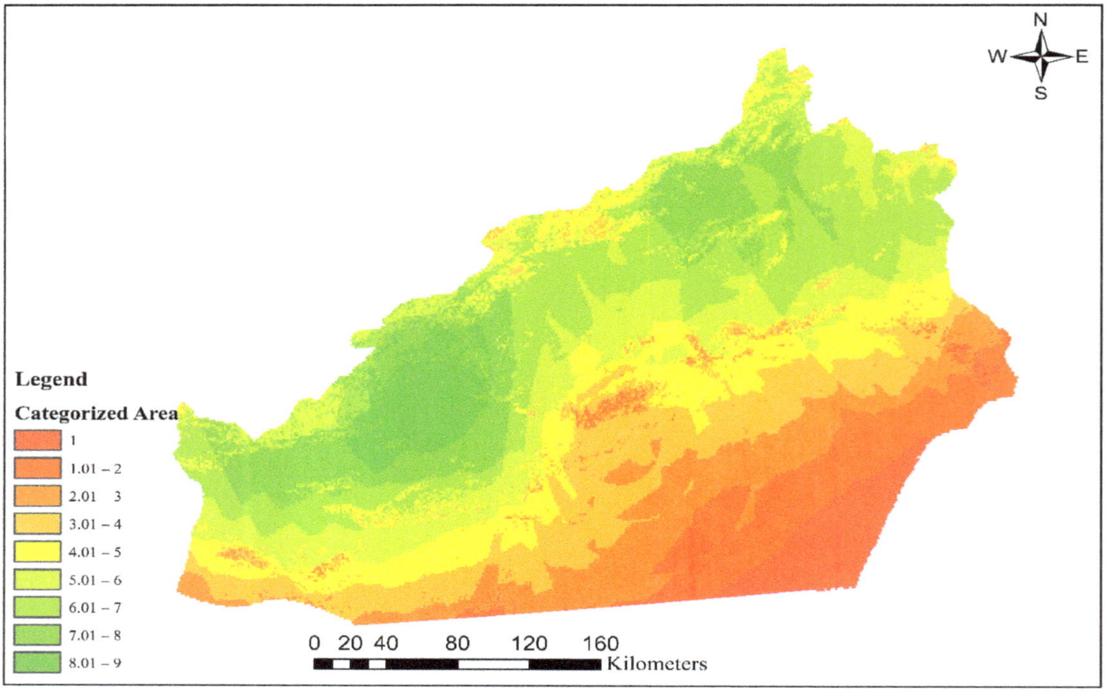

Figure 14. Categorized area based on AHP criteria.

The other essential criteria for this site selection study were the distance from communication routes, substations, and power lines. The remoteness of communication routes, substations, and power lines have adverse effects due to the increase in the cost of installation and maintenance costs. Moreover, the construction of new substations and power transmission lines will cause a sharp increase in the initial and maintenance costs of the whole system. In addition, the increment in the power line's lengths causes more power loss and voltage drop. Therefore, more distant lands have become less of a priority.

The importance of these criteria is of great significance. It has made the northeastern regions of the province, with lower wind speed than the center of the province, have a higher priority for the construction of wind farms. Consequently, the southern regions have the worst conditions due to being deserted and far distance from the roads and power network facilities.

Restrictions were applied to reach the final classified area. The red areas shown in Figure 11 have been removed from the final map based on existing restrictions. Figure 15 shows the final classified map after the restrictions are applied. Due to many residential areas, communication roads, and electricity installations in the northern part of the province, these buffer zones were removed from the final map.

After removing the restricted areas, the best available areas are located on the western side inclined to the province's center and in a small part in the east of the province. According to the classified map, southern areas that are not restricted on account of low population density and lack of cities and protected areas have less priority.

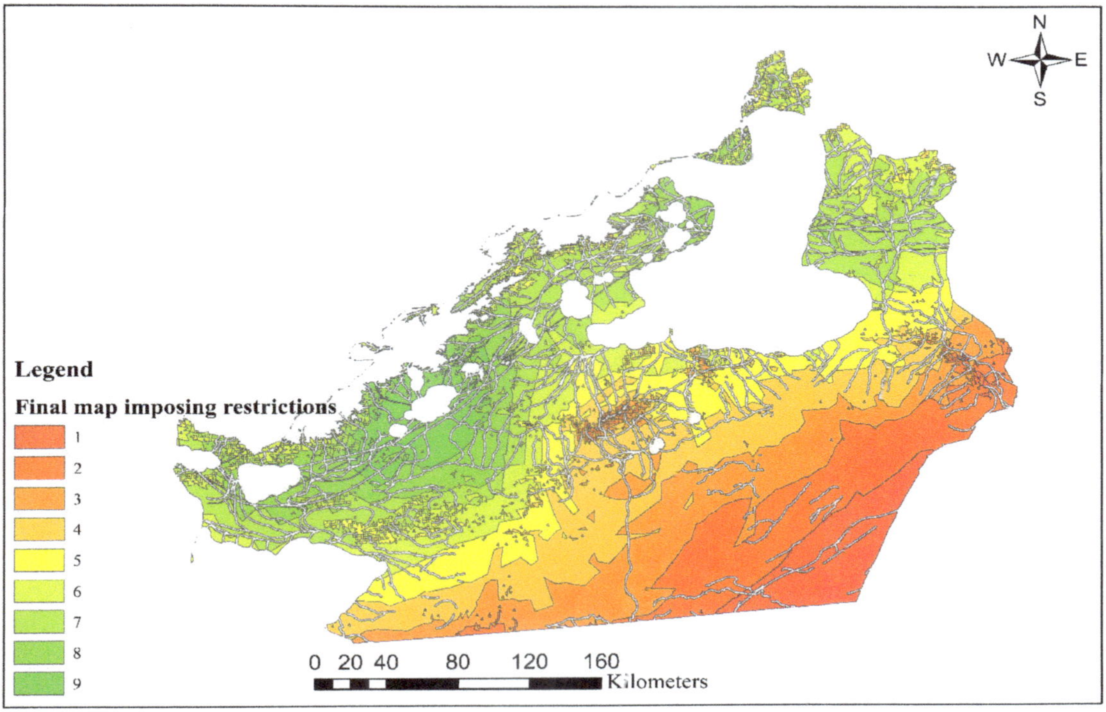

Figure 15. Final categorized map by imposing restrictions.

Figure 16 demonstrates the area of each class with and without restrictions. Classes 9 to 1 are, respectively, the best and worst areas. Class 4 has the largest area (about 16.9%), followed by Class 3 with 14.5%. Only slightly more than 4% of the areas are in Class 9, the best class. The total of the classified areas after removing the buffer zones is a little over 65,000 km^2.

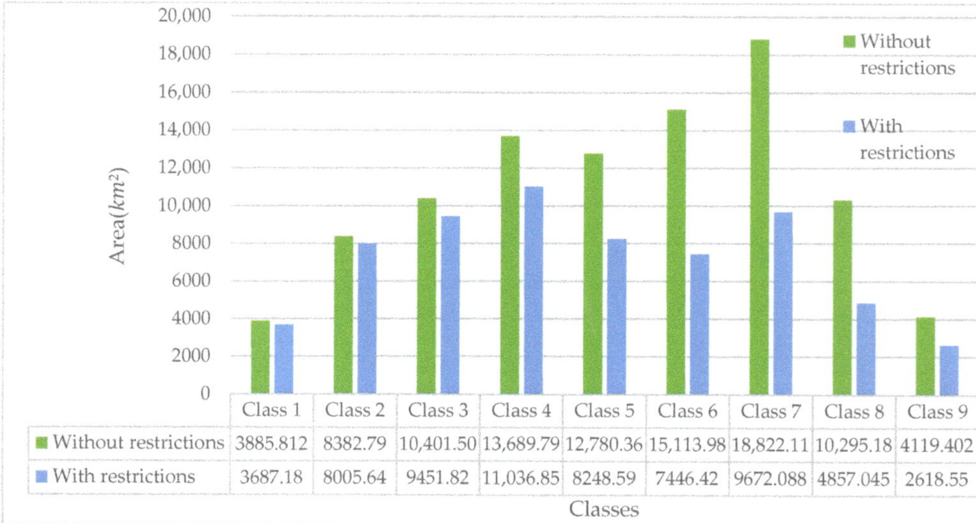

Figure 16. Area of each Class with and without restricted area.

Comparison with Similar Studies

Herein, the wind potential in Semnan province was investigated according to specific criteria. Owing to the increasing popularity of renewable energy, a large number of such articles are placed in various scientific databases annually. However, no similar articles examined Semnan province regarding wind energy potential among the published articles.

In 2011, Mir Hosseini et al. [81] Investigated wind potentials in five locations in Semnan province. Their only criterion to measure wind energy potential was wind speed. To this end, wind data were examined in 3-h intervals between 2003 and 2007. Finally, by comparing wind data, it was concluded that the north area of Damghan has the best conditions among the evaluated areas. The wind was the most important criterion for measuring areas in this paper. According to Figures 2 and 16, the Semnan area can be found among the classified areas, according to which, Damghan is located in the area with grades 7–9, showing good conditions in terms of wind potential.

Moreover, various studies have been conducted to investigate the wind farm potential in the other provinces of Iran. Barzehkar et al. investigated the wind and photovoltaic site potential in Isfahan province, which locates in the south of Semnan. They used a weighted linear combination, AHP, GIS, and fuzzy logic to evaluate the province's area. Their AHP results indicated wind speed and distance from power lines as the most important criteria. According to their final results, almost 15% of the total area had the most potential for wind farm sites, which are mainly located in the province's northeast [82].

Nadizade et al. investigated the multi-renewable energy farms in four eastern provinces of Iran. Three of these provinces locates in the east of Semnan and have a border with this province. According to fuzzy logic and ANP wights, wind speed, distance from urban and protected areas, distance from roads, slope, and elevation were the essential criteria for the wind farm site. Based on their final results, almost 8% of the total study area had a high potential for the wind farm, mostly located in the south part of the study area [83].

In another study, GIS and AHP were used to investigate the wind farm and hydrogen production potential in Yazd, one of Iran's central provinces. According to the AHP weights, the economic criteria such as distance from an urban area and power lines were more important than technical criteria, including slope and wind speed, which is a significant difference between this study and other mentioned studies. Furthermore, they showed that

9. Aubert, A.H.; Bauer, R.; Lienert, J. A review of water-related serious games to specify use in environmental Multi-Criteria Decision Analysis. *Environ. Model. Softw.* **2018**, *105*, 64–78. [CrossRef]
10. Zhang, J.; Zhong, D.; Zhao, M.; Yu, J.; Lv, F. An optimization model for construction stage and zone plans of rockfill dams based on the enhanced whale optimization algorithm. *Energies* **2019**, *12*, 466. [CrossRef]
11. Shao, M.; Han, Z.; Sun, J.; Xiao, C.; Zhang, S.; Zhao, Y. A review of multi-criteria decision making applications for renewable energy site selection. *Renew. Energy* **2020**, *157*, 377–403. [CrossRef]
12. Saaty, T.L. How to make a decision: The analytic hierarchy process. *Eur. J. Oper. Res.* **1990**, *48*, 9–26. [CrossRef]
13. Saaty, T.L.; Vargas, L.G. The analytic network process. In *International Series in Operations Research and Management Science*; Springer: Berlin/Heidelberg, Germany, 2013; Volume 195, pp. 1–40.
14. Ishii, K.; Sugeno, M. A model of human evaluation process using fuzzy measure. *Int. J. Man Mach. Stud.* **1985**, *22*, 19–38. [CrossRef]
15. Zhu, Y.; Tian, D.; Yan, F. Effectiveness of Entropy Weight Method in Decision-Making. *Math. Probl. Eng.* **2020**, *2020*, 3564835. [CrossRef]
16. Alinezhad, A.; Khalili, J. New Methods and Applications in Multiple Attribute Decision Making (MADM). In *International Series in Operations Research and Management Science*; Springer: Cham, Switzerland, 2019; pp. 115–125.
17. Si, S.L.; You, X.Y.; Liu, H.C.; Zhang, P. DEMATEL Technique: A Systematic Review of the State-of-the-Art Literature on Methodologies and Applications. *Math. Probl. Eng.* **2018**, *2018*, 3696457. [CrossRef]
18. Chang, Y.S. Multivariate CUSUM and EWMA control charts for skewed populations using weighted standard deviations. *Commun. Stat. Simul. Comput.* **2007**, *36*, 921–936. [CrossRef]
19. Saaty, T.L.; Ozdemir, M.S. *A Dictionary of Decisions with Dependence and Feedback Based on the Analytic Network Process*; RWS Publication: Pittsburgh, PA, USA, 2005.
20. Wang, T.C.; Chen, Y.H. Applying consistent fuzzy preference relations to partnership selection. *Omega* **2007**, *35*, 384–388. [CrossRef]
21. Deng, H. Multicriteria analysis with fuzzy pairwise comparison. *Int. J. Approx. Reason.* **1999**, *21*, 215–231. [CrossRef]
22. Kahraman, C.; Cebeci, U.; Ulukan, Z. Multi-criteria supplier selection using fuzzy AHP. *Logist. Inf. Manag.* **2003**, *16*, 382–394. [CrossRef]
23. Nguyen, P.H.; Tsai, J.F.; Dang, T.T.; Lin, M.H.; Pham, H.A.; Nguyen, K.A. A hybrid spherical fuzzy MCDM approach to prioritize governmental intervention strategies against the COVID-19 pandemic: A case study from Vietnam. *Mathematics* **2021**, *9*, 2626. [CrossRef]
24. Nguyen, P.H.; Tsai, J.F.; Lin, M.H.; Hu, Y.C. A hybrid model with spherical fuzzy-ahp, pls-sem and ann to predict vaccination intention against COVID-19. *Mathematics* **2021**, *9*, 3075. [CrossRef]
25. Ding, X.; Chong, X.; Bao, Z.; Xue, Y.; Zhang, S. Fuzzy Comprehensive Assessment Method Based on the Entropy Weight Method and Its Application in the Water Environmental Safety Evaluation of the Heshangshan Drinking Water Source Area, Three Gorges Reservoir Area, China. *Water* **2017**, *9*, 329. [CrossRef]
26. Zhao, H.; Guo, S. External benefit evaluation of renewable energy power in China for sustainability. *Sustainability* **2015**, *7*, 4783–4805. [CrossRef]
27. Chen, C.H. A new multi-criteria assessment model combining GRA techniques with intuitionistic fuzzy entropy-based TOPSIS method for sustainable building materials supplier selection. *Sustainability* **2019**, *11*, 2265. [CrossRef]
28. Godlewska, J.; Sidorczuk-Pietraszko, E. Taxonomic assessment of transition to the green economy in Polish regions. *Sustainability* **2019**, *11*, 5098. [CrossRef]
29. Stanujkic, D.; Popovic, G.; Zavadskas, E.K.; Karabasevic, D.; Binkyte-Veliene, A. Assessment of progress towards achieving sustainable development goals of the "Agenda 2030" by using the CoCoSo and the shannon entropy methods: The case of the Eu countries. *Sustainability* **2020**, *12*, 5717. [CrossRef]
30. Zhao, D.-Y.; Ma, Y.-Y.; Lin, H.-L. Using the Entropy and TOPSIS Models to Evaluate Sustainable Development of Islands: A Case in China. *Sustainability* **2022**, *14*, 3707. [CrossRef]
31. Lu, X.; Li, L.Y.; Lei, K.; Wang, L.; Zhai, Y.; Zhai, M. Water quality assessment of Wei River, China using fuzzy synthetic evaluation. *Environ. Earth Sci.* **2010**, *60*, 1693–1699. [CrossRef]
32. Cui, Y.; Feng, P.; Jin, J.; Liu, L. Water resources carrying capacity evaluation and diagnosis based on set pair analysis and improved the entropy weight method. *Entropy* **2018**, *20*, 359. [CrossRef]
33. Ahmadi, M.H.; Dehshiri, S.S.H.; Dehshiri, S.J.H.; Mostafaeipour, A.; Almutairi, K.; Ao, H.X.; Rezaei, M.; Techato, K. A Thorough Economic Evaluation by Implementing Solar/Wind Energies for Hydrogen Production: A Case Study. *Sustainability* **2022**, *14*, 1177. [CrossRef]
34. Tanackov, L.; Badi, I.; Stević, Ž.; Pamučar, D.; Zavadskas, E.K.; Bausys, R. A Novel Hybrid Interval Rough SWARA—Interval Rough ARAS Model for Evaluation Strategies of Cleaner Production. *Sustainability* **2022**, *14*, 4343. [CrossRef]
35. Meng, R.; Zhang, L.; Zang, H.; Jin, S. Evaluation of environmental and economic integrated benefits of photovoltaic poverty alleviation technology in the Sanjiangyuan region of Qinghai province. *Sustainability* **2021**, *13*, 13236. [CrossRef]
36. Ishizaka, A.; Labib, A. Analytic Hierarchy Process and Expert Choice: Benefits and limitations. *OR Insight* **2009**, *22*, 201–220. [CrossRef]

37. Koundinya, S.; Chattopadhyay, D.; Ramanathan, R. Incorporating qualitative objectives in integrated resource planning: Application of analytic hierarchy process and compromise programming. *Energy Sources* **1995**, *17*, 565–581. [CrossRef]
38. Moradi, S.; Yousefi, H.; Noorollahi, Y.; Rosso, D. Multi-criteria decision support system for wind farm site selection and sensitivity analysis: Case study of Alborz Province, Iran. *Energy Strategy Rev.* **2020**, *29*, 100478. [CrossRef]
39. Mosadeghi, R.; Warnken, J.; Tomlinson, R.; Mirfenderesk, H. Comparison of Fuzzy-AHP and AHP in a spatial multi-criteria decision making model for urban land-use planning. *Comput. Environ. Urban Syst.* **2015**, *49*, 54–65. [CrossRef]
40. Khajavi, A.; Reza, M.; Hosseinzadeh, F. Solar PV Power Plant Site Selection Using GIS-FFDEA Based Approach with Application in Iran. *J. Renew. Energy Environ.* **2021**, *8*, 28–43.
41. Xu, Y.; Li, Y.; Zheng, L.; Cui, L.; Li, S.; Li, W.; Cai, Y. Site selection of wind farms using GIS and multi-criteria decision making method in Wafangdian, China. *Energy* **2020**, *207*, 118222. [CrossRef]
42. Colak, H.E.; Memisoglu, T.; Gercek, Y. Optimal site selection for solar photovoltaic (PV) power plants using GIS and AHP: A case study of Malatya Province, Turkey. *Renew. Energy* **2020**, *149*, 565–576. [CrossRef]
43. Castro-Santos, L.; Lamas-Galdo, M.I.; Filgueira-Vizoso, A. Managing the oceans: Site selection of a floating offshore wind farm based on GIS spatial analysis. *Mar. Policy* **2020**, *113*, 103803. [CrossRef]
44. Sayl, K.N.; Mohammed, A.S.; Ahmed, A.D. GIS-based approach for rainwater harvesting site selection. *IOP Conf. Ser. Mater. Sci. Eng.* **2020**, *737*, 012246. [CrossRef]
45. Wu, Y.; Liu, F.; Huang, Y.; Xu, C.; Zhang, B.; Ke, Y.; Jia, W. A two-stage decision framework for inland nuclear power plant site selection based on GIS and type-2 fuzzy PROMETHEE II: Case study in China. *Energy Sci. Eng.* **2020**, *8*, 1941–1961. [CrossRef]
46. Ali, S.A.; Parvin, F.; Al-Ansari, N.; Pham, Q.B.; Ahmad, A.; Raj, M.S.; Anh, D.T.; Ba, L.H.; Thai, V.N. Sanitary landfill site selection by integrating AHP and FTOPSIS with GIS: A case study of Memari Municipality, India. *Environ. Sci. Pollut. Res.* **2021**, *28*, 7528–7550. [CrossRef] [PubMed]
47. Neissi, L.; Albaji, M.; Boroomand Nasab, S. Combination of GIS and AHP for site selection of pressurized irrigation systems in the Izeh plain, Iran. *Agric. Water Manag.* **2020**, *231*, 106004. [CrossRef]
48. Kaya, Ö.; Tortum, A.; Alemdar, K.D.; Çodur, M.Y. Site selection for EVCS in Istanbul by GIS and multi-criteria decision-making. *Transp. Res. Part D Transp. Environ.* **2020**, *80*, 102271. [CrossRef]
49. Ahmadi, S.H.R.; Noorollahi, Y.; Ghanbari, S.; Ebrahimi, M.; Hosseini, H.; Foroozani, A.; Hajinezhad, A. Hybrid fuzzy decision making approach for wind-powered pumped storage power plant site selection: A case study. *Sustain. Energy Technol. Assess.* **2020**, *42*, 100838. [CrossRef]
50. Alhuyi Nazari, M.; Assad, M.E.H.; Haghighat, S.; Maleki, A. Applying TOPSIS Method for Wind Farm Site Selection in Iran. In Proceedings of the 2020 Advances in Science and Engineering Technology International Conferences (ASET), Dubai, United Arab Emirates, 4 February–9 April 2020; p. 4.
51. Ecer, F. Sustainability assessment of existing onshore wind plants in the context of triple bottom line: A best-worst method (BWM) based MCDM framework. *Environ. Sci. Pollut. Res.* **2021**, *28*, 19677–19693. [CrossRef]
52. Mohammadzadeh Bina, S.; Jalilinasrabady, S.; Fujii, H.; Farabi-Asl, H. A comprehensive approach for wind power plant potential assessment, application to northwestern Iran. *Energy* **2018**, *164*, 344–358. [CrossRef]
53. Gao, J.; Guo, F.; Ma, Z.; Huang, X.; Li, X. Multi-criteria group decision-making framework for offshore wind farm site selection based on the intuitionistic linguistic aggregation operators. *Energy* **2020**, *204*, 117899. [CrossRef]
54. Li, M.; Xu, Y.; Guo, J.; Li, Y.; Li, W. Application of a GIS-Based Fuzzy Multi-Criteria evaluation approach for wind farm site selection in China. *Energies* **2020**, *13*, 2426. [CrossRef]
55. Saraswat, S.; Digalwar, A.; Yadev, S.; Kumar, G. MCDM and GIS based modelling technique for assessment of solar and wind farm locations in India. *Renew. Energy* **2021**, *169*, 865–884. [CrossRef]
56. Zalhaf, A.S.; Elboshy, B.; Kotb, K.M.; Han, Y.; Almaliki, A.H.; Aly, R.M.H.; Elkadeem, M.R. A high-resolution wind farms suitability mapping using gis and fuzzy ahp approach: A national-level case study in Sudan. *Sustainability* **2022**, *14*, 358. [CrossRef]
57. Cunden, T.S.M.; Doorga, J.; Lollchund, M.R.; Rughooputh, S.D.D.V. Multi-level constraints wind farms siting for a complex terrain in a tropical region using MCDM approach coupled with GIS. *Energy* **2020**, *211*, 118533. [CrossRef]
58. Erdin, C.; Akbaş, H.E. A comparative analysis of fuzzy TOPSIS and geographic information systems (GIS) for the location selection of shopping malls: A case study from Turkey. *Sustainability* **2019**, *11*, 3837. [CrossRef]
59. Ertuğrul, I.; Karakaşoğlu, N. Comparison of fuzzy AHP and fuzzy TOPSIS methods for facility location selection. *Int. J. Adv. Manuf. Technol.* **2008**, *39*, 783–795. [CrossRef]
60. Rediske, G.; Burin, H.P.; Rigo, P.D.; Rosa, C.B.; Michels, L.; Siluk, J.C.M. Wind power plant site selection: A systematic review. *Renew. Sustain. Energy Rev.* **2021**, *148*, 111293. [CrossRef]
61. Statistical Center of Iran. Available online: https://www.amar.org.ir/english (accessed on 18 May 2022).
62. WorldData. 2021. Available online: WorldData.info (accessed on 19 May 2022).
63. Energy Balance Sheet. Ministry of Energy of Iran. Available online: https://moe.gov.ir/?lang=en-us (accessed on 17 May 2022).
64. Semnan Regional Electricity Company. 2018. Available online: https://semrec.co.ir/page-MyMainEn/en/0 (accessed on 17 April 2022).
65. Semnan Power Distribution Company. 2018. Available online: https://www.semepd.ir/page-enmain/en/0#tabcontrol_15 (accessed on 17 April 2022).

66. Chakraborty, R.; Ray, A.; Dan, P.K. Multi criteria decision making methods for location selection of distribution centers. *Int. J. Ind. Eng. Comput.* **2013**, *4*, 491–504. [CrossRef]
67. Bertolini, M.; Braglia, M.; Carmignani, G. Application of the AHP methodology in making a proposal for a public work contract. *Int. J. Proj. Manag.* **2006**, *24*, 422–430. [CrossRef]
68. Bowen, W.M. Subjective judgements and data envelopment analysis in site selection. *Comput. Environ. Urban Syst.* **1990**, *14*, 133–144. [CrossRef]
69. Kunasekaran, V.; Krishnamoorthy, K. Multi criteria decision making to select the best method for the preparation of solid lipid nanoparticles of rasagiline mesylate using analytic hierarchy process. *J. Adv. Pharm. Technol. Res.* **2014**, *5*, 115–121.
70. Saaty, T.L.; Tran, L.T. On the invalidity of fuzzifying numerical judgments in the Analytic Hierarchy Process. *Math. Comput. Model.* **2007**, *46*, 962–975. [CrossRef]
71. Kablan, M.M. Decision support for energy conservation promotion: An analytic hierarchy process approach. *Energy Policy* **2004**, *32*, 1151–1158. [CrossRef]
72. Satkin, M.; Noorollahi, Y.; Abbaspour, M.; Yousefi, H. Multi criteria site selection model for wind-compressed air energy storage power plants in Iran. *Renew. Sustain. Energy Rev.* **2014**, *32*, 579–590. [CrossRef]
73. Höfer, T.; Sunak, Y.; Siddique, H.; Madlener, R. Wind farm siting using a spatial Analytic Hierarchy Process approach A case study of the Städteregion Aachen. *Appl. Energy* **2016**, *163*, 222–243. [CrossRef]
74. Plan and Budget Organization of Iran. 2022. Available online: https://www.mporg.ir/en (accessed on 18 May 2022).
75. Ameri, M.; Ghadiri, M.; Hosseini, M. Recent advances in the implementation of wind energy in Iran. In Proceedings of the Joint International Conference on Sustainable Energy and Environment (SEE-2), Bangkok, Thailand, 1–3 November 2006; pp 21–23.
76. Bennui, A.; Rattanamanee, P.; Puetpaiboon, U.; Phukpattaranont, P.; Chetpattananondh, K. Site Selection for Large Wind Turbine Using Gis. In Proceedings of the PSU-UNS International Conference on Engineering and Environment—ICEE-2007, Hat Yai, Thailand, 10–11 May 2007; pp. 90–112.
77. Gorsevski, P.V.; Cathcart, S.C.; Mirzaei, G.; Jamali, M.M.; Ye, X.; Gomezdelcampo, E. A group-based spatial decision support system for wind farm site selection in Northwest Ohio. *Energy Policy* **2013**, *55*, 374–385. [CrossRef]
78. Tegou, L.I.; Polatidis, H.; Haralambopoulos, D.A. Environmental management framework for wind farm siting: Methodology and case study. *J. Environ. Manag.* **2010**, *91*, 2134–2147. [CrossRef]
79. Baban, S.M.J.; Parry, T. Developing and applying a GIS-assisted approach to locating wind farms in the UK. *Renew. Energy* **2001**, *24*, 59–71. [CrossRef]
80. Watson, J.J.W.; Hudson, M.D. Regional Scale wind farm and solar farm suitability assessment using GIS-assisted multi-criteria evaluation. *Landsc. Urban Plan.* **2015**, *138*, 20–31. [CrossRef]
81. Mirhosseini, M.; Sharifi, F.; Sedaghat, A. Assessing the wind energy potential locations in province of Semnan in Iran. *Renew. Sustain. Energy Rev.* **2011**, *15*, 449–459. [CrossRef]
82. Barzehkar, M.; Parnell, K.E.; Mobarghaee Dinan, N.; Brodie, G. Decision support tools for wind and solar farm site selection in Isfahan Province, Iran. *Clean. Technol. Environ. Policy* **2021**, *23*, 1179–1195. [CrossRef]
83. Nadizadeh Shorabeh, S.; Argany, M.; Rabiei, J.; Karimi Firozjaei, H.; Nematollahi, O. Potential assessment of multi-renewable energy farms establishment using spatial multi-criteria decision analysis: A case study and mapping in Iran. *J. Clean. Prod.* **2021**, *295*, 126318. [CrossRef]
84. Hosseini Dehshiri, S.S.; Hosseini Dehshiri, S.J. Locating wind farm for power and hydrogen production based on Geographic information system and multi-criteria decision making method: An application. *Int. J. Hydrogen Energy* **2022**, *3*, 83. [CrossRef]

sustainability

MDPI

Article

Study on Combustion Characteristics and Thermodynamic Parameters of Thermal Degradation of Guinea Grass (*Megathyrsus maximus*) in N_2-Pyrolytic and Oxidative Atmospheres

Ayokunle O. Balogun [1], Adekunle A. Adeleke [1,2,*], Peter P. Ikubanni [1,2], Samuel O. Adegoke [3], Abdulbaset M. Alayat [4] and Armando G. McDonald [4]

[1] Department of Mechanical Engineering, College of Engineering, Landmark University, Omu-Aran 251103, Nigeria; balogun.ayokunle@lmu.edu.ng (A.O.B.); ikubanni.peter@lmu.edu.ng (P.P.I.)
[2] SDG-9 (Industry, Innovation, and Infrastructure) Research Group, Landmark University, Omu-Aran 251103, Nigeria
[3] Department of Petroleum Engineering, Faculty of Engineering and Technology, University of Ibadan, Ibadan 200284, Nigeria; sfikayo@gmail.com
[4] Department of Forest, Rangeland and Fire Science, University of Idaho, Moscow, ID 83844-1132, USA; alay0843@vandals.uidaho.edu (A.M.A.); armandm@uidaho.edu (A.G.M.)
* Correspondence: adeleke.kunle@ymail.com

Citation: Balogun, A.O.; Adeleke, A.A.; Ikubanni, P.P.; Adegoke, S.O.; Alayat, A.M.; McDonald, A.G. Study on Combustion Characteristics and Thermodynamic Parameters of Thermal Degradation of Guinea Grass (*Megathyrsus maximus*) in N_2-Pyrolytic and Oxidative Atmospheres. *Sustainability* **2022**, *14*, 112. https://doi.org/10.3390/su14010112

Academic Editors: Farhad Taghizadeh-Hesary and Han Phoumin

Received: 2 November 2021
Accepted: 1 December 2021
Published: 23 December 2021

Abstract: This study provides an extensive investigation on the kinetics, combustion characteristics, and thermodynamic parameters of the thermal degradation of guinea grass (*Megathyrsus maximus*) in N_2-pyrolytic and oxidative atmospheres. A model-fitting technique and three different iso-conversional techniques were used to investigate the kinetics of the thermal process, after which an analysis of the combustion characteristics and thermodynamic parameters was undertaken. Prior to this, experiments on the physico-chemical characterization, thermogravimetric, and spectroscopic analyses were carried out to provide insight into the compositional structure of the guinea grass. The volatile matter, fixed carbon, and total lignin contents by mass were 73.0%, 16.1%, and 21.5%, respectively, while the higher heating value was 15.46 MJ/kg. The cellulose crystallinity index, determined by XRD, was 0.43. The conversion of the GG in air proceeded at a relatively much higher rate as the maximum mass-loss rate peak in a 20 K/min read was −23.1 and −12.3%/min for the oxidative and the pyrolytic, respectively. The kinetics investigation revealed three distinctive stages of decomposition with their corresponding values of activation energy. The average values of activation energy (FWO) at the latter stages of decomposition in the pyrolytic processes (165 kJ/mol) were higher than those in the oxidative processes (125 kJ/mol)—an indication of the distinctive phenomenon at this stage of the reaction. The Coats–Redfern kinetic model revealed that chemical reactions and diffusional models played a predominant role in the thermal decomposition process of the GG. This study showed that the thermodynamic parameters varied with the conversion ratio, and the combustion performance increased with the heating rates. The use of GG as an energy feedstock is recommended based on the findings from this work.

Keywords: physico-chemical characterization; Coats–Redfern model; flammability; integral model; iso-conversional

1. Introduction

The excessive utilization of fossil fuel sources for diverse energy purposes engenders a grave global concern. The combustion of these fuel sources results in the emission of greenhouse gases (GHGs), which have been implicated in global warming and climate change phenomena [1,2]. Again, fossil fuels, which are non-renewables, are being heavily depleted due to an increasing rate of exploitation. Consequently, attention is shifting more toward biomass resources because they are viewed as carbon neutral and abundantly available as inexpensive residues, and they possess widespread availability and huge sustainability potential. It has been reported that energy from biomass has replaced about 14% of the global energy consumption [1]. Diverse biomass materials have been investigated through thermochemical means for probable bioenergy applications and biofuel production. These include rice husk, corn cobs [3], sugarcane straw [4], peanut shells [1], sorghum bicolor glume [5], coffee residues [6], and different woody samples [7–9]. Recently, interest has been turning toward the exploration of different energy crops, which include diverse varieties of grass species, for thermal analysis [10–12].

There are a number of reasons for the recent keen interest in the use of grasses as feedstock materials for biofuel production. The interest may have been prompted by the availability of large expanses of degraded lands, upon which grasses can be cultivated in some reclamation efforts. For instance, Cai et al. [13] stated that about 1107–1411 Mha of degraded land is available globally for the cultivation of energy crops. Grasses have also been identified as short-rotation, non-food crops, as well as low-input, high-yielding biomasses [11,14]. For example, relative to corn feedstock, which has a yield of about 7 Mg/(ha*year), the yield of grasses could reach as high as 40 Mg/(ha*year) [15,16]. This makes them viable substitutes for alleviating the unwholesome competition that arises from the use of food crops, such as corn and sugarcane, for biofuel production. It was noted that, in 2014, 91 billion liters of bioethanol was produced worldwide, mainly through the physicochemical processing of grains and sugarcane—stirring ethical issues from the public [11]. Another attraction to grasses is that though woody biomass residues have been widely investigated, relative to grass their use is characterized by some challenges that include limited land availability, a low annual biomass yield, a slow growth rate, and difficulty in harvesting due to higher energy requirements [14]. Guinea grass (*Megathyrsus maximus*), in particular, has a reputation for being a prolific energy crop in the sub-Saharan region of Africa. It exhibits rapid growth, tolerance to low soil fertility, and resistance to adverse weather conditions [17]. Aside from these characteristics, it is also a lignocellulosic biomass whose polymeric structure is characterized by an intricate matrix of hemicellulose, cellulose, and lignin constituents.

It is important to note that the polymeric structure of grasses differs significantly as a function of certain factors, such as grass variety, maturity stage, and environmental conditions [15]. In terms of climatic conditions, grasses may be broadly classified as tropical (C4) or temperate (C3) region grasses. The major constituents of the former are sucrose and starch, while the latter is predominantly rich in fructose and sucrose [15]. Specifically, it has been noted that *Miscanthus* x *giganteus*, a native Asian grass, has the distinctive feature of a high lignocellulose yield—hemicellulose (20–40%), cellulose (40–60%), and lignin (10–30%) [18]. In contrast, the polymeric constituents of Napier grass (native to Africa) and Bermuda grass (mainly grown in the United States) are comparable to the one mentioned earlier. Switch grass exhibits a slightly different structural composition—the hemicellulose, cellulose, and lignin, respectively, are 25–29%, 37–40%, and 18–25% [15]. Relative to the previous grasses, tall fescue, timothy, yellow flag, and meadow foxtail show markedly different compositions (Table 1). From the foregoing discussion, grasses represent a suitable lignocellulosic feedstock for biofuel production. However, given the wide variation in their structural makeup, it is imperative to undertake detailed characterization analyses prior to deployment for bioenergy purposes. Not only does the information from such analyses provide valuable insight into the chemical character of the feedstock, but it is also profoundly useful in the design of reactors and the modeling and optimization of the associated thermal processes.

Table 1. Polymeric composition of selected grasses [15].

Grasses	Polymeric Composition			Place of Origin
	Cellulose (%)	Hemicellulose (%)	Lignin (%)	
Tall fesecue	25	25	14	Large parts of Europe, Asia, and North Africa
Timothy	28	30	11.5	US, Canada, Europe
Yellow flag	28	10	7	Europe, Western Asia, Northwest Africa
Meadow foxtail	28–31	15–18	11–15	England and Wales

The basic characterization efforts, including proximate and ultimate analyses and higher heating value (HHV) determination, have been utilized extensively in evaluating the composition of diverse biomass feedstock [6,19,20]. The wet chemistry method, typically based on a two-step acidic hydrolysis, is an analysis that has proven reliable in providing insight into biomass composition. However, it is time-consuming and labor-intensive and requires pre-conditioning [21]. Another notable method is infra-red spectroscopic analysis. It is a powerful technique that can be utilized for gathering both quantitative and qualitative data. It is a non-destructive test that is fast and precise. In addition, it is devoid of elaborate sample preparation and the use of expensive and dangerous chemicals. Balogun et al. [22] subjected brewer's spent grain (BSG) to pyrolysis and then undertook the physico-chemical, thermal, and spectroscopic analyses of the BSG and its biochar. They reported a significant variation between the structural configuration of the original BSG and its biochar based on the condensation index and the cellulose crystallinity content. Research has been conducted on the comprehensive characterization of five different biomass samples and revealed that there were distinct differences in the chemical and structural constituents of the samples [23].

Notably, the thermogravimetric analysis (TGA) represents another critical characterization technique that provides a rich source of information regarding the thermal behavior of lignocellulosic biomass. The thermal decomposition of a solid by TGA can be performed isothermally, or otherwise, in an inert or oxidative atmosphere. From the TGA data, deductions can be made on the decomposition mechanism, the kinetic and thermodynamic parameters, and the combustion characteristics of a sample. Though kinetic investigation is suitable for small-sized particles and low-heating-rate processes, it has been widely used because of its high predictability and simplicity. Furthermore, the kinetic data produce sub-models that can be incorporated into complex transport phenomena models to yield practical descriptions of either pyrolysis or combustion processes. The kinetic models, including the one-step global kinetic model, the independent parallel and competitive reaction models, the detailed lumped kinetic model, and the distributed activation energy model, have been formulated and extensively applied [24,25]. Typically, in a kinetic study, the reaction rate is given as a function of temperature and the conversion ratio, and the temperature dependence is expressed as an Arrhenius' equation. The best-fit model, applied to the TGA data, is utilized for the determination of the kinetic parameters and subsequently for simulation. The mathematical approach deployed for solving the rate equation results in two notable techniques, namely iso-conversional and model fitting.

In the model-fitting technique, prior knowledge of the reaction mechanism is required for the selection of an appropriate reaction model. This is achieved by successively fitting different reaction models to the TGA data to select the one with the highest correlation. A popularly utilized model-fitting method is the Coats–Redfern (CR) integral technique. The kinetics of solid-fuel pyrolysis was analyzed using the CR technique and the model identified the probable reaction mechanisms at different stages of pyrolysis [26]. The direct differential and CR methods were deployed to deduce the non-isothermal kinetic parameters of the pyrolysis of pure and crude glycerol, and the distinctive activation energy values were observed [27]. There have been more recent comparative kinetics studies that involved the CR model-fitting method [8,28–31]. The iso-conversional technique, on the other hand, forestalls the need for any foreknowledge of the reaction mechanism. Rather, it relies on the use of several TGA measurements at varied heating rates for the evaluation of the kinetic parameters, and it is based on an approximation technique of the temperature integral. Some kinetic modeling studies have been undertaken through iso-conversional methods, including the Kissinger, Starink, Kissinger–Akahira–Sunose (KAS), and Flynn–Wall–Ozawa (FWO) models [7,25]. It has been demonstrated that thermal degradation of biomass follows a multi-step reaction mechanism because the kinetic parameters vary with the conversion degree [22,30,32].

Globally, energy recovery from biomass is predominantly from combustion processes (about 90%) [33]. Biomass combustion can yield low GHG emissions with efficient monitoring and control. Therefore, the combustion characteristics of specific biomass feedstock need to be quantified for the optimum design and modeling of the combustors and scrubbers. Furthermore, it is also critical to gather information on the feasibility of thermal-conversion processes as well as the energy measurements. This can be achieved by calculating the changes in enthalpy, Gibbs free energy, and entropy from the kinetic parameters [34]. There is limited information on the thermal decomposition of grasses of tropical origin. The objective of this study was to thermally decompose guinea grass in inert and oxidative environments, with the focus on evaluating the kinetic and thermodynamic parameters and the combustion characteristics. The kinetic study will entail the use of model-fitting and iso-conversional techniques, while the feedstock characterization will involve proximate, elemental, compositional, and spectroscopic analyses.

2. Materials and Methods

2.1. Materials

The guinea grass (GG) samples were harvested in an outdoor field (8°7′14″ N; 5°4′56″ E) within the Landmark University premises in June 2020. The grasses (of about 1–1.5 m tall) were initially air dried for 2 weeks and then oven dried at 70 °C for 24 h for ease of pulverization. The sample was pulverized in a ball mill and sieved into 0.6 and 1.18 mm particle sizes with the aid of a mechanical sieve. The 0.6 mm screened particles were used for chemical and TGA characterization. The cellulose (Whatman CF1, Maidstone, England), xylan from corn (TCI America, Portland, OR, USA), softwood kraft lignin (Indulin AT, Meadwestvaco, Charleston, SC, USA), and hardwood organosolv lignin (Lignovate LLC, Fayetteville, AR, USA) were used as received.

2.2. Biomass Characterization

The HHV was determined using a Parr oxygen bomb calorimeter (model 1261, Modline, IL, USA) on densified GG samples (1.0 g, 6 mm diameter using a Carver laboratory press (Wabash, IN, USA) at 68 MPa) according to ASTM D5865-04. The ash content, volatile matter (VM), and fixed carbon (FC) for GG were evaluated based on proximate analysis (ASTM E870-82). The elemental analysis was conducted on a Costech ESC 4010 elemental analyzer (Valencia, CA, USA) to obtain the C and N contents.

The GG sample (4.0 g) was submitted to Soxhlet extraction using CH_2Cl_2 (150 mL) for 16 h, and the extractives content was determined gravimetrically, according to ASTM D1108-96. The CH_2Cl_2 extract was analyzed for lipid profiles as their fatty acid methyl ester (FAME) derivatives after acidic methanolysis (2 mL of $CH_3OH/H_2SO_4/CHCl_3$ (1.7:0.3:2.0 $v/v/v$) at 90 °C for 90 min) and subsequent gas chromatography-mass spectrometry analysis (Thermoscientific ISQ-Trace1300 (Madison, WI, USA); Phenomenex (Torrance, CA, USA) ZB5 30 m x 0.25 mm column; 40 °C (1 min) to 280 °C at 5 °C/min) [22]. The extractive-free GG (200 mg) was subsequently analyzed for lignin and carbohydrate contents by acid hydrolysis [72% H_2SO_4 (2 mL), 60 min, 30 °C], followed by secondary hydrolysis [4% H_2SO_4, 30 min, 121 °C] in an autoclave, according to ASTM D 1106-96. The Klason lignin content was evaluated gravimetrically after filtration. Acid soluble lignin was determined at 205 nm of the filtered hydrolysate (250 mL), using an extinction coefficient of 110 L g^{-1} cm^{-1} (Genesys 50, ThermoScientific, Hanover Park, IL, USA). The hydrolysis filtrate (5 mL) was subjected to carbohydrate analysis according to ASTM E 1758-01. The monosaccharides were quantified by HPLC (two Rezex RPM columns, 7.8 mm × 300 mm, Phenomenex, Torrance, CA, USA) at 85°C on elution with water (0.5 mL min^{-1}) using differential refractive index detection (Waters model 2414, Milford, MA, USA). All analyses were performed in duplicate.

FTIR spectroscopy was conducted on an iS5 spectrometer (ThermoNicolet, Madison, WI, USA) in the single bounce attenuated total reflection (ATR) mode (iD5, ZnSe). The determination of the lignin syringyl/guaiacyl (S/G) ratio was conducted at the relative band intensities at 1462 and 1508 cm^{-1}. The relative band intensities at 1370 and 2900 cm^{-1} were used in determining cellulose crystallinity as the total crystalline index (TCI). [7]. The cellulose lateral order index (LOI) was determined from band intensity ratios at 1430 and 897 cm^{-1} [35]. XRD was carried out on a Siemens D5000 diffractometer (Karlsruhe, Germany) (2θ from 5 to 50° with steps of 0.2°). The diffractogram was peak fitted using Origin software prior to determination of the cellulose crystallinity index (CCI = (1 − (I_{am}/I_{002})), where I_{am} is the intensity of the peak at 2θ = 15° and I_{002} is the maximum intensity of the (002) plane diffraction at 2θ = 22°) [22].

2.3. *Thermogravimetric Analysis (TGA)*

The GG sample with an initial mass of 5.44 ± 0.25 mg was subjected to dynamic heating experiments in a Perkin Elmer TGA-7 (Waltham, MA, USA) instrument in either an N_2 or dry air environment at a flow rate of 30 mL/min. The heating temperature was raised from ambient conditions (29.15 ± 0.64 °C) to 900 °C at three heating rates (5, 10, and 20 °C/min), and the data obtained were analyzed using the Pyris v11 software. The experiments were performed in duplicate.

2.4. *Kinetic Modelling*

The rate equation for a single-step global kinetic model for solid-state degradation under isothermal heating is given as Equation (1).

$$\frac{d\theta}{dt} = Aexp^{\left(-\frac{E}{RT}\right)} f(\theta) \tag{1}$$

where R is the universal gas constant (8.314 J/(mol*K), $f(\theta)$ is the differential decomposition model, A = pre-exponential frequency factor, and θ is the conversion degree expressed as Equation (2).

$$\theta = \frac{W - W_i}{W_f - W_i} \tag{2}$$

The W, W_i, W_f, respectively, are sample mass (%) at temperature T, initial mass, and residual mass. Inserting the constant linear heating rate, $\beta = \frac{dT}{dt}$, into Equation (1) yields the dynamic heating condition (Equation (3)):

$$\frac{d\theta}{dT} = \frac{A}{\beta} exp^{\left(-\frac{E}{RT}\right)} f(\theta) \tag{3}$$

2.4.1. Coats–Redfern (CR) Method

The ordinary differential equation in Equation (3) can be handled by integration by separation of the variable to obtain a temperature integral function as shown in Equation (4). However, an analytical solution is not attainable.

$$g(\theta) = \int_0^{\theta} \frac{d\theta}{f(\theta)} = \int_{T_0}^{T} \frac{A}{\beta} exp^{\left(-\frac{E}{RT}\right)} dT \tag{4}$$

Note that $g(\theta)$ represents an integral decomposition model that represents the reaction mechanism that relates to the solid-state degradation. A couple of such models are given in Table 2. The logarithmic transformation of Equation (3) alongside Equation (4) yields the CR model for the derivation of the kinetic parameters, as shown in Equation (5):

$$\ln\left(\frac{g(\theta)}{T^2}\right) = \ln\frac{AR}{\beta E}\left(1 - 2\frac{RT}{E}\right) - \frac{E}{RT} \tag{5}$$

Table 2. Empirical correlations for $g(\theta)$ on different reaction mechanisms [28,29].

Mechanism Model	$g(\theta)$
Power Law (n = 1, 2, 3)	$\theta^{1/n}$
Nucleation Reaction Models	
Avarami-Eroféve (n = 1.5, 2, 3)	$[-\ln(1-\theta)]^{1/n}$
Contracting sphere	$1-(1-\theta)^{1/2}$
Contracting cylinder	$1-(1-\theta)^{1/3}$
Diffusional Models	
1-D diffusion	θ^2
2-D diffusion	$[(1-\theta)\times\ln(1-\theta)]+\theta$
3-D diffusion-Jander	$\left[1-(1-\theta)^{1/3}\right]^2$
3-D diffusion-GB	$1-\frac{2\theta}{3}-(1-\theta)^{2/3}$
Chemical Reaction Models	
1st order	$-\ln(1-\theta)$
n-th order	$\left[1-(1-\theta)^{1-n}\right]/(1-n)$

The plot of the left-hand side of Equation (5) against the reciprocal of temperature yields, approximately, a linear curve from whose slope the activation energy can be obtained. It is assumed that $RT \ll E$; therefore, the intercept is given as $intercept = \ln\frac{AR}{\beta E}$, from which the pre-frequency factor is computed.

2.4.2. Differential Friedman Method (DFM)

If the natural logarithm is applied to Equation (3), it yields Equation (6), which is commonly referred to as the differential Friedman's kinetic model.

$$\ln\left[\frac{d\theta}{dt}\right] = \ln\left[\beta\left(\frac{d\theta}{dT}\right)\right] = \ln[Af(\theta)] - \frac{E}{RT} \tag{6}$$

In Friedman's relation, the conversion function, $f(\theta)$, is assumed constant. This implies that the solid-state degradation is primarily dependent on the mass-loss rate and is independent of the temperature. The linear plot of $\ln\left[\frac{d\theta}{dt}\right]$ against $\frac{1}{T}$ is generated for different heating rates, and the activation energy is determined from the slope $\left(slope = -\frac{E}{R}\right)$. It is important to note that the use of the derivative conversion data makes the DFM prone to noise sensitivity and numerical instability, and therefore, caution must be exercised in the data interpretation [36].

2.4.3. Flynn–Wall–Ozawa (FWO) Method

The FWO model takes the apparent activation energy to be constant during the thermal decomposition process and engages Doyle's relation to approximate the temperature integral function. Taking the logarithm of the integral function and inserting Doyle's approximation yields Equation (7).

$$\log\beta = \log\left(A\frac{E}{Rg(\theta)}\right) - 2.315 - 0.4567\frac{E}{RT} \tag{7}$$

A plot of $\log\beta$ against $\frac{1}{T}$ for different heating rates produces straight lines. Again, the activation energy can be evaluated from the slope of the lines as $\left(slope = -0.4567\frac{E}{R}\right)$.

2.4.4. Starink (STK) Method

The Starink method is based on the optimization of two iso-conversional methods, namely the Flynn–Wall–Ozawa and the Kissinger–Akhira–Sunose (KAS), and it is expressed as Equation (8).

$$\ln\left(\frac{\beta}{T^{1.92}}\right) = C_s - 1.0008\frac{E}{RT} \tag{8}$$

A plot of $\ln\left(\frac{\beta}{T^{1.92}}\right)$ against the reciprocal of temperature generates linear curves and the activation energy can be computed from their slopes. It was noted that Starink's model presented an accuracy of an order of magnitude higher than the FWO and KAS.

2.4.5. Combustion Characteristics Indices

The indices, ignition temperature (T_i), the temperature at the maximum DTG (T_{max}), the burnout temperature (T_b), the corresponding time (t_i, t_{max}, t_b), and the maximum and average DTG ($-R_p$ and $-R_v$), can be obtained from the TGA data [1,8]. These were subsequently used to monitor the combustion characteristics, comprehensive combustibility (S), flammability (C), ignition (D_i), and burnout (D_b), according to the relations in Equations (9)–(12).

$$S = \frac{-R_p \times -R_v}{T_i^2 \times T_b} \tag{9}$$

$$C = \frac{-R_p}{T_i^2} \tag{10}$$

$$D_i = \frac{-R_p}{t_i \times t_b} \tag{11}$$

$$D_b = \frac{-R_p}{\Delta t_{1/2} \times t_p \times t_b} \tag{12}$$

2.4.6. Thermodynamic Analysis

The thermodynamic parameters [change in enthalpy (ΔH, J/mol), Gibbs free energy (ΔG, J/mol), and entropy (ΔS,J/((mol*K)))] were deduced as functions of conversions from the kinetic parameters, as shown in Equations (13)–(15).

$$\Delta H = E_\theta - RT \tag{13}$$

$$\Delta G = E_\theta + RT_{max}\ln\left(\frac{k_B T_{max}}{hA_\theta}\right) \tag{14}$$

$$\Delta S = \frac{\Delta H - \Delta G}{T_{max}} \tag{15}$$

where k_B, and h are the Boltzmann constant (1.381×10^{-23} J/K) and the Planck constant (6.626×10^{-34} J s), respectively.

3. Results and Discussion

3.1. Characterization of Guinea Grass

Table 3 shows the results of the proximate, elemental, compositional, and calorific value analyses of the GG. It is shown that the GG contained 40% C, 1.3% N, and 5.1% ash. The ash content was well within the range reported for other grass species, such as Elephant grass [37], Camel grass (6.31%) [38], and *Echinochloa stagnina* (6.31%) [23], as well as other biomass wastes, such as jackfruit peel (5.56%) and seeds (6.64%) [39]. In comparison to low-rank coals, the lower ash content of biomass makes it more suitable for combustion processes [39]. This is due to technical problems, such as slagging and fouling, which impede heat and mass transfer. The estimated protein content of 8.1% was in the range (5.3–8.8%) for fresh GG, as documented by Aganga and Tshwenyane [40]. An FC value of 16.1% was in the range (8.5–16.9%) for Napier grass [41]. The calorific value obtained for the GG was 15.5 MJ kg^{-1}, and is comparable to Napier grass (16.2–18.1 MJ kg^{-1}) [41], tamarind residues (17.5 MJ kg^{-1}) [42], smoked cigarette butts (18.5 MJ kg^{-1}) [43], and jackfruit wastes (16.3–17.2 MJ kg^{-1}) [39]. An extractives content of 1.4% agreed with the literature [40,44]. A lignin content of 21.5%, which was nearly twice that reported by Ratsamee et al. [45] for GG, using the acetyl bromide method, was recorded, while Mohammed et al. [41] obtained a lignin content for Napier grass of 24%. The higher lignin value is attributable to protein interference in the Klason lignin determination of grasses [46]. Detailed carbohydrate analysis showed mainly glucan (34%) and xylan (18%), together with galactan (1.2%) and arabinan (6.4%). The total carbohydrate value is lower than that reported by Ratsamee et al. (hemicellulose (27.1%) + cellulose (41.7%)) for GG [45], but higher than for the other grasses listed in Table 1 [15]. The variations observed in some parameters may be due to differences in genetics and/or environmental conditions.

Table 3. Proximate, Elemental, and Compositional data for GG sample.

Parameter	GG
Proximate Analysis	
Volatile matter (VM) (%)	73.0 ± 0.3
Fixed carbon (FC) (%)	16.1 ± 0.8
Ash (%)	5.09 ± 0.01
Elemental Analysis	
C (%)	40.1 ± 0.4
N (%)	1.30 ± 0.01
Protein (N * 6.25) (%)	8.12 ± 0.08
Compositional Analysis	
CH_2Cl_2 extractives (%)	1.41 ± 0.03
Acid soluble lignin (%)	3.4 ± 0.2
Klason lignin (%)	18.1 ± 0.5
Total lignin (%)	21.5 ± 1.5
Glucan (%)	33.6 ± 0.5
Xylan (%)	17.7 ± 0.5
Galactan (%)	1.2 ± 0.4
Arabinan (%)	6.4 ± 0.4
Total Neutral sugar (%)	58.9
HHV (MJ kg^{-1})	15.46 ± 0.16

Fatty acids are an important source of unsaturated acids in grasses for foraging animals [47]. The fatty acid profile of GG extractives was determined as FAME derivatives and given in Table 4. The fatty acids were from C_{12} (lauric acid) to C_{24} (lignoceric acid), with the most abundant being palmitic acid (74 mg/g extract), linoleic (39 mg/g extract), and oleic (24 mg/g extract) acids. Lauric (C12) to stearic (C18) acids, saturated and unsaturated, have been observed in several types of forage grass [47].

Table 4. Fatty acid profile of GG extract.

FAME	RT (min)	M^+ (m/z)	Concentration (mg/g Extract)
Lauric acid (C12:0)	24.15	214	8.13
Myristic (C14:0)	28.69	242	7.16
Pentadecanoic acid (C15:0)	30.79	256	2.22
Palmitelaidic acid (16:1)	32.31	268	3.26
Palmitic acid (C16:0)	32.83	270	74.4
Heptadecanoic acid (C17:0)	34.83	284	3.51
Linoleic acid (C18:2)	35.84	294	39.4

Table 4. *Cont.*

FAME	RT (min)	M^+ (m/z)	Concentration (mg/g Extract)
Oleic acid (C18:1)	36.00	296	23.9
Stearic acid (C18:0)	36.54	298	6.93
Arachidic acid (C20:0)	40.00	326	3.18
Behenic acid (C22:0)	43.14	354	4.10
Tricosanoic acid (C23:0)	44.82	368	2.08
Lignoceric acid (C24:0)	46.85	382	4.69

Cellulose crystallinity is a key factor in the biological or thermal degradability of biomass. The XRD analysis of the GG (Figure 1a) showed a typical diffractogram of cellulose I, with 2θ peaks at 15° and 22°, which were assigned to the cellulose planes of (101) and (002), respectively [22]. The CCI was determined after peak fitting at 0.431, and it was found to be higher than that of Napier grass, having a CCI of 0.327 [46].

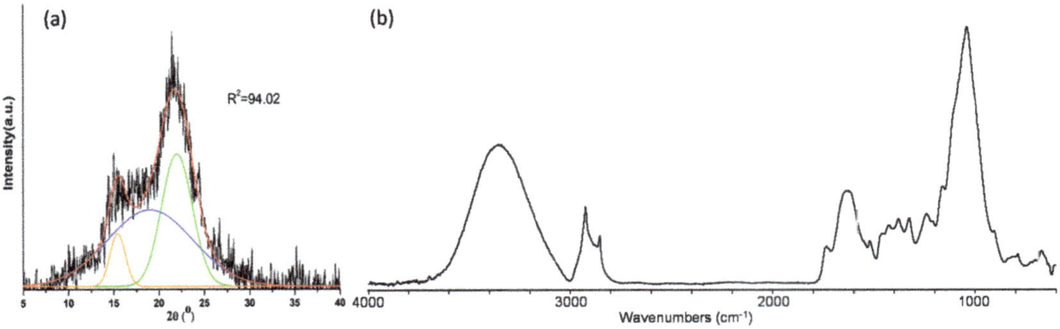

Figure 1. (**a**) X-ray diffractogram and (**b**) FTIR spectrum of guinea grass.

The GG was analyzed for chemical properties by FTIR spectroscopy (Figure 1b). The vibrational band assignments in GG by Balogun et al. [22] were used. An O-H stretching band around 3300 cm^{-1} was assigned to the polysaccharides and lignin. The C–H stretching band at 2920 cm^{-1} was assigned to the aliphatic structures, while the carbonyl band around 1735 cm^{-1} was assigned to the acetyl and uronic acid groups in xylan. The presence of lignin was confirmed by the distinct bands at 1514 and 1604 cm^{-1}, assigned to the aromatic skeletal vibrations. The large band centered at 1037 cm^{-1} was assigned to the C-O stretching in the cellulose, hemicellulose, and lignin polymers. As mentioned earlier, cellulose degradation is associated with its crystallinity. Cellulose crystalline information was determined by its TCI (crystallinity) and LOI (cellulose I), and the values obtained were both 1.1 for the GG. The values of TCI and LOI for Napier grass were 1.25 and 0.53, respectively [48], while sorghum glume has a lower value of LOI (0.75) [49]. The glass transition temperature (softening point) and reactivity of lignin are influenced by its S/G ratio [34]. The lignin S/G ratio of GG was calculated at 1.2, and it was higher than straw soda lignin (1.05) [49]. Sun et al. [34] used Raman spectroscopy to determine S/G ratios for switchgrass (0.92) and maize (1.1).

3.2. Thermal Degradation Characteristics of GG at Different Heating Rates

Figure 2a,b present the thermograms of GG in the N_2 and air atmospheres, respectively, at different heating rates.

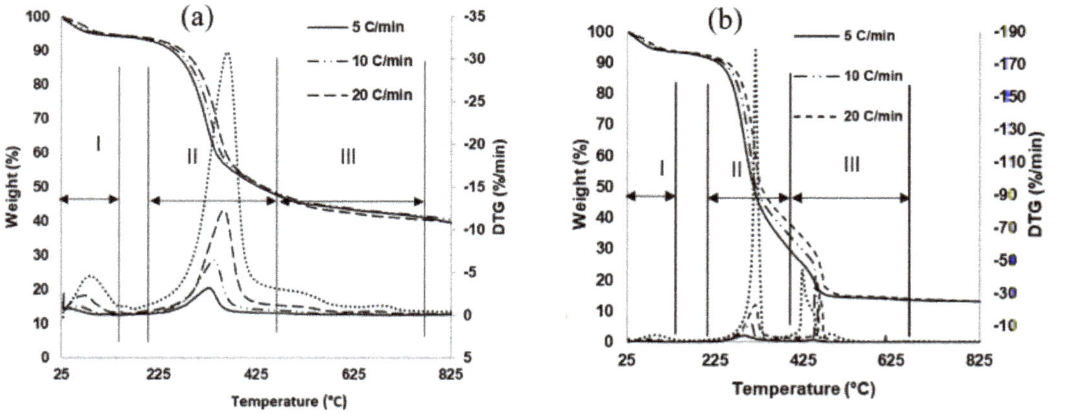

Figure 2. Thermograms of mass loss and mass-loss rate (DTG) for GG in (**a**) N_2 and (**b**) air atmospheres.

The thermograms for the thermal degradation of GG in N_2 and air reveal that the degradation trend consists mainly of three regions, namely dehydration (I), devolatilization (II), and solid burnout or char formation (III). The first region is often ascribed to moisture loss and occurs below 100 °C for all heating rates. Shortly after, around 220 °C, is a characteristic "shoulder" that is indicative of hemicellulose degradation. Of the polymeric constituents, it is the least stable and thus the most reactive. Prior to the "shoulder", between 140 and 200 °C, a mass loss of less than 1% is noted. This may be due to the breakdown of some low-molecular organics with very weak bonds. At around 290 and 370 °C, a sharp and conspicuous DTG peak emerges. This is typically assigned to cellulose degradation because, as it has been noted, at temperatures above 250 °C cellulose is totally decomposed [22]. The devolatilization region is where substantial parts of the bonds in hemicellulose and cellulose and partly in lignin are deconstructed, leading to a release of large amounts of volatiles. This notion is supported by other researchers who have also observed that this region is suggestive of a simultaneous decomposition of the hemicellulose, cellulose, and lignin components [8]. Lignin decomposes over a wide range of temperatures, and thus, there is the trailing effect that extends to much higher temperatures. At about 600 °C, a barely visible peak appears—this may be indicative of the deconstruction of lignin's strongest bonds and/or the thermal cracking of some of the condensed lignin structures formed from the previous primary reactions [50].

Figure 3 depicts the thermograms of the isolated polymeric constituents (xylan, cellulose, and lignin (softwood and hardwood)), and the GG decomposition under N_2 conditions at 20 °C/min. This provides insight into the sequence of the decomposition of the lignocellulosic biomass constituents. It is shown that hemicellulose (xylan) is the most reactive with a DTG peak at about 270 °C. Then, cellulose appears, with a prominent DTG peak at about 400 °C, which is closely followed by lignin. There are quantitative variations in terms of the DTG peaks (height and temp) between the different constituents, and these affirm the discussion on the trend observed for GG thermal decomposition. The apparent discrepancy may be due to the fact that the isolated constituents are devoid of the complexity associated with the lignocellulosic macromolecular structure in their natural matrix.

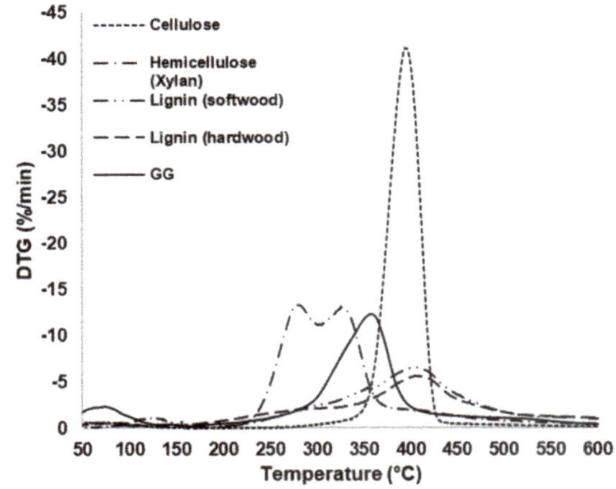

Figure 3. Thermograms of mass-loss rate (DTG) of cellulose, xylan, lignin, and GG in N_2 atmosphere at 20 °C/min.

Table 5 presents the thermal characteristics data from the decomposition of the GG in both atmospheres. A marginal difference exists in the residual mass as the heating rate rises for the decomposition in both the N_2 and the air atmosphere—this underscores the independence of the mass-loss trend on the heating rate. Similar trends have been shown in previous findings for peanut shells [1] and other biomass residues [3] in N_2 and air atmospheres. It is also demonstrated that the DTG peaks increase in height and shift to higher temperatures with an increase in the heating rate. For instance, the maximum DTG peaks in absolute terms for N_2 at 5, 10, and 20 °C/min read 3.21%/min (327 °C), 6.45%/min (339 °C), and 12.3%/min (358 °C), respectively. This is consistent with findings from the published literature for the pyrolysis and combustion of tobacco wastes [51]. This could be due to the heat transfer limitations that arise from the successive rise in the thermal gradient. This phenomenon usually occurs when there is rapid heating at the exterior relative to the inner parts of the sample, thus pushing the decomposition temperature to higher values [3].

Table 5. Heating rate, maximum peak temperature, residual mass, and mass-loss rate of GG decomposition under N_2 and air atmosphere.

	N_2			Air		
β	T_{max} (°C)	DTG (%/min)	R_w (%)	T_{max} (°C)	DTG (%/min)	R_w (%)
	40.2	−0.80		48.5	−0.64	
	326.9	−3.21		289.7	−4.15	
5			28.7	447.4	−1.42	13.2
	617.7	−0.15		638.0	−0.14	
	58.0	−1.09		64.0	−1.08	
	338.5	−6.45		307.3	−11.2	
10			28.8	451.9	−40.6	13.2
	642.3	−0.26		649.3	−0.20	
	74.2	−2.33		74.8	−1.96	
20	358.3	−12.3	29.8	316.2	−23.1	13.1
				462.9	−29.4	
Average			29.1 ± 0.6			13.2 ± 0.1

β = heating rate, T_{max} = maximum peak temperature, R_w = residual mass.

Specifically, the residual mass after thermal degradation in the air is far less than (above 100%) in N_2. This agrees with the findings from the published literature [3]. Thermal degradation in the air is an oxidative reaction (combustion), and it is expected to possess a comparatively higher decomposition rate. This is further attested to by the mass-loss rate peak (DTG), which is shown to be much higher in the air atmosphere for the corresponding stages of degradation. For example, at this 20 °C/min the DTG peaks for the air and N_2 environments, respectively, read −23.1 and −12.3%/min. It is important to note that under air, a prominent peak appears at about 450 °C, as opposed to a very small peak under N_2. This conforms with the published data [8,52,53]. These observations suggest a thermal degradation phenomenon that is related to the combustion of char as char formation is predominant in the latter stage of oxidative degradation processes [8,52].

3.3. *Kinetic Modeling*

3.3.1. Model-Free Technique

Figure 4a,b, respectively, show the plots of the linear curves derived from the application of Equations (6)–(8) under N_2 and air conditions. The DFM, FWO, and STK models have proven suitable in predicting the kinetic parameters because of the high correlation ($R^2 > 0.9$) shown in both atmospheres. The plot was limited to a conversion degree, θ, ($0.15 \leq \theta \leq 0.8$) as this region is where the chemical reactions are predominant, and the kinetic models are more likely to produce realistic results. For the conversion range of 0.2 to 0.6, particularly for the integral methods (FWO and STK), the lines of best fit were approximately parallel. This may be an indication of a similar kinetic behavior in which the same reaction mechanism is exhibited for the specified range. In some cases, the non-parallel nature of the lines was largely restricted to either the earlier or the later part of the conversion. Perhaps, it is a pointer to a dissimilarity in the reaction mechanism that characterizes these decomposition stages. It has been suggested that the non-parallel nature of the linear fits could be an indication of a change in the reaction mechanism at a higher decomposition temperature [54,55]. Wang et al. [56] attributed it to the heterogeneity of the solid produced at the latter stages of degradation. The reaction mechanism at this stage is considered to involve a complex intertwine of diffusion, secondary reactions, and in situ catalysis of metals. It is also important to note that the non-parallel trend demonstrated by the DFM technique affirms the complexity of biomass decomposition, which arises from its intrinsically heterogenous character [55].

The dependence of activation energy, E_θ, on conversion ratio, θ, for the three iso-conversional methods is displayed in Figure 5a,b for the N_2 and the air atmosphere, respectively. Table 6 also presents the E_θ, θ, and the coefficient of determination, R^2, data for both scenarios. These values were determined at an increment of 0.05 for θ, from 0.15 to 0.8. The E_θ had a significant positive correlation with θ (r = 0.27; $p < 0.05$). It has been noted that a significant variation in the apparent activation energy with conversion underscores the complexity associated with the kinetic process [57]. The effects of the decomposition atmospheres were also statistical analyzed, and it was shown that the E_θ was negatively correlated for air (r = −0.25; $p < 0.05$), while it was positively correlated for N_2 (r = 0.84; $p < 0.05$). The trend in the inert environment is different from that in the oxidative. In the inert scenario, the trajectories of the curves were similar, especially for the FWO and STK that are modeled according to the temperature integral approximation. The DFM is a differential technique that is not based on the integral approximation of the temperature function, and thus, it is relatively more accurate [36]. The E_θ versus θ plots show three distinct stages (Figure 5a). The first decomposition stage, θ = 0.15–0.2 for DFM, FWO, and STK presents average values of E_θ as 105, 101, and 98 kJ/mol, respectively. The degradation of the hemicellulose fractions is the predominant process at this stage due to its high reactivity relative to the other polymeric fractions. This may have accounted for the relatively low E_θ values [22]. Following a similar sequence, the average values of E_θ were evaluated as 137, 129, and 127 kJ/mol for θ = 0.25–0.6. At this stage, the combined contribution of the three polymeric constituents is more likely, albeit in different proportions. However, cellulose may be expected to play a significant role given its highly ordered crystalline nature that makes it stable thermally [22,51]. Again, this is corroborated by the earlier observation that cellulose degradation occurs within this period. In the last reaction stage (θ = 0.65–0.8), the E_θ surges to much higher levels, recording average values of 189, 165, and 164 kJ/mol with the widest variability. Other researchers have reported similar trends for other lignocellulosic biomass [36,53]. The final decomposition stage for the pyrolytic processes is typically characterized by lignin and secondary product decomposition as well as char formation [22,51].

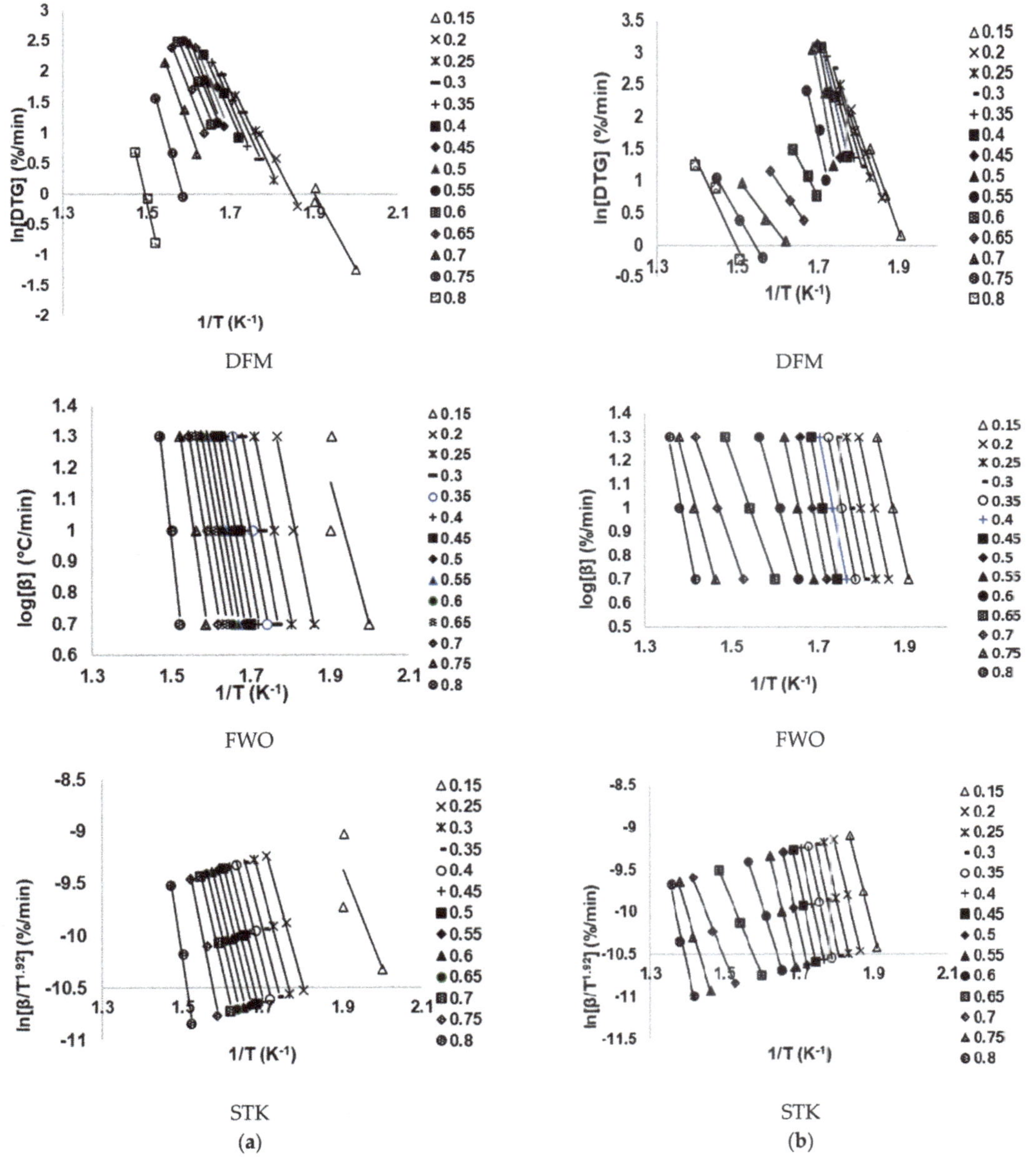

Figure 4. Plots of linear curves for DFM, FWO, STK models in (**a**) N_2 and (**b**) air environments.

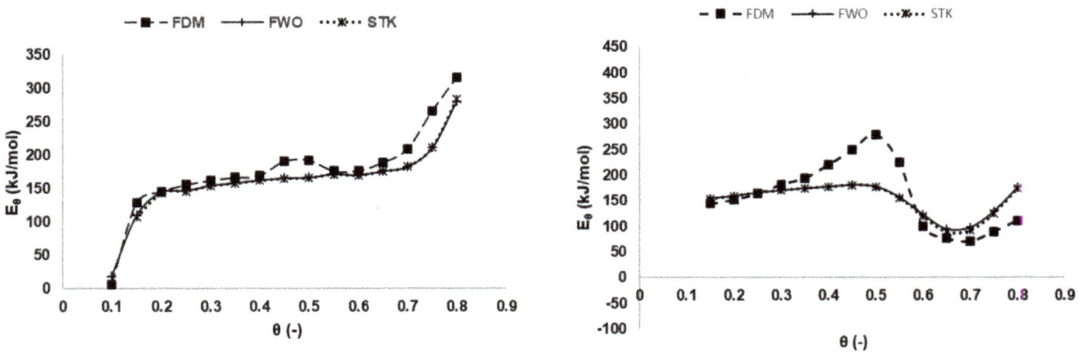

(a) (b)

Figure 5. Plots of apparent activation energy, E_θ, against conversion ratio, θ, (**a**) N_2 and (**b**) air environments.

Table 6. Values of E_θ (kJ/mol), and R^2 for DFM, FWO, and STK models under N_2 and air environments.

	θ	N_2 DFM		N_2 FWO		N_2 STK		Air DFM		Air FWO		Air STK	
		E_θ (kJ/mol)	R^2	E_θ (kJ/mol)	R^2	E_θ (kJ/mol)	R^2	E_θ (kJ/mol)	R^2	E_θ (kJ/mol)	R^2	E_θ (kJ/mol)	R^2
STAGE I	0.15	105.3	0.969	84.7	0.916	80.9	0.694	143.8	0.997	155.2	0.999	154.7	0.999
	0.2	106.3	0.980	118.3	0.916	115.6	0.996	151.8	0.999	159.7	0.999	159.2	0.999
Average		105.8 ± 0.5		101.5 ± 16.8		98.2 ± 17.3		147.8 ± 5.2		157.5 ± 6.4		156.5 ± 6.5	
STAGE II	0.25	122.3	0.982	117.5	0.997	114.4	0.998	163.8	0.999	166.5	0.999	166.2	0.999
	0.3	128.2	0.983	123.3	0.992	120.3	0.994	181.6	0.999	170.6	0.999	170.4	0.999
	0.35	131.3	0.979	127.2	0.988	124.3	0.989	193.9	0.999	174.5	0.999	174.4	0.999
	0.4	132.8	0.977	130.9	0.986	128.1	0.987	219.5	0.993	177.2	0.998	177.2	0.997
	0.45	147.5	0.919	131.0	0.982	128.1	0.983	249.1	0.979	181.1	0.997	181.1	0.995
	0.5	152.1	0.913	132.9	0.982	130.0	0.983	279.1	0.935	177.4	0.997	177.1	0.996
	0.55	138.1	0.975	136.1	0.979	133.3	0.980	224.8	0.928	157.06	0.998	155.5	0.997
	0.6	139.9	0.973	135.8	0.976	132.9	0.977	100.2	0.992	122.9	0.998	119.3	0.999
Average		136.5 ± 9.3		129.3 ± 6.0		126.7 ± 6.1		201.5 ± 51.6		165.9 ± 17.7		162.0 ± 18.8	
STAGE III	0.65	147.2	0.976	138.9	0.974	136.1	0.974	76.3	0.999	94.8	0.999	89.3	0.999
	0.7	163.4	0.982	143.4	0.973	140.7	0.974	69.9	0.991	97.9	0.999	92.2	0.998
	0.75	202.7	0.996	163.6	0.980	161.7	0.981	88.8	0.998	129.1	0.989	124.5	0.983
	0.8	241.5	0.985	214.9	0.981	215.3	0.982	110.8	0.941	176.1	0.983	173.7	0.977
Average		188.7 ± 36.6		165.2 ± 30.2		163.5 ± 31.5		86.4 ± 15.6		124.5 ± 32.7		119.6 ± 33.9	

For the oxidative scenario, the three methods initially show a positive slope, and then plunge into a deep valley mid-way into the conversion (Figure 5b). A similar pattern has been reported in the literature [51]. At θ = 0.15–0.2, a gentle slope with a very high correlation of $R^2 > 0.99$ and comparable values of E_θ (148, 158, 157 kJ/mol) was observed for the three methods. These values compare well with the published data for the E_θ of *Lentinula edodes* pileus in a similar decomposition range [50]. Wu et al. [51] suggested the reactions here involve the oxidative breakdown of hemicellulose, pectin, and N-containing compounds. Although the activation energies for FWO and STK are about 160 kJ/mol, DFM displays a sharp rise in slope from 164 to 279 kJ/mol. It is noteworthy that the average value of E_θ in stage II for DFM was much higher than the other stages as well as the other methods. This may be another attestation to the complexity involved in the thermal degradation processes as this stage involves the simultaneous oxidative decomposition of hemicellulose, cellulose, and lignin [51]. Just above θ = 0.5, E_θ decreases to a minimum of around 0.68, then increases. This trend highlights the fact that this is a complex thermal process and the activation energy values obtained are simply "apparent". Therefore, it is not uncommon for the values of E_θ to exhibit a marked variation from the intrinsic kinetic parameters of an individual step [58]. Unlike the pyrolytic process, the last stage of the oxidative decomposition process involves the combustion of char [51].

Generally, the average values of E_θ for air, for the first and second stages of decomposition, are much higher than for N_2. For instance, the Friedman model at the second stage had E_θ = 202 and 137 kJ/mol for air and N_2, respectively. This observation has been previously observed [1]. The implication of this lower energy barrier is required for the reaction to proceed in an inert environment. Significantly, the average values of E_θ in the final decomposition stage for N_2 were higher than in air. For instance, E_θ for the FWO method in air and N_2 gave 125 and 165 kJ/mol, respectively. This may be due to the distinct phenomenon in an inert environment resulting in char formation, while in air it is combusted [51,53]. These results show that both the pyrolytic and the oxidative processes involve complicated, multi-step reaction mechanisms.

3.3.2. Model-Fitting Technique

Table 7 shows the values of E_θ, R^2, and A at different heating rates derived from the CR model for the GG samples. These were deduced from the slope of linear plots of $\ln\left(\frac{g(\theta)}{T^2}\right)$ against the reciprocal of temperature. The reaction order, n, may be taken as a positive or negative integer. However, it is more practicable to define it as Equation (9) [58].

$$0 \leq n \leq 3 \tag{16}$$

There is a postulation that compares the average value of E_θ obtained from the CR method with that of an iso-conversional technique such as FWO. This provides a means to choose an appropriate decomposition mechanism [8,59]. The closest E_θ among the given integral models is believed to represent a probable reaction mechanism. In this investigation, this postulation was employed. Some non-realistic values were obtained for some models and stages of decomposition; therefore, the values of E_θ that had the same order of magnitude as the model-free kinetic data were the only ones considered. Only the data from stage II decomposition satisfy this criterion and therefore are presented for discussion in Table 7. A strong correlation ($R^2 > 0.9$) was demonstrated for all the heating rates in both conditions—implying that the models fairly approximate the decomposition process. Of all the models, as listed in Table 2, it was demonstrated that only two of them, namely chemical reaction and diffusional, represent the probable mechanism that predominates the thermal process. This validates the assertion from the iso-conversional approximation regarding the multi-step reaction pathways and the associated complexities of GG thermal decomposition. The activation energy and the pre-frequency exponential factor, for the integral model, increase with the heating rates. For the chemical reaction model, they increase with an increase in reaction order.

For β 5, 10, 20, and °C/min under N_2, the values of E_θ for the second-order reaction model were 88.3, 102, and 103 kJ/mol, respectively, while the third-order reactions were, respectively, 109, 126, and 127 kJ/mol. In both (N_2 and air) conditions, the diffusional model presents the highest average value of 216 kJ/mol (air) and 137 kJ/mol (N_2). For the second stage of decomposition, the closest average value of E_θ for the FWO (129 kJ/mol) and CR (130 kJ/mol) models in the N_2 atmosphere represents the diffusional model. This suggests the critical role of diffusion in this stage of decomposition for the tropical grass being investigated. It has been reported in the literature that where the mobility of the reactant constituents depends on the lattice defects, the solid-state reactions mostly occur between either the molecules penetrating the lattices or the crystal lattices [59].

Table 7. Values of E_θ (kJ/mol), R^2, and A (min^{-1}) for N_2 and air atmospheres (stage II) based on Coats–Redfern model.

N_2	5 °C/min			10 °C/min			20 °C/min			Average		
$g(\theta)$	E_θ	R^2	A	E_θ	R^2	A	E_θ	R^2	A	E_θ	R^2	A
Chemical reaction model												
$(1-(1-\theta))^{-1}/(-1)$	88.3	0.966	1.04×10^9	101.9	0.990	2.83×10^{10}	102.8	0.991	3.99×10^{10}	99.8	0.985	2.31×10^{10}
$(1-(1-\theta))^{-2}/(-2)$	108.7	0.954	1.22×10^{11}	126.2	0.983	6.67×10^{12}	127.3	0.983	8.42×10^{12}	123.4	0.976	5.07×10^{12}
Diffusional model												
θ^2	102.6	0.985	5.40×10^9	118.0	0.999	1.97×10^{11}	119.1	0.999	2.58×10^{11}	115.7	0.996	1.53×10^{11}
$(1-\theta) * \ln(1-\theta) + \theta$	111.4	0.982	2.1×10^{10}	128.3	0.999	1.02×10^{12}	129.4	0.999	1.28×10^{12}	125.7	0.994	7.74×10^{11}
$\left(1-(1-\theta)^{1/3}\right)^2$	121.4	0.979	4.78×10^{10}	140.1	0.997	3.24×10^{12}	141.3	0.997	3.83×10^{12}	137.2	0.993	2.37×10^{12}
$1-(2/3)\theta-(1-\theta)^{2/3}$	114.7	0.981	1.01×10^{10}	132.2	0.998	5.50×10^{11}	133.4	0.998	6.74×10^{11}	129.5	0.994	4.11×10^{11}
Air												
Chemical reaction model												
$(1-(1-\theta))^{-1}/(-1)$	105.7	0.997	7.24×10^{10}	124.1	0.989	6.42×10^{12}	121.9	0.976	5.09×10^{12}	156.7	0.985	3.86×10^{12}
$(1-(1-\theta))^{-2}/(-2)$	131.0	0.999	3.05×10^{13}	155.1	0.991	7.23×10^{15}	152.5	0.980	4.65×10^{15}	196.2	0.986	3.97×10^{15}
Diffusional model												
θ^2	120.4	0.984	4.84×10^{11}	142.2	0.974	7.47×10^{13}	139.1	0.953	4.52×10^{13}	180.8	0.973	4.01×10^{13}
$(1-\theta) * \ln(1-\theta) + \theta$	131.3	0.986	3.13×10^{12}	155.2	0.977	7.27×10^{14}	151.9	0.958	3.99×10^{14}	197.4	0.976	3.76×10^{14}
$\left(1-(1-\theta)^{1/3}\right)^2$	143.7	0.990	1.29×10^{13}	170.1	0.980	4.82×10^{15}	166.6	0.962	2.37×10^{15}	216.4	0.978	2.40×10^{15}
$1-(2/3)\theta-(1-\theta)^{2/3}$	135.4	0.988	1.84×10^{12}	160.1	0.978	4.99×10^{14}	156.7	0.959	2.65×10^{14}	203.7	0.977	2.55×10^{14}

3.3.3. Combustion Characteristics

Table 8 presents the combustion characteristics of GG at different heating rates based on the TGA data and Equations (9)–(12). It is shown that the ignition and burnout temperatures rise with the increasing heating rate, while the corresponding time decreases. The combustion performance indices, S, C, D_i, and D_b, all increased with the heating rate. This highlights the fact that the combustion efficiency increases with the heating rate. This same trend has been observed in the published literature [8]. A correlation analysis was conducted using Pearson's coefficient, r, for the combustion atmosphere (n = 3). It was shown that the heating rate had a significantly positive correlation with the $-Rp$ ($p \leq 0.03$), Rv ($p \leq 0.0001$), and C ($p \leq 0.03$), whereas it showed no significant positive correlation with T_i, T_b, S, D_i, and D_b ($p > 0.05$). A similar finding, with the exception of comprehensive combustibility, S, was observed in the combustion study of tea leaves and waste teas [57]. The values of the indices obtained were of the same order of magnitude for nearly all the heating rates that were reported by Chen et al. [8] and Cai et al. [57]. In comparison with other lignocellulosic biomass in relation to their ignition temperatures at 10 °C/min (Pinus: 243 °C, 12 min [8]; Jackfruit peel: 213 °C, 18 min; Jackfruit seed: 246 °C, 21 min [39]; Tamarind husk: 240 °C, 21 min; Tamarind seed: 250 °C, 22 min [42]), the ignition temperature and time of the GG were higher. This implies that the ignition process for the GG may be relatively more difficult.

Table 8. Combustion characteristics of GG at different heating rates.

β (K/min)	Temperature (K) T_i	T_{max}	T_b	Time (min) t_i	t_p	t_b	$\Delta t_{1/2}$	Combustion Characteristics $-R_p$ (%/min)	$-R_v$ (%/min)	$S \times 10^{-7}$ (%^{2}/(min^2*K^3))	$C \times 10^{-5}$ (%/(min*K^2))	$D_i \times 10^{-2}$ (%/min^3)	$D_b \times 10^{-3}$ (%/min^4)
5	528.6	562.8	732.1	47.05	54.05	90.20	60.00	4.15	0.50	0.10	1.49	0.16	0.014
10	544.7	580.4	734.2	25.00	28.60	45.15	26.20	11.15	1.00	0.51	3.76	1.56	0 33
20	555.7	589.4	773.6	13.13	14.80	24.63	22.82	23.11	2.00	1.93	7.49	11.9	2 78

3.3.4. Thermodynamics Analysis

Table 9 shows the thermodynamic parameters for the thermal decomposition of GG at 20 K/min in N_2 and air atmospheres. These values were calculated from the apparent activation energy E_θ based on DFM and a first-order differential model. The choice of DFM was informed by the fact that the temperature function was not modeled as an integral approximation, and the result may thus be expected to be relatively more accurate. Though the global reaction mechanism was assumed here, it is noteworthy that the underlying principle for iso-conversional methods supposes that the description of the process kinetics is based on multiple single-step kinetic equations, in which each equation is associated with a given conversion extent over a narrow temperature range, thus validating the use of the DFM model. The model was used for the computation of the pre-exponential frequency factor (A), which gives insight into the frequency of collisions between reactants [60]. It has been shown that for pre-exponential factor values with orders of magnitude $\leq 10^9$ min^{-1}, the surface reaction pathway is predominantly manifested [61]—this pathway was clearly noticeable for $\theta = 0.6 - 0.7$, in the air atmosphere. The ΔH was negatively correlated with the conversion for air (r = -0.27; $p < 0.05$), while it was positively correlated for N_2 (r = 0.84; $p < 0.05$). The average value of ΔH for decomposition in the air (156 kJ/mol) is higher than for N_2 (142 kJ/mol)—an indication of a relatively higher reactivity in an oxidative reaction, as earlier noted. Zou et al. [50] observed that the ΔH is a reflection of the exchange of heat between complex activated species and reactants so that the higher the value, the higher the reactivity and the faster the rate of reaction. It has been noted that a positive value of ΔH indicates an endothermic reaction [8]. By implication, the pyrolytic and oxidative processes were endothermic throughout the conversion range. The average values of ΔH obtained in this study for both thermal conditions are higher than that for peanut shells, (74.8 kJ/mol; air) and (29.3 kJ/mol; N_2) [1]. The values of ΔG for both atmospheres were positively correlated with conversion. The ΔG is an indication of the total energy increase for the reaction in a thermal degradation process and portrays the difficulty and direction of the reactions. Chen et al. [8] stated that large values of ΔG are suggestive of the low possibility of reactions, and positive values indicate non-spontaneity in reactions. The mean values of ΔG for thermal degradation in N_2 and air are 143 and 135 kJ/mol, respectively. This conforms with the findings in the literature [1]. This suggests a less favorable reaction of GG in N_2 relative to air. In addition, it may be inferred that both the oxidative and pyrolytic processes of GG proceeded in a non-spontaneous manner. Ahmad et al. [60] opined that ΔG is a measure of the amount of energy that is available from the thermal degradation of a given biomass. In comparison to peanut shells and red pepper waste, the thermal decomposition of GG will generate more energy [1,62], whereas it will liberate relatively less energy than Camel grass and Napier grass [38,63]. Generally, the ΔS provides information on the degree of disorder for the reactions taking place in the thermal decomposition process. Unlike in air (r = −0.33; $p < 0.05$), the values of ΔS for N_2 (r = 0.84; $p < 0.05$) were positively correlated with conversion. The trend for ΔS was quite similar to that reported by Cai et al. [57] for tea leaves in a N_2/O_2 atmosphere and that by Ahmad et al. [64] for Para grass, particularly with respect to the appearances of positive and negative values, whereas in air, at $\theta \leq 0.55$, which corresponds mainly to hemicellulose and cellulose decomposition, ΔS had positive values and ΔS in N_2, at $\theta \leq 0.40$, had negative values. This may suggest that the earlier stages of the pyrolytic process had a relatively lower degree of disorder for the product formation, while the converse was true for the oxidative process. Significantly, the dominancy of ΔS positive values, which attests to a high degree of disorder for the thermal processes, has been observed in the literature [57]. The average value of ΔS in N_2 (0.043 kJ/(mol*K) is lower than in air (0.1132 kJ/(mol*K).

Table 9. Thermodynamic parameters for GG thermal decomposition at 20 K/min based on DFM.

	N_2				Air			
θ	A (min^{-1})	ΔH (kJ/mol)	ΔG (kJ/mol)	ΔS (kJ/(mol∗K))	A (min^{-1})	ΔH (kJ/mol)	ΔG (kJ/mol)	ΔS (kJ/(mol∗K))
0.15	3.44×10^{10}	100.96	121.35	−0.0569	2.83×10^{14}	139.21	128.38	0.0134
0.20	2.28×10^{10}	101.61	123.56	−0.0613	1.44×10^{15}	147.17	128.47	0.0317
0.25	5.98×10^{11}	117.47	129.85	−0.0346	1.72×10^{16}	159.03	128.27	0.0522
0.30	1.83×10^{12}	123.21	132.35	−0.0255	7.04×10^{17}	176.86	127.96	0.0830
0.35	3.14×10^{12}	126.47	133.92	−0.0212	8.41×10^{18}	189.12	128.10	0.1035
0.40	3.82×10^{12}	127.90	134.83	−0.0198	1.47×10^{21}	214.60	128.31	0.1464
0.45	6.10×10^{13}	142.37	141.25	0.0031	5.57×10^{23}	244.24	128.90	0.1957
0.50	1.36×10^{14}	146.94	143.48	0.0097	1.80×10^{26}	274.13	130.50	0.2437
0.55	8.32×10^{12}	132.88	137.79	−0.0137	1.13×10^{21}	219.84	134.97	0.1440
0.60	1.03×10^{13}	134.64	138.97	−0.0121	4.21×10^{9}	95.115	139.29	−0.0750
0.65	3.29×10^{13}	141.88	142.79	−0.0025	1.82×10^{7}	71.052	142.09	−0.1205
0.7	4.54×10^{14}	158.00	151.15	0.0191	2.85×10^{6}	64.410	144.75	−0.1363
0.75	2.54×10^{17}	197.27	171.65	0.0715	5.87×10^{7}	83.028	148.80	−0.1116
0.8	3.82×10^{17}	235.88	195.51	0.1126	7.78×10^{25}	104.80	152.79	−0.0814
Average	4.55×10^{16}	141.97	142.75	0.0432	1.85×10^{25}	155.90	135.11	0.1132

4. Conclusions

Guinea grass (*Megathyrsus maximus*) was subjected to thermal degradation in a non-isothermal TGA (N_2 inert and dry air atmosphere) at multiple heating rates of 5, 10, and 20 K/min. The Coats–Redfern, Flynn–Wall–Ozawa, and Starink techniques were utilized to evaluate the kinetic parameters, and these were subsequently used in the combustion characteristics and thermodynamics analyses. A couple of integral models, representative of various decomposition mechanisms, was tested in the model-fitting technique. This was with the primary objective of evaluating the bioenergy potential of GG.

The thermal profile of the GG proceeded distinctly under the different heating scenarios—the average residual mass after the GG decomposition was 29.1% and 13.2%, respectively, for the N_2 and air environments. The model-fitting technique suggested that the chemical reaction and diffusional models play critical roles in the thermal decomposition processes of GG, both in the N_2 and the air atmospheres. The kinetics model also revealed three distinctive stages of decomposition, which correspond to moisture evaporation, devolatilization, and solid residue burnout/char formation. On average, the activation energy for the decomposition in the air is relatively higher than in the N_2 atmosphere. The decomposition process in air showed relatively higher reactivity, with an average value enthalpy change being 156 kJ/mol, as opposed to 141 kJ/mol for the N_2 environment. According to the change in Gibbs free energy, it was also shown that both processes proceeded in a non-spontaneous manner. It may be concluded from the foregoing that the thermal decomposition process of GG, either in an N_2-pyrolytic or oxidative environment, follows a complex pathway that involves parallel and successive reactions. It has been demonstrated that GG would be a suitable feedstock for biofuel production and bioenergy purposes. The information derived from the combustion characteristics would be useful in the development and application of combustion technology for the GG.

Author Contributions: A.O.B., A.A.A. and P.P.I. conceived the research idea; A.O.B., A.A.A., P.P.I. and S.O.A. provided the materials for the research; A.O.B., A.A.A. and P.P.I. did the preliminary sample preparation in the laboratory; A.M.A. and A.G.M. carried out the further laboratory experiments on the prepared samples to obtain the required data; A.O.B. did the data analyses and wrote the first draft of the manuscript; A.A.A., P.P.I., S.O.A., A.M.A. and A.G.M. contributed to the scientific discussion of the manuscript. All authors have read and agreed to the published version of the manuscript.

Funding: This research received no external funding.

Data Availability Statement: The data presented in this study are available on request from the corresponding author.

Conflicts of Interest: The authors declare no conflict of interest.

References

1. Xu, Z.; Xiao, X.; Fang, P.; Ye, L.; Huang, J.; Wu, H.; Tang, Z.; Chen, D. Comparison of combustion and pyrolysis behavior of the peanut shells in air and N_2: Kinetics, thermodynamics and gas emissions. *Sustainability* **2020**, *12*, 464. [CrossRef]
2. Li, Q.; Zhao, C.; Chen, X.; Wu, W.; Li, Y. Comparison of pulverized coal combustion in air and in O_2/CO_2 mixtures by thermo-gravimetric analysis. *J. Anal. Appl. Pyrolysis* **2009**, *85*, 521–528. [CrossRef]
3. Sattar, H.; Muzaffar, I.; Munir, S. Thermal and kinetic study of rice husk, corn cobs, peanut crust and Khushab coal under inert (N_2) and oxidative (dry air) atmospheres. *Renew. Energy* **2020**, *149*, 794–805. [CrossRef]
4. Rueda-Ordóñez, Y.J.; Tannous, K. Thermal decomposition of sugarcane straw, kinetics and heat of reaction in synthetic air. *Bioresour. Technol.* **2016**, *211*, 231–239. [CrossRef] [PubMed]
5. Lasode, O.A.; Balogun, A.O.; McDonald, A.G. Torrefaction of some Nigerian lignocellulosic resources and decomposition kinetics. *J. Anal. Appl. Pyrolysis* **2014**, *109*, 47–55. [CrossRef]
6. Lisseth, C.; Martinez, M.; Patrícia, E.; Rocha, A.; Cassia, A.d.C.O.; Carneiro, O.; José, F.; Gomes, B.; Aparecida, L.; Batalha, R.; et al. Characterization of residual biomasses from the coffee production chain and assessment the potential for energy purposes. *Biomass Bioenergy* **2019**, *120*, 68–76.
7. Balogun, A.O.; Lasode, O.A.; McDonald, A.G. Devolatilisation kinetics and pyrolytic analyses of Tectona grandis (teak). *Bioresour. Technol.* **2014**, *156*, 57–62. [CrossRef] [PubMed]
8. Chen, R.; Li, Q.; Xu, X.; Zhang, D.; Hao, R. Combustion characteristics, kinetics and thermodynamics of Pinus Sylvestris pine needle via non-isothermal thermogravimetry coupled with model-free and model-fitting methods. *Case Stud. Therm. Eng.* **2020**, *22*, 100756. [CrossRef]
9. Azeez, A.M.; Meier, D.; Odermatt, J. Temperature dependence of fast pyrolysis volatile products from European and African biomasses. *J. Anal. Appl. Pyrolysis* **2011**, *90*, 81–92. [CrossRef]
10. Adesemuyi, M.F.; Adebayo, M.A.; Akinola, A.O.; Olasehinde, E.F.; Adewole, K.A.; Lajide, L. Preparation and characterisation of biochars from elephant grass and their utilisation for aqueous nitrate removal: Effect of pyrolysis temperature. *J. Environ. Chem. Eng.* **2020**, *8*, 1–12. [CrossRef]
11. Mohammed, I.Y.; Abakr, Y.A.; Mokaya, R. Biofuel and valuable products recovery from Napier grass pre-processing: Process design and economic analysis. *J. Environ. Chem. Eng.* **2019**, *7*, 102962. [CrossRef]
12. Awasthi, A.; Singh, G.; Dhyani, V.; Kumar, J.; Reddy, Y.S.; Adarsh, V.P.; Puthiyamadam, A.; Mullepureddy, K.K.; Sukumaran, R.K.; Ummalyma, S.B.; et al. Co-pyrolysis of phumdi and para grass biomass from Loktak Lake. *Bioresour. Technol.* **2019**, *285*, 1–6. [CrossRef]
13. Cai, X.; Zhang, X.; Wang, D. Land availability for biofuel production. *Environ. Sci. Technol.* **2011**, *45*, 334–339. [CrossRef] [PubMed]
14. Kumar, S.; Ghosh, P. Sustainable bio-energy potential of perennial energy grass from reclaimed coalmine spoil (marginal sites) of India. *Renew. Energy* **2018**, *123*, 475–485. [CrossRef]

15. Mohapatra, S.; Mishra, C.; Behera, S.S.; Thatoi, H. Application of pretreatment, fermentation and molecular techniques for enhancing bioethanol production from grass biomass—A review. *Renew. Sustain. Energy Rev.* **2017**, *78*, 1007–1032. [CrossRef]
16. Tan, K.T.; Lee, K.T.; Mohamed, A.R. Role of energy policy in renewable energy accomplishment: The case of second-generation bioethanol. *Energy Policy* **2008**, *36*, 3360–3365. [CrossRef]
17. Egwu, U.; Sallis, P.; Oko, E. Study of the impacts of supplements on the specific methane production during anaerobic digestion of the West African Gamba and Guinea Grass. *Fuel* **2021**, *285*, 1–10. [CrossRef]
18. Brosse, N.; Dufour, A.; Meng, X.; Sun, Q.; Ragauskas, A. Miscanthus: A fast-grwoing crop for biofuels and chemicals production. *Biofuels Bioprod. Biorefining* **2012**, *6*, 580–598. [CrossRef]
19. Balogun, A.O.; Lasode, O.A.; McDonald, A.G. Thermo-analytical and physico-chemical characterization of woody and non-woody biomass from an agro-ecological zone in Nigeria. *BioResources* **2014**, *9*, 5099–5113. [CrossRef]
20. Ajimotokan, H.A.; Ehindero, A.O.; Ajao, K.S.; Adeleke, A.A.; Ikubanni, P.P.; Shuaib-Babata, Y.L. Combustion characteristics of fuel briquettes made from charcoal particles and sawdust agglomerates. *Sci. Afr.* **2019**, *6*, 1–9. [CrossRef]
21. Xu, F.; Yu, J.; Tesso, T.; Dowell, F.; Wang, D. Qualitative and quantitative analysis of lignocellulosic biomass using infrared techniques: A mini-review. *Appl. Energy* **2013**, *104*, 801–809. [CrossRef]
22. Balogun, A.O.; Sotoudehniakarani, F.; Mcdonald, A.G. Thermo-kinetic, spectroscopic study of brewer's spent grains and characterisation of their pyrolysis products. *J. Anal. Appl. Pyrolysis* **2017**, *127*, 8–16. [CrossRef]
23. Singh, Y.D.; Mahanta, P.; Bora, U. Comprehensive characterization of lignocellulosic biomass through proximate, ultimate and compositional analysis for bioenergy production. *Renew. Energy* **2017**, *103*, 490–500. [CrossRef]
24. Hameed, S.; Sharma, A.; Pareek, V.; Wu, H.; Yu, Y. A review on biomass pyrolysis models: Kinetic, network and mechanistic models. *Biomass Bioenergy* **2019**, *123*, 104–122. [CrossRef]
25. Papari, S.; Hawboldt, K. A review on the pyrolysis of woody biomass to bio-oil: Focus on kinetic models. *Renew. Sustain. Energy Rev.* **2015**, *52*, 1580–1595. [CrossRef]
26. Masnadi, M.S.; Habibi, R.; Kopyscinski, J.; Hill, J.M.; Bi, X.; Lim, C.J.; Ellis, N.; Grace, J.R. Fuel characterization and co-pyrolysis kinetics of biomass and fossil fuels. *Fuel* **2014**, *117*, 1204–1214. [CrossRef]
27. Almazrouei, M.; Janajreh, I. Model-fitting approach to kinetic analysis of non-isothermal pyrolysis of pure and crude glycerol. *Renew. Energy* **2020**, *145*, 1693–1708. [CrossRef]
28. Ashraf, A.; Sattar, H.; Munir, S.; Husk, R. A comparative applicability study of model-fitting and model-free kinetic analysis approaches to non-isothermal pyrolysis of coal and agricultural residues. *Fuel* **2019**, *240*, 326–333. [CrossRef]
29. Haykırı-açma, H.; Yaman, S. Kinetic modelling of RDF pyrolysis: Model-fitting and model-free approaches. *Waste Manag.* **2016**, *48*, 275–284.
30. Ali, I.; Raza, S.; Bahadar, A. Kinetic analysis of Botryococcus braunii pyrolysis using model-free and model fitting methods. *Fuel* **2018**, *214*, 369–380. [CrossRef]
31. Amoo, L.M. Computational fluid dynamics simulation of Lafia-Obi bituminous coal in a fluidized-bed chamber for air- and oxy-fuel combustion technologies. *Fuel* **2015**, *140*, 178–191. [CrossRef]
32. Wilk, M.; Magdziarz, A.; Gajek, M.; Zajemska, M.; Jayaraman, K. Combustion and kinetic parameters estimation of torrefied pine, acacia and Miscanthus giganteus using experimental and modelling techniques. *Bioresour. Technol.* **2017**, *243*, 304–314. [CrossRef]
33. Bach, Q.V.; Tran, K.Q.; Skreiberg, Ø. Combustion kinetics of wet-torrefied forest residues using the distributed activation energy model (DAEM). *Appl. Energy* **2017**, *185*, 1059–1066. [CrossRef]
34. Sun, L.; Varanasi, P.; Yang, F.; Loqué, D.; Simmons, B.A.; Singh, S. Rapid determination of syringyl: Guaiacyl ratios using FT-Raman spectroscopy. *Biotechnol. Bioeng.* **2012**, *109*, 647–656. [CrossRef] [PubMed]
35. Balogun, A.O.; McDonald, A.G. Decomposition kinetic study, spectroscopic and pyrolytic analyses of Isoberlinia doka and Pinus ponderosa. *Biomass Convers. Biorefinery* **2016**, *6*, 315–324. [CrossRef]
36. Cai, J.; Xu, D.; Dong, Z.; Yu, X.; Yang, Y.; Banks, S.W.; Bridgwater, A.V. Processing thermogravimetric analysis data for isoconversional kinetic analysis of lignocellulosic biomass pyrolysis: Case study of corn stalk. *Renew. Sustain. Energy Rev.* **2018**, *82*, 2705–2715. [CrossRef]
37. Martínez-Ramírez, S.; Frías, M.; Nakanishi, E.Y.; Savastano, H. Pozzolanic Reaction of a Biomass Waste as Mineral Addition to Cement Based Materials: Studies by Nuclear Magnetic Resonance (NMR). *Int. J. Concr. Struct. Mater.* **2019**, *13*, 1–8. [CrossRef]
38. Mehmood, M.A.; Ye, G.; Luo, H.; Liu, C.; Malik, S.; Afzal, I.; Xu, J.; Ahmad, M.S. Pyrolysis and kinetic analyses of Camel grass (Cymbopogon schoenanthus) for bioenergy. *Bioresour. Technol.* **2017**, *228*, 18–24. [CrossRef] [PubMed]
39. Alves, J.L.F.; da Silva, J.C.G.; Mumbach, G.D.; di Domenico, M.; da Silva Filho, V.F.; de Sena, R.F.; Machado, R.A.F.; Marangoni, C. Insights into the bioenergy potential of jackfruit wastes considering their physicochemical properties, bioenergy indicators, combustion behaviors, and emission characteristics. *Renew. Energy* **2020**, *155*, 1328–1338. [CrossRef]
40. Aganga, A.A.; Tshwenyane, S. Potentials of Guinea Grass (Panicum maximum) as Forage Crop in Livestock Production. *Pakistan J. Nutr.* **2004**, *3*, 1–4.
41. Mohammed, I.Y.; Abakr, Y.A.; Kazi, F.K.; Yusup, S.; Alshareef, I.; Chin, S.A. Comprehensive Characterization of Napier Grass as a Feedstock for Thermochemical Conversion. *Energies* **2015**, *8*, 3403–3417. [CrossRef]
42. Alves, J.L.F.; da Silva, J.C.G.; Mumbach, G.D.; di Domenico, M.; de Sena, R.F.; Machado, R.A.F.; Marangoni, C. Demonstrating the Suitability of Tamarind Residues to Bioenergy Exploitation Via Combustion Through Physicochemical Properties, Performance Indexes, and Emission Characteristics. *Bioenergy Res.* **2020**, *13*, 1308–1320. [CrossRef]

43. Alves, J.L.F.; da Silva, J.C.G.; Mumbach, G.D.; da Silva Filho, V.F.; Di Domenico, M.; de Sena, R.F.; Bclzan, A.; Machado, R.A.F.; Marangoni, C. Thermo-kinetic investigation of the multi-step pyrolysis of smoked cigarette butts towards its energy recovery potential. *Biomass Convers. Biorefinery* **2020**, 1–15. [CrossRef]
44. Gandra, J.R.; de Oliveira, E.R.; de Goes, R.H.T.B.; de Oliveira, K.M.P.; Takiya, C.S.; Del Valle, T.A.; Araki, H.M.C.; Silveira, K.; SIlva, D.; da Silva Pause, A.G. Microbial inoculant and an extract of Trichoderma longibrachiatum with xylanase activity effect on chemical composition, fermentative profile and aerobic stability of guinea grass (Pancium maximum Jacq.) silage. *J. Anim. Feed Sci.* **2017**, *26*, 339–347. [CrossRef]
45. Ratsamee, S.; Akaracharanya, A.; Leepipatpiboon, N.; Srinorakutara, T.; Kitpreechavanich, V.; Tolieng, V. Purple guinea grass: Pretreatment and ethanol fermentation. *BioResources* **2012**, *7*, 1891–1906. [CrossRef]
46. Detmann, E.; de Oliveira Franco, M.; Íris Gomes, D.; Medrado Barbosa, M.; de Campos Valadares Filho, S. Protein contamination on Klason lignin contents in tropical grasses and legumes. *Pesq. Agropec. Bras* **2014**, *49*, 994–997. [CrossRef]
47. Clapham, W.M.; Foster, J.G.; Neel, J.P.S. Fatty Acid Composition of Traditional and Novel Forages. *J. Agric. Food Chem.* **2005**, *53*, 10068–10073. [CrossRef] [PubMed]
48. Luengnaruemitchai, A.; Anupapwisetkul, C. Surface morphology and cellulose structure of Napier grass pretreated with the ionic liquid 1-ethyl-3-methylimidazolium acetate combined with either water or dimethyl sulfoxide as a co-solvent under microwave irradiation. *Biomass Convers. Biorefinery* **2020**, *10*, 435–446. [CrossRef]
49. Li, H.; McDonald, A.G. Fractionation and characterization of industrial lignins. *Ind. Crops Prod.* **2014**, *62*, 67–76. [CrossRef]
50. Zou, H.; Evrendilek, F.; Liu, J.; Buyukada, M. Combustion behaviors of pileus and stipe parts of Lentinus edodes using thermogravimetric-mass spectrometry and Fourier transform infrared spectroscopy analyses: Thermal conversion, kinetic, thermodynamic, gas emission and optimization analyses. *Bioresour. Technol.* **2019**, *288*, 121481. [CrossRef] [PubMed]
51. Wu, W.; Mei, Y.; Zhang, L.; Liu, R.; Cai, J. Kinetics and reaction chemistry of pyrolysis and combustion of tobacco waste. *Fuel* **2015**, *156*, 71–80. [CrossRef]
52. Ding, Y.; Huang, B.; Wu, C.; He, Q.; Lu, K. Kinetic model and parameters study of lignocellulosic biomass oxidative pyrolysis. *Energy* **2019**, *181*, 11–17. [CrossRef]
53. Greenhalf, C.E.; Nowakowski, D.J.; Bridgwater, A.V.; Titiloye, J.; Yates, N.; Riche, A.; Shield, I. Thermochemical characterisation of straws and high yielding perennial grasses. *Ind. Crops Prod.* **2012**, *36*, 449–459. [CrossRef]
54. Alvarenga, L.M.; Xavier, T.P.; Barrozo, M.A.S.; Bacelos, M.S.; Lira, T.S. Analysis of reaction kinetics of carton packaging pyrolysis. *Procedia Eng.* **2012**, *42*, 113–122. [CrossRef]
55. Gogoi, M.; Konwar, K.; Bhuyan, N.; Chandra, R. Assessments of pyrolysis kinetics and mechanisms of biomass residues using thermogravimetry. *Bioresour. Technol. Rep.* **2018**, *4*, 40–49. [CrossRef]
56. Wang, X.; Hu, M.; Hu, W.; Chen, Z.; Liu, S.; Hu, Z.; Xiao, B. Thermogravimetric kinetic study of agricultural residue biomass pyrolysis based on combined kinetics. *Bioresour. Technol.* **2016**, *219*, 510–520. [CrossRef] [PubMed]
57. Cai, H.; Zou, H.; Liu, J.; Xie, W.; Kuo, J.; Buyukada, M.; Evrendilek, F. Thermal degradations and processes of waste tea and tea leaves via TG-FTIR: Combustion performances, kinetics, thermodynamics, products and optimization. *Bioresour. Technol.* **2018**, *268*, 715–725. [CrossRef]
58. Vyazovkin, S.; Burnham, A.K.; Criado, J.M.; Pérez-Maqueda, L.A.; Popescu, C.; Sbirrazzuoli, N. ICTAC Kinetics Committee recommendations for performing kinetic computations on thermal analysis data. *Thermochim. Acta* **2011**, *520*, 1–19. [CrossRef]
59. Ding, Y.; Ezekoye, O.A.; Lu, S.; Wang, C.; Zhou, R. Comparative pyrolysis behaviors and reaction mechanisms of hardwood and softwood. *Energy Convers. Manag.* **2017**, *132*, 102–109. [CrossRef]
60. Ahmad, M.S.; Mehmood, M.A.; Taqvi, S.T.H.; Elkamel, A.; Liu, C.G.; Xu, J.; Rahimuddin, S.A.; Gull, M. Pyrolysis, kinetics analysis, thermodynamics parameters and reaction mechanism of Typha latifolia to evaluate its bioenergy potential. *Bioresour. Technol.* **2017**, *245*, 491–501. [CrossRef]
61. Badshah, S.L.; Shah, Z.; Alves, J.L.F.; da Silva, J.C.G.; Noreen, N.; Iqbal, A. Kinetic and thermodynamics study of the pyrolytic process of the freshwater macroalga, Chara vulgaris. *J. Appl. Phycol.* **2021**, *33*, 2511–2521. [CrossRef]
62. Maia, A.A.D.; de Morais, L.C. Kinetic parameters of red pepper waste as biomass to solid biofuel. *Bioresour. Technol.* **2016**, *204*, 157–163. [CrossRef] [PubMed]
63. Manatura, K. Napier Grass Pyrolysis: Kinetic and Thermodynamics Analysis. *RMUTI J. Sci. Technol.* **2019**, *12*, 1–13.
64. Ahmad, M.S.; Mehmood, M.A.; Al Ayed, O.S.; Ye, G.; Luo, H.; Ibrahim, M.; Rashid, U.; Arbi Nehdi, I.; Qadir, G. Kinetic analyses and pyrolytic behavior of Para grass (Urochloa mutica) for its bioenergy potential. *Bioresour. Technol.* **2017**, *224*, 708–713. [CrossRef] [PubMed]

sustainability

MDPI

Article

Environmental Assessment of Latent Heat Thermal Energy Storage Technology System with Phase Change Material for Domestic Heating Applications

Daniel Chocontá Bernal [1], Edmundo Muñoz [2], Giovanni Manente [3], Adriano Sciacovelli [3], Hossein Ameli [4] and Alejandro Gallego-Schmid [5,*]

1 Pariser Building, School of Mechanical, Aerospace and Civil Engineering, The University of Manchester, Sackville Street, Manchester M13 9PL, UK; Daniel.chocontabernal@postgrad.manchester.ac.uk
2 Center for Sustainability Research, Universidad Andres Bello, República 440, Santiago 8370251, Chile; edmundo.munoz@unab.cl
3 Birmingham Centre for Energy Storage, School of Chemical Engineering, University of Birmingham, Edgbaston, Birmingham B15 2TT, UK; G.Manente@bham.ac.uk (G.M.); a.sciacovelli@bham.ac.uk (A.S.)
4 Department of Electrical and Electronic Engineering, Imperial College London, Exhibition Rd, South Kensington, London SW7 2BU, UK; h.ameli14@imperial.ac.uk
5 Pariser Building, Tyndall Centre for Climate Change Research, School of Mechanical, Aerospace and Civil Engineering, The University of Manchester, Sackville Street, Manchester M13 9PL, UK
* Correspondence: alejandro.gallegoschmid@manchester.ac.uk

Citation: Chocontá Bernal, D.; Muñoz, E.; Manente, G.; Sciacovelli, A.; Ameli, H.; Gallego-Schmid, A. Environmental Assessment of Latent Heat Thermal Energy Storage Technology System with Phase Change Material for Domestic Heating Applications. *Sustainability* **2021**, *13*, 11265. https://doi.org/10.3390/su132011265

Academic Editors: Fausto Cavallaro and Valeria Palomba

Received: 1 September 2021
Accepted: 7 October 2021
Published: 13 October 2021

Abstract: The emissions generated by the space and water heating of UK homes need to be reduced to meet the goal of becoming carbon neutral by 2050. The combination of solar (S) collectors with latent heat thermal energy storage (LHTES) technologies with phase change materials (PCM) can potentially help to achieve this goal. However, there is limited understanding of the environmental sustainability of LHTES technologies from a full life cycle perspective. This study assesses for the first time 18 environmental impacts of a full S-LHTES-PCM system from a cradle to grave perspective and compares the results with the most common sources of heat in UK homes. The results show that the system's main environmental hotspots are the solar collector, the PCM, the PCM tank, and the heat exchanger. The main cause of most of the impacts is the extensive consumption of electricity and heat during the production of raw materials for these components. The comparison with other sources of household heat (biomass, heat pump, and natural gas) indicates that the S-LHTES-PCM system generates the highest environmental impact in 11 of 18 categories. However, a sensitivity analysis based on the lifetime of the S-LHTES-PCM systems shows that, when the lifetime increases to 40 years, almost all the impacts are significantly reduced. In fact, a 40-year S-LHTES-PCM system has a lower global warming potential than natural gas.

Keywords: thermal energy storage (TES); latent heat thermal energy storage (LHTES); circular economy; environmental sustainability; life cycle assessment (LCA); climate change

1. Introduction

The main goal of the Paris Agreement is to limit global warming to well below 2 °C, preferably to 1.5 °C, compared to pre-industrial levels [1]. The UK is one of the 192 signatories and has already taken some steps towards low carbon growth. For example, the UK electricity generation sector produced 48.5% of the electricity with low carbon technologies in 2019 [2]. Nevertheless, to meet the goals of the Paris Agreement and become carbon neutral by 2050 [3], the emissions generated by the heating of homes and industry need to be reduced, as they currently account for almost a third of all the total current UK emissions [4]. As stated by the Committee on Climate Change [5], to meet the mentioned net-zero target, the UK has to move entirely to a low-carbon heating system by

2050, which implies that by 2035, the installation of new gas boilers needs to be phased out and replaced by low-carbon heating systems.

Thermal energy storage (TES) is a form of energy storage that can store heat or cold to be used later [6]. This energy storage mechanism is a possible solution to reduce environmental impacts by balancing the energy demand and supply on a daily, monthly, or seasonal basis [6]. Likewise, the implementation of TES can facilitate the integration of heat pumps or solar collectors into the energy network because the combination of both can reduce the cost of distributed heating for consumers by taking advantage of time-of-use electricity rates. The idea behind this strategy is to store heat in off-peak price periods and release that heat in on-peak price periods [7]. Currently, there are three TES technologies with different readiness levels: sensible heat storage (SHS), thermochemical heat storage (THS), and latent heat thermal energy storage (LHTES). In the case of SHS, heat is stored or released due to a temperature change of the stored material (normally water). The main benefits of SHS are relatively low prices and the use of non-harmful materials. However, the low energy density of this technology causes the use of large quantities of stored material. For example, the energy stored density of an SHS system using water is 84 MJ/m^3, whereas for an LHTES system using salt hydrates is 300 MJ/m^3 [6]. Hence, the deployment of these systems requires vast areas, making it viable only at industrial scale. On the other hand, THS technologies are promising due to their highest energy densities and lower heat losses. These technologies charge heat during an endothermic reaction and discharge it during an exothermic reaction. Nevertheless, these technologies are still not commercially available [8]. Lastly, LHTES stores energy through the phase change of the storage medium. Its energy density is significantly higher than SHS, meaning the system can be more compact, and it absorbs and releases heat at a constant temperature, which makes the process more efficient and with less thermal loss. The storage media for LHTES are phase change materials (PCM) with high latent heat of fusion, which allows them to store large amounts of heat when the material changes phase [9]. Nonetheless, these materials have low thermal conductivity, making the charging and discharging of the TES system slow, unless high conducting material such as graphite powder is added. Moreover, incongruently melting PCMs (materials that do not melt uniformly), like many salt hydrates, tend to suffer from phase separation phenomenon, which reduces the heat storage capacity over repeated heating and cooling storage cycles. However, this problem can be solved by selectively adding thickening agents, which limit the distance that the phases can separate by increasing the viscosity of the PCM mixture [10]. Finally, "supercooling" prevents the heat of fusion from being released during the discharge process when the melting point of the PCM is reached, and it is usually avoided by using various nucleating agents. However, stable supercooling is sought in long-term LHTES to minimize the heat losses to the environment, as described in Dannemand et al. [11].

LHTES systems with PCM are technologically ready to be implemented commercially [12]. However, the environmental sustainability of LHTES systems with PCM has been scarcely analysed in the scientific literature. Jungbluth [13] performed a life cycle assessment (LCA) on only the production stage of a phase change material (sodium acetate) for energy storage. The study is based in Germany and is focused only on evaluating the global warming potential and cumulative energy demand of sodium acetate. The results showed that the total greenhouse gas emissions for PCM production are about 5 kg CO_2 eq. per kg of sodium acetate produced. Moreover, this study concluded that the incineration of the PCM and electricity consumption during its production are the main environmental hot spots in terms of global warming potential. Noël et al. [14] performed another LCA focused on the embodied energy and CO_2 emissions of an organic biosourced PCM (Dodecanoic acid) for energy storage. Results showed that this PCM is feasible in energy terms, since less than two years are needed to pay back the embodied energy. Additionally, this PCM could reduce up to 16 t CO_2 eq. the greenhouse emissions in a typical house in the USA (over a 10-year period). Hence, to date, there are no studies that consider the whole life cycle of the LHTES systems, a wide range of environmental impacts (and potential trade-offs) and the comparison with other sources of heat at the house level.

Regarding the assessment of the combination of solar energy and TES systems, Lamnatou et al. [15] reviewed existing literature on the environmental impact of different storage systems used for building-integrated photovoltaic (BIPV) and building-integrated photovoltaic/thermal (BIPVT) installations. The storage systems analyzed were batteries, PCM, and water tanks in different countries such as Canada and the USA. This review concluded that the environmental impact of a configuration depends on the PCM used and the climate conditions, and found that PCM components present a high environmental impact in human toxicity and ecotoxicity categories. Nevertheless, it is worth noting that the studies focused on paraffins and salt-based PCMs used as envelopes of a building and not as the storage medium of an LHTES system, and considered only a limited amount of environmental impact categories. Moreover, none of the reviewed studies analyzed the environmental impacts of the whole heating system of the building (PV panels, heat exchangers, piping, storage system, pumps, valves, etc.).

Therefore, this study investigates, for the first time, 18 environmental impacts and main hotspots of a solar-powered LHTES (S-LHTES) system using sodium acetate trihydrate (SAT) as PCM for short- and long-term heat storage integrated in a UK household combined heating system (hot water and space heating). This approach gives a much more holistic perspective of the environmental sustainability of the S-LHTES-PCM system, observing potential trade-offs between different impacts and proposing improvements based on the hotspots. SAT was selected due to the suitable melting temperature, the high latent heat of fusion and the ability to supercool consistently down to temperatures well below the ambient temperature. The environmental performance of this system is compared, also for the first time, with the most common heat systems in UK households (natural gas, biomass, and heat pumps). Firstly, the methodology used to perform the LCA and calculate the impact categories is described in Section 2. Secondly, the results for the S-LHTES-PCM system, a sensitivity analysis related to the extension of the lifetime of the system and the comparison with other common sources of heat are explained and discussed in Section 3. Finally, the main conclusions are summarised in Section 4.

2. Methods

The life cycle environmental impacts of the combined solar heating system using SAT heat storage have been calculated using the ISO 14040/44 guidelines [16,17]. In compliance with these standards, the study followed the four LCA phases: (i) goal and scope definition; (ii) inventory analysis; (iii) impact assessment, and (iv) interpretation of the results. These stages are described in the next sections.

2.1. Goal and Scope

The goal of this study is to assess the environmental impacts of an S-LHTES-PCM system and compare it with other common sources of domestic heat in the UK using an attributional LCA. The LHTES system will store heat produced by solar collectors using SAT as the PCM and its purpose is to supply heating and domestic hot water to a household in the UK.

To accomplish the stated aim, the functional unit (FU) selected for the study was 1 kWh of heat produced by the novel S-LHTES-PCM system. This unit was selected because the main purpose of the system is to provide heat. Moreover, this FU allows to compare the results obtained with other energy storage technologies or sources of heat, irrespective of the size of the systems. For example, Oró et al. [18] performed a comparative LCA of three different thermal energy storage systems for solar power plants, and the same FU was selected.

The system boundaries of the study illustrated in Figure 1 are from cradle to grave, including the following activities and stages:

- Production of materials:

- ○ Phase change material mixture: SAT (which is the PCM) and Carboxymethyl Cellulose (CMC, thickening agent);
- ○ Evacuated tube solar collector, which includes glass, copper, propylene glycol, and stone wool;
- ○ Water and PCM storage tanks, made of foam and steel;
- ○ Stainless steel (used in the hot water tank, the PCM storage tanks, heat exchanger, and pumps)
- ○ Other materials: rubber (used in solar collectors), butyl acrylate (used in expansion vessels) and cast iron, aluminium, and polyvinylchloride (all used in pumps).

- Manufacturing of:
  - ○ Hot water tanks, PCM tanks, solar collectors, PCM, piping, pumps, expansion vessels, heat exchangers, valves, stratifier, and the crystallization activation device. On the other hand, the hydronic circuits with the radiators and taps/showers were not included because they are not part of the S-LHTES-PCM system per se and are components that a typical UK household already has.
- Use:
  - ○ Electricity consumption of the pumps to circulate the water.
- End of life:
  - ○ Disposal of components after they have reached their life expectancy.
- Distribution:
  - ○ Transport of all the components from production to the household in the UK and to the treatment plant at the end of life.

2.2. *Inventory Data*

The life expectancy of the whole system (20 years) and corresponding components was obtained from existing literature and warranty documents produced by manufacturers (Table 1). A sensitivity analysis has been performed to assess the effect on the environmental impacts of changes in the life expectancy of the S-LHTES-PCM system (see Section 2.3). The S-LHTES-PCM system reaches a solar fraction of heat supply equal to 56% [19]. Since the total annual energy consumption of the house is 3723 kWh [11,19], this means 2085 kWh are provided by the S-LHTES-PCM system (41,700 kWh in 20 years). It is worth noting that the annual energy consumption of 3723 kWh is typical of a low-energy single-family house built according to the passive house standard located in a Danish climate (similar weather conditions as in the UK). The total energy consumption is the sum of two contributions: the space heating demand (2031 kWh) and the domestic hot water demand (1692 kWh) [11,19]. To calculate the domestic hot water demand, a daily water consumption equal to 99 L/day was considered (33 L draw off at 7.00, 12:00, and 18:00), assuming a supply temperature of 50 °C and a cold-water temperature of 10 °C. The space heating demand was calculated on an hourly basis using a building energy simulation tool and the weather data of the Danish climate (similar weather conditions as in the UK). The space heating system was a low-temperature system (floor heating).

The inventory data for the S-LHTES-PCM system is detailed in Table 2. Scientific articles have been used as the main source of primary production data, including the amount and type of raw materials and the electricity and heat consumption during production [11,19,20]. Where there was no information available in the literature, the manufacturer's product catalogues were consulted [21–23]. The database Ecoinvent v3.7 [24] was used as the primary source for background data, while the NREL USLCI database [25] was used to fill data gaps. The following sections describe the inventory data in more detail.

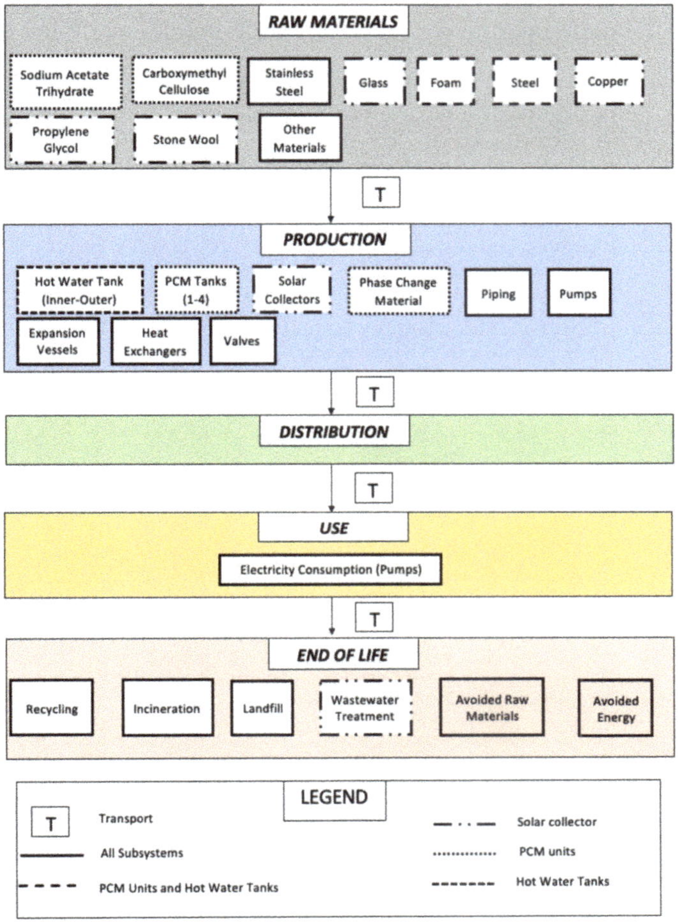

Figure 1. System boundaries for a latent heat thermal energy storage system with phase change materials (PCM), including solar collectors and hot water tanks. (Other materials include rubber (solar collector) butyl acrylate (expansion vessel), acrylonitrile butadiene styrene (stratifier and activation device) and cast iron, aluminium, and polyvinylchloride (pumps)).

Table 1. Lifetime expectancy of all the components used in a solar energy and latent heat thermal energy storage system with phase change material.

Component	Lifetime	Reference
Phase change material	10 years	[26]
Heat exchanger	30 years	[27]
Inner tank	25 years	[28]
Solar collector	20 years	[21]
Heat transfer fluid	3 years	[23]
Water pump	5 years	[24]
Expansion vessels	15 years	[29]
Piping	Copper 40 years Steel 40 years	[30]
Valves	40 years	[31]

Table 2. Inventory data for a solar energy and latent heat thermal energy storage (S-LHTES) system with phase change material (PCM). All values are expressed per 1 kWh of heat produced by the system.

Life Cycle Stage	S-LHTES-PCM	Reference
Raw materials		
Sodium acetate trihydrate (PCM) (g)	41.8	[10,11,32]
Carboxymethyl cellulose (PCM) (g)	0.4	
Stainless steel (hot water tank, PCM tanks, heat exchanger, solar collector, and pumps) (g)	12.7	[21,22,24,28,32,33]
Glass tube (solar collector) (g)	7.6	[21,24]
Foam (hot water tank, PCM-tanks) (g)	5.8	[20]
Carbon steel (hot water tank, piping, heat exchanger, expansion vessel, and valves) (g)	19.6	[22,24,28,32]
Copper (solar collector, piping, and pumps) (g)	1.5	[21,24]
Propylene glycol (solar collector) (g)	2	[23]
Stone wool (solar collector) (g)	1.1	[21,24]
Other materials (solar collector, expansion vessel, and pumps) (g) [a]	0.9	[21,24]
Production		
Electricity (all elements) (kJ)	99.7	[13,24,34]
Heat (all elements) (kJ)	29	
Transport		
Freight lorry (all elements) (kg·km)	2.1	[24]
Freight train (all elements) (kg·km)	0.02	
Use		
Electricity Consumption (kJ)	12.9	[32]
End of Life		
Recycling: Plastics (g)	1.8	[35]
Recycling: Metals (g)	27	[36,37]
Incineration with energy recovery: Plastics (g)	3.4	[35]
Incineration with energy recovery: PCM (g)	42.2	[13]
Incineration with energy recovery: Stone wool (g)	0.7	[38]
Landfilling: Plastics (g)	1.3	[35]
Landfilling: Glass (g)	7.6	[39]
Landfilling: Metals (g)	5.1	[36,37]
Landfilling: Stone wool (g)	0.4	[38]
Wastewater treatment (l)	0.005	[24]

[a] Other materials: Rubber (solar collector) 0.4 g; butyl acrylate (expansion vessel) 0.2 g; cast iron (pumps) 0.4 g; aluminium (pumps) 0.006 g; and polyvinchloride (pumps) 0.01 g and acrylonitrile butadiene styrene (activation device and stratifier) 0.04g.

2.2.1. Raw Materials

The whole S-LHTES-PCM system is divided into three subsections as illustrated in Figure 2: (i) the solar collector unit, (ii) the PCM storage unit, and (iii) the hot water tank unit.

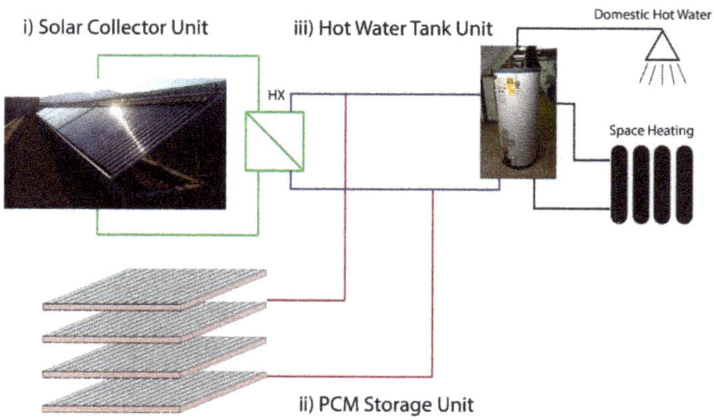

Figure 2. Diagram of a solar energy and latent heat thermal energy storage (S-LHTES) system with phase change material (PCM). Adapted from Englmair et al. [19].

The solar collector subsystem includes the solar collectors, a plate heat exchanger, the heat transfer fluid, and a copper piping section. Seven evacuated tubular collectors with an aperture area of 3.2 m^2 each have been considered. Specifically, the Thermomax HP 450 tubular collectors from Kingspan Solar were selected due to their low heat loss coefficient [21]. This low coefficient enables higher temperatures of the heat transfer fluid and, in turn, the achievement of the conditions required for stable supercooling of SAT [11]. According to the data reported by Kingspan Solar [21], the absorber is made of copper with a $TiNO_X$ coating reaching 95% absorbance. Each collector is composed of thirty low-iron glass tubes, which maintain a vacuum condition [21]. Additionally, stainless steel and ethylene propylene diene monomer (EPDM) rubber are needed for the mounting frames and clips. A layer of stone wool is also needed to increase insulation. The Ecoinvent dataset [24] for a home solar collector system with evacuated tube collectors was adapted according to the design characteristics mentioned above. The heat transfer fluid considered is TYFOCOR-LS liquid, which is an aqueous solution of propylene glycol [23]. As stated by Kingspan Solar [21], each solar collector needs 1.7 L of fluid, so taking into account the life expectancy stated in Table 1, and the density of propylene glycol, 116.9 kg of propylene glycol have been considered (2.0 g/FU). The plate heat exchanger is composed of 2.5 kg of stainless steel (0.06 g/FU) and 31 kg of carbon steel (0.74 g/FU) [22].

The raw materials for the PCM unit subsystem were obtained from LHTES system studies [10,11,32]. According to these studies, four separate modules of 150 L each, composed the PCM storage unit. Considering the life expectancies in Table 1, 11,742.4 kg of SAT (41.8 g/FU) and 17.8 kg (0.4 g/FU) of carboxymethyl cellulose are needed as PCM and thickening agent, respectively. To store this quantity of PCM mixture, four stainless steel tanks are needed to avoid corrosion [33]. Each tank features an insulation layer of 10 cm of rigid foam to minimize thermal losses, which is the most common type of advanced heating insulation material used in industry [20].

A parallel channel heat exchanger made of steel is used to achieve a high heat exchange capacity rate in the PCM unit, which is an important factor for the system performance. The channels have an internal height of 4 mm and a width of 130 mm separated by 20 mm spacers, as reported by Englmair et al. [32]. In total, there are 16 parallel channels at the bottom of the PCM chamber and 14 on the top. The weight of the heat exchanger (13.8 g/FU) was obtained by estimating the difference between the total weight of each empty unit and the weight of the PCM container.

The dimensions for the hot water tank unit were gathered from Englmair et al. [10], where a "tank in tank" technology provides both domestic hot water and space heating for the house. The outer tank and inner tanks have dimensions of 0.8 m × 1.6 m × 0.002 m and 0.45 m × 1.1 m × 0.002 m, respectively. To obtain the raw materials used in the "tank in tank", data provided by a hot water tank manufacturer called ACV was used [28]. According to ACV, the outer tank is made of steel with a 0.05 m insulation layer of rigid foam, while the inner tank is made of stainless steel. Additionally, the inner tank has a citric acid coating to improve the corrosion resistance properties of the material. Englmair et al. [32] also state that the hot water tank needs a polymeric inlet stratifier (for properly admitting the hot water). It has been assumed that the stratifier is made of acrylonitrile butadiene styrene (ABS) and that it is a cylinder with an inner diameter of 16 mm, an outer diameter of 20 mm and a height of 1.6 m.

The S-LH-PCM system also needs extra components such as piping sections, pumps, expansion vessels, and valves. A total number of 11 valves, five 50 L expansion vessels and three 50 W pumps are needed [32]. For these components, the Ecoinvent dataset [24] was used. Regarding the piping used for this heating system, the weight per unit of length was obtained from the inner and outer diameters and the type of material reported in Aste and Groppi [40] (see Table S1 in Supporting Materials).

2.2.2. Production

All the components were assumed to be manufactured in the UK unless stated differently. The Ecoinvent 3.7 dataset [24] was used to obtain the amount of electricity and heat consumption during the production of most of the components. However, for the PCM, the data was obtained from a chemical company (SGL Carbon, 2021) and the PCM was assumed to be produced in Germany [13]. For all the storage tanks, the consumption of energy during the production was obtained from an LCA of household water tanks [34].

2.2.3. Use

The electricity consumption of the pumps was estimated from the power rating of the devices. There were three pumps with a power consumption of 50 W each and operating for 3000 h/yr. Therefore, the total electricity consumption of the system is 150 kWh/yr and 12.9 kJ/FU (Table 2).

2.2.4. Transport

The transportation distances of the different raw materials and components from the market processes of the Ecoinvent 3.7 database [24] have been considered due to the lack of specific data from the suppliers. Regarding the transportation devices, the hot water tanks and PCM tanks were assumed to be transported from the production site to the household by a 16–32 t Euro 6 lorry. For the PCM, 16–32 t Euro 6 lorries and freight trains were considered. Lastly, for the waste treatment, a distance of 50 km in a 16–32 t Euro 6 lorry was assumed for all the materials used.

2.2.5. End of Life

Table 3 illustrates the type of treatment, values, and literature references considered for each raw material. Average UK values for recycling, incineration, and landfilling were used for each raw material. The wastewater treatment associated with the end of life of the heat transfer fluid (propylene glycol) has also been considered. The "net scrap" approach was applied for recycling [41], meaning that only the percentage of recycled material that exceeds the recycled content in the original raw materials has been credited to the system. The environmental impact associated with the recycling process has also been included and, in the case of incineration, the amount of energy produced during the process (heat and electricity) has been attributed to the system. The Ecoinvent 3.7 database [24] has been considered for all waste and wastewater treatment methods.

Table 3. End of life treatment for different raw materials in the UK.

Material	Recycle	Landfill	Incineration	Reference
Plastics	28%	20%	52%	[35]
Steel and cast iron	85%	15%		[36]
Stone wool		39%	61%	[38]
Tempered glass			100%	[39]
Copper	65%	35%		[37]
Phase change material			100%	[13]

2.3. Sensitivity Analysis

A sensitivity analysis was performed to assess the effect on expanding the life expectancy of the S-LHTES-PCM system. The lifetime of each part of the system is highly uncertain as it depends on variable aspects such as the level of maintenance. S-LHTES-PCM systems are a new and under development technology, and there is the potential to increase the life expectancy of the different parts of the system with appropriate maintenance, in line with the circular economy strategy of slowing [42–44]. For this reason, it has been estimated that the heat exchanger and inner tank can increase their useful life up to 40 years (from the initial 20 years; see Table 1), while the solar collector, expansion vessels, and PCM useful life can be increased to 25, 20, and 15 years, respectively (from the initial

20, 15, and 10 years; see Table 1). For pipes, valves, and water pumps, the useful life was not changed. Considering these premises, the sensitivity analysis considers different levels of lifetime extension (25, 30, 35, and 40 years of operation) and compares them with the current situation (20 years).

2.4. Comparison with Current Sources of Heat in the UK

Current heat generation scenarios for households in the UK were analyzed and compared with the S-LHTES-PCM system. Data was collected from Ecoinvent 3.7 [24] and adapted to the UK conditions when possible (e.g., UK electricity grid and UK natural gas). The following heat generation scenarios were evaluated: (i) biomass, (ii) heat pump, (iii) natural gas, (iv) S-LHTES-PCM system, and (v) S-LHTES-PCM system with 40 years of lifetime (see Section 2.3). Natural gas was selected because it is the main source of heat in UK households (76% of the total) and biomass and heat pumps because they are considered low-carbon sources [45]. The biomass system includes the production of natural wood from the forest, infrastructure, air emissions, electricity required for operation, and ash disposal. The Ecoinvent database [24] considers wood-fired furnaces for domestic use with nominal capacities of less than 15 kW. For the heat pump, the Ecoinvent database was adapted to European conditions. The system has a heat capacity of 10 kW and a lifetime of 20 years. The heat pump system delivered approximately 20,000 kWh in 2000 operating hours and Ecoinvent 3.7 [24] includes emissions of the refrigerant R134a during operation. For the natural gas system, the Ecoinvent database includes a mix of central and small-scale gas boilers, natural gas production, the energy requirements (electricity, heat, and burnt natural gas), and the emissions of the high-pressure distribution network from the UK.

2.5. Impact Assessment

SimaPro 9.1 software has been used to model the system and the impact categories have been calculated according to the "Recipe 2016 midpoint (H)" methodology [46]. The following impact categories have been considered: global warming potential (GWP), stratospheric ozone depletion potential (ODP), ionizing radiation potential (IRP), ozone formation potential, human health (OFPh), fine particulate matter formation potential (PMP), ozone formation potential, terrestrial ecosystems (OFPt), terrestrial acidification potential (TAP), freshwater eutrophication potential (FEP), marine eutrophication potential (MEP), terrestrial ecotoxicity potential (TEP), freshwater ecotoxicity potential (FETP), human carcinogenic toxicity potential (HTPc), human non-carcinogenic toxicity potential (HTPnc), agricultural land occupation potential (ALOP), mineral depletion potential (MDP), fossil depletion potential (FDP), water consumption potential (WDP), and cumulative energy demand (CED).

3. Results

3.1. Life Cycle Assessment

As shown in Figure 3, the system's main environmental hotspots are the solar collector in nine of the 18 impact categories (contributions between 35% in OFPh and 73% in FETP), the PCM in five categories (between 26% in GWP and 65% in WDP), the PCM tank in three categories (between 35% in MEP and 49% in ODP), and the heat exchanger is the main contributor to HTPc (30% of the total). Altogether, these parts contribute over 83% in all the 18 impact categories evaluated.

For solar collectors the evacuated tube collector is the leading environmental hotspot, representing between 67% (water consumption, WDP) and 90% (land use, ALOP) of solar collector-related impacts in all categories evaluated. These impacts are mainly associated with the raw materials (particularly copper and borosilicate glass) in the evacuated tube collector and the production processes. For example, over 90% of the impacts of the evacuated tube collector in the TEP, FEP, and HTPnc categories are associated with its copper content. The borosilicate glass is the primary environmental hotspot of the evaluated tube collector in the GWP, IRP, OFPh, ALOP, and FDP impact categories, with contributions over

35%. In GWP, the impact is mainly associated with the electricity and heat consumption in the borosilicate glass production plant. Electricity consumption is also responsible for 82% of the solar collector-related IRP impact. In the OFPh category, NO_x emissions generated in the glass tube production plant are responsible for 54% of the impact of the evacuated tube collector. Heat and electricity used in the production contribute, respectively, 40% and 35% of the FDP impact of the evacuated tube collector.

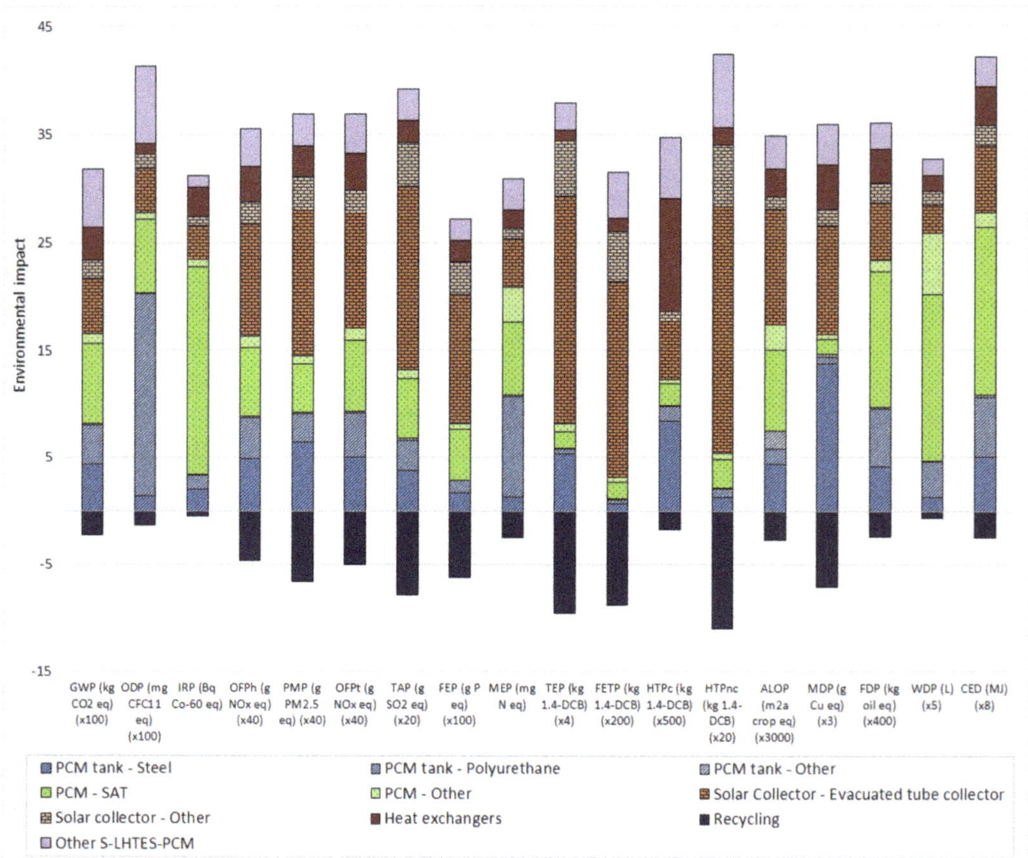

Figure 3. Environmental contribution analysis of a solar domestic system with latent heat thermal energy storage technology with phase change material (S-LHTES-PCM). (All impacts are expressed per kWh of heat produced. GWP: global warming potential; ODP: ozone depletion potential; IRP: ionising radiation potential; OFPh: ozone formation potential, human health; PMP: fine particulate matter formation potential; OFPt: ozone formation potential, terrestrial ecosystems; TAP: terrestrial acidification potential; FEP: freshwater eutrophication potential; MEP: marine eutrophication potential; TEP: terrestrial ecotoxicity potential; FETP: freshwater ecotoxicity potential; HTPc: human carcinogenic toxicity potential; HTPnc: human non-carcinogenic toxicity potential; ALOP: agricultural land occupation potential; MDP: mineral depletion potential; FDP: fossil depletion potential; WDP: water depletion potential; CED: Cumulative energy demand; SAT: Sodium acetate trihydrate).

With regards to the PCM subsystem, SAT contributes to the highest environmental burden, accounting for 68% (MEP and TEP) and 96% (IRP) of the impact of the PCM. As illustrated in Figure 3, SAT is the leading environmental contributor in the whole S-LHTES-PCM system in the GWP (23% of the total), IRP (62%), FDP (35%), WDP (48%), and CED (37%) categories. The main environmental hotspots in the SAT life cycle are associated with the electricity used in the production of the components (sodium hydroxide and acetic acid), and to a lesser extent, in the production process of sodium acetate trihydrate. As for the PCM tank, steel and polyurethane (foam) represent the major environmental impacts. These parts of the PCM tank represent 68% (ALOP) and 100% (ODP) of the impact of the PCM system. Polyurethane is one of the main environmental hotspots of the PCM tank in the categories of ODP (93% of the contribution of the whole PCM system), MEP (87%), FDP (57%), WDP (71%), and CED (52%). These impacts are mainly due to the use of methylene diphenyl diisocyanate (MDI) in polyurethane production. Nitrate emissions to water in MDI production are responsible for 80% of the MEP impact of the PCM system. The other categories mentioned above (ODP, FDP, WDP, and CED) are strongly influenced by the use of aniline in the MDI production process. If we focus on the PCM tank, steel is the main environmental hotspot in 13 out of the 18 impact categories analysed. In the GWP category, heat production from hard coal industrial furnaces and electricity production are the steel main environmental hotspots. Emissions of particulate matter < 2.5 μm in the ferrochromium production process is the main environmental hotspot in the PMP category, accounting for 47% of the impact of steel on the PMP category. In the TEP category, air emissions of copper in ferronickel production are responsible for 60% of the steel impact. Air emissions of chromium VI associated with ferrochromium production are responsible for 48% of the HTPc impact, while nickel ore consumption in ferronickel production is responsible for 85% of the impact in the MDP category.

End-of-life stages generate both environmental benefits and impacts. Steel recycling generates higher environmental benefits in eight impact categories (GWP, OFPh, OFPt, MEP, ALOP, MDP, FDP, and CED), while copper does so in ten (ODP, IRP, PMP, TAP, FEP, TEP, FETP, HTPc, HTPcn, and WDP). Overall, recycling reduces the total environmental impacts of the solar energy and S-LHTES-PCM system by up to 28% (FETP).

3.2. Sensitivity Analysis: Lifetime Extension

A sensitivity analysis was performed based on the lifetime of the S-LHTES-PCM system (see Section 2.3). As shown in Figure 4, the increase of life expectancy of the S-LHTES-PCM system to 25 years significantly reduces the environmental impacts in all the evaluated categories. However, when the lifetime increases to 30 years, the environmental impacts of five categories (FETP, TEP, HTPnc, FEP, and TAP) increase over the impacts of the original system (20 years lifetime). This is because the S-LHTES-PCM system requires the replacement of the solar collector, which has reached its lifetime at 25 years. The solar collector is the leading environmental hotspot of the S-LHTES-PCM system in the five categories mentioned above. Because of this, and given that the system will have a lifetime of 30 years, the new solar collector will only be used for five years, and, therefore, not being able to amortize its environmental burden over its potential lifetime, which increases the impacts of the system.

As illustrated in Figure 4, when the lifetime increases from 30 to 40 years, most of the impacts decrease because the environmental burden of the leading environmental hotspots of the system (PCM and solar collector) are distributed throughout their lifetime. FETP is the only impact category that did not decrease, but had a negligible increase (0.6%). The TEP and HTPnc impact categories showed minor decreases when the lifetime increased to 40 years (3% and 4%, respectively). These small variations can be a result of the inherent uncertainty associated with the data. However, these minor variations can be justified because the solar collector is the primary environmental hotspot in these categories (FETP, TEP, and HTPnc). The solar collector initially has a lifetime of 20 years and, although it increases its lifetime to 25 years, two solar collectors are used when the useful life of the S-LHTES-PCM system increases to 40 years, which does not affect these categories. However, these categories are affected to a lesser extent by the PCM tank that doubles its lifetime, which generates the environmental benefit observed in the TEP and HTPnc categories. On the other hand, the HTPc, ODP, and GWP categories showed the most significant decreases with 53%, 40%, and 36% reduction, respectively. HTPc presents the highest reduction in all the environmental impacts, because its main environmental hotspots are PCM tank (28%) and heat exchanger (30%). Both parts of the S-LHTES-PCM system increase from 20 years to 40 years, which influences the environmental benefits of the HTPc category. For ODP, the main environmental hotspot is the PCM tank (49%), and, therefore, increasing its lifetime decreases the impact of this category. In GWP, the PCM tank (26%) and the heat exchanger (10%) are two of the main environment hotspots, and increasing their lifetime decreases the global warming impact. This will imply a reduction from 0.29 kg CO_2 eq./kWh to 0.19 kg CO_2 eq./kWh when the lifetime of the S-LHTES-PCM system increases to 40 years. However, it should be noted that increasing the lifetime of the system could potentially increase the cost of repairs and maintenance. Therefore, the economic impact of increasing the lifetime of the system should be analyzed in future studies.

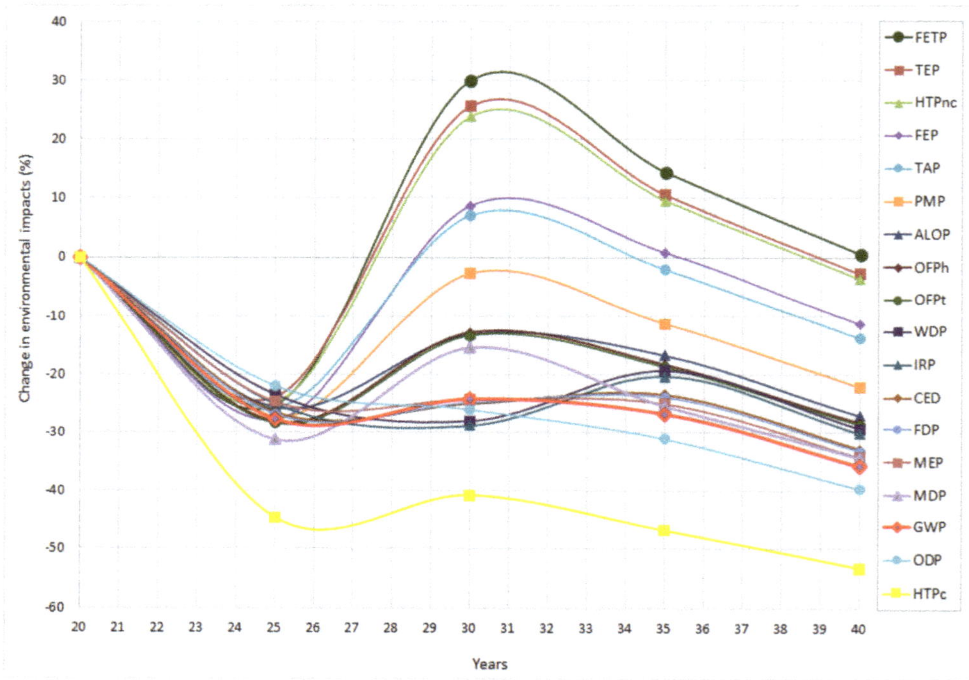

Figure 4. Percentual variation of the environmental impacts for different lifetimes of a solar domestic system with latent heat thermal energy storage technology with phase change material (S-LHTES-PCM). For impact nomenclature, see Figure 3.

3.3. *Comparison with Natural Gas, Biomass and Heat Pumps*

For the scenario analysis, the generation of 1 kWh of heat from biomass, heat pump, natural gas, and two S-LHTES-PCM systems (with regular and extended lifetimes of 20 and 40 years) was considered (see Section 2.3). As illustrated in Figure 5, the S-LHTES-PCM system (20 years) generates the highest environmental impacts in 11 of the 18 categories evaluated (GWP, ODP, TAP, FEP, MEP, TEP, FETP, HTPc, HTPnc, MDP, and WDP). The main cause of these impacts is the emissions from the extraction of raw materials and processes associated with the production of the components of the S-LHTES-PCM system, especially solar collector, PCM, PCM tanks, and heat exchanger (see Section 3.1). In the case of GWP, the S-LHTES-PCM system generates 0.30 kg CO_2 eq./kWh. However, when the life expectancy of the S-LHTES-PCM system is expanded to 40 years, the GWP value is reduced to 0.19 kg CO_2 eq./kWh, becoming the third-best option and with similar values to the second (heat pumps with 0.17 kg CO_2 eq./kWh). Natural gas technology generates the second-highest impacts in GWP with 0.27 kg CO_2 eq./kWh, mainly due to direct emissions from the natural gas combustion process, while biomass technology generates the lowest environmental impacts in the GWP category (0.03 kg CO_2 eq./kWh). However, biomass presents the highest value in OFPh, OFPt, ALOP, and CED. Although there is a significant amount of biomass in the UK [47], the indigenous biomass resources and energy crops only could service, in the best scenario, up to 44% of UK energy demand by 2050 [48]. Bioenergy is a key renewable energy technology targeted to provide options for decarbonizing heat, power, and transport energy in the UK. However, there are growing demands for bioenergy for different energy vectors. Therefore, there will likely be growing competition within the bioenergy sector for feedstock [47]. In addition, biomass resources (crop residues, forestry products, waste, and land) also compete with food production, conservation, animal feed, animal bedding, construction material, panel industry, sawmills, pulp and paper industry, among others [47], leading to uncertainty about their future availability in significant amounts for household heating. The heat pump technology has the highest impacts in the IRP category and the second-highest impacts in another ten (TAP, FEP, MEP, TEP, FETP, HTPc, HTPnc, ALOP, MDP, and WDP). The impacts of this technology are mainly associated with the use of energy from the UK electricity system, representing between 60% to 99% of all impact categories evaluated.

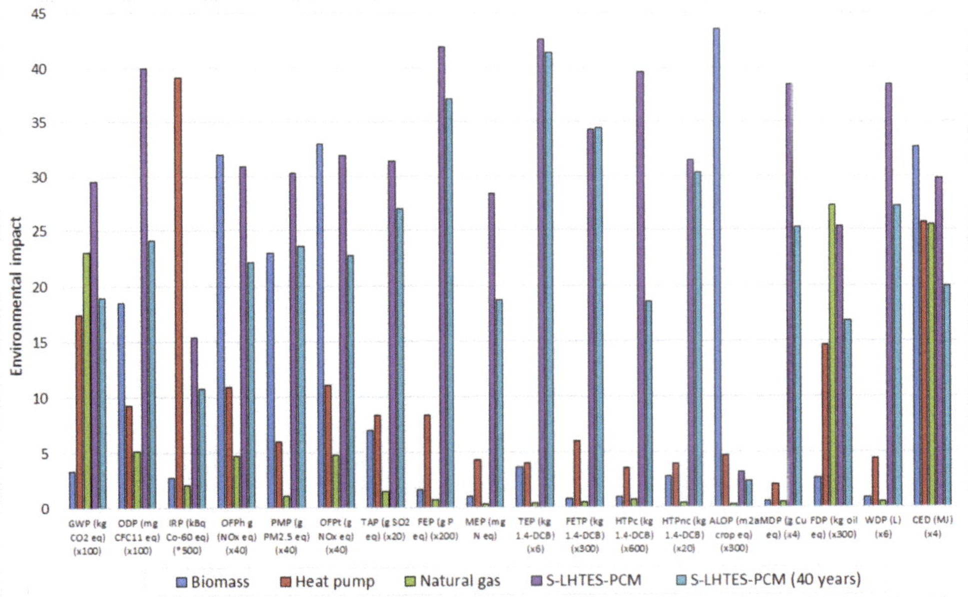

Figure 5. Comparative environmental impact analysis of different types of household energy sources. For acronyms and impact nomenclature, see Figure 3.

Finally, it is essential to highlight that if the lifetime of the S-LHTES-PCM system increases to 40 years, its use in households can minimize most of the environmental impacts. For example, as mentioned above, an S-LHTES-PCM system with a 40 years life expectancy has significantly lower values of GWP compared to natural gas, which is the primary source of energy in households in the UK [45,49]. This is relevant considering that the S-LHTES-PCM system is a technology still under development with potential room for improvement. In this sense, improving the efficiency of the S-LHTES-PCM system, applying the circular economy principles regarding maintenance and lifetime expansion, and decarbonizing the UK electricity are options that can improve the environmental performance of the system and minimize all the impacts, including climate change.

4. Conclusions

This study analyses, for the first time, 18 environmental impact categories for an S-LHTES-PCM system. The environmental performance of the system is compared against traditional heat sources in the UK (biomass, heat pumps, and natural gas). Moreover, a sensitivity analysis was performed based on the lifetime of the S-LHTES-PCM system.

Thanks to the implementation of the life cycle assessment methodology, it was possible to identify the main environmental hotspots of the S-LHTES-PCM system: the solar collector, PCM, PCM tank, and heat exchanger. Altogether, these parts contribute over 83% in all the 18 impact categories evaluated. The environmental impacts are mainly associated with the system's raw materials and the energy consumption in the production processes. For this reason, extending the lifetime of the systems according to circular economy principles improves the system's environmental performance. In this sense, thanks to a sensitivity analysis, it has been demonstrated for the first time that when the S-LHTES-PCM system lifetime is increased to 40 years (from the initial 20 years considered), the environmental performance improves and can be a competitive option from an environmental perspective if compared with other traditional household energy sources, like natural gas, biomass, and heat pumps.

The results of this study will work as a baseline to identify where LHTES systems need to improve before being implemented commercially in the UK. Even more, the general conclusions and hotspots identified can be applied to the implementation of similar systems in other countries. Future research should investigate the social and economic aspects to have a holistic vision that considers the three pillars of sustainable development. The results obtained could be of special interest to stakeholders of the construction and energy sectors and policymakers interested in potential solutions to achieve a more sustainable delivery of space and water heat in households.

Supplementary Materials: The following are available online at https://www.mdpi.com/article/10.3390/su132011265/s1, Table S1. Main features of the piping in the phase change material (PCM), collector and water tank circuits. Adapted from Englmair et al. (2020).

Author Contributions: Conceptualization, D.C.B., E.M., G.M., A.S. and A.G.-S.; Data curation, D.C.B. and E.M.; Formal analysis, A.G.-S.; Funding acquisition, A.S., H.A. and A.G.-S.; Investigation, D.C.B., E.M. and A.G.-S.; Methodology, A.G.-S.; Project administration, A.S., H.A. and A.G.-S.; Resources, G.M. and A.G.-S.; Software, D.C.B. and E.M.; Supervision, A.S. and A.G.-S.; Validation, A.G.-S.; Writing—original draft, D.C.B., E.M. and A.G.-S.; Writing—review & editing, D.C.B., A.S., H.A. and A.G.-S. All authors have read and agreed to the published version of the manuscript.

Funding: The authors would like to acknowledge the financial support received by the EPSRC Supergen Energy Network (EP/S00078X/1) through the Flex Funding of the project 'Whole system analysis impact of thermal energy storage technologies in future UK energy networks'.

Institutional Review Board Statement: Not applicable.

Informed Consent Statement: Not applicable.

Data Availability Statement: The data presented in this study are available on request from the corresponding author.

Acknowledgments: The images in Figure 2 were obtained from Wikimedia Commons. The authors would like to thank Xi Zhang for his contribution to the article with data regarding the main sources of heat in UK households.

Conflicts of Interest: The authors declare no conflict of interest.

Acronyms

ABS	Acrylonitrile butadiene styrene
ALOP	Agricultural land occupation potential
BIPV	Building integrated photovoltaic
BIPVT	Building integrated photovoltaic/thermal

CED	Cumulative energy demand
CMC	Carboxymethyl cellulose
EPDM	Ethylene propylene diene monomer
FDP	Fossil depletion potential
FEP	Freshwater eutrophication potential
FETP	Freshwater ecotoxicity potential
FU	Functional unit
GWP	Global warming potential
HTPc	Human carcinogenic toxicity potential
HTPnc	Human non-carcinogenic toxicity potential
IRP	Ionizing radiation potential
LCA	Life cycle assessment
LHTES	Latent heat thermal energy storage
MDI	Methylene diphenyl diisocyanate
MDP	Mineral depletion potential
MEP	Marine eutrophication potential
ODP	Stratospheric ozone depletion potential
OFPh	Ozone formation potential human health
OFPt	Ozone formation potential terrestrial ecosystems
PCM	Phase change material
PMP	Fine particulate matter formation potential
SAT	Sodium acetate trihydrate
SHS	Sensible heat storage
S-LHTES	Solar power latent heat thermal energy storage
S-LHTES-PCM	Solar power latent heat thermal energy storage with phase change material
TAP	Terrestrial acidification potential
TEP	Terrestrial ecotoxicity potential
TES	Thermal energy storage
THS	Thermochemical heat storage
WDP	Water consumption potential

References

1. United Nations Climate Change the Paris Agreement. Available online: https://unfccc.int/process-and-meetings/the-paris-agreement/the-paris-agreement (accessed on 10 June 2021).
2. National Grid Balancing the Grid. Available online: https://www.nationalgrid.com/britain-hits-historic-clean-energy-milestone-zero-carbon-electricity-outstrips-fossil-fuels-2019 (accessed on 28 July 2021).
3. Department for Business Energy & Strategy UK Becomes First Major Economy to Pass Net Zero Emissions Law. Available online: https://www.gov.uk/government/news/uk-becomes-first-major-economy-to-pass-net-zero-emissions-law (accessed on 1 August 2021).
4. The Parliamentary Office of Science and Technology Carbon Footprint of Heat Generation. Available online: http://researchbriefings.files.parliament.uk/documents/POST-PN-0523/POST-PN-0523.pdf (accessed on 28 May 2021).
5. Committee on Climate Change Reducing UK Emissions. Available online: https://www.theccc.org.uk/publication/reducing-uk-emissions-2020-progress-report-to-parliament/ (accessed on 27 May 2021).
6. Cabeza, L.F.; Martorell, I.; Miró, L.; Fernández, A.I.; Barreneche, C. Introduction to thermal energy storage (TES) systems. In *Advances in Thermal Energy Storage Systems*; Woodhead Publishing Limited: Cambridge, UK, 2015; pp. 1–28. [CrossRef]
7. Zhao, B.C.; Li, T.X.; Gao, J.C.; Wang, R.Z. Latent heat thermal storage using salt hydrates for distributed building heating: A multi-level scale-up research. *Renew. Sustain. Energy Rev.* **2020**, *121*. [CrossRef]
8. Kerskes, H. Thermochemical Energy Storage. In *Storing Energy: With Special Reference to Renewable Energy Sources*; Elsevier Inc: Amsterdam, The Netherlands, 2016; pp. 345–372. [CrossRef]
9. Wu, S. Heat energy storage and cooling in buildings. In *Materials for Energy Efficiency and Thermal Comfort in Buildings*; Woodhead Publishing: New York, NY, USA, 2010; pp. 101–126. [CrossRef]
10. Englmair, G.; Moser, C.; Furbo, S.; Dannemand, M.; Fan, J. Design and functionality of a segmented heat-storage prototype utilizing stable supercooling of sodium acetate trihydrate in a solar heating system. *Appl. Energy* **2018**, *221*, 522–534. [CrossRef]
11. Dannemand, M.; Schultz, J.M.; Johansen, J.B.; Furbo, S. Long term thermal energy storage with stable supercooled sodium acetate trihydrate. *Appl. Therm. Eng.* **2015**, *91*, 671–678. [CrossRef]
12. Cabeza, L.F. Advances in thermal energy storage systems: Methods and applications. In *Advances in Thermal Energy Storage Systems*, 2nd ed.; Woodhead Publishing Limited: Cambridge, UK, 2021; pp. 37–54. [CrossRef]
13. Jungbluth, N. Life Cycle Inventory of Sodium Acetate and Expanded Graphite. Available online: https://www.osti.gov/etdeweb/servlets/purl/22119615 (accessed on 18 June 2021).

14. Noël, J.A.; Allred, P.M.; White, M.A. Life cycle assessment of two biologically produced phase change materials and their related products. *Int. J. Life Cycle Assess.* **2015**, *20*, 367–376. [CrossRef]
15. Lamnatou, C.; Notton, G.; Chemisana, D.; Cristofari, C. Storage systems for building-integrated photovoltaic (BIPV) and building-integrated photovoltaic/thermal (BIPVT) installations: Environmental profile and other aspects. *Sci. Total Environ.* **2020**, *699*, 134269. [CrossRef] [PubMed]
16. ISO. ISO 14044:2006 Environmental Management–Life Cycle Assessment–Requirements and Guidelines. Available online: https://www.iso.org/standard/38498.html (accessed on 1 September 2021).
17. ISO. ISO 14040:2006 Environmental Management–Life Cycle Assessment–Principles and Framework. Available online: https://www.iso.org/standard/37456.html (accessed on 1 September 2021).
18. Oró, E.; Gil, A.; de Gracia, A.; Boer, D.; Cabeza, L.F. Comparative life cycle assessment of thermal energy storage systems for solar power plants. *Renew. Energy* **2012**, *44*, 166–173. [CrossRef]
19. Englmair, G.; Moser, C.; Schranzhofer, H.; Fan, J.; Furbo, S. A solar combi-system utilizing stable supercooling of sodium acetate trihydrate for heat storage: Numerical performance investigation. *Appl. Energy* **2019**, *242*, 1108–1120. [CrossRef]
20. Zhang, H. Heat-insulating Materials and Sound-absorbing Materials. *Build. Mater. Civ. Eng.* **2011**, 316–423. [CrossRef]
21. Kingspan Solar Thermomax HP400 New Slimline Design Heat Pipe Vacuum Tube Solar Collector. Available online: https://www.midlandheating.ie/docs/solar-1-kingspan.pdf (accessed on 7 July 2021).
22. Armstrong Efficient Heat Transfer in a Compact Design. Available online: https://docplayer.net/27537357-Phe-plate-heat-exchangers.html (accessed on 15 July 2020).
23. Tyfo TYCOR-LS. Available online: https://www.plumbase.co.uk/link/1/h030224_31294_t.pdf (accessed on 15 August 2021).
24. Ecoinvent Center the Ecoinvent Database. Available online: https://www.ecoinvent.org/database/database.html (accessed on 6 June 2021).
25. Federal LCA Commons USLCI Database. Available online: https://www.lcacommons.gov/lca-collaboration/search/page=1&group=National_Renewable_Energy_Laboratory (accessed on 7 June 2021).
26. De Falco, M.; Capocelli, M.; Losito, G.; Piemonte, V. LCA perspective to assess the environmental impact of a novel PCM-based cold storage unit for the civil air conditioning. *J. Clean. Prod.* **2017**, *165*, 697–704. [CrossRef]
27. Adolfsson, M.; Rashid, S. Life Cycle Assessment and Life Cycle Cost of Heat Exchangers A Case for Inter Terminals Sweden AB Located in Port of Gothenburg. Master's Thesis, Chalmers University of Technology, Gothenburg, Sweden, 2016.
28. ACV Smart 600-Tank in Tank. Available online: https://www.acv.com/belgie/customer/product/06619301-393/smart-600 (accessed on 12 July 2020).
29. Complete Pumpsupplies Flamco Flexcon Premium 50 Litre Expansion Vessel. Available online: https://www.completepumpsupplies.co.uk/flamco-flexcon-top-50-litre-expansion-vessel (accessed on 16 June 2021).
30. StoneMark Construction Management Pipes Are Not a Lifetime Component. Available online: https://stonemarkcm.com/blog/pipes-not-lifetime-component/ (accessed on 5 August 2021).
31. EPA Table 1: Typical Equipment Life Expectancy. Available online: https://www.waterboards.ca.gov/drinking_water/certlic/drinkingwater/documents/tmfplanningandreports/Typical_life.pdf (accessed on 21 July 2021).
32. Englmair, G.; Kong, W.; Brinkø Berg, J.; Furbo, S.; Fan, J. Demonstration of a solar combi-system utilizing stable supercooling of sodium acetate trihydrate for heat storage. *Appl. Therm. Eng.* **2020**, *166*, 114647. [CrossRef]
33. Masteel Applications of Grade 304 Stainless Steel. Available online: https://masteel.co.uk/type-304-stainless-steel/ (accessed on 2 July 2021).
34. Shah, K.N.; Varandani, N.S.; Panchani, M. Life Cycle Assessment of Household Water Tanks—A Study of LLDPE, Mild Steel and RCC Tanks. *J. Environ. Prot.* **2016**, *7*, 760–769. [CrossRef]
35. Plastics Europe. Plastics—The Facts 2019 an Analysis of European Plastics Production, Demand and Waste Data. Available online: https://www.plasticseurope.org/en/resources/market-data (accessed on 2 July 2021).
36. Bowyer, J.; Bratkovich, S.; Fernholz, K.; Frank, M.; Groot, H.; Howe, J.; Pepke, E. *Understanding Steel Recovery and Recycling Rates and Limitations to Recycling*; Dovetail Partners Inc.: Minneapolis, MN, USA, 2015; pp. 1–12.
37. Copper Alliance A Global View of Production and Trade. Available online: https://sustainablecopper.org/global-pathways/ (accessed on 15 August 2021).
38. OECD. OECD.stat. Available online: https://stats.oecd.org/# (accessed on 28 July 2021).
39. Recovery Recycling Worldwide Glass Recycling-Current Market Trends. Available online: https://www.recovery-worldwide.com/en/artikel/glass-recycling-current-market-trends_3248774.html (accessed on 20 June 2021).
40. Aste, N.; Groppi, F.; del Pero, C. The first installation under the Italian PV Rooftop Programme: A performance analysis referred to five years of operation. In Proceedings of the 2007 International Conference on Clean Electrical Power, Capri, Italy, 21–23 May; IEEE: Milan, Italy, 2007; pp. 360–365.
41. Bergsma, G.; Sevenster, M. *End-of-Life Best Approach for Allocating Recycling Benefits in LCAs of Metal Packaging*; CE Delft: Delft, The Netherlands, 2013; p. 25.
42. Gallego-Schmid, A.; Chen, H.-M.; Sharmina, M.; Mendoza, J.M.F. Links between circular economy and climate change mitigation in the built environment. *J. Clean. Prod.* **2020**, *260*, 121115. [CrossRef]
43. Heyes, G.; Sharmina, M.; Mendoza, J.M.F.; Gallego-Schmid, A.; Azapagic, A. Developing and implementing circular economy business models in service-oriented technology companies. *J. Clean. Prod.* **2018**, *177*, 621–632. [CrossRef]

44. Mendoza, J.M.F.; Sharmina, M.; Gallego-Schmid, A.; Heyes, G.; Azapagic, A. Integrating Backcasting and Eco-Design for the Circular Economy: The BECE Framework. *J. Ind. Ecol.* **2017**, *21*, 526–544. [CrossRef]
45. Sansom, R. Decarbonising Low Grade Heat for Low Carbon Future. PhD. Thesis, Imperial College London, London, UK, 2014.
46. Huijbregts, M.A.J.; Steinmann, Z.J.N.; Elshout, P.M.F.; Stam, G.; Verones, F.; Vieira, M.; Zijp, M ; Hollander, A.; van Zelm, R. ReCiPe2016: A harmonised life cycle impact assessment method at midpoint and endpoint level. *Int. J. Life Cycle Assess.* **2017**, *22*, 138–147. [CrossRef]
47. Welfle, A.; Thornley, P.; Röder, M. A review of the role of bioenergy modelling in renewable energy research & policy development. *Biomass Bioenergy* **2020**, *136*. [CrossRef]
48. Welfle, A.; Gilbert, P.; Thornley, P. Increasing biomass resource availability through supply chain analysis. *Biomass Bioenergy* **2014**, *70*, 249–266. [CrossRef]
49. Strbac, G.; Pudjianto, D.; Sansom, R.; Djapic, P.; Ameli, H.; Shah, N.; Brandon, N.; Hawkes, A. Analysis of Alternative UK Heat Decarbonisation Pathways. Available online: https://www.theccc.org.uk/wp-content/uploads/2018/06/Imperial-College-2018-Analysis-of-Alternative-UK-Heat-Decarbonisation-Pathways.pdf (accessed on 29 July 2021).

sustainability

MDPI

Article

Sustainable Energy-Related Infrastructure Development in the Mekong Subregion: Key Drivers and Policy Implications

Han Phoumin [1,*], Sopheak Meas [2] and Hatda Pich An [2]

[1] Economic Research Institute for ASEAN and East Asia (ERIA), Jakarta 10270, Indonesia
[2] Mekong River Commission Secretariat (MRCS), Vientiane 01000, Laos; sopheak@mrcmekong.org (S.M.); hatda@mrcmekong.org (H.P.A.)
* Correspondence: han.phoumin@eria.org

Abstract: Many players have supported infrastructure development in the Mekong Subregion, bridging the missing links in Southeast Asia. While the influx of energy-related infrastructure development investments to the region has improved the livelihoods of millions of people on the one hand, it has brought about a myriad of challenges to the wider region in guiding investments for quality infrastructure and for promoting a low-carbon economy, and energy access and affordability, on the other hand. Besides reviewing key regional initiatives for infrastructure investment and development, this paper examines energy demand and supply, and forecasts energy consumption in the subregion during 2017–2050 using energy modeling scenario analysis. The study found that to satisfy growing energy demand in the subregion, huge power generation infrastructure investment, estimated at around USD 190 billion–220 billion, is necessary between 2017 and 2050 and that such an investment will need to be guided by appropriate policy. We argue that without redesigning energy policy towards high-quality energy infrastructure, it is very likely that the increasing use of coal upon which the region greatly depends will lead to the widespread construction of coal-fired power plants, which could result in increased greenhouse gas and carbon dioxide emissions.

Keywords: connectivity; energy infrastructure; fossil fuels; emissions; Mekong Subregion

Citation: Phoumin, H.; Meas, S.; An, H.P. Sustainable Energy-Related Infrastructure Development in the Mekong Subregion: Key Drivers and Policy Implications. *Sustainability* **2021**, *13*, 5720. https://doi.org/10.3390/su13105720

Academic Editor: Carlos Oliveira Cruz

Received: 13 March 2021
Accepted: 11 May 2021
Published: 20 May 2021

1. Introduction

The Mekong Subregion is linked by common energy challenges. There are challenges in maintaining economic growth and ensuring energy security, while curbing climate change and reducing air pollution. At the intersection of these challenges is the corresponding need to rapidly develop and deploy energy efficiency, low-emissions coal technology, and double the share of renewables in the energy mix towards more inclusive and sustainable growth, as the region's energy demand is expected to rise significantly over the next 30 years [1]. Such an increase is bringing both opportunities and challenges, including climate change, which is a result of fossil fuels. Despite significant progress in recent decades in terms of energy poverty alleviation, countries such as Cambodia and Myanmar are still struggling to provide energy access to their rural populations.

The coronavirus disease (COVID-19) pandemic is another major challenge of our time. It has caused a global economic downturn, with economic output set to contract by 2.5% in 2020. This economic impact has also brought about low energy demand in all sectors. As a result, daily global emission levels fell by 17% in the first quarter of 2020 [2]. However, as governments begin lifting restrictions and business activities resume, so too will the demand for energy. Economic recovery could see levels of carbon dioxide (CO_2) emissions bounce back very quickly. Indeed, global data from late May 2020 show record levels of CO_2 as countries started reopening their economies [3]. The post-COVID-19 economic recovery will drive increased energy demand, which emphasises the need to secure investment to fill infrastructure gaps.

Quality infrastructure, connectivity, and innovation are considered key for the region to ensure prosperity and sustainable development. In fact, fast connectivity—along with high-quality infrastructure and human resources development in the Southeast Asian region—has already resulted in opportunities for growth. These developments have also lifted living standards through income generation and employment opportunities. They have enabled the region to participate in the production network at different degrees and made it ready to benefit from the global value chain in the near future as improved connectivity attracts more investment, cuts logistics costs, and creates synergies and location advantages [4]. The region is arguably fortunate to have different stakeholders supporting infrastructure improvement that has bridged the missing links in Southeast Asia. However, the influx of investment, particularly in energy infrastructure development, has raised questions about both sustainability and quality, and the identification of partners the region should prioritize working with to promote long-term development sustainability, quality, and innovation in the Mekong Subregion. This report aims to review and analyze major initiatives that drive energy-related infrastructure development in the subregion; conducts energy modeling and estimation for energy demand and supply in the subregion during 2017–2050; and, from there, draws key policy implications that guide high-quality, energy-related infrastructure development.

This paper comprises seven sections. The second section discusses the study's approaches. The third section reviews regional platforms and initiatives for infrastructure development related to the Mekong Subregion by engaging relevant literature. The fourth section examines economic impacts brought by connectivity. The region's energy landscape, the required investment to meet the rising energy demand in the region for the foreseeable future, and the region's energy transition are discussed in the fifth and sixth sections. The final section concludes with policy implications.

2. The Study's Approach

This study employs several approaches to gathering data and information. Data on economic investment, in particular energy for the Mekong Subregion, are available in different forms and for time periods. The study relies on several past studies conducted by the Economic Research Institute for ASEAN and East Asia (ERIA) for the economic impacts brought by infrastructure connectivity in the Mekong Subregion. For project infrastructure investment, the study uses data and information from past projects and studies conducted by the Asian Development Bank (ADB). For the energy data and analysis, we conducted our own energy modeling and estimation for energy demand and supply for the Mekong Subregion. We also reviewed key regional initiatives for infrastructure investment and development platforms, such as quality infrastructure initiated by Japan at the G20 in Osaka; China's Belt and Road Initiative; the United States (US) Blue Dot Network (BDN); the Free and Open Indo-Pacific (FOIP); and other subregional initiatives, such as the Mekong River Commission, Lancang–Mekong Cooperation, and Mekong–Japan Cooperation.

Our analysis of the economic impacts brought by Mekong Subregion connectivity involves the quantitative assessment of existing and proposed infrastructure development up to 2030. The ERIA study on economic impact assessment employed a geographical simulation model (GSM), which was developed to track the progress on quality infrastructure development in the Association of Southeast Asian Nations (ASEAN) and East Asia. Jointly developed by the ERIA and the Institute of Developing Economies in 2007, the model calculates the proposed infrastructure-related projects for connectivity and innovation and includes a sophisticated level of information on infrastructure development status to facilitate any assessment.

For energy demand and supply in the Mekong Subregion, we employ energy modeling using the Long-Range Energy Alternative Planning System (LEAP) software, an accounting system used to develop projections of energy balance tables based on final energy consumption and energy input and output in the transformation sector. Final energy consumption is forecast using energy demand equations by energy and sector and future macroeconomic assumptions. For consistency, the historical energy data in the Mekong Subregion used in this analysis came from the energy balances of the International Energy Agency (IEA) for the Organization for Economic Cooperation and Development (OECD) and non-OECD countries [5]. Energy demand and supply has two scenarios: the business-as-usual (BAU) scenario, reflecting each country's current goals, action plans, and policies; and the alternative policy scenario (APS), which includes additional goals, action plans, and policies that countries could achieve with their best efforts given energy policy reforms and technological development. The APS consists of assumptions such as more efficient final energy consumption, more efficient thermal power generation, and higher consumption of new and renewable energy and biofuels.

The study also quantifies the required investment for power generation demand from 2017 to 2050, using the following formula:

$$Investment(i) = GenCapacity(i) \times Unit\ Cost\ (\$/\mathrm{GW})$$

$$GenCapacity\ (i) = \frac{GWh(i)}{[24\ hours \times 365\ days \times CapF(i)]}$$

where (*i*) is the fuel type, such as coal, gas, hydropower, and renewables; *investment* (*i*) is the required investment amount of fuel type (*i*); *GenCapacity* (*i*) is the generation capacity of fuel type (*i*) in gigawatts; and *CapF*(*i*) is the capacity factor of fuel type (*i*).

The study does not consider other required investments in the power grid or connectivity costs. It only estimates the required generation to meet the growing demand from 2017 to 2050.

3. Review on Regional Initiatives for Infrastructure Development

3.1. Initiatives for Quality Infrastructure

The region is arguably very fortunate to have different stakeholders supporting infrastructure improvement in a manner that bridges the missing links in the wider ASEAN region. However, quality is far more critical than quantity if the region is to develop sustainably. The region and particularly ASEAN, therefore, should focus on key development partners that promote long-term development sustainability, especially those that promote quality infrastructure, build responsible human resources, and bring new knowledge and innovation to the region. Some of the key players driving quality infrastructure in Southeast Asia are briefly discussed below.

3.1.1. G20 Principles for Quality Infrastructure Investment

Japan has been pioneering and promoting quality infrastructure for many years to empower Asia as a growth center to drive the global economy. Most importantly, at the G20 in Osaka in June 2019, Japan successfully launched an initiative, known as the G20 Principles for Quality Infrastructure Investment, as a key to promoting investment for sustainable development. According to the Ministry of Finance, Japan [6], the principles took into account many aspects of sustainability to ensure that quality infrastructure is in harmony with local environments, communities, and people's livelihoods through generating local employment and facilitating technology transfer. So far, Japan has committed USD 110 billion for quality infrastructure in Asia from 2015 to 2020 [7]. Such a commitment will accelerate financial resource mobilization into the region from private companies around the globe. This is in line with Japan's global commitment to promote high-quality infrastructure investment to address sustainable economic growth and reduce poverty and disparity.

Japan's promotion of quality infrastructure in Southeast Asia can be seen in the country's efforts to enhance ASEAN's connectivity through core land and maritime corridors and soft infrastructure development. The land corridors are high-quality hard infrastructure developments. They connect the South China Sea and the Indian Ocean; develop the Southern Economic Corridor that connects Ho Chi Minh City, Phnom Penh, Bangkok, and Dawei; and establish the East–West Economic Corridor (EWEC) that extends from Da Nang to Mawlamyaing in Myanmar as a trading center and seaport, connecting Southeast Asia to India and beyond. Another hard infrastructure development is the Maritime Economic Corridor, which consolidates connectivity through the development of port and port-associated industries as well as energy and information and communication technology networks, in major cities. This allows the Mekong Subregion to connect to Brunei Darussalam, Indonesia, Malaysia, the Philippines, and Singapore, thus enhancing connectivity across ASEAN.

3.1.2. Belt and Road Initiative

In recent years, China has also invested enormously in Asian infrastructure through its Belt and Road Initiative (BRI). The BRI is a major Chinese strategy aiming to push China's economic links to Southeast Asia, South Asia, Central Asia, Pacific Oceania, Africa, and the Baltic region (Central and Eastern Europe) through various infrastructure and development projects [8]. The BRI has been officially renamed several times since 2013 when Chinese President Xi Jinping announced the policy. It was previously called One Belt, One Road; the Silk Road Economic Belt; and the 21st-Century Maritime Silk Road. The policy was more fully articulated in 2015 as a vision statement, and numerous supporting policy documents have since been produced to support the implementation of the vision statement.

The BRI is expected to involve over USD 1 trillion in investments, largely in infrastructure development, for ports, roads, railways, and airports, as well as power plants and telecommunications networks [9]. Financing sources will include those typical of Chinese overseas investments, such as Chinese banks (commercial and policy), bonds, state-owned enterprises, private Chinese equity, private/public partnerships, the Asian Infrastructure Investment Bank, and others. However, it is expected that Chinese banks will continue to be the main source of financing for Chinese overseas projects, including those along BRI routes. Numerous projects have been proposed or are already in development. According to data from the Ministry of Commerce, China [10], from January to August 2016, Chinese companies signed almost 4000 project contracts in 61 countries. The value of these projects amounted close to USD 70 billion (Data on BRI investments are known to vary, particularly since it is unclear if existing projects are retroactively categorized by the Chinese government as BRI investments. This figure from the Ministry of Commerce is considered official).

There are growing concerns from recent experiences of BRI megaprojects that have come under a host of criticism. There is fear that the BRI could be a *debt trap* due to the high interest rates associated with some of the BRI's projects, as in the notorious case of Sri Lanka's Hambantota Port [11–13]. There are concerns that projects under the BRI are not transparent and that the BRI itself will be damaging to the environment [14] because it does not offer explicit guidelines on how Chinese investors should regard environmental protection or civil society [15]. There is also fear that the BRI is modern *Chinese colonialism*, often taking as an example the Chinese presence in Africa, and connecting to the long-standing *yellow peril* phobia (see, for example, [16,17]). There is another fear that, despite its effectiveness in relation to construction speed [13], the projects under the BRI are not sustainable but are the cause of environmental and social issues [9]. China's official responses have been mostly on the defensive, trying to delink the BRI from geopolitical or hegemonic ambitions, arguing that BRI projects 'benefit the local population' and are opportunities for 'shared development' (see, for example, [18]).

The BRI is considered as a second wave of Chinese overseas investment and should be seen as a renewed version of the Chinese policy, also known as China's 'Go Global' strategy [15]. This policy was the first to call on Chinese enterprises and industries to 'go out' and invest abroad. It is also seen as the key driver to advance China's interests overseas, and demonstrates its growing influence as a rule-shaper in the economic governance of the region and beyond [8], something that countries in the Mekong Subregion need to deal with carefully. However, if the BRI is to be successful, the Principles for Quality Infrastructure Investment initiative will need to be considered in all infrastructure investments, and local communities developing BRI projects will have to play an active role. In addition, host-country stakeholders will need to improve the quality of their governance systems.

3.1.3. Blue Dot Network

In November 2019, the US, Australia, and Japan came together to establish what is now known as a trilateral BDN to help develop and promote quality infrastructure in the Indo-Pacific region and around the world. Focusing on transparency and sustainability, the BDN aims to set a standard of excellence in infrastructure development. Hansbrough [19] argued that the BDN is primarily a vision of what global infrastructure should look like. In the eyes of many observers, the BDN is also seen as an alternative to China's BRI, or a counter to the rising debt traps and low-quality infrastructure that boost quantitative and nontransparent aspects of the projects see, for example, [12,20–23].

According to the US Department of State [24], the BDN is a multistakeholder initiative seeking to bring together governments, the private sector, and civil society to encourage the adoption of trusted standards for quality global infrastructure development in an open and inclusive framework. It also encourages responsible construction and lending practices through international norms. Infrastructure projects have to follow the G20 Principles for Quality Infrastructure Investment, aimed at sustainable lending and borrowing; the G7 Charlevoix Commitment on Innovative Financing for Development; and the Equator Principles, which mandate financial institutions to assess and manage environmental and social risks in a given project. Projects that aim for certification under the BDN will have to give an undertaking that they adhere to these principles. The undertaking will then be scrutinized. Certification by the BDN means that a project is high-quality and has transparent origins, much like an 'organic' label for produce. Likewise, a country that agrees to follow BDN standards signifies that its government values high-quality infrastructure that benefits local communities.

The BDN plans to certify projects around the world (whose investment totals an estimated USD 94 trillion) that meet high-quality infrastructure standard over the next two decades [25]. This will meet the projected infrastructure investment need identified by ADB (2017) up to 2040 [4]. In Asia alone, the investment will require approximately USD 26 trillion from 2016 to 2030, or USD 1.7 trillion per year, if the region is to maintain its growth momentum, eradicate poverty, and respond to climate change [26].

The BDN looks promising for the Mekong Subregion and for the world, as it seeks to build the robust, resilient infrastructure essential to a country's growth and its people's wellbeing [27]. But this remains to be seen. The initiative has not been fully defined and project financing facilities are amongst the many details that have to be clarified [22,25].

3.1.4. Free and Open Indo-Pacific

The region has also witnessed another initiative called the FOIP, as a mechanism complementary to other initiatives for infrastructure investment. In Japan, former Prime Minister Shinzo Abe unveiled the FOIP concept in August 2006, just before his first term as Japan's leader, and formally laid it down as a strategy in 2016 [28,29]. In late 2017, the US also launched a new FOIP [30], but it was not until 2019 that the concept was actually formalized [31].

Extending from Japan in the east to India in the west, the FOIP involves middle and major powers such as Japan, the US, Australia, and India; and other regional partners. It seeks to build a vision for Asia established around the concept of a strong coalition of like-minded regional democracies. However, a host of scholars and analysts have viewed the FOIP as a mechanism that provides the region with alternatives to China's BRI [32–35] or for countering China's influence [30,36–39]. The Government of Japan, nevertheless, views this differently. The FOIP is an inclusive concept that ultimately aims to incorporate China and other powers in an inclusive political and economic system in the Indo-Pacific [28].

It is also a comprehensive framework or vision for Japanese regional policies, mostly its economic and development cooperation, such as infrastructure development and support for regional connectivity [40–42].

Despite different views, the Mekong countries welcomed the FOIP. For example, they welcomed Japan's commitment to support their efforts made in line with ASEAN's Outlook on the Indo-Pacific [41,43]. Perhaps they saw this as another option for quality infrastructure projects. As Swaine [38] argued, infrastructure development initiatives under the FOIP could prove instrumental for both engaging and challenging China by advancing common principles for economic development and enabling developing countries to choose their own economic paths free from coercion. In this respect, the cooperative and competitive elements of the China challenge could merge as the allies pursue dialogue with Beijing on rules and norms while attempting to dilute its influence.

3.1.5. The Mekong River Commission

The Mekong River Commission (MRC) is another key driving force behind quality energy infrastructure development in the region. As the only treaty-based river basin organization in the region, the 25-year-old MRC has put in place two crucial strategies to guide its four member countries—Cambodia, Laos, Thailand, and Vietnam—in assessing and developing hydropower projects in the Lower Mekong River Basin (LMB) to optimise transboundary benefits while minimising adverse cross-national impacts.

One of them is the basin-wide Sustainable Hydropower Development Strategy (SHDS) for the LMB adopted in 2001 by the MRC Council of Ministers, the organization's highest governing body. The SHDS recognizes that while each member country has the full responsibility and right to plan and implement hydropower projects nationally, the MRC is tasked with striking a balance between regional and basin needs, and economic development and environmental protection [44]. The SHDS thus sets out strategic priorities and actions at the basin level to address hydropower opportunities and risks and strengthens basin-wide cooperation and sustainable development [45]. It also draws a close linkage between the energy and water sectors because the need for linked planning between the energy and water sectors is now more critical than ever before in the Mekong Region.

The Preliminary Design Guidance for Proposed Mainstream Dams in the LMB is another key strategic guidance resource. Adopted in 2009, it provides performance targets and principles for the design and operation of mainstream dams to help avoid, minimize, and mitigate harmful effects and limit the potential for substantial damage [46]. It seeks to establish a common design and operational approach, aiming to meet common objectives and mitigate commonly understood risks, and making it possible for developers to plan for and undertake the assessments and designs for mitigation and management measures as early as possible in the project cycle.

However, both documents are aging and need to be revisited. With rapid development in the basin, especially in the hydropower sector, it is important that the documents are updated, taking into account major changes the basin has faced over the last two decades. Studies by the MRC and others (see [47–53]) have indicated that hydropower dams constructed on the mainstream in the upper part in China where the river is called the Lancang and on the lower reaches where the river is called the Mekong and on tributaries in the LMB had changed the natural flow regime of the river, yielding both opportunities and risks on hydropower development now and in the future. Gathering the significant economic and greenhouse gas (GHG) reduction benefits offered through hydropower development should not come at the expense of the unique and abundant ecosystem services and biodiversity on which so many communities in the basin depend. In addition, although the MRC has a critical role to play in water diplomacy and energy infrastructure development in the region, this and its wider role have not received sufficient credit [54]. Thus, the Mekong River Commission (MRC) needs to evolve, and its founding member countries need to empower it further if the Mekong River is to develop sustainably and responsibly [55,56].

3.1.6. Lancang–Mekong Cooperation

The Lancang–Mekong Cooperation (LMC), despite its relatively young age, is one of the most rapidly progressive and notable platforms in the Mekong Subregion. In 2012, Thailand proposed an initiative for sustainable development of the Mekong Subregion, which received a positive response from China. At the 17th China–ASEAN Summit held in November 2014, Chinese Premier Li Keqiang proposed the establishment of the LMC Framework, which was welcomed by the other five Mekong countries. In March 2016, China and the other five Mekong countries held their first LMC Leaders' Meeting, which released the Sanya Declaration and officially launched the LMC mechanism [57].

Although the LMC seeks to promote many aspects of cooperation on security, economic, cultural, agriculture, and poverty reduction issues [57–59], the major driving force is seen through its emphasis on infrastructure development for the region. Some of the major examples are Myanmar's Kyaukpyu Port and gas pipeline, the Laos and Thailand's high-speed railway projects, Cambodia's irrigation systems and transport infrastructure, and more plans to develop better capacity for navigation along the Mekong River [60].

As a subregional cooperation mechanism connecting the six countries along the Mekong River, the LMC has seen China emerge as a willing investor and guarantor as part of its wider BRI. While a comprehensive list of LMC projects is not publicly available, the LMC has provided financial support for at least 132 projects in the Mekong Region as of 2018 [61]. During the LMC Ministerial Meeting in 2019, the LMC proposed 101 additional projects, all of which were considered fast-track—to be carried out in one year or less—in the six Mekong countries [62] to respond to "socioeconomic demands and water related challenges" [63] (p. 2). The LMC, like the BRI, is often promoted as an effective platform that offers countries in the Mekong Subregion the resources they need for development (see, for example, [64–68]).

Critics, however, have voiced strongly that China is using the LMC to build its regional strategic influence and that the LMC per se does not promote good governance. China's strong interest in driving the development of the LMC stemmed from gaining substantial control over the Mekong Subregion, delimiting the influence of external actors such as the US and Japan, and pushing forward its neighborhood diplomacy [69,70]. While the LMC can be a building block for stronger regional multilateralism, it can also work against the advancement of broader ASEAN regional cooperation and marginalize other Mekong Subregion bodies [60]. Amongst all the seemingly unchecked development that has flourished as a result of the LMC, perhaps none has had such an impact on local communities and the environment as the dams that have sprouted up across the region, where China has taken the role of developer or funding agency [61]. While Chinese investment in infrastructure development through the LMC is a welcome source of capital for Mekong countries, Southeast Asia should approach it more critically to avoid development that later becomes a debt trap, does not last, and only benefits the few.

3.1.7. Mekong–Japan Cooperation

Mekong–Japan connectivity is another important dimension for the Mekong Region. It aims to promote infrastructure development in the region and to enhance institutional connectivity through the improvement of systems, development of Special Economic Zones (SEZs) and other industrial bases, industrial promotion measures, improvement of customs procedures, and people-to-people connectivity to ensure that the whole region benefits from growth [71]. Key pillars of cooperation in the development of infrastructure are to fill the missing links of the East–West and Southern Economic Corridors. Once the links are filled, they will connect the corridors more smoothly through the improvement of systems such as customs procedures; they will also promote land development along the corridors (e.g., the development of industrial parks, industrial promotion measures, and so on) and improve access from neighboring areas to corridors so that the region can develop as a whole. Finally, they will help to promote the development of industrial human resources that will support growth in the region and strengthen people-to-people networks.

It can be argued that the Mekong Subregion has benefited significantly from the infrastructure improvement brought by official development assistance support from Japan, with high-quality roads, bridges, and other hard and soft infrastructure.

4. The Economic Impacts of Connectivity and Infrastructure Investment

4.1. The Economic Impacts of Connectivity in the Mekong Subregion

The coordinated development of soft and hard infrastructure is also essential to maintain growth in the region. The new international division of labor calls for a novel approach to infrastructure development, in which the Mekong Subregion is prepared to participate actively in the promotion of economic corridors: the Southern Economic Corridor, the EWEC, and the North–South Economic Corridor. These economic corridors—together with the fast acceleration of domestic infrastructure development including SEZ, urban amenities, and other economic activities—have already promoted regional participation in the production network by reducing the cost of service links that connect remote locations. Mekong Subregion connectivity is just one piece of the puzzle in ASEAN connectivity with the rest of the world. China's BRI is another very large 'connectivity for development' strategy, linking China to Eurasian countries and the rest of Asia.

As the region embarks on rapid infrastructure development, quality infrastructure, connectivity, and innovation are key to ensure prosperity and sustainable development. Infrastructure development and stages of economic development can be explained by the development of recent economic theories: fragmentation theory and new economic geography [72]. The theory classifies infrastructure projects into three tiers. Tier 1 includes projects that serve countries/regions that are already in production networks and have started forming industrial agglomerations. Tier 2 consists of projects supporting countries/regions that are about to participate in production networks. Tier 3 consists of projects in remote areas where participation in production networks is difficult in the short run, but where better and more reliable connectivity can generate new business models in agriculture, mining, tourism, and other industries. Thus, the ultimate aim of quality infrastructure and services development is in tier 1, in which some ASEAN Member States are experiencing and enjoying quality growth, particularly Singapore and to some extent Brunei Darussalam. Malaysia and Thailand are also doing well, with the quality of infrastructure in tier 2 possibly moving to tier 1 in the near future. The Mekong Subregion has achieved lower middle-income status, improving infrastructure quality from tier 3 and possibly joining tier 2 in the near future. Indonesia and the Philippines have achieved middle-income status and infrastructure development is in the early stage of tier 2, likely catching Malaysia and Thailand in the near future.

By and large, connectivity and innovation promote agglomeration forces and dispersion forces generated by production–consumption interactions in both internal and external economies in which people and ideas can move easily. Agglomeration forces mean that economic activities and people are attracted to the core, where positive agglomeration effects are found in the form of the ease of finding business partners and proximity to the market, etc. On the other hand, dispersion forces generate movements of economic activities and people from the core to the periphery. One source of dispersion forces is negative agglomeration effects or 'congestion' in the core, which includes wage increases, land price hikes, traffic congestion, and environmental pollution [72].

One practical example of new economic geography creating 'location advantages' through connectivity and innovation is Cambodian labor force migration. Currently, about 1 million (out of a population of 16 million) Cambodians are in Thailand working in unskilled labor-intensive sectors and the informal sector rather than in Phnom Penh. The question is: How can Phnom Penh attract labor from rural areas and, at the same time, attract production blocks from Thailand? If the wage gap between Bangkok and Phnom Penh is too large, people will not move to Phnom Penh; however, at the same time, production blocks may be motivated to move. On the other hand, if the wage gap is too small, production blocks will not move even though people may flow into Phnom Penh.

Then, how can Phnom Penh attract both production blocks and people? The answer is the improvement of location advantages and liveability in Phnom Penh.

Another example is the Mekong–India Economic Corridor (MIEC)/EWEC connecting Ho Chi Minh (HCM) City, Phnom Penh, Bangkok Metropolitan Area, and Dawei. This has great potential to become a major manufacturing corridor in the near future. However, the question is how to attract labor and investment to Dawei. In this regard, the MIEC will need to have at least three projects implemented at the same time—industrial estates, highway connection to Thailand, and a deep seaport. According to Han [4], the road situation between Phnom Penh and HCM City was relatively bad in 1999. Before the road was upgraded, travel time from Phnom Penh to HCM City was about 9–10 h, and cross-border trade at Moc Bai (Vietnam)–Bavet (Cambodia) was worth about USD 10 million per year. However, the situation completely changed in 2014 after implementing hardware and software infrastructure between Phnom Penh and HCM City. The travel time was reduced to 5–6 h, and cross-border trade at Moc Bai–Bavet grew to USD 708 million per year [72]. Further, connectivity promoted other economic development corridors, such as investment brought to Trang Bang Industrial Park (in Moc Bai), consisting of 41 projects with USD 270 million in new investment, creating about 3000 jobs.

The top 10 beneficiaries from the MIEC, based on ERIA (2015), are Dawei, Phnom Penh, Dong Nai, Kawthoung, HCM City, Kandal, Sihanoukville, Banteay Meanchey, Svay Rieng, and Battambang. For Phnom Penh, it was estimated that the connectivity would increase gross domestic product (GDP) by almost 400% as a cumulative impact during 2021–2030. ERIA also estimated the remainder of the economic corridor in the Mekong Subregion, and found significant impacts for all participating countries in the connectivity.

For power connectivity in the Greater Mekong Subregion (GMS), ERIA's study on energy markets in ASEAN and East Asia examined the power trade and development in the subregion for the foreseeable future [73]. The study showed that the 2030 Scenario (in which the GMS realizes the potential of hydropower) will provide both economic and environmental benefits. The GMS at large will benefit by about USD 40 billion and reduce CO_2 emissions by almost 70 million tons per year. For ASEAN power connectivity as a whole, the study estimated that ASEAN would save USD 25 billion over 20 years by substituting hydropower for fossil fuels.

4.2. Infrastructure Investment Projects in the Mekong Subregion

The GMS regional investment framework 2014–2022 (RIF 2022) pipeline projects consist of 143 investment projects requiring USD 65.7 billion and 84 technical assistance projects requiring USD 295 million (see Table 1). Of the total 227 prioritized projects, which require investment of about USD 66 billion, there are financing gaps for 121 projects amounting to USD 27 billion (about 40% of the total investment). Of the projects currently identified with available financing, 70% have government financing, 18% have ADB financing, 6% have financing through other development partners, and 6% have private sector investment or public–private partnerships [74].

The RIF 2022 is heavily skewed towards transportation sector projects, as the table shows. However, intersectoral linkages, such as tourism supported through transport networks, are more prominent in the RIF 2022. Furthermore, there is an increase in transportation subsectors, with new projects in ports and waterways, logistics, and border crossings, which were missing or underrepresented in earlier pipelines. Railway infrastructure, because of its greenfield nature and extensive civil works, continues to make up the bulk of the required investment costs in the RIF 2022. Some railway projects have commenced, with domestic budgets and bilateral assistance from China. The GMS Railway Association is assessing which railway lines to prioritize for the subregion and examining alternative modalities to address the vast financing needs for rail infrastructure [74,75]. In addition to projects in new transport subsectors in the RIF 2022, projects in border area or border zone development involve multisectoral interventions such as road and/or border infrastructure, trade facilitation, technical and vocational education and training, schools, urban infrastructure, and tourism. The GMS Tourism Infrastructure for Inclusive Growth projects also use this multisectoral approach.

Table 1. Regional Investment Framework 2022 summary by sector.

	Number of Projects			Cost Estimates (USD, Million)		
Sector	Investment	TA	Total	Investment	TA	Total
Transport	85	12	97	55,753	10	55,763
Energy	11	8	19	2230	15	2245
Agriculture	9	10	19	1695	96	1791
Environment	3	4	7	560	13	573
Health and other HRD	4	7	11	702	22	724
Urban development	7	6	13	1147	10	1157
Others/BEZ	6	6	12	2085	8	2093
Tourism	12	17	29	1430	83	1513
TTF	3	9	12	91	17	108
ICT	3	5	8	28	22	50
Total	143	84	227	65,722	296	66,017

BEZ = border economic zone, HRD = human resources development, ICT = information and communication technology, TA = technical assistance, TTF = transport and trade facilitation. Source: ADB [74].

Of the total transport sector investment projects, as shown in Table 1, railways took 62% of the total (about USD 35 billion investment in the RIF 2022), followed by roads and bridges at 36% (about USD 20 billion). If the railway, road, and bridge projects under construction and potential new projects are realized in the near future, the GMS will be a region of connectivity by rail and road, which will play out very well for connectivity to Malaysia and Singapore. Thus, the flows of goods and services could see potential increases in volume, positively affecting economic growth in the region.

5. Energy Landscape in the Mekong Subregion

The data sources presented in Figures 1–5 are provided by country experts in the Mekong subregion who are the members of the ASEAN and East Asia Energy Outlook managed by ERIA. One of the co-authors of this paper is also the coeditor of the East Asia Energy Outlook. The presented numbers are also derived from using the Long-range Energy Alternatives Planning (LEAP) system software, an accounting system used to develop projections of energy balance tables based on final energy consumption and energy input/output in the transformation sector. LEAP is also an energy modeling software capable of conducting a variety of analyses of energy systems including Demand Analysis, Transformation Analysis, Resource Analysis, and Environmental Analysis. The data structures in a LEAP are organized using a hierarchical tree. The types of data entered at each branch depend on the type of branch, and its position in the tree (for example whether it is a Demand or Transformation branch). At the core of LEAP is the concept of scenario analysis. Scenarios are self-consistent storylines of how a future energy system might evolve over time in a particular demographic and socioeconomic setting and under a particular set of policy conditions. Using LEAP, we can build scenarios and then compare them to assess their energy requirements, social costs and benefits, and environmental impacts. All scenarios start from a common base year. We can use scenarios to ask an unlimited number of 'what if' questions, such as: what if more efficient appliances are introduced, what if different electric generation capacity expansion plans are pursued, what if indigenous reserves of oil and gas are discovered, what if renewable energy technologies are introduced, and so on. From here, the results presented in Figures 1–5 are the authors' calculations based on the country data provided by the regular members of ERIA's energy outlook and energy-saving potential panel of experts.

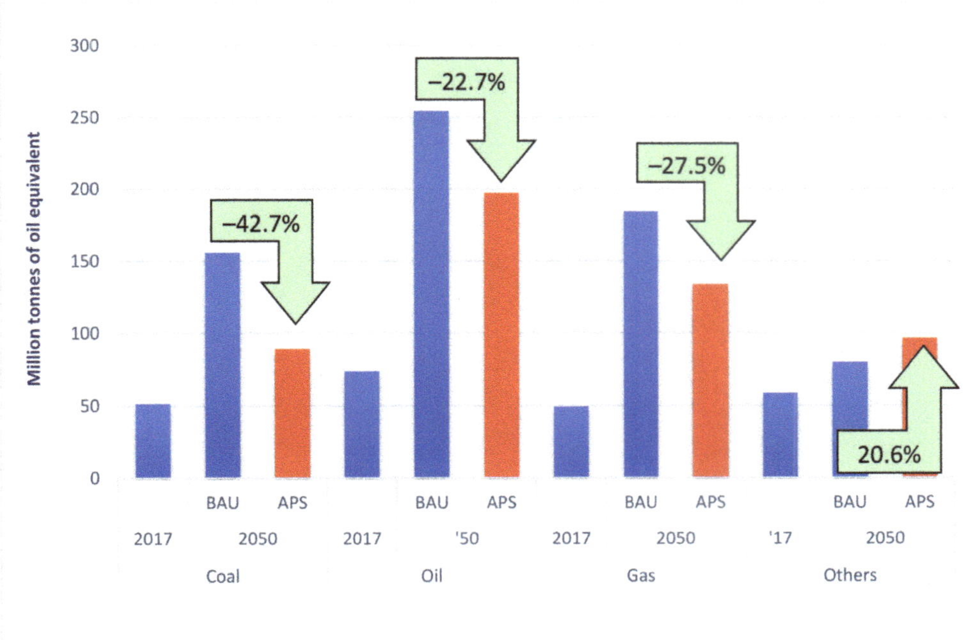

Figure 1. TPES by energy source, BAU (in blue) vs. APS (in red). APS = alternative policy scenario, BAU = business as usual, TPES = total primary energy supply. Source: authors' calculations.

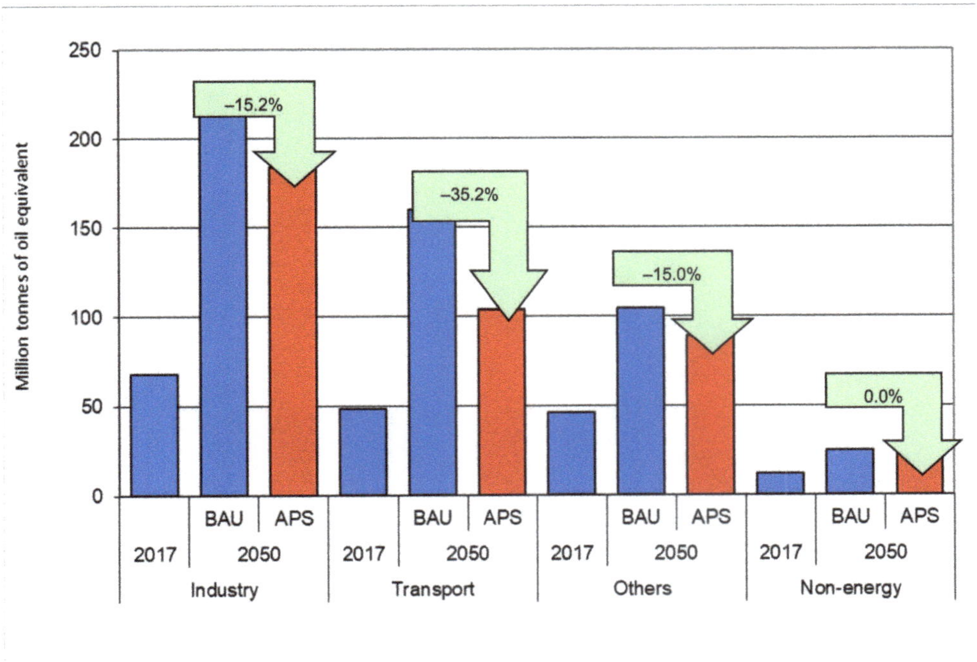

Figure 2. TFEC by sector, BAU (in blue) vs. APS (in red). APS = alternative policy scenario, BAU = business as usual, TFEC = total final energy consumption. Source: authors' calculations.

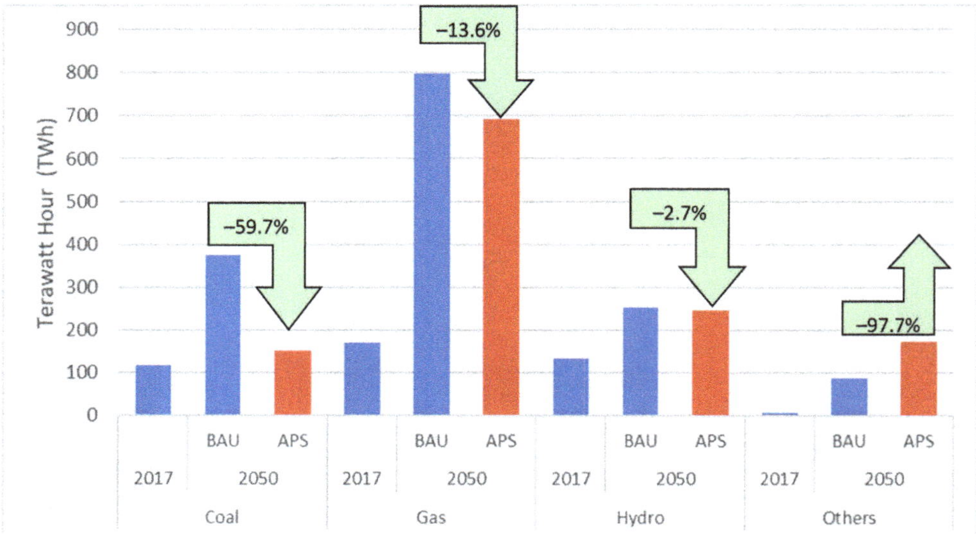

Figure 3. Total power generation (TFEC) by energy source, BAU (in blue) vs. APS (in red). APS = alternative policy scenario. BAU = business as usual, TFEC = total final energy consumption. Source: authors' calculations.

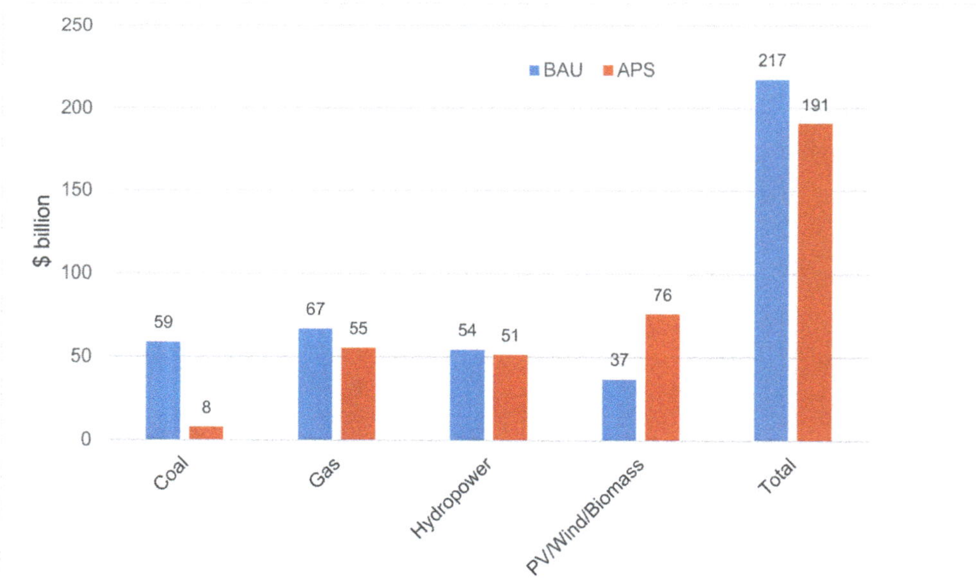

Figure 4. Investment in power generation by energy source, BAU vs. APS. APS = alternative policy scenario, BAU = business as usual, PV = photovoltaic. Source: authors' calculations.

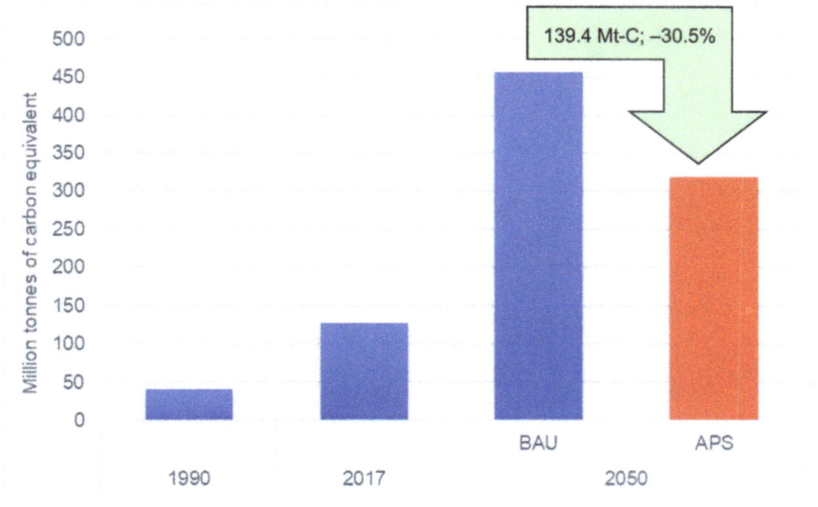

Figure 5. CO_2 emissions in the Mekong Subregion, BAU (in blue) vs. APS (in red). APS = alternative policy scenario, BAU = business as usual, CO_2 = carbon dioxide, Mt-C = million tonnes of carbon equivalent. Source: authors' calculations.

5.1. Energy Supply in the Mekong Subregion

The total primary energy supply (TPES) in the Mekong Subregion (Cambodia, Laos, Myanmar, Thailand, and Vietnam) is projected to increase by 189% in the BAU scenario, and by 121% in the APS from 2017 to 2050. In actual amounts, it will increase from 234 million tonnes of oil equivalent (Mtoe) in 2017 to 675 Mtoe in the BAU scenario, and to 516 Mtoe in the APS by 2050. It is observed that the Mekong Subregion depends heavily on fossil fuel consumption (oil, coal, and gas). Based on the baseline data from 2017, the fossil fuel share in the energy supply is around 75% of the total in the Mekong Subregion. It is projected that the Mekong Subregion will see growing dependency on fossil fuels in the future. In this regard, the study results showed that by 2050, the share of fossil fuels in the energy supply will be about 88% in the BAU scenario and 81% in the APS. In actual amounts, the combined coal, oil, and gas in the energy supply is expected to increase from 175 Mtoe in 2017 to 595 Mtoe in the BAU scenario and to 420 Mtoe in the APS by 2050. Oil is the dominant energy source in the energy supply, followed by natural gas and coal (see Figure 1). Oil is expected to increase from 74 Mtoe in 2017 to 255 Mtoe for the BAU scenario and to 197 Mtoe for the APS in 2050. Natural gas is expected to increase from 49.3 Mtoe in 2017 to 184.3 Mtoe for the BAU scenario and to 133.6 Mtoe for the APS in 2050. Coal will increase from 51.6 Mtoe to 155.8 Mtoe for the BAU scenario and to 89.3 Mtoe for the APS in 2050. Other sectors, including biomass, wind, solar, and electricity, will see increases from 58.8 Mtoe in 2017 to 80.0 Mtoe for the BAU scenario and to 96.5 Mtoe for the APS by 2050.

The difference between the BAU scenario and the APS is the energy saving potential in the TPES. Coal will see the largest energy saving, with potential of 42.7%, followed by 27.5% for natural gas and 22.7% for oil. These large energy savings are expected from the implementation of energy efficiencies, with improved efficiency in thermal power plants and energy efficiency in end-use sectors such as transportation, industry, commercial, and residential sectors. The Mekong Subregion is expected to see an increase in renewables of about 20.6% in the energy supply mix by 2050 (see Figure 1).

5.2. Final Energy Consumption in the Mekong Subregion

In the total final energy consumption (TFEC), industry accounts for the largest share, followed by transportation, and other commercial and residential sectors, as Figure 2 shows. Energy consumption in the industrial sector is expected to increase from 68 Mtoe in 2017 to

217 Mtoe for the BAU scenario and to 184 Mtoe for the APS by 2050. Energy consumption in the transport sector is predicted to increase from 48 Mtoe in 2017 to 160 Mtoe for the BAU scenario and to 104 Mtoe for the APS by 2050. For other sectors, including the commercial and residential sectors, energy consumption is expected to increase from 46 Mtoe in 2017 to 105 Mtoe for the BAU scenario and to 89 Mtoe for the APS by 2050. The nonenergy sector (naphtha) is also used in the TFEC, especially for the refinery and petrochemical industries, with its use remaining the same for the BAU scenario and the APS in 2050.

Energy saving is expected to be highest for the transportation sector at 35.2%, 15.2% for the industrial sector, and 15.0% for the commercial and residential sectors, as indicated in Figure 2. The reduction in energy consumption in the final energy sector will derive from fuel efficiencies in the transportation, industry, commercial, and residential sectors (e.g., the introduction of more efficient heat and power, a shift to electric vehicles, hybrid and fuel cell vehicles, more efficient electric appliances, and energy-saving buildings).

5.3. Power Generation Mix in the Mekong Subregion

In the power sector, remarkable progress has been made in the subregion over the past two decades. This includes rural electrification access, rapid provision of large-scale and high-volume national grid systems, successful mobilization of indigenous resources, the adoption of new technologies, the gradual share of renewables into energy mix, and the beginnings of cross-country trade. However, the future energy landscape in the Mekong Subregion will rely on today's actions/policies and investment to change course towards a cleaner energy system.

Natural gas is the dominant fuel source in power generation, followed by coal and hydropower, as Figure 3 shows. Natural gas is expected to increase from 170.4 megawatt-hours (MWh) in 2017 to 798.7 MWh by 2050 in the BAU scenario and to 690.3 MWh in the APS by 2050. Electricity from coal-fired power generation will increase from 116 MWh in 2017 to 374 MWh in the BAU scenario and 150 MWh in the APS by 2050. Electricity from hydropower is expected to increase from 133 MWh in 2017 to 252 MWh in the BAU scenario and to 245 MWh in the APS by 2050.

Electricity from 'others' (including biomass, wind, and solar) will see a large increase from 6.2 MWh in 2017 to 87.2 MWh in the BAU scenario and to 172.4 MWh in the APS by 2050. Significant energy saving is expected in coal-fired power generation (59.7% saving, a reduction from BAU to the APS) followed by the gas combined cycle (13.6%). Energy saving in power generation is expected due to the introduction of high thermal efficiency. Electricity from renewables such as biomass, wind, and solar is expected to increase sharply by 97.7% due to upscaling renewables in the power mix in the APS scenario compared with the BAU scenario.

5.4. Required Power Generation Investment to Meet Rising Demand in the Mekong Subregion

To satisfy growing energy demand in the Mekong Subregion, huge power generation infrastructure investment is necessary from 2017 to 2050, as indicated in Figure 4. This study estimates that USD 191 billion–217 billion will be needed for cumulative investment in power generation in coal, gas, and hydropower. The investment in natural gas combined cycle power generation will require USD 55 billion–67 billion for the BAU scenario and APS from 2017 to 2050. Coal-fired power generation will require around USD 59 billion in the BAU scenario. However, coal-fired power plant (CPP) capacity may be reduced in the APS, depending on the Mekong Subregion's energy policy. In this case, the estimate for coal-fired power investment could drop to about USD 8 billion from 2017 to 2050. For renewables such as solar photovoltaic (PV), wind, and biomass, the required investment is expected to increase from USD 37 billion in the BAU scenario to USD 76 billion in the APS. More broadly, at the ASEAN level, the *Energy Outlook* projects that USD 2.1 trillion will be required for oil, gas, coal, and power supply [76]. More than 60% of investment goes to the power sector, with transmission and distribution accounting for more than half.

Thus, the huge potential for energy infrastructure related investment will need to be guided by the appropriate policy to promote quality infrastructure and resilience in the Mekong Subregion for growth and sustainability.

5.5. Carbon Dioxide Emissions in the Mekong Subregion

The region will continue to rely on fossil fuel consumption in the foreseeable future (see Figure 5). This is mainly because of the presence of the high combined share of fossil fuels in the power generation mix of the Mekong Subregion, at 67% in 2017 and 78% in the BAU scenario by 2050, as well as the high share of fossil fuel use in the TFEC. CO_2 emissions rose from 42 million tonnes of carbon equivalent (Mt-C) in 1990 to 127 Mt-C in 2017. These emissions are expected to rise to 457 Mt-C in the BAU scenario and to 318 Mt-C in the APS by 2050.

Thus, the clean use of fossil fuels through clean technology deployment is indispensable in decarbonizing the Mekong Subregion's emissions, as also recently shown in a study by Han et al. [77]. Further, natural gas should be promoted as a transitional fuel to bridge towards more renewable energy in the future.

6. Energy Transition in the Mekong Subregion

The Mekong Subregion faces mounting challenges in matching its energy demand with sustainable energy supply. This is because the regional reliance on fossil fuel consumption is projected to last at least until 2050. The transition to a lower-carbon economy will require the region to develop and deploy greener energy sources and clean use of fossil fuels through innovative technology such as high-efficiency, low emissions (HELE technologies) technologies. Coal-use patterns in the region reflect the rising demand for electricity to power and steer economic growth. Hence, building low-efficiency CPPs is an obvious choice for power-hungry emerging Southeast Asia due to lower capital costs. However, such plants cause more environmental harm and health issues due to air pollution, CO_2, and other GHG emissions. Widespread coal power plant construction could also point to the low environmental standards for coal-fired power generation in the Mekong Subregion [78]. The Mekong Subregion countries have relatively high allowable emissions in terms of sulfur oxides (SOx), nitrogen oxides (NOx), and particulate matter (PM) (see Figure 6). This means that countries in the subregion have lower emissions standards than advanced countries such as Germany, the Republic of Korea, and Japan, where HELE coal fired power plants are mandatory.

Major harmful air pollutants, such as SOx, NOx, and PM, come from fossil fuel and biomass power plants, which therefore need to be carefully regulated. It is known that short-term exposure to sulfur dioxide (SO_2) can harm the human respiratory system and make breathing difficult.

Thus, the Mekong Subregion's leaders may need to consider more strongly the promotion of HELE technologies, higher standards or stringent environmental regulation for CPPs, and effective enforcement. This may push investors to select more advanced technologies, especially ultrasupercritical technology, for CPPs. Such plants are considered clean power because they use coal more efficiently and cleanly than traditional subcritical CPPs. Furthermore, supporting frameworks to ensure that developing countries can afford HELE technologies are urgent because the up-front investment costs of HELE technologies are much higher than those of traditional CPP technologies.

The role of natural gas in the energy transition cannot be overlooked. This is because it can be used as a bridging fuel between high emissions fuels, such as coal and oil, to cleaner energy systems in which renewables and clean fuels take the major share in the energy supply mix. The prospects for using natural gas in the Mekong Subregion are good, with demand likely to quadruple depending on the future stability of gas and liquefied natural gas (LNG) prices in the market; whether a competitive gas/LNG market can be created in Southeast Asia; and the role of gas/LNG from Australia, the US, and other sources. The region is expected to be a key market for future gas demand, thus gas infrastructure investment, such as gas pipelines and LNG terminals, will be crucial in supporting the demand for gas in the region [79].

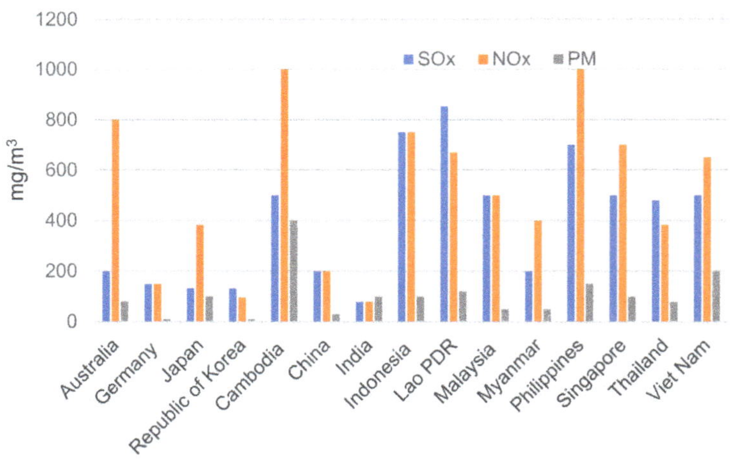

Figure 6. Emissions standards for newly constructed CPPs in selected countries (SOx, NOx, and PM). CPP = coal-fired power plant, mg/m^3 = milligram per cubic meter, SOx = sulfur oxides, NOx = nitrogen oxides, PM = particulate matter. Source: Mitsuru et al. [78].

In the current situation, hydropower accounts for quite a large share of the energy mix in the Mekong Subregion. However, as energy demand is expected to increase further, hydropower sources will be fully utilized. Thus, the share of renewables, such as wind, solar, and biomass, will play a critical role in the future clean energy system in the Mekong Subregion. The lower cost of these renewables will make it possible for a higher share of wind and solar in the energy mix [80]. Since electricity from wind and solar sources is variable and intermittent, there is a need to invest in grid infrastructure with smart grids, using the Internet of Things (IoT) and other technology to predict electricity production.

The Mekong Subregion may benefit enormously from the development of renewable hydrogen, as the region has large hydropower potential and the possibility of a higher share of solar and wind power see [77]. Thus, electricity from wind and solar plus other unused electricity during low-demand hours should be converted to hydrogen as stored energy. Fast-moving technological development will drive down the cost of hydrogen production in the future and give hydrogen a larger role in the clean energy future [81]. Thus, the Mekong Subregion may need to prepare a roadmap for rolling out a hydrogen plan in the future.

7. Conclusions and Policy Implications

The Mekong Subregion's fast connectivity—including rail, road, port, aviation, and energy infrastructure—has integrated the region further in terms of compressing time and space for the movement of goods and services. However, the wider ASEAN region faces challenges in guiding investments for long-term sustainability, especially on quality infrastructure. In the region, key players channel their investments through regional and subregional initiatives and platforms such as China's BRI and LMC, the US EDN, the FOIP, the MRC, and Mekong–Japan Cooperation. Although there is a clear need for resilience and quality infrastructure in the Mekong Subregion, policy measures and actions undertaken in each country towards high-quality infrastructure vary, reflecting the differences in socioeconomic, political, and geographical contexts. Thus, this makes it difficult for the region to promote sustainable growth and a low-carbon economy, energy access and affordability, and resilient and sustainable quality infrastructure.

As the Mekong Subregion continues to rely on fossil fuels, its energy transition will need to consider cleaner use through clean technology investment such as HELE technologies and other high-quality energy infrastructure. Currently, investment in renewable energy and clean technologies is unstable and high in cost. These challenges need to be addressed through political commitment to ensure that an energy technology development and deployment support framework can scale up the share of renewables and clean fuels. Without redesigning energy policy towards high-quality energy infrastructure, it is very likely that the increasing use of coal will lead to the widespread construction of CPPs, which, without the employment of the best available HELE technologies, will result in increased GHG [1,77].

The investment opportunities for energy-related infrastructure are huge. This study estimates that around USD 190 billion–220 billion will be required from 2017 to 2050 for power generation alone. However, this estimate does not include the transmission and distribution network, LNG terminals, and refineries. The challenge will be to ensure quality infrastructure to promote sustainability in the region. Energy sustainability in the Mekong Subregion requires an increase in the share of renewables in the energy mix. Currently, it is dominated by coal, gas, and hydropower. Although intermittent renewables (solar and wind) comprise the most abundant energy resources in the region, they have so far taken a minimal share of the power mix.

As this study shows what countries in the Mekong Subregion will need, as development accelerates and climate change intensifies, is an environmentally friendly, logistically feasible, and economically responsible alternative energy source and infrastructure. Derived from this study, the following key policy implications are provided with this consideration in mind.

First, the region will need to promote quality infrastructure investment. Given the region's vulnerability to climate change, resilient and high-quality infrastructure will play a key role in the region's long-term sustainability. Thus, regional and subregional platforms and initiatives such as the BRI, quality infrastructure by Japan, the BDN, and other subregional initiatives will need to promote high-quality infrastructure investment. For instance, the region should and will need to discuss the quality and standards that can guide investment to meet the need for high-quality infrastructure. Willingness to pay could be a barrier because of the high cost of quality infrastructure. Thus, a mechanism to reduce costs through innovative financing will be important for the successful deployment of high standards in the region.

Second, the current climate narrative and policy approach of banning coal use will need to be reviewed to assist emerging Asia to afford HELE technologies. This is primarily because there are less available alternative energy options in the medium term to meet energy demand. Treating HELE technologies as a technological solution in the energy transition will be a win–win solution for a climate-friendly world as Asia faces energy accessibility and affordability. Emerging Southeast Asia will rely on whatever HELE technologies are available in the market at affordable prices. The up-front costs of such ultrasupercritical technology or advanced ultrasupercritical technology are higher than supercritical and subcritical technologies. Thus, it is necessary to lower the up-front costs through policies such as attractive financial/loan schemes or a strong political institution to deliver public financing for HELE technologies in the region.

Third, there is a need for public consultation on and local participation in the potential impacts of any selected power plant infrastructure and technologies. However, for the Mekong Subregion, the government institutions have not emphasized such local participation strongly enough, just yet. Thus, an active organization or mechanism is needed to disseminate information on the potential harm resulting from less efficient CPPs.

Fourth, the region will see a rise in LNG imports to meet demand. Thus, the region's leaders will need to consider energy policy to increase the use of LNG in the future as a bridging fuel towards a clean energy future. Redesigning policy to promote LNG use will, to some extent, reduce coal use in the power mix. The countries in the Mekong Subregion should investigate the LNG infrastructure gap to develop policy to promote investment. This includes LNG terminals, pipelines, regasification plants, transportation, and storage.

Fifth, the region will need to prepare for a sharp increase in renewable energy from wind, solar, and biomass in the energy supply mix; and at the same time, promote the use of clean fuels and clean technologies. It will also need to look wider in terms of power grid connectivity. In this case, investment in 'hard' quality infrastructure will need to be connected to ASEAN.

Finally, the Mekong Subregion should boldly increase the portion of funding in the economic recovery package on green energy investment, as it will promote jobs, environmental protection, and social benefits for long-term sustainability. Governments and financial institutions may need to promote the financing of green projects through green bonds or other financial instruments. Of course, the region will also need to work on carbon credits in the future, as this will promote renewable and clean technology development.

Author Contributions: Conceptualization, H.P. and S.M.; methodology, H.P.; software, H.P.; validation, H.P., S.M., and H.P.A.; formal analysis, H.P.A. and S.M.; investigation, H.P.A.; resources, H.P.A.; data curation, H.P.A.; writing—original draft preparation, H.P.A.; writing—review and editing, S.M., H.P.A., and H.P.A.; visualization, H.P.A.; supervision, H.P.A.; project administration, H.P.A.; funding acquisition, H.P.A. All authors have read and agreed to the published version of the manuscript.

Funding: This research received no external funding.

Institutional Review Board Statement: Not applicable.

Informed Consent Statement: Not applicable.

Data Availability Statement: Not applicable.

Conflicts of Interest: The authors declare no conflict of interest.

References

1. Han, P. The Need for Quality Infrastructure to Meet Rising Energy Demand in the ASEAN Region. In *Powering Southeast Asia: Meeting the Region's Electricity Needs*; Phoumin, H., Herberg, M.E., Tsafos, N., Weatherby, C., Eds.; The National Bureau of Asian Research: Washington, DC, USA, 2020; pp. 31–41. Available online: https://www.iea.org/reports/world-energy-outlook-2017 (accessed on 10 September 2020).
2. Han, P. Embracing Renewable, Clean Technologies to Decarbonize Emissions in ASEAN Post-COVID-19′. *The Korea Times*, 13 June 2020.
3. 2° Institute. Global CO2 Levels. 2020. Available online: https://www.co2levels.org (accessed on 9 December 2020).
4. Han, P.; Perspectives on Mekong–Japan Cooperation for Inclusive Growth and Mutual Benefits. In Asia Pathways: The Blog of the Asian Development Bank Institute, 16 November 2018. Available online: https://bit.ly/386NWap (accessed on 5 November 2020).
5. IEA World Energy Model. 2019. Available online: https://www.iea.org/reports/world-energy-outlook-2019 (accessed on 17 September 2020).
6. Ministry of Finance, Japan. G20 Principles for Quality Infrastructure Investment 2019, G20 Osaka Summit 2019, 28–29 June 2019. Available online: https://www.mof.go.jp/english/international_policy/convention/g20/annex6_1.pdf (accessed on 9 December 2020).
7. Han, P. Quality Energy Infrastructure for Sustainability. *The ASEAN Post*, 13 August 2020.
8. Yu, H. China's Belt and Road Initiative and Its Implications for Southeast Asia. *Asia Policy* **2017**, *24*, 117–122. [CrossRef]
9. OECD. China's Belt and Road Initiative in the Global Trade, Investment and Finance Landscape. In *OECD Business and Finance Outlook 2018*; Organisation for Economic Co-operation and Development: Paris, France, 2018; pp. 3–44.
10. Ministry of Commerce, China. Investment and Cooperation Statistics about Countries along 'Belt and Road' in January–August of 2016′, Article, 23 September 2016. Available online: http://english.mofcom.gov.cn/article/statistic/foreigntradecooperation/201610/20161001410655.shtml (accessed on 15 November 2020).
11. Abi-Habib, M. How China Got Sri Lanka to Cough Up a Port. *The New York Times*, 25 June 2018.
12. Geraci, M. An Update on American Perspectives on the Belt and Road Initiative. *Issue Brief, 23 January 2020*. Available online: https://chinaus-icas.org/research/an-update-on-american-perspectives-on-the-belt-and-road-initiative/ (accessed on 12 November 2020).
13. Sultana, G. Sri Lanka after Rajapaksa: Can it Ignore China? *Strat. Anal.* **2016**, *40*, 1–10. [CrossRef]
14. Russel, D.R.; Berger, B. *Navigating the Belt and Road Initiative*; Asia Society Policy Institute: New York, NY, USA, 2019.
15. Friends of the Earth US. *China's Belt and Road Initiative: An Introduction*; Friends of the Earth US: Washington, DC, USA, 2016.

16. Grammaticas, D. Chinese Colonialism? *BBC*, 19 July 2012.
17. Wu, A. Fear of 'Yellow Peril' Lives On. *South China Morning Post*, 6 May 2013.
18. Cheong, D. Belt and Road Initiative Not Debt-Trap Diplomacy, Says China. *The Straits Times*, 9 March 2019.
19. Hansbrough, J.V. From the Blue Dot Network to the Blue Dot Marketplace: A Way to Cooperate in Strategic Competition. In *Hindsight, Insight, Foresight: Thinking About Security in the Indo-Pacific*; Vuving, A.L., Ed.; Asia-Pacific Center for Security Studies: Honolulu, HI, USA, 2020; pp. 179–189.
20. Panda, J.P. India, the Blue Dot Network, and the 'Quad Plus' Calculus. *Air Force J. Indo-Pac. Aff.* **2020**, *3*, 4–22.
21. Lyn, J. US Blue Dot Network to Counter China's BRI. *Asia Times*, 21 May 2020.
22. McCawley, P. Connecting the Dots on the Blue Dot Network. *The Interpreter*, 12 November 2019.
23. Heydarian, R.J. *The Indo-Pacific: Trump, China, and the New Struggle for Global Mastery*; Palgrave Macmillan: Singapore, 2020.
24. US Department of State. Blue Dot Network: Frequently Asked Questions. Available online: https://www.state.gov/blue-dot-network-frequently-asked-questions/ (accessed on 15 November 2020).
25. Kuo, M.A. Blue Dot Network: The Belt and Road Alternative. *The Diplomat*, 7 April 2020.
26. ADB. *Meeting Asia's Infrastructure Needs*; Asian Development Bank: Manila, Philippines, 2017.
27. Basol, P.K.; Basar, B. *A New Global (?) Infrastructure Initiative: The Blue Dot Network*; Hergüner Bilgen Özeke: Levent, Istanbul, Turkey, 2020; pp. 20–22.
28. Satake, T. Japan's 'Free and Open Indo-Pacific Strategy' and Its Implications for ASEAN. In *Southeast Asian Affairs 2019*; Singh, D., Cook, M., Eds.; ISEAS–Yusof Ishak Institute: Singapore, 2019; pp. 69–82.
29. Szechenyi, N.; Hosoya, Y. Working toward a Free and Open Indo-Pacific. In *Alliance Policy Coordination Brief*; Carnegie Endowment for International Peace: Washington, DC, USA, 2019.
30. Arase, D. Free and Open Indo-Pacific Strategy Outlook. In *Trends in Southeast Asia*; No. 12; ISEAS–Yusof Ishak Institute: Singapore, 2019.
31. US Department of State. *A Free and Open Indo-Pacific: Advancing a Shared Vision*; US Department of State: Washington, DC, USA, 2019.
32. Berkofsky, A.; 'Free and Open Indo-Pacific': Tokyo's Plans and Priorities. Italian Institute for International Political Studies, Commentary, 4 June 2018. Available online: https://www.ispionline.it/en/pubblicazione/free-and-open-indo-pacific-tokyos-plans-and-priorities-20690 (accessed on 15 November 2020).
33. Brewster, D. A 'Free and Open Indo-Pacific' and What It Means for Australia. *The Interpreter*, 7 March 2018.
34. Maslow, S.; Japan's 'Pivot to Asia': Tokyo Discovers the Indo-Pacific. Asia and the Pacific Policy Society, 1 August 2018. Available online: https://www.policyforum.net/japans-pivot-asia/ (accessed on 22 November 2020).
35. Herberg, M.E. High-Quality Infrastructure and the Free and Open Indo-Pacific Vision. In *Powering Southeast Asia: Meeting the Region's Electricity Needs*; Han, P., Herberg, M.E., Tsafos, N., Weatherby, C., Eds.; National Bureau of Asian Research: Washington, DC, USA, 2020; pp. 23–30.
36. Ford, L.W. *The Trump Administration and the Free and Open Indo-Pacific*; Brookings Institution: Washington, DC, USA, 2020.
37. Kawashima, S. A New Phase for the Free and Open Indo-Pacific. *The Diplomat*, 16 November 2020.
38. Swaine, M.D. *Creating an Unstable Asia: The U.S. 'Free and Open Indo-Pacific' Strategy*; Carnegie Endowment for International Peace: Washington, DC, USA, 2018; Available online: https://carnegieendowment.org/2018/03/02/creating-unstable-asia-u.s.-free-and-open-indo-pacific-strategy-pub-75720 (accessed on 20 November 2020).
39. Valencia, M.J. What Does a 'Free and Open Indo-Pacific' Actually Mean? *The Diplomat*, 30 March 2018.
40. Ministry of Foreign Affairs, Japan. Address by Prime Minister Shinzo Abe at the Opening Session of the Sixth Tokyo International Conference on African Development (TICAD VI). Nairobi, 27 August 2016. Available online: https://www.mofa.go.jp/afr/af2/page4e_000496.html (accessed on 15 November 2020).
41. Ministry of Foreign Affairs, Japan. *Priority Policy for Development Cooperation FY2017*; Ministry of Foreign Affairs of Japan: Tokyo, Japan, 2017.
42. Editorial Desk of the Gaiko (Diplomacy). Interview: Foreign Minister Kono Taro—Japan's Diplomatic Landscape in 2018: Pushing North Korea toward Denuclearization by Applying Continued Pressure. Discuss Japan—Japan Foreign Policy Forum, Diplomacy, No. 47. 2018. Available online: https://www.japanpolicyforum.jp/diplomacy/pt20180529193613.html (accessed on 17 November 2020).
43. Ministry of Foreign Affairs, Japan. *Joint Statement of the 11th Mekong–Japan Summit*; Ministry of Foreign Affairs of Japan: Bangkok, Thailand, 4 November 2019.
44. MRC. *Integrated Water Resources Managed-Based Basin Development Strategy for the Lower Mekong Basin, 2016–2020*; Mekong River Commission Secretariat: Vientiane, Laos, 2016.
45. MRC. *Basin-Wide Strategy for Sustainable Hydropower Development for the Lower Mekong Basin*; Mekong River Commission Secretariat: Vientiane, Laos, 2001.
46. *MRC, Preliminary Design Guidance for Proposed Mainstream Dams in the Lower Mekong Basin*; Mekong River Commission Secretariat: Vientiane, Laos, 2009.
47. MRC. *State of the Basin Report 2018*; Mekong River Commission Secretariat: Vientiane, Laos, 2019.

48. MRC; Basist, A.; Williams, C. *Understanding the Mekong River's Hydrological Conditions: A Brief Commentary Note on the 'Monitoring the Quantity of Water Flowing through the Upper Mekong Basin under Natural (Unimpeded) Conditions*; Mekong River Commission Secretariat: Vientiane, Laos, 2020.
49. MRC. *Situation Report on Hydrological Conditions in the Lower Mekong River Basin in January–July 2020*; Mekong River Commission Secretariat: Vientiane, Laos, 2020.
50. MRC. *Macro-Economic Assessment Report: The Study on the Sustainable Management and Development of the Mekong River Basin Including Impacts of Mainstream Hydropower Projects*; Mekong River Commission Secretariat: Vientiane, Laos, 2018.
51. Kummu, M.; Varis, O. Sediment-related impacts due to upstream reservoir trapping, the Lower Mekong River. *Geomorphology* **2007**, *85*, 275–293. [CrossRef]
52. Kondolf, G.M.; Rubin, Z.K.; Minear, J.T. Dams on the Mekong: Cumulative sediment starvation. *Water Resour. Res.* **2014**, *50*, 5158–5169. [CrossRef]
53. Kuenzer, C.; Campbell, I.; Roch, M.; Leinenkugel, P.; Tuan, V.Q.; Dech, S. Understanding the impact of hydropower developments in the context of upstream–downstream relations in the Mekong river basin. *Sustain. Sci.* **2013**, *8*, 565–584. [CrossRef]
54. Kittikhoun, A.; Staubli, D.M. Water diplomacy and conflict management in the Mekong: From rivalries to cooperation. *J. Hydrol.* **2018**, *567*, 654–667. [CrossRef]
55. Sok, S.; Meas, S.; Chea, S.; Chhinh, N. Regional cooperation and benefit sharing for sustainable water resources management in the Lower Mekong Basin. *Lakes Reserv. Res. Manag.* **2019**, *24*, 215–227. [CrossRef]
56. An, H.P.; Kittikhoun, A.; Meas, S. 25 Years of Mekong Cooperation: Achievements, Challenges and the Way Ahead. *Mekong Connect* **2020**, *2*, 1–4.
57. LMC. A Brief Introduction of Lancang–Mekong Cooperation. 13 December 2017. Available online: http://www.lmcchina.org/eng/n3/2020/0904/c416294-9755278.html (accessed on 10 September 2020).
58. Gong, X. China's Economic Statecraft: The Belt and Road in Southeast Asia and the Impact on the Indo-Pacific. *Secur. Chall.* **2020**, *16*, 39–48.
59. Zhang, H.; Li, M. China's water diplomacy in the Mekong: A paradigm shift and the role of Yunnan provincial government. *Water Int.* **2020**, *45*, 347–364. [CrossRef]
60. Busbarat, P. Grabbing *the Forgotten*: China's Leadership Consolidation in Mainland Southeast Asia through the Mekong–Lancang Cooperation. *ISEAS Perspect.* **2018**, *7*, 1–8.
61. The ASEAN Post. Lancang–Mekong Cooperation: Blessing or Curse? *The ASEAN Post*, 3 April 2019.
62. LMC. *List of Proposed Projects on Lancang–Mekong Water Resources Cooperation*; LMC Water Center: Beijing, China, 2019.
63. LMC. *Joint Statement of the Ministerial Meeting of Lancang–Mekong Water Resources Cooperation*; LMC Water Center: Beijing, China, 2019; pp. 1–8.
64. Liena, F.; Juan, L.; Pengfei, M. Spotlight: Lancang–Mekong Cooperation Boosts Development of China–ASEAN Community of Shared Future. *Xinhau*, 1 October 2018.
65. Qingrun, S. Opinion: Lancang–Mekong Cooperation Promotes Regional Development. *CGTN*, 25 September 2018.
66. Commentary: Hyping Mekong Water Resources Issue a Political Farce. *Xinhau*, 13 September 2020.
67. Chinese State Councilor Lauds Progress in Lancang–Mekong Cooperation. *Xinhau*, 21 February 2020.
68. Xing, W. Lancang-Mekong River Cooperation and Trans-Boundary Water Governance: A Chinese Perspective. *China Q. Int. Strat. Stud.* **2017**, *3*, 377–393. [CrossRef]
69. Biba, S. China's 'Old' and 'New' Mekong River Politics: The Lancang–Mekong Cooperation from a Comparative Benefit-Sharing Perspective. *Water Int.* **2018**, *43*, 622–641. [CrossRef]
70. Middleton, C.; Allouche, J. Watershed or Powershed? Critical Hydropolitics, China and the 'Lancang–Mekong Cooperation Framework'. *Int. Spect.* **2016**, *51*, 100–117. [CrossRef]
71. Verbiest, J.-P.A. Regional Cooperation and Integration in the Mekong Region. *Asian Econ. Policy Rev.* **2013**, *8*, 148–164. [CrossRef]
72. ERIA. *The Comprehensive Asian Development Plan 2.0 (CADP 2.0): Infrastructure for Connectivity and Innovation*; Economic Research Institute for ASEAN and East Asia: Jakarta, Indonesia, 2015.
73. Han, P.; Kimura, F. Introduction. In *Energy Market Integration in East Asia: Energy Trade, Cross Border Electricity, and Price Mechanism*; Kimura, F., Han, P., Eds.; Economic Research Institute for ASEAN and East Asia: Jakarta, Indonesia, 2014; pp. 1–12.
74. GMS Secretariat. *Greater Mekong Subregion Economic Cooperation Program—Regional Investment Framework 2022: First Progress Report and Update*; Asian Development Bank: Manila, Philippines, 2019.
75. ADB. Greater Mekong Subregion Statistics on Growth, Infrastructure, and Trade. In *Greater Mekong Subregion Eight Economic Corridors Forum, Phnom Penh, 3–4 August 2016*; Asian Development Bank: Manila, Philippines, 2016.
76. IEA. *World Energy Outlook 2017*; International Energy Agency: Paris, France, 2017.
77. Phoumin, H.; Kimura, F.; Arima, J. Potential Renewable Hydrogen from Curtailed Electricity to Decarbonize ASEAN's Emissions: Policy Implications. *Sustainability* **2020**, *12*, 10560. [CrossRef]
78. Mitsuru, M.; Jongkyun, L.; Ichiro, K.; Han, P. (Eds.) *Improving Emission Regulations for Coal-Fired Power Plants in ASEAN*; Economic Research Institute for ASEAN and East Asia: Jakarta, Indonesia, 2017.
79. Kobayashi, Y.; Han, P. (Eds.) *Natural Gas Mater Plan for Myanmar*; Economic Research Institute for ASEAN and East Asia: Jakarta, Indonesia, 2018.

80. Denholm, P.; Cochran, J. *Wind and Solar on the Power Grid: Myths and Misperceptions*; USAID Office of Global Climate Change and National Renewable Energy Laboratory: Washington, DC, USA, 2015. Available online: https://www.nrel.gov/docs/fy15osti/63045.pdf (accessed on 11 November 2020).
81. IRENA. *Innovation Landscape Brief: Renewable Power-to-Hydrogen*; International Renewable Energy Agency: Abu Dhabi, United Arab Emirates, 2019. Available online: https://bit.ly/3rcOd38 (accessed on 11 November 2020).

sustainability

MDPI

Article

Analyzing the Characteristics of Green Bond Markets to Facilitate Green Finance in the Post-COVID-19 World

Farhad Taghizadeh-Hesary [1,*], Naoyuki Yoshino [2] and Han Phoumin [3]

1 Social Science Research Institute, Tokai University, Hiratsuka-shi 259-1292, Japan
2 Faculty of Economics, Keio University, Tokyo 108-8345, Japan; yoshino@econ.keio.ac.jp
3 Economic Research Institute for ASEAN and East Asia (ERIA), Jakarta 10270, Indonesia; han.phoumin@eria.org
* Correspondence: farhad@tsc.u-tokai.ac.jp or farhadth@gmail.com

Abstract: The COVID-19 pandemic and the global recessions have reduced the investments in green projects globally that would endanger the achievement of the climate-related goals. Therefore, the post-COVID-19 world needs to adopt the green financial system by introducing new financial instruments. In this regard, green bonds—a type of debt instrument aiming to finance sustainable infrastructure projects—are growing in popularity. While the literature does not contest their effectiveness in fighting climate change, research highlights the high level of risks and low returns associated with this instrument. This study analyzes the green bond markets in different regions with a focus on Asia and the Pacific. It aims to fill the gap in the literature by conducting a comparative study of the characteristics, risks, and returns of green bonds based on the region. The study is based on theoretical background and empirical analysis using the data retrieved from Bloomberg New Energy Finance and the Climate Bonds Initiative. The empirical results are based on several econometrics tests using panel data analysis estimation methods, namely pooled ordinary least squares and generalized least squares random effects estimator. Our findings prove that green bonds in Asia tend to show higher returns but higher risks and higher heterogeneity. Generally, the Asian green bonds market is dominated by the banking sector, representing 60% of all issuance. Given that bonds issued by this sector tend to show lower returns than average, we recommend policies that could increase the rate of return of bonds issued by the banking sector through the use of tax spillover. In the era of post-COVID-19, diversification of issuers, with higher participation from the public sector and de-risking policies, could also be considered.

Keywords: green bonds; post-COVID-19 era; Asia and the Pacific; green finance; sustainable development

Citation: Taghizadeh-Hesary, F.; Yoshino, N.; Phoumin, H. Analyzing the Characteristics of Green Bond Markets to Facilitate Green Finance in the Post-COVID-19 World. *Sustainability* **2021**, *13*, 5719. https://doi.org/10.3390/su13105719

Academic Editor: Chia-Lin Chang

Received: 7 April 2021
Accepted: 9 May 2021
Published: 20 May 2021

1. Introduction

Since the beginning of the century, the world has been consistently growing at around 3%, without following a sustainable path. The past decade (2010–2020) has been marked by rising environmental awareness and demand for the promotion of renewable energy sources. Alarming reports from the Intergovernmental Panel on Climate Change have shown that climate change is a pressing matter that needs to be addressed, and in 2015, United Nations members agreed on keeping global warming below 2 °C through Nationally Determined Contributions. The United Nations also acknowledged the matter by including 'Climate Action' in the Sustainable Development Goals (SDGs). Yet, the Intergovernmental Panel on Climate Change and the United Nations Environment Program reports highlight that further actions need to be taken to reach this goal and fulfill the SDGs. Several SDGs are directly and indirectly related to green and low-carbon energy developments and the environment. SDG 7 (affordable and clean energy) and SDG13 (climate action) are directly related. SDG3 (good health and well-being), SDG14 (life below water), and SDG15 (life on land) are indirectly related. This means that the UN global agenda clarified the importance of green energy and reducing pollutions (CO_2 or NO_x).

One of the biggest barriers in the development of renewable energy is the low level of investment [1]. As of 2018, the majority of the world's investment in energy still went to carbon-emitting sources—fossil fuels. For instance, while 39% of investments in power supply generation went to renewable energy, they only represented 19% of total investments in the energy sector [2]. In comparison, fossil fuels received about 60% of total investments in the same year [2], with the remainder going to nuclear, biofuels, or battery storage, which are still, to a lesser extent, sources of greenhouse gas (GHG) emissions.

In 2020–2021 due to the COVID-19 pandemic and the global economic recession, the ongoing investment in renewable power, energy efficiency, and other green projects fell drastically. The COVID-19 pandemic and the economic downturns resulted in a drastic reduction in fossil fuel prices. Low fossil fuel price is harmful to developing renewable energy projects, making solar, wind, and other renewable energy resources less competitive electricity sources. This reduces investors' interest in clean fuels that would threaten the Paris Agreement on climate change and several SDGs.

Funding green infrastructure projects remains an issue. In general, these projects require large borrowings, as they are capital-intensive [3]. In addition, green projects are usually associated with 'high risk and low returns at the initial research and development stage' [4]. Difficulties in accessing finance for green projects is especially the case in Asia, whose financial sector is dominated by banks; hence, banks are the main source of funding [1]. Venture capitalists are scarce in Asia, including East and Southeast Asia [3], although they are more likely to provide funds for green projects, while banks generally deem green projects risky [4]. In addition to risk overvaluation, the authors of [5] highlighted the existence of a maturity mismatch between bank loans, which are generally short-term, and green projects, which are thought to be medium- to long-term projects. Thus, banks are not usually well-suited to providing loans for green projects. Second-level financial institutions (e.g., insurance or pension funds) may provide funds for longer-term projects as they hold long-term money but are reluctant to invest in electricity projects whose tariffs are generally regulated by the public sector [5]. Overall, traditional finance is failing to provide enough funding for green projects, so there is a need for innovative finance or the establishment of a green financial system to fill this gap. Introducing new financial instruments such as green bonds besides the conventional banking system will help fill the green finance gap required to achieve the related SDGs.

This research aims at analyzing green bonds—a special type of green finance instrument and an essential part of the green financial system.

Green bonds are fixed-income securities whose popularity has increased significantly in the past few years. While their definition varies, they are usually understood as a form of debt instrument used to finance green projects, such as renewable energy infrastructure or projects that comprise an energy efficiency dimension. The Asia and the Pacific regions have been increasing the use of this instrument to bridge the gap between infrastructure projects and access to financing. In 2018, Asia and the Pacific achieved the highest regional growth of green bond issuance, with an annual rate of 35% [6]. The region has consistently been the second-largest issuer of green bonds by volume since 2016 and accounts for the most diverse pool of issuers in the world, with 345 different institutions [7]. While this new instrument may be favored in Asia, one cannot help but wonder how the peculiar nature of the Asian financial sector, which is dominated by traditional forms of banking, may affect the characteristics of green bonds issued in the region, in terms of associated returns and risks. The recent literature on the topic has shown that green bonds tend to show lower returns than their conventional counterparts [8–11]. The lower return of the green bonds is due to their intrinsic characteristics. Green bonds are financial instruments designed to fund green infrastructure and green technologies. Green technologies and generally green projects (such as renewable energy technologies) are often earlier in the development stage and not always commercially viable compared to the brown projects (such as fossil fuel-based energy projects) field, where many of them date back to 100 years ago [12,13]. This makes green technologies more expensive and riskier ventures. As a result, the rate of return of green bonds is expected to lower compared to conventional bonds. In addition to this, other reasons make green projects more expensive. According to the OECD Companion to the Inventory of Support Measures for Fossil Fuels 2015, the production or consumption of fossil fuels is supported by almost 800 individual policies [14]. Another form of subsidy, an indirect one, takes place when fossil fuel companies are not taxed efficiently [15].

Refer. [16,17] showed that the green bond market was more volatile and hence riskier than the conventional bond market. However, the studies mentioned above conducted global analyses of green bonds, even though issuers' regional characteristics may play a crucial role in determining the risks and returns of these instruments.

There are several reasons behind the hypothesis that the characteristics of green bonds may depend on the region of issuance. First, economic theory and empirical research confirm that the performance of fixed-income instruments is highly dependent on macroeconomic variables such as changes in financial markets, economic uncertainty, or daily economic activity [18]. Therefore, it is likely that the performance and associated risks of green bonds vary depending on the region's economic activity or the investors' uncertainty evaluation and risk aversion. A second rationale for this hypothesis comes from the difference in the inherent characteristics of financial markets, based on the region, as previously explained. The research questions of this study are (i) how do green bonds in Asia compare in terms of size and time to maturity with green bonds issued in other regions? Do they differ because of the characteristics of Asian financial markets? (ii) In terms of return and risk, how do green bonds in Asia compare with green bonds from the rest of the world? (iii) Does the type of issuer affect the performance of the bond? In particular, does the influence of the issuer depend on the region?

This research aims to fill the gap in the literature by conducting a comparative study of the characteristics of green bonds, based on the region. In particular, we seek to determine whether the domination of traditional banking has an impact on the return of green bonds issued in Asia and the Pacific.

The study is organized as follows: Section 2 presents a literature review, which discusses green finance and recent academic debates related to green bonds. Section 3 introduces the dataset used in this study and discusses our methodology. Section 4 shows the empirical results of this research, and Section 5 concludes this paper and provides policy recommendations.

2. Literature Review

2.1. An Introduction to Green Finance and Green Bonds

The concept of green finance emerged in the 2010s and can be defined as 'a type of future-oriented finance that simultaneously pursues the development of financial industry, improvement of the environment, and economic growth' [4]. Green finance is a broad concept that includes sustainable finance for socially inclusive green projects, environmental finance to promote environmental protection, carbon finance, targeting a reduction in GHG emissions, and climate finance, focusing on climate change adaptation and mitigation [4]. The term 'green finance' also covers a wide range of instruments, from private loans to insurance, and includes equity, derivatives, and fiscal or investment funds [4].

Increasing green finance, climate finance, and low-carbon investments are directly and indirectly related to various SDGs. Investments in green energy projects are crucial to achieve the SDGs and meet the Paris Agreements [1]. Fiscal policy has an essential role in assuring the sustainable use of resources and keeping the environment for meeting the related SDGs. This applied to both sides of the government budget. In the revenue side, carbon taxation adjusted with greenness efforts and green bonds are two essential tools, and their importance is increasing. Various fiscal measures could help green-specific priority sectors. Green-adjusted tax on polluting gases can help generate revenue for environmental purposes and redirect the flow of investments from brown to green and low-carbon sectors by introducing green floating rate bonds [5,13]. Another fiscal measure is global taxation on CO_2, NO_X, and other pollutions, for bringing back optimal portfolio allocation in green investments.

In this research, we focus on green bonds. Since their creation in 2007, USD754 billion worth of green bonds have been issued—primarily in the United States, China, and France—in compliance with the Green Bond Principles [7]. Green bonds can be issued by central and local governments, banks, or corporations, and include any debt format [6]. Since 2014, Asia-Pacific's bond issuance has been growing at 35%, placing the region second in terms of green bond volume [6]. Figure 1 shows the evolution of the amount issued for green bonds, per region of issuance. The graph clearly shows that green bonds are a relatively new form of financial instrument, as their issuance started timidly in the early 2010s and skyrocketed after 2015. Europe is the leading issuer of green bonds, although Asia-Pacific has witnessed steady growth in recent years.

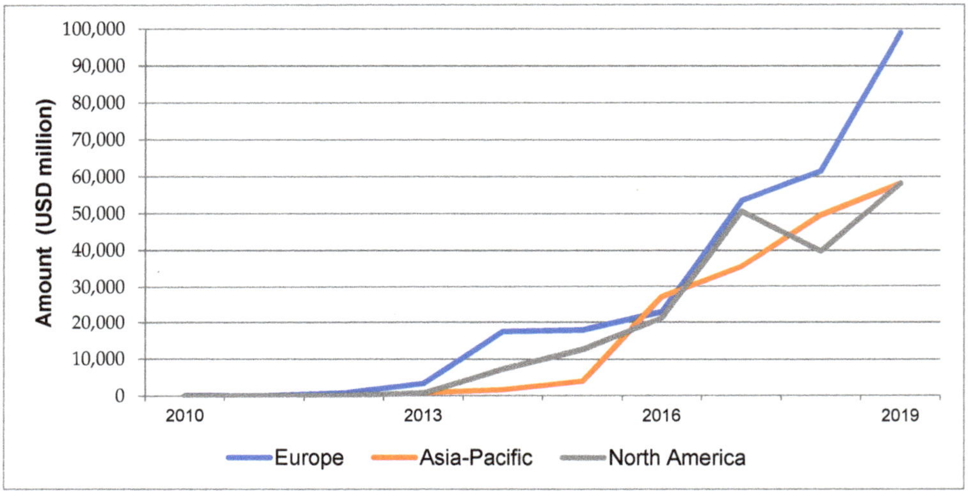

Figure 1. Evolution of the amount issued for green bonds per region. Note: 'North America' includes issuers from other regions of the world, apart from Europe and Asia-Pacific. However, the majority of the issuance in this category is from the United States and Canada. Source: Authors' compilation, using data from [7].

Increasing awareness of climate change could be the reason behind the surge in popularity of this instrument. Typically, green bonds are a form of fixed-income finance which can be applied to many debt formats, such as private placements, securitizations, and covered bonds, as well as green loans [7]. The particularity of this form of finance is their target, as the term only encompasses finance for climate change solutions whose proceeds go to green assets [7]. To clarify which bonds could be qualified as such, a consortium of investment banks established the Green Bonds Principles in 2014, based on four main components: (i) the use of the proceeds, (ii) the process for project evaluation and selection, (iii) the management of the proceeds, and (iv) reporting [19]. The principles do not define what is 'green' about the bonds, but merely list target sectors in which green bonds are considered valid. However, these principles simply have an indicative value, and were only agreed on by the investment banks that created them. To date, there is no general taxonomy for green bonds, although the European Union has proposed including one in the upcoming European Green Deal [7].

2.2. Characteristics and Challenges of Green Bonds

The increasing popularity of this instrument has attracted the attention of academic researchers. Studies have provided some empirical proof that green bonds can be useful in fighting climate change [20]. The main academic debate regarding green bonds is the existence of the 'green premium', also called 'greenium', defined as 'a discount that makes green bonds funded cheaper than other bonds from the same issuer' [8,21]. Many recent studies have attempted to compare the yields of green bonds with those of conventional bonds, and the results vary depending on the methodology used. The authors of [11] conducted a global study, matching green bonds with similar conventional bonds and applying a two-step regression method, and concluded that green bonds had lower yields, on average. This effect was especially pronounced for bonds issued by the financial sector and low-rated bonds [11]. This conclusion is shared by recent studies such as [8,10,11].

Other studies, however, tend to have mixed results. For instance, [22] showed that the green premium was actually positive, meaning that matched green bonds had higher yields than their closest brown counterparts. The authors explain their results by arguing that the sign of the green premium depends on the issuer, and that privately issued bonds generally have a positive premium [22]. Similarly, [23] found that the sign of the green premium was not obvious, and depended on the rating achieved by the bond. In particular, highly rated green bonds consistently showed higher returns, which, the authors argued, could make up for the external costs of issuance [23,24]. Finally, [25] could not find statistically significant evidence of the existence of the green premium, even though they used several methodologies such as matching with difference-in-differences and traditional panel techniques (fixed effect). Due to the lack of consensus regarding the green premium, [26] provided a comprehensive literature review on the topic, detailing the methodology of each paper. The authors concluded that the majority of the studies on the topic prove the existence of a green premium in secondary markets.

Interestingly, there does not appear to be a consensus on the riskiness of green bonds either. While [22] found that green bonds had lower variance than conventional bonds, the results of [16], who studied the volatility of the green bond market using a multivariate GARCH approach, contradict this theory. The authors of [16] proved that the market of labeled green bonds was highly volatile—far more so than the unlabeled market of conventional bonds. There is a close link between green bonds and fixed-income and currency markets, with the latter's green bonds receiving price spillover from the latter [27]. Generally, green bonds are strongly affected by changes in stock, changes in energy, and high-yield corporate bond markets [28], as well as the liquidity risk of the bond market [29].

Apart from their generally low returns and high risks, green bonds also represent a challenge for their issuers. Both [23] and [27] highlighted that issuing green bonds tends to be more expensive than issuing a conventional bond due to additional costs arising from the certification, reporting, and administrative burden of the proceeds. The authors of [27] also pointed out the need to bridge the informational gap between issuers and investors and offer clear and unified green criteria to provide assurance of the green nature of the investment [27]. The major issue faced by green bonds is generally the lack of uniform definition and labeling. While the Green Bonds Principles are a major step towards this direction, they remain an informal form of labeling that was only generated by a handful of private actors. Hence, it does not have global legitimacy.

A review of the literature has revealed the evolution and contribution of green bonds. As fixed-income instruments, green bonds can be useful in fighting climate change and bridging the investment gap for green projects. At the same time, these bonds are characterized by lower returns and higher risks than their conventional counterparts. Administrative costs arising from certification and lack of uniform taxonomy have added to their relative lack of attractiveness. Nevertheless, there is ongoing debate regarding the characteristics of green bonds, particularly the existence of a green premium, while results tend to vary depending on the bond rating and issuer [22,23].

A literature review shows that green bonds are essential financial instruments for financing ecological and green projects, and that their importance is increasing. However, we could not find any study that provides a comparative study of the characteristics of green bonds, based on the region. In particular, we could not find any study that determines whether the domination of traditional banking impacts the return of green bonds issued in Asia and the Pacific. Hence, from this aspect, this study is novel and contributes to the literature.

3. Methodology and Data Description

In this section, we detail the approach taken in this study to determine the regional characteristics of green bonds, with a specific focus on those issued in Asia and the Pacific.

3.1. Data and Description of Variables

The study combined two datasets from Bloomberg New Energy Finance (BNEF) and the Climate Bonds Initiative (CBI). The BNEF database only provides bonds with an issued amount of at least \$100. Both sources are considered authorities on data related to green finance and have been employed in many recent studies (e.g., [9,11,22–24]). In this research, we only focus on green bonds with a minimum of \$100 in size, issued from 2017 to 2020. Hence, this study presents an analysis of unbalanced panel data of 1014 bonds, from 2017 to 2020, for a total of 1174 observations. To be precise, since we are missing many observations of the rate of return of bonds in 2017 and 2018, the length of the panel is about two time periods. A description of the variables used in the study is provided in Table 1.

Table 1. Description of variables.

Name of Variable	Observations	Unit	Description	Source
Rate of return	1174	%	Rate of return on investment, as measured on 10 January each year.	Bloomberg NEF
Days to maturity	1174	Days	Remaining days before the principal of a security is due and payable.	Bloomberg NEF
Amount issued	1174	\$	Cumulative amount issued from the original security pricing date through to the current date for debt securities. The amount will include taps/increases or reopenings.	Bloomberg NEF
Coupon rate	1174	%	Current interest rate of the security.	Bloomberg NEF
Issuer name	1174	/	Name of the issuing entity.	Bloomberg NEF
Region of issuance	1174	/	Set of dummy variables, with possible values being Asia and the Pacific, Europe, and North America/Others.	Bloomberg NEF
Sector of issuance	1174	/	Set of dummy variables, with possible values being banking and finance, public, manufacturing, power and utilities, construction, and others.	Authors' compilation, based on issuer name provided by Bloomberg NEF

Source: Authors' compilation.

3.2. Methodology

To determine the characteristics of green bonds, we propose several methods, each assessing different dimensions of bonds. First, an analysis of the distribution of issuers, maturity, and issued size is proposed, to determine whether green bonds issued in Asia present an inherent difference in their nature. We then move on to a mean-variance analysis, distinguishing between regions and sectors of issuance, to discuss how Asian green bonds compare with their counterparts in terms of risks and returns. Finally, the latter part of the empirical analysis is devoted to investigating the impact of the sector of issuance on the performance of green bonds, as measured by the rate of return, depending on the region.

To this end, we develop an econometric model, which is given by the following equation:

$$Return_{i,t} = \alpha + \sum_{i=1}^{4} \beta_i Sector_i + \sum_{t=2018}^{2020} \gamma_t Year_t + \chi_1 Coupon_i + \chi_2 Maturity_{i,t} + \epsilon_{i,t} + u_{i,t}, \quad (1)$$

where $Return_{i,t}$ denotes the rate of return of bond i at year t, $Sector_i$ is a set of dummy variables denoting the bond i's issuing sector, $Year_t$ is a set of dummy variables for time fixed-effects, $Coupon_i$ is the bond i's coupon rate, $Maturity_{i,t}$ denotes the number of days until the bond i reaches maturity at year t, and $\epsilon_{i,t}$ and $u_{i,t}$ are idiosyncratic and time-varying error terms, respectively.

While many studies use yield as a dependent variable [11,22,23], we decided to use the rate of return of the bonds as our dependent variable, as an approximation of the bond's performance, due to limitations on data availability. Since this study aims to determine the impact of the type of issuer on the bond's performance, we also include a set of four dummy variables, representing the issuer's sector, constructed based on the issuer name provided by BNEF. Sectors analyzed in the study are grouped into five categories: public, banking and finance, manufacturing, power and utilities, and other issuers. Public issuers are generally state and regional development banks and international organizations, but we do not include state-owned enterprises in this category. Banking and finance are essentially composed of national and local banking institutions, but investment banks and insurance are also considered. Finally, manufacturing in our sample is mostly composed of information technology and paper companies, while other issuers are dominated by companies belonging to real estate and construction.

The choice of remaining control variables is based on existing literature on the topic. The coupon rate, issued size, and maturity are often used in studies tackling the existence of the green premium, as they are essential components for matching green and brown bonds [11,22] or as control variables in regression [8,23,24]. Furthermore, we decided to include year fixed-effects to control for variation over time since our other control variables describe fixed characteristics of bonds. The variable *Maturity* is time-dependent, but its variation is fixed over time so it cannot fully capture changes in time periods. We aim to capture the effects of changes in the financial market and economic policy uncertainty through these dummy variables, as these macroeconomic variables were shown to have a significant impact on green bonds' returns. The authors of [18,25] took a similar approach by including year fixed-effects as a control variable in their regression.

4. Empirical Analysis

This section presents the results of the empirical analysis and is divided into three parts. The study will first discuss the characteristics of green bonds, based on summary statistics and a general description of the dataset, and will move on to mean-variance and regression analysis.

4.1. Summary Statistics

Since this study aims at identifying the regional characteristics of green bonds, we first delve into the description of our dataset. To this end, we present summary statistics in Table 2, while Figure 2 presents the distribution of issuers per sector and per region. Note that the summary statistics of Table 2 are constructed using data from [19] due to the larger amount of bonds in their database. Summary statistics of the dataset from BNEF are presented in Appendix A, for reference.

Table 2. Summary statistics.

	Amount Issued (USD Million)			Time to Maturity		
Item	**Asia-Pacific**	**Europe**	**North America**	**Asia-Pacific**	**Europe**	**North America**
Observations	624	835	3899	608	823	3886
Mean	288.28	349.73	49.71	8505.14	4624.98	4502.73
Standard deviation	443.31	628.25	129.91	46,149.90	25,339.42	2067.03
Minimum	0.99	0.38	0.02	161	19	24
Maximum	4355.1	7558.6	2250.0	364,635.0	364,877.0	36,594.0

Source: Authors' compilation.

There are already several takeaways regarding the regional characteristics of green bonds, based on Table 2. The number of green bonds issued in North America is a little less than three times the amount of bonds issued in Asia-Pacific and Europe combined. However, North American bonds are characterized by their small issued amount, which explains why the region is lagging the Asia-Pacific and Europe in the overall green bonds market, as shown in Figure 1. The dominance of small green bonds also explains the small share of North American bonds in the BNEF sample size. While bonds issued in the Asia-Pacific are comparable in size to their European counterparts, they are characterized by a long-term orientation, as the number of days before reaching maturity is almost twice that of European and North American bonds. Nonetheless, it is essential to note that

Asian bonds are far more diverse in terms of maturity, and to a lesser extent, size, than bonds issued in other regions of the world. Therefore, it might be challenging to reach an overall conclusion on the characteristics of Asian bonds, solely based on an analysis of summary statistics.

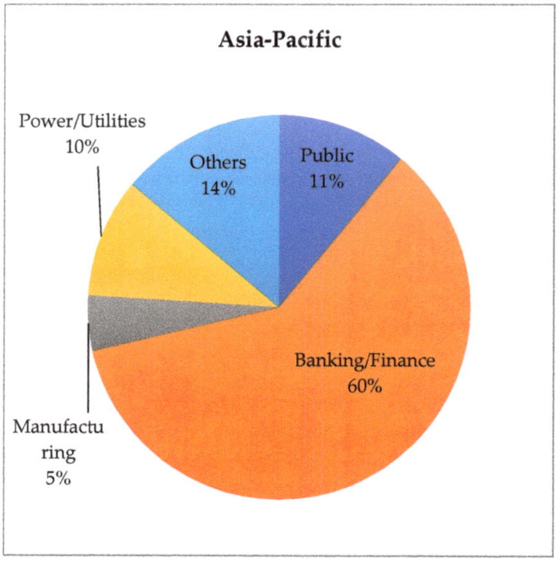

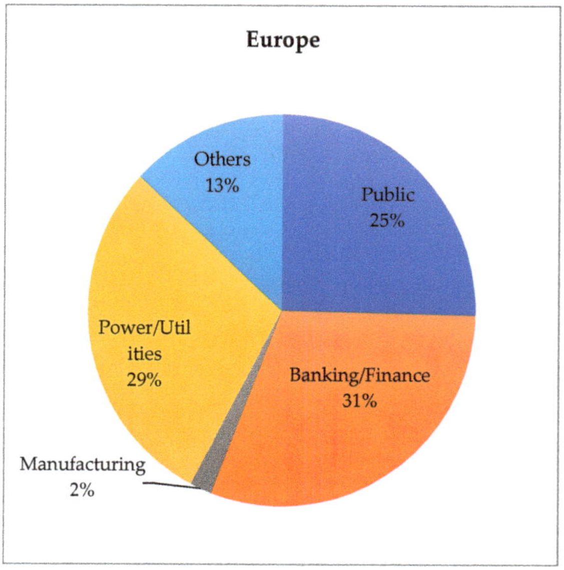

Figure 2. *Cont.*

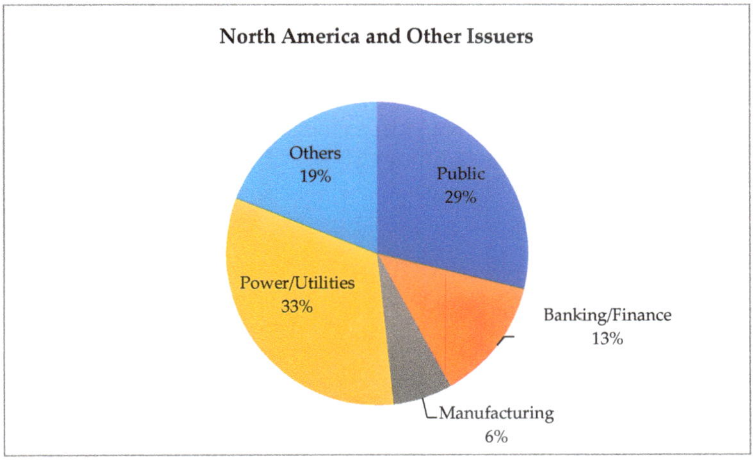

Figure 2. Distribution of green bonds issuers by sector and by region Source: Authors' compilation.

Since this study aims to determine whether green bonds present different characteristics based on the region of issuance, we use the Kolmogorov–Smirnov test to check whether distribution in the sample differs, depending on the region. Results of the test are presented in Table 3. The test only compares two samples. Hence, each region was tested against the remaining two. Four variables were used, all of which define some crucial characteristics of bonds, namely issued amount, time to maturity, coupon rate, and rate of return, based on the data provided by BNEF. Regardless of the variable, the results suggest a rejection of the null hypothesis of identical distribution. Each region possesses a unique distribution when it comes to essential characteristics of green bonds.

Table 3. Kolmogorov–Smirnov test for equality of distribution.

	Asia–Europe vs. North America	Asia–North America vs. Europe	Europe–North America vs. Asia
Rate of Return	0.43 *** (0.00)	0.66 *** (0.00)	0.52 *** (0.00)
Amount Issued	0.12 *** (0.00)	0.43 *** (0.00)	0.41 *** (0.00)
Time to Maturity	0.26 *** (0.00)	0.23 *** (0.00)	0.43 *** (0.00)
Coupon Rate	0.39 *** (0.00)	0.63 *** (0.00)	0.46 *** (0.00)

Source: Authors' compilation. Associated probability in parenthesis. *** denotes significance at the 1% level.

The sectoral distribution of green bond issuers provides another insight into the particular nature of Asian green bonds. While the share of issuers in Europe, North America, and the rest of the world is quite balanced between the public, utilities, and banking categories, the banking and finance sector share in the Asia-Pacific represents almost two-thirds of the total issuance. Regardless of the region, however, issuance from manufacturers, real estate, construction, and other types of firms is relatively uniform. The imbalance observed in the Asia-Pacific comes from the low shares of the public and utility sectors, with the amount of bonds issued even lower than that of real estate, construction, and other sectors (Figure 2). This observation confirms our initial hypothesis of the dominance of traditional forms of banking in Asia. As the literature review showed, the Asian financial sector is mostly composed of traditional banking institutions [3], but this result confirms that this trend is also passed on to green finance instruments such as green bonds. Due to the risk of overvaluation and maturity mismatch in traditional forms of banking [5], banking dominance likely has a significant impact on the performance of green bonds.

4.2. Mean-Variance Analysis

Since the dominance of banking and finance—as the issuers of green bonds in Asia—may affect the bonds' performance, we present a mean-variance analysis of the rate of return of bonds, based on the region of issuance and the type of issuer. The results of this analysis on the overall sample are presented in Figure 3, and numerical values for mean and variance are provided in Appendix A.

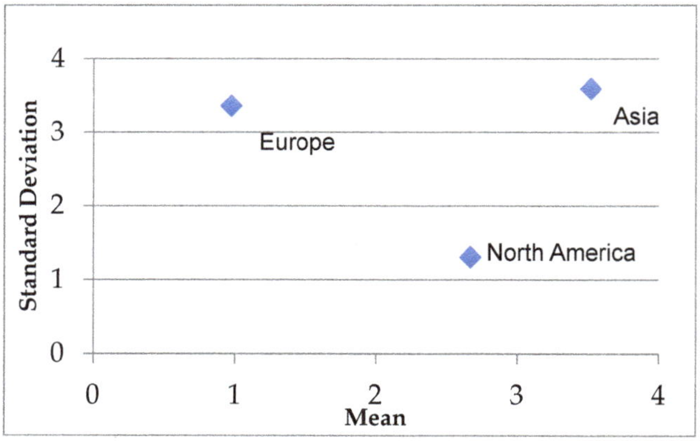

Figure 3. Mean-variance analysis of the rate of return of green bonds. Source: Authors' compilation.

The overall analysis of the mean and variance of the returns of bonds shows high variation between regions of issuance. The relatively high variance of Asian bonds reflects the diversity of these bonds (confirmed in Section 4.1). In Europe, the bonds issued appear to have higher risks, with relatively low returns, and, in comparison with Asian and North American bonds, do not seem to be appealing to investors.

Figure 4 represents the main focus of our mean-variance analysis, as it provides a sectoral analysis of the risks and returns of green bonds, based on the region of issuance. As in Figure 3, specific numerical values for the mean and variance are reported in Appendix A.2. First, the mean and variance values for the manufacturing sector stand out, as they are twice as large as those of other sectors, especially in the case of European bonds. These extreme values could be explained by the size of this particular subsample, as manufacturers represent around 5% of all issuance on average. With the notable exception of the manufacturing sector, Asian bonds tend to offer higher returns than those issued in Europe and North America, but also come with higher risks. It is interesting to note that bonds issued by companies in banking and finance in Asia do not present a striking difference with those issued by other sectors, contrary to what our hypothesis would suggest. On the other hand, bonds issued by power and utilities stand out due to their high variance, compared with other sectors. This feature could explain the small share of issuance of power and utilities in the Asia-Pacific, especially as their low risk characterizes bonds issued by power and utilities companies in Europe and North America. Indeed, if bonds issued by power and utilities are deemed risky, then it is not surprising that they attract few investors, hence their relatively low share. Generally, European bonds are characterized by low returns but have low associated risks, with both the mean and variance around 1. This could explain the dominance of Europe in the green bond markets, as they could be considered more reliable assets by investors.

4.3. Regression Analysis

The core of our empirical findings lies in the regression analysis. While summary statistics and mean-variance analysis can highlight the characteristics and features of data on sectoral issuers and the difference in performance depending on the region and type of issuer, it cannot provide a conclusion on the relationship between the issuer and performance, nor can it help elucidate the significance of the difference in performance, depending on the region and issuer.

To answer these questions, the study introduces a regression analysis, estimated based on the equation provided in sub-Section 3.2, whose results are presented in Table 4. Equations are estimated on the full sample (using dummy variables to represent each region), as well as on each of the three regional subsamples, using White robust standard errors to control for model misspecifications, such as heteroskedasticity. The relatively short length of the panel (t = 2 for most observations) exempts us from additional time series testing on the data. Therefore, we use traditional panel data analysis estimation methods: pooled ordinary least squares (OLS) and generalized least squares (GLS) random effects (RE) estimator. The lack of time-varying independent variables precludes the use of a fixed-effect (FE) estimator. Indeed, the inclusion of a cross-sectional FE dummy variable (for each bond) does not allow us to determine the impact of the bonds' characteristics, such as the sector of issuance. Instead, adding both FE and sectorial dummy variables provokes issues of multicollinearity, as individual characteristics are both captured by FE dummy and sectorial dummy variables. Therefore, the study prefers the RE estimator, in line with [17]. Since we are interested to see the effect of the banking sector on green bonds, we further include interaction terms between each region and the banking dummy variable. Regional and interaction dummy variables for North America are used as references and excluded so as not to cause a multicollinearity issue.

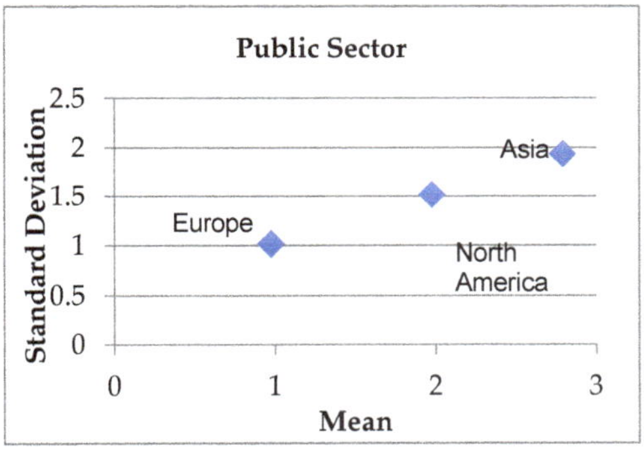

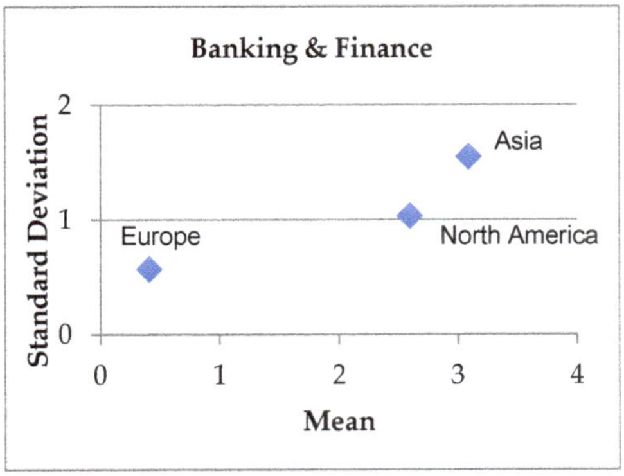

Figure 4. *Cont.*

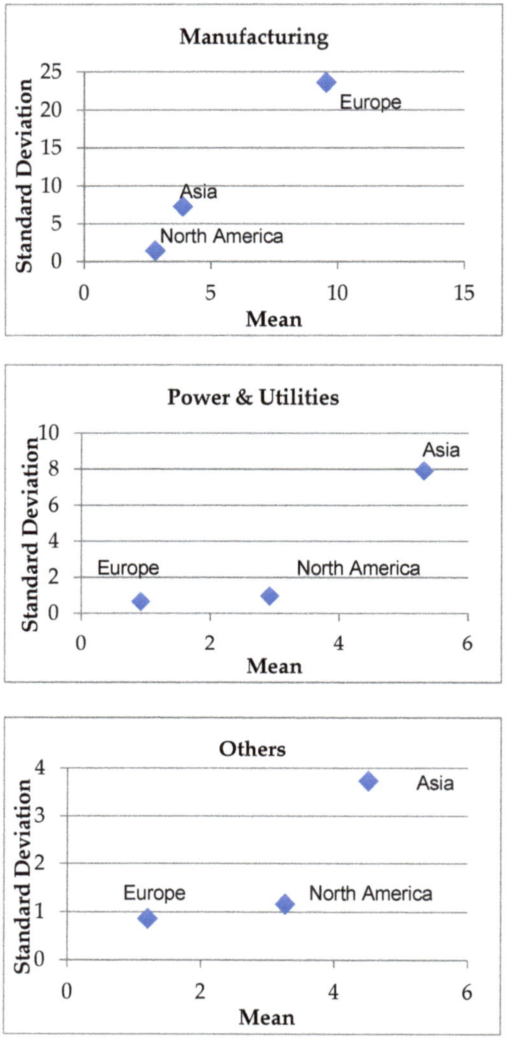

Figure 4. Mean-variance analysis of rate of return of green bonds, by sector of issuance. Source: Authors' calculation.

The regression analysis provides further information on the characteristics of green bonds, as the level of significance of the variables tends to vary depending on the region of issuance. It is interesting to note the difference in significance, depending on the analyzed sample. First, and regardless of the sample, the level of significance of the control variables is in line with the literature on the topic. For instance, the coupon rate was also found to be a significant variable in [22,24]. Similarly, maturity is often used as a control variable in studies assessing bonds' performance but is generally not found to be significant [23,24]. However, apart from these control variables, no sectorial dummy, or regional dummy, or even their interaction terms, appears to be significant. This is all the more surprising that, when conducting regressions on regional sample, sectorial dummy variables show significance, to an extent. This could potentially be explained and improved by using a larger sample of analysis.

Table 4. Regression results.

Item	Full Sample		Asia-Pacific		Europe		North America	
	Pooled OLS	**GLS Regression (Random Effect)**	**Pooled OLS**	**GLS Regression (Random Effect)**	**Pooled OLS**	**GLS Regression (Random Effect)**	**Pooled OLS**	**GLS Regression (Random Effect)**
Days to maturity	-9.78×10^{-7} (1.07×10^{-6})	-9.78×10^{-7} (1.15×10^{-6})	8.80×10^{-6} (7.28×10^{-5})	4.97×10^{-6} (8.85×10^{-5})	-1.15×10^{-6} (1.74×10^{-6})	-1.15×10^{-6} (1.51×10^{-6})	3.87×10^{-5} *** (1.08×10^{-5})	4.85×10^{-5} *** (1.28×10^{-5})
Coupon Rate	1.20 *** (0.20)	1.20 *** (0.20)	1.14 *** (0.13)	1.14 *** (0.16)	1.57 ** (0.69)	1.57 *** (0.56)	0.78 *** (0.04)	0.76 *** (0.04)
Banking	−0.05 (0.20)	−0.05 (0.27)	−0.62 *** (0.19)	−0.57 ** (0.24)	0.33 (0.38)	0.33 (0.31)	−0.07 (0.14)	−0.04 (0.18)
Manufacturing	2.27 (1.82)	2.27 (1.85)	1.84 (1.20)	1.79 (1.50)	5.30 (4.87)	5.30 (4.70)	−0.06 (0.11)	0.05 (0.13)
Power/Utilities	−0.13 (0.18)	−0.13 (0.21)	1.04 (1.00)	0.97 (1.05)	−0.30 * (0.15)	−0.30 (0.20)	−0.07 (0.11)	0.03 (0.15)
Others	−0.45 ** (0.23)	−0.45 * (0.25)	−0.52 (0.34)	−0.42 (0.45)	−0.35 (0.33)	−0.35 (0.33)	−0.04 (0.12)	0.05 (0.16)
2018	0.11 (0.11)	0.11 (0.08)	0.06 (0.23)	0.25 * (0.13)	0.07 (0.22)	0.07 (0.15)	0.10 (0.13)	−0.00496 (0.06)
2019	0.25 ** (0.11)	0.25 *** (0.08)	−0.05 (0.25)	−0.01 (0.19)	0.33 * (0.18)	0.33 *** (0.11)	0.27 ** (0.12)	0.36 *** (0.07)
2020	−0.20 (0.17)	−0.20 (0.16)	−0.63 ** (0.24)	−0.60 *** (0.17)	0.08 (0.30)	0.08 (0.29)	−0.66 *** (0.10)	−0.69 *** (0.05)
Asia	0.65 * (0.38)	0.65 (0.43)						
Europe	0.49 (0.56)	0.49 (0.55)						
Asia—Banking	−0.76 * (0.41)	−0.76 (0.51)						
Europe—Banking	0.08 (0.24)	0.08 (0.31)						
Constant	−0.90 (0.70)	−0.90 (0.70)	0.06 (0.33)	5.83×10^{-3} (0.40)	−1.11 (0.96)	−1.11 (0.77)	0.51 *** (0.13)	0.50 *** (0.13)
Observations	1174	1174	366	366	603	603	205	205
R-squared	0.41	0.61	0.47	0.56	0.26	0.49	0.86	0.87

GLS = generalized least squares, OLS = ordinary least squares. Notes: Standard errors, presented in parentheses, are obtained using the White method, robust with heteroskedasticity and serial correlation. ***, **, and * denote significance at the 1%, 5%, and 10% levels, respectively. Source: Authors' calculation.

The majority of sectoral dummy variables show a lack of significance, with the notable exception of banking and finance in the Asia-Pacific. Our results prove that bonds issued by companies in the banking and financial sector consistently display lower rates of return. Not only does this sector issue low-performing bonds, but the size of the associated coefficient (0.62 or 0.57, depending on the method of estimation) is relatively large, as the average return of Asian bonds is 3.52. Even when using the full sample, being a green bond issued from the banking sector in Asia is shown to have a slightly significant negative sign. This is all the more striking as it appears that no other sectoral dummy variable shows such high levels of significance in other regions. This result confirms that the dominance of traditional forms of banking in the Asian financial sector has an impact on the characteristics of green bonds, specifically on the performance of bonds.

The significance of year dummy variables also provides a few other takeaways from this study. As the rate of return is measured on 10 January each year, each dummy captures the state of the market at the beginning of the year. Keeping this in mind, it comes as no surprise that bonds performed relatively poorly at the beginning of 2020 in the Asia-Pacific. As the majority of Asian bonds were issued in China, their performance took a severe hit from the outbreak of the coronavirus disease (COVID-19) at the end of 2019, as shown by the negative and large coefficient linked with the 2020 dummy variable. The negative sign of the same variable in the North American sample could reflect the level of dependence of the United States economy on China: the negative expected performance of Asian bonds could therefore bring down American bonds as well.

4.4. Test and Diagnostics

This section provides a discussion of the results of the tests and diagnostics to assess the quality of the results presented in the previous section. The results of the poolability test are shown in Table 5, while Table 6 displays the diagnostics, and more specifically, the distribution of standard errors between idiosyncratic and time-invariant terms.

Table 5. Misspecification tests.

Item	Regional Sample	Test Statistic	Probability
	Full Sample	2.22	0.00 ***
Poolability test	Asia and the Pacific	5.00	0.00 ***
	Europe	1.08	0.24
	North America and other issuers	31.13	0.00 ***

Source: Authors' calculation. *** denotes significance at the 1% level.

Table 6. Diagnostics.

Item	Full Sample	Asia-Pacific	Europe	North America
Idiosyncratic error term, $\epsilon_{i,t}$	2.67	2.03	3.31	0.28
Time − invariant error, $u_{i,t}$	<0.00	1.52	<0.00	0.33
Fraction of variance due to individual heterogeneity	<0.00	0.36	<0.00	0.59

Source: Authors' calculation.

Table 5 presents the results of the poolability test, related to model misspecification, and allows us to decide between the pooled OLS estimates and the FE/RE estimates. The test results suggest that results from RE are more reliable in the case of Asia and the Pacific and North America. In the case of Europe, however, the test seems to favor pooled OLS, even though the model showed a lower R-squared overall. Overall, the results of the misspecification tests confirm the validity of our results.

Finally, we introduce empirical estimates of $\epsilon_{i,t}$ and $u_{i,t}$, time-varying and idiosyncratic error terms, in Table 6. As one would expect, the size of the idiosyncratic error term is rather large in all models. It is worth noting that, for European bonds, the majority of the unobserved terms are captured by time-varying factors, meaning that European bonds are quite homogenous in terms of risks. This was already observed by the mean-variance analysis of European bonds. As for the region of interest in this study, it appears that variance due to heterogeneity across bonds accounts for 36% of unobserved factors determining performance, thereby confirming the high risks associated with Asian green bonds. Indeed, if the performance of Asian bonds has such high variation, they are naturally considered less reliable by investors in general.

To prove this last point, we also provide ratings of Asian green bonds by major agencies, namely S&P and Moody's. Due to the lack of data availability, we only provide ratings for 48 bonds, all issued in Asia. The distribution of ratings is shown in Figure 5. This figure shows the heterogeneity of Asian green bonds, as no rating category dominates the sample. That being said, one can also see that the majority of bond ratings are mid-tier (A+, A, and A− for S&P, and A1, A3, Baa1, and Baa2 for Moody's). Furthermore, 5% to 8% of Asian green bonds are below BBB and Baa3, and hence considered risky investments, which is a relatively high percentage. Of course, agencies are not infallible, but Asian green bonds could be seen as relatively risky investments, based on their ratings.

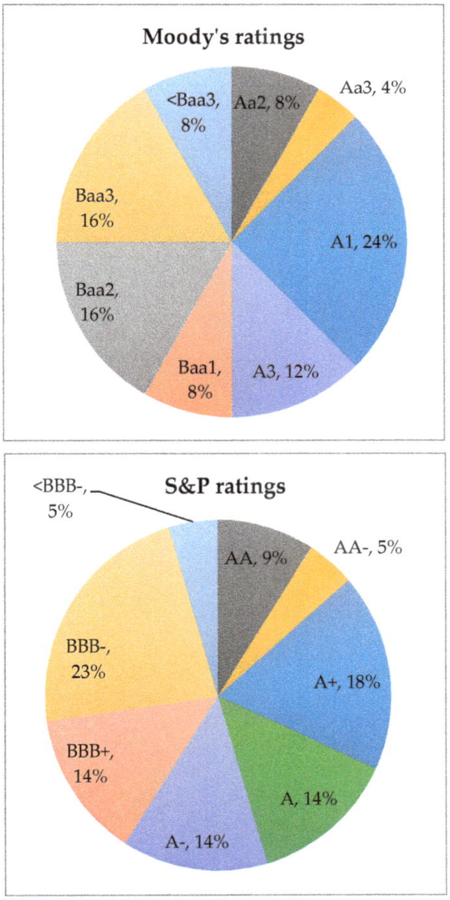

Figure 5. Probability of default of Asian green bonds—ratings distribution. Source: Authors' compilation. (Aa3, Aa2, A1, A3, Baa1, Baa2, Baa3, <Baa3) And (AA, AA-, A+, A, A-, BBB+, BBB-, <BBB-) are the credit ranges by the two credit rating agencies.

5. Conclusions and Policy Implications

5.1. Conclusions and Further Steps

The increasing prominence of green bonds as a financial tool to fight climate change has sparked the interest of many researchers in recent years. While it has been recognized that green bonds can be useful for climate policy, the existence of a green bond premium—meaning that green bonds show a lower rate of return than their brown or conventional counterparts—remains open to academic debate. Furthermore, researchers seem to have reached a consensus that green bonds tend to be riskier assets. However, research on green bonds provides general conclusions on the global green bonds market. No study so far has looked at the regional characteristics of green bonds, based on the place of issuance, with the focus on Asia and the Pacific.

The financial sector in Asia and the Pacific is dominated by traditional banking, with venture capitalists being quite scarce [3]. However, [5] highlighted that traditional banking is not necessarily an appropriate source of funding for green bonds due to maturity mismatch and the conservative approach of banking. Indeed, the study argued that maturity mismatch occurs as bank liabilities are short- to medium-term, while infrastructure projects are more long-term oriented, leading to risk overvaluation. Therefore, this study aimed to provide a comparative analysis of regional characteristics and green bonds' performance.

Using data from both BNEF and CBI, we gathered panel data composed of a total of 1174 observations and divided them into regional subsamples. Then, the study combined summary statistics as well as mean-variance and regression analysis to reach its conclusion. The results of this research are summarized in Table 7.

Table 7. Regional characteristics of green bonds.

Item	Asia and the Pacific	Europe	North America
Risks	High	Low	Moderate
Return	High	Low	Moderate
Homogeneity between bonds	Heterogeneous	Homogenous	Heterogeneous
Sector of issuance	Dominated by banking and finance	Well-balanced between public, utilities, and banking and other issuers	Well-balanced, between public, utilities, and banking and other issuers
Size	Large	Large	Small
Maturity	Long-term	Medium-term	Medium-term

Source: Authors' compilation.

Based on the empirical results, we were able to show that green bonds issued in Asia and the Pacific had different characteristics from those issued in Europe and North America. Specifically, Asian bonds proved to have higher returns, but also higher associated risks, as these bonds showed higher levels of heterogeneity than their European or North American counterparts. In the sample, bonds from Asia and the Pacific were generally issued in the long term, as their time to maturity was almost twice as long as that of bonds issued in other regions. However, the summary statistics revealed the dominance of the banking and finance sector in Asia—a trend that is not found in other regions. The empirical analysis proved that bonds issued by banks in Asia consistently showed lower returns; hence, there is an urgent need for diversification of issuers in Asia and the Pacific.

5.2. Policy Implications

As restrictions of the COVID-19 are easing and economies are opening, governments are beginning to unveil their economic recovery plans. However, there is a lack of motivation to strengthen the green agenda in recovery plans. This is because the recovery outlook seems to follow the 'growth first and green when possible' approach of existing development plans. This will endanger meeting the SDGs and the Paris Agreement on climate change.

Therefore, in the current insufficient investment level in the green sector, especially in the post-COVID-19 era, imperative financial and fiscal policy reforms, such as global or regional carbon taxation, regulations, and strategies on green financing, supporting policies for facilitating the issuance of green bonds, the establishment of green credit rating to measure the greenness of the projects, targeting the energy subsidies, reducing the direct and indirect subsidies to fossil fuels, and introducing public de-risking tools such as a green credit guarantee scheme for reducing the risk of green investments, are required. In other words, the world is required to establish a green financial system in order to facilitate the public and private financing of the green projects.

A major takeaway from this study is the relatively high risk and return associated with bonds issued in Asia and the Pacific. Most importantly, the research showed that bonds issued by banks in Asia were associated with lower returns. Thus, the study proposes several policy recommendations to address each of the weaknesses of Asian bonds, and eventually encourage their issuance, as green bonds are useful tools against climate change.

First, this study proposes using tax spillover to increase the rate of return of green bonds issued by banking and finance. Since this sector represents 60% of issuance in Asia, it is likely that traditional banking will keep playing a decisive role in green finance in the region. While green infrastructure requires high up-front costs, these projects create employment and revenue in the long term. Subsidizing green bonds in the early stages of project development could be a solution, as in the long term, these subsidies could be repaid to the public sector through tax spillover generated by employment and increased economic activity. A similar idea is developed by [5], although not applied to green bonds in particular. Figure 6 displays how an increase in the rate of return can directly impact investors' portfolios and contribute to making Asian green bonds more attractive. Detailed calculations behind this policy recommendation are provided in Appendix A.3.

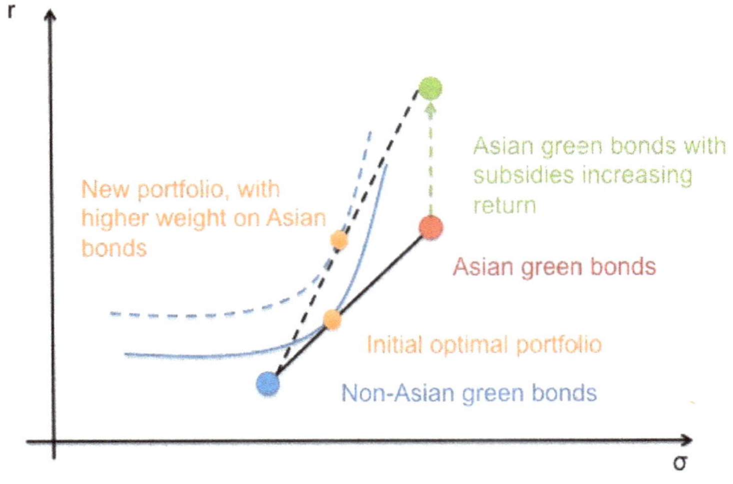

Figure 6. Effect of tax spillover on rate of return of green bonds. Source: Authors' depiction.

Since bonds issued by the banking and finance sector in Asia are shown to have lower returns, another solution to increase their attractiveness would simply be to encourage the diversification of issuers, and generally by promoting the involvement of the public sector. As shown in Figure 4, bonds issued by the public sector in Asia have high associated risks and relatively high returns. Diversification is not necessarily limited to the sector of issuance, however, and [30] highlighted the possibility of increased financial connectivity between Asian and European public institutions in financing green infrastructure.

Finally, a last remedy to increase the amount of green bonds issued in Asia and the Pacific is to reduce the risks associated with these instruments. Several studies have highlighted the risks associated with green infrastructure projects and proposed de-risking approaches for policymakers. The authors of [31] suggested a simplification of administrative procedures linked with project developments. They also proposed the establishment of agreements with local governments or companies, as green infrastructure projects are often more oriented towards the long term. The authors of [32] also proposed a wide array of de-risking solutions—ranging from general measures such as the unbundling of the electricity market, corruption control mechanisms, or reforms of fossil fuel subsidies, to financial de-risking measures such as credit guarantees or guaranteed power prices and the establishment of public–private partnerships to reduce political risks generally associated with green policies. Specifically, [33] proposed a model green credit guarantee scheme, where a public entity absorbs the risks related to green infrastructure projects by providing a credit guarantee. As many companies involved in green projects tend to be small- and medium-sized enterprises, credit guarantee schemes can allow these firms to receive higher funding, as the public entity acts as a form of collateral. Utilizing tax spillover to increase the rate of return of green bonds, diversifying sectors and regions, and de-risking policies could surely contribute to increasing the attractiveness of Asian green bonds and help to accelerate the fight against climate change in the region [13].

In order to have a well-developed green bond market, it is crucial to have a clear definition of what green is. This means an unambiguous definition of green bond is needed. In the meantime, green labeling has helped somewhat, but it is not enough. Currently, 80% brown and 20% green is called green, and 90% green and 10% brown is also called green. There are many different definitions of greenness that are all called green, and green bonds are used for financing them. Therefore, we need a clear greenness credit rating to show the ratio of greenness. Nowadays, satellite photos can show how much CO_2 is exposed by each company or each project, and it is possible to detect and measure the emissions that would be used to assess the greenness of the projects. Globally, having unified green rating agencies rather than having different standards for each country is required [26,34–36].

Finally, in bank-oriented financial systems such as in Asia and several other regions, just relying on green bond issuance might not be an adequate solution to fill the green finance gap. Green bond is a complementary financial instrument that needs to be used besides banking solutions. There are several mechanisms and instruments that can help to bridge the green finance gap for meeting SDGs. These mechanisms include the modification of the collateral framework, changes in capital adequacy ratios, a market of SDG lending certificates, the introduction of rediscounting policies, the establishment of a green credit guarantee scheme, green credit rating, etc. [37,38].

Another fundamental problem is 'decoupling,' i.e., the fact that green bonds apply a financial logic to solve an ecological issue, which is created in the first place by the economic system and its financial indicators. As a result, green bond investment strategies prioritize ecosystem services generating the most significant and most stable payment flows to the detriment of other invisible ecosystem services, but just as essential. The author of [39,40] uses the term a financial "logos" (defined as a structuring discourse integrated into financial practices' management tools and belief systems) to describe this problem. His article argues that any ecological finance theory devised to fit the SDGs needs a paradigm shift in the morphology of randomness underlying financial risk modeling by integrating the characteristics of "nature" and sustainability into the modeling carried out. Most recently, the authors of [41] have proposed a strategy to incorporate ecological issues into financial economics. They used the concepts of resilience, diversity, self-thinning, self-regulated mitosis, and ecological transparency from biology and introduced them to the field of financial economics.

In addition, public financial institutions (PFIs)—or those publicly created and/or mandated financial institutions that have often been created to correct for the lack of market-based finance through the provision of missing financial services—have a potentially vital role to play to scale-up private sector investments in green projects for meeting SDGs. However, there are four critical points for the involvement of PFIs in green projects: (1) They need to provide long-term financing (long-term loans) compared to private commercial banks, (2) setting up the interest rate lower than private banks, stable and fixed, and (3) avoid harmful effects of government lending through PFIs. This means avoid increasing the government's role in the economy and avoid crowding out private deposits and loans. (4) Make loans by PFIs, where the private sector cannot make loans.

Author Contributions: Conceptualization, F.T.-H.; Formal analysis, F.T.-H. and H.P.; Methodology, N.Y.; Project administration, H.P.; Software, F.T.-H.; Supervision, H.P., Validation, N.Y.; Writing—original draft, F.T.-H.; Writing—review and editing, N.Y. and H.P. All authors have read and agreed to the published version of the manuscript.

Funding: Economic Research Institute for ASEAN and East Asia (ERIA) financially supported this project and the article processing fee.

Institutional Review Board Statement: Not applicable.

Informed Consent Statement: Not applicable.

Data Availability Statement: Data sharing not applicable.

Acknowledgments: This paper results from a research project titled 'Energy Sustainability and Climate Change in ASEAN and East Asia' for the Economic Research Institute for ASEAN and East Asia (ERIA). The authors are grateful to Fukunari Kimura and Jun Arima and all members of the working group in the aforementioned project for their valuable comments and inputs, which significantly contributed to improving the quality of this study. The authors are also grateful to the Climate Bonds Initiative (CBI) for providing part of the data used in this research. The authors are grateful to Aline Mortha for her research assistance throughout this project.

Conflicts of Interest: The authors declare no conflict of interest.

Appendix A.

Appendix A.1. Summary Statistics with the Reduced Sample, Using Data from Bloomberg New Energy Finance

Table A1. Asian subsample.

Item	Observations	Mean	Std. Dev.	Min	Max
Rate of return	366	3.515615	3.595249	−0.216	48.955
Days to maturity	760	1806.713	1212.08	145	11,217
Amount issued	760	4.38×10^8	5.52×10^8	9.98×10^7	4.33×10^9
Coupon rate	760	3.425405	2.185087	0	15.5
Private	760	0.9052632	0.293044	0	1
Banking	760	0.5578947	0.4969639	0	1
Manufacturing	760	0.0578947	0.2336981	0	1
Power/Utilities	760	0.1263158	0.3324237	0	1
Others	760	0.1631579	0.369753	0	1

Std. dev. = standard deviation. Source: Authors' compilation.

Table A2. European subsample.

Item	Observations	Mean	Std. Dev.	Min	Max
Rate of return	603	0.9731144	3.362978	−0.572	80.075
Days to maturity	1140	5578.874	30,351.63	147	36,6305
Amount issued	1140	6.66×10^8	4.68×10^8	1.00×10^3	4.46×10^9
Coupon rate	1140	1.162737	0.9079918	0	7.125
Private	1140	0.7894737	0.4078614	0	1
Banking	1140	0.322807	0.4677548	0	1
Manufacturing	1140	0.0210526	0.1436228	0	1
Power/Utilities	1140	0.2877193	0.4528983	0	1
Others	1140	0.1578947	0.3648023	0	1

Std. dev. = standard deviation. Source: Authors' compilation.

Table A3. North American and Other Issuers subsample.

Item	Observations	Mean	Std. Dev.	Min	Max
Rate of return	205	2.666659	1.306911	−0.3	7.72
Days to maturity	432	4122.398	3229.278	245	13,655
Amount issued	432	5.26×10^8	3.72×10^8	9.51×10^7	2.25×10^9
Coupon rate	432	2.958718	1.474915	0	8
Private	432	0.7222222	0.4484225	0	1
Banking	432	0.1388889	0.3462315	0	1
Manufacturing	432	0.0555556	0.229327	0	1
Power/Utilities	432	0.3333333	0.4719511	0	1
Others	432	0.1944444	0.3962313	0	1

Std. dev. = standard deviation. Source: Authors' compilation.

Appendix A.2. Summary Statistics by Sector

Table A4. Asian subsample.

Item		Observations	Mean	Std. Dev.	Min	Max
Public sector	Rate of return	40	2.78625	1.938086	−0.216	9.03
	Days to maturity	72	2501.5	1105.893	555	4815
	Amount issued	72	6.47×10^8	4.60×10^8	1.10×10^8	2.24×10^9
	Coupon rate	72	2.273056	1.815014	0	7.125
Banking/Finance	Rate of return	221	3.086752	1.548352	−0.059	6.95
	Days to maturity	424	1466.811	820.2578	145	4797
	Amount issued	424	5.05×10^8	6.85×10^8	1.02×10^8	4.33×10^9
	Coupon rate	424	3.296255	1.611012	0	6.5
Manufacturing	Rate of return	17	3.903	7.30048	−0.184	26.092
	Days to maturity	44	1914.045	939.1773	472	4305
	Amount issued	44	2.69×10^8	1.71×10^8	1.00×10^8	7.05×10^8
	Coupon rate	44	2.105455	2.437531	0	7.5
Power/Utilities	Rate of return	37	5.32027	7.92042	0.744	48.955
	Days to maturity	96	1922.677	1069.983	218	4723
	Amount issued	96	2.85×10^8	1.42×10^8	9.98×10^7	5.90×10^8
	Coupon rate	96	3.845875	1.855951	0.85	7.9
Others	Rate of return	51	4.507686	3.735481	0.231	17.395
	Days to maturity	124	2437.669	1938.155	174	11,217
	Amount issued	124	2.67×10^8	1.44×10^8	1.00×10^8	6.00×10^8
	Coupon rate	124	4.678968	3.273824	0.09	15.5

Std. dev. = standard deviation. Source: Authors' compilation.

Table A5. European subsample.

Item		Observations	Mean	Std. Dev.	Min	Max
Public sector	Rate of return	153	0.9720131	1.020399	−0.556	3.263
	Days to maturity	240	2667.183	1878.802	147	11,266
	Amount issued	240	8.56×10^8	7.43×10^8	1.16×10^8	4.46×10^9
	Coupon rate	240	1.16585	0.8552492	0	3.3
Banking/Finance	Rate of return	184	0.4063315	0.5703005	−0.572	2.615
	Days to maturity	368	2601.359	1844.092	151	12,251
	Amount issued	368	6.35×10^8	3.32×10^8	1.06×10^8	1.74×10^9
	Coupon rate	368	0.6791848	0.5777155	0	2.5
Manufacturing	Rate of return	11	9.553909	23.62808	0.221	80.075
	Days to maturity	24	2298.333	1084.292	753	4692
	Amount issued	24	3.65×10^8	2.55×10^8	1.08×10^8	8.37×10^8
	Coupon rate	24	2.371333	2.296941	0.5	7.125

Table A5. *Cont.*

Item		Observations	Mean	Std. Dev.	Min	Max
Power/Utilities	Rate of return	176	0.9256023	0.6572373	−0.224	3.602
	Days to maturity	328	12,709.1	55,942.4	473	366,305
	Amount issued	328	6.79×10^8	3.68×10^8	1.09×10^8	1.93×10^9
	Coupon rate	328	1.374195	0.7278362	0	4.496
Others	Rate of return	79	1.206405	0.8654128	−0.202	4.732
	Days to maturity	180	2993.033	1406.762	888	9954
	Amount issued	180	4.93×10^8	2.89×10^8	1.00×10^8	1.14×10^9
	Coupon rate	180	1.600711	0.9940294	0.1	5

Std. dev. = standard deviation. Source: Authors' compilation.

Table A6. North American and Other Issuers subsample.

Item		Observations	Mean	Std. Dev.	Min	Max
Public sector	Rate of return	59	1.973627	1.518473	−0.3	7.72
	Days to maturity	120	3083.367	2605.471	265	12,148
	Amount issued	120	4.47×10^8	3.33×10^8	1.00×10^8	1.20×10^9
	Coupon rate	120	2.060917	1.905424	0	8
Banking/Finance	Rate of return	27	2.594	1.033445	−0.004	4.205
	Days to maturity	60	2902.767	2313.379	245	10,886
	Amount issued	60	7.24×10^8	6.14×10^8	1.10×10^8	2.25×10^9
	Coupon rate	60	2.847733	1.344723	0.25	5.25
Manufacturing	Rate of return	13	2.812	1.428624	0.05	5.219
	Days to maturity	24	3279.167	1107.982	1140	5422
	Amount issued	24	9.77×10^8	3.39×10^8	4.50×10^8	1.50×10^9
	Coupon rate	24	2.633333	1.938997	0	5.5
Power/Utilities	Rate of return	67	2.92809	0.9791559	0.068	4.512
	Days to maturity	144	5643.694	3752.187	705	12,103
	Amount issued	144	4.31×10^8	1.68×10^8	9.51×10^7	7.50×10^8
	Coupon rate	144	3.449056	0.7427922	1	4.6
Others	Rate of return	39	3.267821	1.166826	0.789	7.079
	Days to maturity	84	4110.881	3026.646	894	13,655
	Amount issued	84	5.33×10^8	5.33×10^8	1.00×10^8	1.23×10^9
	Coupon rate	84	3.572952	0.8995293	1.625	5.875

Std. dev. = standard deviation. Source: Authors' compilation.

Appendix A.3. Theoretical Framework for Policy Recommendation

Policy implications for this research are based on a theoretical framework, detailed below. Since Asian bonds are characterized by higher relative risks and returns, we derive the optimal portfolio of a theoretical investor, who can choose to assign a weight α on green bonds not issued in Asia and a weight $(1-\alpha)$ on Asian bonds.

The rate of return and associated variance of this portfolio is given by Equations (A.1) and (C.2), respectively:

$$r = \alpha r_{NA} + (1-\alpha) r_A \tag{A1}$$

$$\sigma^2 = \alpha^2 \sigma_{NA}^2 + (1-\alpha)^2 \sigma_A^2 + 2\alpha(1-\alpha)\sigma_{NA/A} \tag{A2}$$

where r, r_{NA}, and r_A denote the rate of return of portfolio, non-Asian bonds, and Asian bonds respectively, and σ^2, σ_{NA}^2, σ_A^2, and $\sigma_{NA/A}$ denote the variance of portfolio, non-Asian bonds, Asian bonds, and covariance between Asian and non-Asian bonds.

Then, the theoretical investor aims at maximizing the utility derived from their portfolio. This study assumes that their utility function is given by:

$$U(r,\sigma) = r - \beta\sigma \tag{A3}$$

Substituting (A1) and (A2) into (A3), we obtain:

$$U(r,\sigma) = \alpha r_{NA} + (1-\alpha) r_A - \beta\left\{\alpha^2 \sigma_{NA}^2 + (1-\alpha)^2 \sigma_A^2 + 2\alpha(1-\alpha)\sigma_{NA/A}\right\} \tag{A4}$$

Thus,

The investor's utility maximization problem is given by Equation (A5):

$$\max_{\alpha} U(r,\sigma) \tag{A5}$$

The first-order condition, with respect to α, is:

$$\frac{\partial U}{\partial \alpha} = (r_{NA} - r_A) - \beta\left\{2\alpha^*\sigma_{NA}^2 - 2(1-\alpha^*)\sigma_A^2 + (2-4\alpha^*)\sigma_{NA/A}\right\} = 0 \tag{A6}$$

Solving this equation for α^*, we obtain the optimal weight the investor can put on non-Asian bonds:

$$\alpha^* = \frac{\frac{1}{\beta}(r_{NA} - r_A) - \left(2\sigma_{NA/A} - 2\sigma_A^2\right)}{2\sigma_{NA}^2 + 2\sigma_A^2 - 4\sigma_{NA/A}} \tag{A7}$$

To change this optimal weight, policymakers in Asia and the Pacific can act on parameters of this utility maximization problem, namely on r_A and σ_A^2.

For instance, one can increase the weight put on Asian bonds by increasing the rate of return, r_A, by subsidising bonds through tax spillover, denoted by θ_{tax}. The new rate of return of this subsidised portfolio, denoted by $r_{spillover}$, is given by Equation (A8):

$$r_{spillover} = \alpha r_{NA} + (1-\alpha)(r_A + \theta_{tax}),\ where\ \theta_{tax} \geq 0 \tag{A8}$$

Then, the investor's utility becomes:

$$U(r,\sigma)_{spillover} = \alpha r_{NA} + (1-\alpha)(r_A + \theta_{tax}) - \beta\left\{\alpha^2\sigma_{NA}^2 + (1-\alpha)^2\sigma_A^2 + 2\alpha(1-\alpha)\sigma_{NA/A}\right\} \tag{A9}$$

Solving the utility maximization problem, we obtain the new optimal weight for this investor:

$$\alpha_{spillover}^* = \frac{\frac{1}{\beta}(r_{NA} - (r_A + \theta_{tax})) - \left(2\sigma_{NA/A} - 2\sigma_A^2\right)}{2\sigma_{NA}^2 + 2\sigma_A^2 - 4\sigma_{NA/A}} \tag{A10}$$

Note that

$$\alpha_{tax}^* \leq \alpha^* \tag{A11}$$

where the equality holds if and only if $\theta_{tax} = 0$.

Since α^* denotes the optimal portfolio weight attributed to bonds not issued in Asia and the Pacific, policymakers can make green bonds more attractive for investors by using spillover from tax returns.

References

1. Sachs, J.; Woo, T.W.; Yoshino, N.; Taghizadeh-Hesary, F. *Handbook of Green Finance : Energy Security and Sustainable Development*; Springer Singapore and Asian Development Bank Institute: Singapore, 2019.
2. Paris: International Energy Agency. World Energy Investment 2019—Data Tables. Available online: https://www.iea.org/reports/world-energy-investment-2019 (accessed on 22 November 2020).
3. Peimani, H. Financial Barriers for Development of Renewable and Green Energy Projects in Asia. In *Handbook of Green Finance: Energy Security and Sustainable Development*; Sachs, J., Wing, T.W., Yoshino, N., Taghizadeh-Hesary, F., Eds.; Springer Singapore and Asian Development Bank Institute: Singapore, 2019; pp. 15–34.
4. Noh, H.J. Financial Strategies to Accelerate Green Growth. In *Handbook of Green Finance: Energy Security and Sustainable Development*; Sachs, J., Wing, T.W., Yoshino, N., Taghizadeh-Hesary, F., Eds.; Springer Singapore and Asian Development Bank Institute: Singapore, 2019; pp. 37–62.
5. Yoshino, N.; Taghizadeh-Hesary, F.; Nakahigashi, M. Modelling the Social Funding and Spillover Tax for Addressing the Green Energy Financing Gap. *Econ. Model.* **2019**, *77*, 34–41. [CrossRef]
6. London: Climate Bonds Initiative. Green Bonds: The State of the Market. 2018. Available online: https://www.climatebonds.net/files/reports/cbi_gbm_final_032019_web.pdf (accessed on 22 November 2020).
7. London: Climate Bonds Initiative. Green Bonds: Global State of the Market. 2019. Available online: https://www.climatebonds.net/resources/reports/green-bonds-global-state-market-2019 (accessed on 22 November 2020).

8. Agliardi, E.; Agliardi, R. Financing Environmentally-Sustainable Projects with Green Bonds. *Environ. Dev. Econ.* **2019**, *24*, 608–623. [CrossRef]
9. Baker, M.; Bergstresser, D.; Serafeim, G.; Wurgler, J. Financing the Response to Climate Change: The Pricing and Ownership of U.S. Available online: https://www.nber.org/system/files/working_papers/w25194/w25194.pdf (accessed on 22 November 2020).
10. Gianfrate, G.; Peri, M. The Green Advantage: Exploring the Convenience of Issuing Green Bonds. *J. Clean. Prod.* **2019**, *219*, 127–135. [CrossRef]
11. Zerbib, O. The Effect of Pro-Environmental Preferences on Bond Prices: Evidence from Green Bonds. *J. Bank. Financ.* **2019**, *98*, 39–60. [CrossRef]
12. Zou, C.; Zhao, Q.; Zhang, G.; Xiong, B. Energy revolution: From a fossil energy era to a new energy era. *Nat. Gas Ind B* **2016**, *3*, 1–11. [CrossRef]
13. Taghizadeh-Hesary, F.; Yoshino, N. Sustainable Solutions for Green Financing and Investment in Renewable Energy Projects. *Energies* **2020**, *13*, 788. [CrossRef]
14. OECD. *OECD Companion to the Inventory of Support. Measures for Fossil Fuels 2015*; OECD Publishing: Paris, France, 2015. [CrossRef]
15. Coady, D.; Parry, I.; Sears, L.; Shang, B. *How Large Are Global Energy Subsidies? IMF Working Paper No. WP/15/105*; International Monetary Fund: Washington, DC, USA, 2015.
16. Pham, L. Is It Risky to Go Green? A Volatility Analysis of the Green Bond Market. *J. Sustain. Financ. Invest.* **2016**, *6*, 263–291. [CrossRef]
17. Naeem, M.A.; Nguyen, T.T.H.; Nepal, R.; Ngo, Q.-T.; Taghizadeh-Hesary, F. Asymmetric relationship between green bonds and commodities: Evidence from extreme quantile approach. *Financ. Res. Lett.* **2021**, *101983*. [CrossRef]
18. Broadstock, D.; Cheng, L. Time-Varying Relation Between Black and Green Bond Price Benchmarks: Macroeconomic Determinants for the First Decade. *Financ. Res. Lett.* **2019**, *29*, 17–22. [CrossRef]
19. CBI. Green Bond Principles & Climate Bonds Standard. Available online: https://www.climatebonds.net/market/best-practice-guidelines (accessed on 22 November 2020).
20. Flaherty, M.; Gevorkyan, A.; Radpour, S.; Semmler, W. Financing Climate Policies Through Climate Bonds—A Three Stage Model and Empirics. *Res. Int. Bus. Financ.* **2017**, *42*, 468–479. [CrossRef]
21. Hyun, S.; Park, D.; Tian, S. Pricing of Green Labeling: A Comparison of Labeled and Unlabeled Green Bonds. *Financ. Res. Lett.* **2020**, *101816*. [CrossRef]
22. Bachelet, M.J.; Becchetti, L.; Manfredonia, S. The Green Bonds Premium Puzzle: The Role of Issuer Characteristics and Third-Party Verification. *Sustainable* **2019**, *11*, 1098. [CrossRef]
23. Hachenberg, B.; Schiereck, D. Are Green Bonds Priced Differently from Conventional Bonds? *J. Asset Manag.* **2018**, *19*, 371–383. [CrossRef]
24. Chiesa, M.; Barua, S. The Surge of Impact Borrowing: The Magnitude and Determinants of Green Bond Supply and Its Heterogeneity Across Markets. *J. Sustain. Financ. Invest.* **2019**, *9*, 138–161. [CrossRef]
25. Tang, D.; Zhang, Y. Do Shareholders Benefit from Green Bonds? *J. Corp. Financ.* **2020**, *61*. [CrossRef]
26. MacAskill, S.; Roca, E.; Liu, B.; Stewart, R.A.; Sahin, O. Is There a Green Premium in the Green Bond Market? Systematic Literature Review Revealing Premium Determinants. *J. Clean. Prod.* **2021**, *280 Pt 2*. [CrossRef]
27. Sartzetakis, E. Green Bonds as an Instrument to Finance Low Carbon Transition. *Econ. Chang. Restruct.* **2020**, 1–25. [CrossRef]
28. Reboredo, J.; Ugolini, A. Price Connectedness Between Green Bond and Financial Markets. *Econ. Model.* **2020**, *88*, 25–38. [CrossRef]
29. Febi, W.; Schäfer, D.; Stephan, A.; Sun, C. The Impact of Liquidity Risk on the Yield Spread of Green Bonds. *Financ. Res. Lett.* **2018**, *27*, 53–59. [CrossRef]
30. Yoshino, N.; Hossain, M.; Taghizadeh-Hesary, F. Enhancing Financial Connectivity Between Asia and Europe: Implications for Infrastructure Convergence Between the Two Regions. *Asian Econ. Pap.* **2020**, *19*, 84–101. [CrossRef]
31. Komendantova, N.; Schinko, T.; Patt, A. De-Risking Policies as a Substantial Determinant of Climate Change Mitigation Costs in Developing Countries: Case Study of the Middle East and North African Region. *Energy Policy* **2019**, *127*, 404–411. [CrossRef]
32. Carafa, L.; Frisari, G.; Vidican, G. Electricity Transition in the Middle East and North Africa: A De-Risking Governance Approach. *J. Clean. Prod.* **2016**, *128*, 34–47. [CrossRef]
33. Taghizadeh-Hesary, F.; Yoshino, N. The Way to Induce Private Participation in Green Finance and Investment. *Financ. Res. Lett.* **2019**, *31*, 98–103. [CrossRef]
34. Revelli, C. Re-embedding Financial Stakes Within Ethical and Social Values in Socially Responsible Investing (SRI). *Res. Int. Bus. Financ.* **2016**, *38*, 1–5. [CrossRef]
35. Paranque, B.; Revelli, C. Ethico-economic Analysis of Impact Finance: The Case of Green Bonds. *Res. Int. Bus. Financ.* **2019**, *47*, 57–66. [CrossRef]
36. Willmott, H. Creating 'Value' Beyond the Point of Production: Branding, Financialization and Market Capitalization. *Organization* **2010**, *17*, 517–542. [CrossRef]
37. Lagoarde-Segot, T. Financing the Sustainable Development Goals. *Sustainability* **2020**, *12*, 2775. [CrossRef]
38. Yoshino, N.; Taghizadeh-Hesary, F.; Otsuka, M. Covid-19 and Optimal Portfolio Selection for Investment in Sustainable Development Goals. *Financ. Res. Lett.* **2020**, 101695. [CrossRef]

39. Chiapello, E. The work of Financialisation. In *The Making of Finance*; Lenglet, M., Chambost, I., Tadjeddine, Y., Eds.; Routledge: London, UK, 2019; pp. 192–200.
40. Walter, C. Sustainable Financial Risk Modelling Fitting the SDGs: Some Reflections. *Sustainability* **2020**, *12*, 7789. [CrossRef]
41. Lagoarde-Segot, T.; Martínez, E.A. Ecological finance theory: New foundations. *Int. Rev. Financ. Anal.* **2021**, *75*, 101741. [CrossRef]

sustainability

MDPI

Article

Optimal Allocation of Gas Resources to Different Consumption Sectors Using Multi-Objective Goal Programming

Ieva Meidute-Kavaliauskiene [1,*], Vida Davidaviciene [1], Shahryar Ghorbani [2] and Iman Ghasemian Sahebi [3]

1 Department of Business Technologies and Entrepreneurship, Vilnius Gediminas Technical University, Saulėtekio al. 11, 10223 Vilnius, Lithuania; vida.davidaviciene@vilniustech.lt
2 Department of Production Management, University of Sakarya, 54050 Sakarya, Turkey; mg.shahryar@gmail.com
3 Department of Industrial Management, University of Tehran, Tehran 1417466191, Iran; iman.ghasemian@ut.ac.ir
* Correspondence: ieva.meidute-kavaliauskiene@vilniustech.lt

Abstract: Natural gas is a main source of energy in Iran, and optimal allocation to different sectors is crucial, based on realities, geopolitical considerations, and national security concerns. In this paper, a multi-objective goal programming model is developed to study the optimal allocation of this resource to various consumption sectors, including household business, industry, petrochemical industry, power plants, injection to oil fields, and export from Iran for the horizon of 2025. In this research, the energy security index is prioritized over other indicators. Two objective functions are considered: the first is maximizing the energy security index (minimizing the cost of energy security), and the second is minimizing the relative weight of different consumption sectors. The goal share of the various gas sectors, the amount of consumption of the various consumption sectors, and the allocatable and predicted amount for each year is calculated. Household business, power plants, petrochemical industries, industry, and export aid injection to oil fields are the most consuming sectors in 2025, respectively. Also, based on cost minimization, power plants, petrochemical industries, and industries in general are the more consuming sectors, respectively.

Keywords: natural gas; multi-objective; goal programming; optimization; allocation

Citation: Meidute-Kavaliauskiene, I.; Davidaviciene, V.; Ghorbani, S.; Sahebi, I.G. Optimal Allocation of Gas Resources to Different Consumption Sectors Using Multi-Objective Goal Programming. *Sustainability* **2021**, *13*, 5663. https://doi.org/10.3390/su13105663

Academic Editors: Farhad Taghizadeh-Hesary and Han Phoumin

Received: 23 February 2021
Accepted: 12 May 2021
Published: 18 May 2021

1. Introduction

Nowadays, natural gas (NG) is the main source of energy in many countries. Being more efficient and having less carbon, NG is increasingly used in different sectors rather than other fossil fuels and non-renewable energy sources [1]. While the global portion of oil consumption in the energy sector fell from 45% in 1970 to 43.6% in 2020, NG experienced an increase from 17.2% to 33.7% in a similar period [2]. Iran is known to be among the main suppliers of NG in the world and its own region. Having been explored and found to have more than 33.5 trillion cubic meters of NG reserves in 2015, Iran is the second-biggest owner of NG reserves globally, and it is estimated to own almost 18% of all explored NG on the Earth (Figure 1).

Importantly, explorations by Iran over the past two decades increased its global proportion by 2% [2]. Energy use and consequent NG consumption in Iran has dramatically increased over the past 10 years. Despite all mentioned facts, and its increased extraction, Iran is currently an importer of NG from Turkmenistan, and in search of new NG exporters [3]. It is worth mentioning that there is a reliable planning and allocation sector which is associated with population and consumption, and that forecasting is crucial.

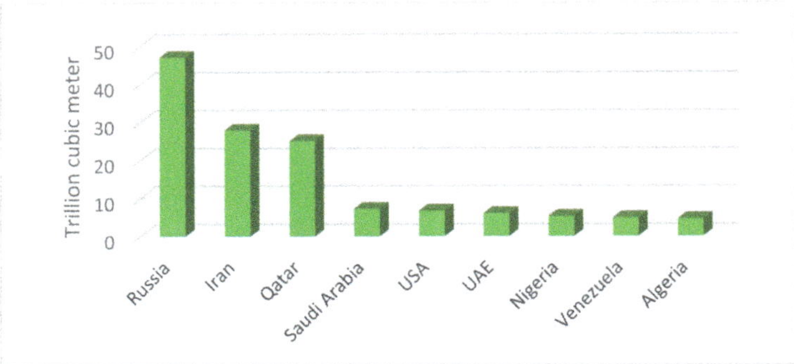

Figure 1. Country natural gas reserves (Data from eoearth.org, 2020, access on 3 October 2020).

Because of its operational features, NG can be distributed to different sectors of energy consumers. NG is the main fuel used by petrochemical and refinery industries to gain added value. However, a significant proportion of NG is consumed in transportation, domestic heating and cooking, and industry.

NG allocation should be associated with the expansion of social welfare over time, and this measure must be considered as the opportunity cost of gas distribution to various sectors. Consequently, the cost of allocation between the NG exploitation time and various sectors' uses should be minimized. Hence, a function of the profit and costs of gas allocation to the different sectors should be considered [4]. In fact, optimally allocating limited sources of NG to different consumers is a crucial political and economic challenge. Hence, this paper aims to study the optimal allocation of NG to distinct sectors in Iran by developing an optimization model.

One of the major factors playing a key role in optimally allocating resources is predicting future consumptions in order to make it applicable. There are several approaches to forecast energy demand, e.g., artificial neural network, data-driven model, time-series analysis such as ARIMA, etc. [5–7]. However, in this paper, forecasts of future energy consumptions in Iran are extracted from energy balance sheets by simple statistical methods. As well, different approaches and optimization algorithms, such as fuzzy goal programming [8], weighted goal programming [9], and mixed integer programming [10], are employed by operations research scholars to optimize energy resource allocation.

In this paper, a multi-objective goal programming method is employed to optimally allocate NG to the different Iranian consumers for the horizon of 2025. The rest of the paper is organized as follows: in Section 2 the literature is reviewed, in Section 3 the methodology and the mathematical model are presented, in Section 4 the results are demonstrated and discussed, and, finally, there is a conclusion on the research and the results.

2. Literature Review

Due to economic growth, technological advances, and increasing demand, planning for energy is now a complex multi-variable, multi-objective problem. Accordingly, a variety of models are developed to solve the problem based on a different point of view worldwide [11]. While they have pros and cons, many of them cannot be considered as decision-making assistance tools. Also, some of them do not adequately reflect energy policies. For instance, they do not take into account the policies which the World Energy Council has proposed: e.g., by 2050, new technologies should generate about 37% of the total energy in the world [12].

Pollution and environmental problems caused by overuse of fossil fuels, especially for transportation, have exacerbated the need for alternative fuels. Romm [13] thoroughly investigated alternative fuels for transportation systems in the future. Arslan et al. [14] reviewed possible scenarios of supplying energy for cars rather than fossil fuels in Turkey. Also, Babtista et al. [15] studied short-term and long-term resources and road consumption scenarios in Portugal and found alternative fuels necessary for longer horizons. Sehatpour et al. [16] made a comprehensive research on fossil-fuel alternatives for light-duty vehicles and, based on a multi-criteria evaluation, concluded NG and biogas are superior options for the mid-term in Iran. Santisirisomboon et al. [17] studied policies of carbon taxation to study the competitiveness of biomass energy with fossil fuels in Thailand.

Due to the importance of the problem, there are many decision support tools and simulation models available, such as the Vienna Automated System Planning Package, MESSAGE (Model for Energy Supply Strategy Alternatives and their General Environmental Impact) [18], the Long-range Energy Alternatives Planning system (LEAP), the MARKAL-EFOM Integrated System (TIMES), and the Energy PLAN [19]. These models allocate energy based on minimizing costs and priorities of demands. Environmental concerns, planning policies, and availability of energy resources can be defined as constraints [20].

A different application of these models is recorded in the literature. Strachan & Kannan [21] employed MARKAL-Macro (M-M) to study the long-term reduction of CO_2 emissions in the UK energy sector. Liu et al. [22] applied the energy model of MESSAGE-China to study the trend of novel energy technologies and their contributions to GHG reduction in China. Ball et al. [23] employed the energy system model MOREHyS to plan spatially and temporally a set-up of a hydrogen-based transport infrastructure system in Germany for the horizon of 2030. Chiodi et al. [24] analyzed the competing demands for land-use, import dependency, availability of sustainable bioenergy, and economics under the framework of an Irish energy systems model of TIMES. Tavakoli et al. [25] and Valinejad et al. [26] found the energy model system as a convenient and user-friendly approach to analyze energy policies.

Goal programming (GP) is a current multi-objective optimization method, which can address multi-criteria decision analysis (MCDA) problems. Jayaraman et al. [27] used a GP model for efficient allocation of labor resources considering the criteria of economic, energy, and environment in the United Arab Emirates using the approach of prioritizing areas for strategic planning and resource allocation for the sustainability of the strategies. They presented mathematical and economic indicators in order to digitize criteria. Kumar et al. [28] developed an insight into the application of various multiple criteria decision-making (MCDM) methods in the renewable energy sector. Zografidou et al. [29] programmed a GP model with all possible weight combinations to analyze energy allocation and budgeting in Greece and provided a multi-dimensional decision-makers' framework to determine the optimal budgeting mix to attract investors and guarantee the success of the venture. Kumar et al. [30] optimized priorities among suppliers considering the three dimensions of economic, social, and environmental sustainability in India. They integrated fuzzy AHP and fuzzy multi-objective linear programming approaches. Other extensions to GP are also applied to energy suitability problems, e.g., stochastic goal programming [7], weighted goal programming [31], fuzzy goal programming [32], and fractional goal programming [33]. Flisberg et al. [34] kept a schedule of the harvest and chipping operations of forest fuels in Sweden and studied alternatives. They employed indicators for all operations and solved them by a decision support system. Mekonnen et al. [35] explored the commutation between domestic and other applications of biomass energy sources in Ethiopia by employing a no-separable-farm household model in which labor energy is also considered in the stages of collection and farming. They concluded that the application of dung as a domestic fuel source negatively affects the value of harvested crops, however, the application of on-farm fuelwood is compromised with an increase in the value of agricultural output. Chong et al. [36] analyzed factors which had an impact on the energy consumption growth in Guangdong Province by employing the logarithmic mean Divisia index I (LMDI) based on the physical processes of energy utilization from the source to end-users. They concluded GDP and population are the most influential factors in energy consumption.

Atabaki and Aryanpur [37] developed a multi-objective linear planning model to analyze Iran's long-term power sector development from economic, environmental, social, and sustainable perspectives. Three objective functions in this study are included: minimizing the cost, carbon production, and maximizing the job market. To assess expert-based weights, analytical hierarchy process (AHP) methods are employed and, moreover, to support the decision-makers, the MESSAGE model as a planning tool is used to define different scenarios for developing clean technologies. The results show that a sustainable scenario leads to high technology diversification. Furthermore, the combined cycle would be the dominant option in Iran's long-term generation mix. In addition, power generation from non-hydro renewables, solar PV in particular, should grow faster than the total electricity demand. The findings indicate that the economic scenario fulfills Iran's commitment to 4% reduction of emissions compared to the current trend. However, the sustainable and environmental scenarios would achieve the superior 12% reduction goal. Multi-objective analysis shows that moving away from one's objective optimum value leads to significant improvements in other objective values. Adnan et al. [38] formulated a multi-objective scheduling problem to optimize the allocation of renewable energy resources and electric vehicle (EV) charging stations.

2.1. Comparative Analysis of Conventional Method

Investigating the appropriateness of different fuels and technologies, including renewable energy, is a practical decision-making practice for policymakers. Renewable energy sources can help to increase energy supply as well, to reduce emissions of greenhouse gas pollutants. Due to limitations of supplying the total energy from new sources of energy for now, finding an optimal combination of supply from both renewable and non-renewable energy sources is favorable. However, the rapid development and rising capacity of new technologies in different areas should be taken into account in planning future and targeting [39]. Ehsan and Yang [40] comprehensively investigated optimization methods employed in distributed generation in the power distribution networks. The Pros and cons of each method are shown in Table 1. The literature was investigated with a systematic review. Criteria for selecting articles included: publication dates between 1990 and 2020, use of a quantitative allocation technique, presence of a case study, use of optimization method, and those published in reputable journals.

Table 1. Comparative analysis of conventional techniques and metaheuristic algorithms [40].

Techniques Category	Planning Method	Disadvantages	Advantages
Conventional technique	Analytical techniques	Inaccuracies in case of complex problems	Low power consumption, ease of use, non-repetitiveness
	Comprehensive analysis	Computationally inefficient	understandable
	Linear integer programming of mixtures	Possibly incorrect results	Easily applicable, relatively flexible
	Nonlinear integer programming of mixtures	Difficult to make calculations, requiring decision variables	very accurate
	Optimal power flow	Difficult to troubleshoot some parameters	very accurate
	Possible methods	Too much data is required	Suitable for the model generation of renewable DGs
Metaheuristic algorithm	Genetic Algorithm	The possibility of early convergence	Suitable for discrete parameters and complex issues
	Particle Swarm Optimization	Difficulty in designing basic parameters, the possibility of early convergence	Easy to run, low parameters are required to adjust
	Taboo Search	Need to repeat	Suitable for continuous and discrete variables

Alternate energy sources cannot guarantee the continuity and reliability of the power supply. Vega-Garita et al. [41] and Calpa et al. [42] analyzed the effects of high PV penetration as the main source of energy for the Spanish electric grid. Renewable resources have many technical and economic benefits. Its technical benefits include reducing system losses, improved voltage, quality, high reliability, and economic benefits including low maintenance costs and fuel costs. The most common distribution generation (DG) systems today are residential solar technology, small wind turbines, and fuel cells. Some research work focusing on energy resource allocation optimization is listed in Table 2.

Table 2. Previous research in the field of energy resource allocation optimization.

Authors	Targets	Findings/Contribution	Limitations
Rowse [4]	Allocation of gas resources to domestic uses	Calculates a social welfare function to find the optimal route of the amount of export and domestic consumption	Limiting the supply from new sources
Ja'fari and Dehghani [43]	Optimal allocation of natural gas to various uses such as exports, petrochemical	The order of priority of gas projects is gas exports, gas injection, and petrochemical projects, respectively	Lack of comparability with similar research
Rowse [3]	Allocating gas resources using the hyperbolic discount rate	Estimates the previous model (1986) with different hyperbolic discount rates.	Not estimating the discount rate directly
Renani et al. [44]	Prioritization for the use of gas reserves in Iran to domestic uses	Gas injection is preferred to gas exports at low discount rates. Nevertheless, at the higher rates, this priority does not exist and the optimum gas quantity is determined simultaneously	Not using the exponential discount
Mohaghar et al. [45]	Prioritizing the sectors for allocation of natural gas	Injection into the oil fields is the first priority	Using a simple MCDM method
Lo and Schober [46]	Optimizing electrical energy allocation	Using computer simulation methods to solve the model	Not considering the budget limitation
Hutagalung [47]	NG allocation priority to domestic sectors	The priority of sectors for NG allocating determine as: industry, petrochemical, oil production, and electricity and power plant	Not considering the effect of NG price
Orlov [48]	Allocating NG to various sectors	The priority of sectors for NG allocation determined as injection, export, and power plant	Not using uncertainties in utility function
Zhang et al. [49]	Analyzing the effect of domestic NG price increase on the overall economy index	Chemical industry will be most influenced by any NG price increase	Lack of comparability with similar research
Daneshzand et al. [50]	Sustainability of domestic NG supply on providing financial capital	Energy price is one of the main variables directly influencing energy demand and supply	Not using pricing policy tools
Daneshzand et al. [51]	Optimal allocation of natural gas to various demand sectors	Residential sector should have a much smaller share and the export sector a much larger share of the consumption sector	Lack of comparability with similar research
Alavi et al. [52]	Optimal allocation of NG	Gas exports do not maximize social welfare.	Not using pricing policy tools and subsidies in modeling

2.2. Research Gap

Hashemipour et al. [53] presented a mathematical model to optimize the NG allocation to an oil field of Iran by using genetic algorithms. The objective function included (1) maximizing the production rate and (2) maximizing the profit. Alikhani and Rshidi [54] proposed a stochastic programming model for NG allocation with an energy security cost approach. They concluded that the priorities of NG resources allocations are as follows: domestic and commercial sections, power plants, industries, gas reinjection, exports, refinery, road transport, and agriculture, respectively. Kazemi et al. [55] comprehensively studied the problem of NG allocation models by AHP approach and 13 models were ranked in this study. Using multi-objective goal programming, Chedid et al. [56] suggested a model for NG allocation in Lebanon. Borges and Antunes [57] formulated energy allocation to the domestic sector using fuzzy multi-objective programming. Hatagalung [47] analyzed the optimization allocation of sustainable energy source by a non-linear mathematical model. Li et al. [58] examined the energy allocation problem by minimizing errors of budget. Maroufmashat and Sattari [59] presented a linear programing model to allocate NG resources to Iran's various demand sectors. Alikhani and Azar [60] employed a fuzzy goal programming model for allocating NG to different sectors. It was shown that NG injection, export, and road transportation are the most important sectors, respectively.

Some other studies investigated the NG prices' effects on the economics of importers and exporters since this influences the cycle of energy demand and supply [61–63]. Orlov [64] studied the Russian government's policy to reduce domestic NG price regulation and concluded that the domestic NG price should be 55% of the export netback price. Wang and Lin [61] presented a dynamic model to analyze the effect of NG price increase on the economy indexes (GDP, imports, and household income) in China. The effects of NG pricing were not presented or researched in any of the studies analyzed by the authors (or in systematic literature review made by the authors). Hence, some literature with their features is listed in Table 3 for a better idea of research trends needed in this field.

Table 3. Modeling approaches.

Author(s)	Modelling		Proposed Model				Case Study
	Method	Dynamic	Domestic NG Development	Consumption Priorities		Effect of NG Price	
				Optimal Allocation	Different Demand Sectors		
Rowse [4]	NLP	√	√	√			Canada
Arab et al. [65]	LP	√	√				Nigeria
Boucher and Smeers [66]	NLP	√	√				Indonesia
Hutagalung [47]	CGE	√	√		√	√	Indonesia
Orlov [67]	CGE			√		√	Russia
Alikhani and Azar [60]	FGP				√		Iran
Maroufmashat and Sattari [59]	LP			√	√		Iran
Zhu et al. [68]	FGP				√	√	China
Orlov [64]	CGE	√			√		Russia
Zhang et al. [49]	CGE				√	√	China
Salehi and Gazijahani [69]	SLP	√		√			Iran
Daneshzand et al. [50]	SD	√	√			√	Iran
Daneshzand et al. [51]	SD	√		√		√	Iran
Alavi et al. [52]	DP	√		√		√	Iran
This Research	GP	√	√	√	√	√	Iran

Note: NLP: none-linear programing, LP: linear programing, CGC: computational general equilibrium, FGP: fuzzy goal programming, SLP: stochastic linear programing, SD: system dynamic, DP: dynamic programing, and GP: goal programing.

In this paper, the optimal allocation of natural gas resources to different sectors of consumption is investigated by applying multi-objective goal programming decision-making techniques. Therefore, the objective function is the goal and system constraints of the multi-goal programming decision-making technique based on the relative weights assigned to different sectors of consumption.

3. Methods and Material

3.1. Multi-Objective Function Model—Scenario 1 (Minimizing Gas Consumption in Different Sectors)

Simplex in linear programming is standard method for solving an optimization problem, typically one involving a function and several constraints expressed as inequalities. The inequalities define a polygonal region, and the solution is typically at one of the vertices. The simplex method is a systematic procedure for testing the vertices as possible solutions. In this research the simplex method was used for model formulation based on goal programming.

Let us take X_{ij} the amount of gas which must be allocated in year t to the sectors $i = 1, 2, \ldots, 6$, which represent the household business, industry, petrochemicals, power plants, injection into oil fields, and exports, respectively, and j is the number of the year (2018–2025—$j = 1, 2, \ldots, 8$).

Definition of indexes and parameters are as follows:

j: time (year);
i: The number of the consumer sector;
Z_i: Objective I;
P_i: The priority of objective I;
d_i^+: Positive deviation from objective I;
d_i^-: Negative deviation from objective I;
P_{ij}: Price of gas consumed in sector i in year j;
AC_{ij}: Minimum natural gas consumption of each sector; and
FC_{ij}: The goal portion of various gas sectors.

The main proposed model is the multi-objective function based on goal programming, as follows:

$$Min\ P_k d_k^+; \quad \forall\, k \in i \tag{1}$$

Subject to:

$$\sum_{k=1}^{i} \sum_{l=1}^{j} P_{kl} \cdot X_{kl} = Z_k; \quad \forall\, k \in i \text{ and } \forall\, l \in j \tag{2}$$

$$\sum_{k=1}^{i} \sum_{l=1}^{j} P_{kl} \cdot X_{kl} \leq E_k; \quad \forall\, k \in i \text{ and } \forall\, l \in j \tag{3}$$

$$AC_{ij} \leq X_{ij} \leq FC_{ij}; \quad \forall\, i = 1, \ldots, 6 \text{ and } \forall\, j = 1, \ldots, 8$$

$$d_k^+, d_k^- = 0; \quad \forall\, k \in i \tag{4}$$

$$X_{ij},\ d_k^+,\ d_k^- \geq 0; \quad \forall\, k \in i \tag{5}$$

As it is clear, this model consists of six objectives. The prior objective is to allocate gas to the export sector in different production years. Therefore, in the equation of the multi-objective model, the related deviations are considered as positive and so for the other sectors. Also, each of the objective functions are assigned and continued to minimize deviations from the goal that are included in the final objective function constraints. Also, since a target function cannot have values greater than or less than the goal, six final constraints are used in this regard. Discussed target functions are presented as below:

3.2. Objective Function Model Allocation of Household-Commercial Gas Surplus to Other Sectors

F_j: Gas deficit rate in year j

E_j: Gas household surplus in j year

$$Min\ Z = \sum_{i=2}^{6}\sum_{j=1}^{8} P_{ij}\cdot X_{ij} \tag{6}$$

$$\sum_{i=2}^{6} X_{ik} \leq E_k,\ k = 1, \ldots, 8 \tag{7}$$

$$X_{ij} \geq 0; \tag{8}$$

3.3. Objective Function Model Optimizing Energy Security Cost

Energy security is a factor that is associated with national security and the availability of natural resources. In this context, the security of the energy supply is of the utmost importance. It is worth mentioning that energy security plays the most important role for decision-makers, that is to say, this factor is highly influential on the price of energy resources, and political power and relation among exporter and importer counties.

The cost of energy security should be quantized and have a measurement. All applications to secure energy should be taken into account, e.g., increasing fuel costs from a source to reduce consumption from an unsafe and insecure source, costs of infrastructure construction to create new and safe systems, and the political costs of securing and storing energy resources. Accordingly, in a country which holds energy policy as its priority, circumstances are different to those where cheaper energy is the objective.

Threats endangering the transmission and consumption of fuel in any of the subsectors can be categorized into seven general groups:

Investment threats: This threat is considered as a physical threat that affects the external or physical nature of fuel transmission. Delay or lack of investment due to the administrative bureaucracy and the complexity of the country's structure in absorbing domestic and foreign capital, legal restrictions, international sanctions, and taxes, uncertainty about the fate of investment, the high risk of investing in the particular project, and the lack of return on investment are among the investment threats.

Technical threats: This threat can disrupt the supply. Technical problems in the process of the production can result in waste or no fuel extraction, however, those problems related to gas transmission pipelines are likely to be solved by storing it or bypassing to injection fields.

The threats of demand management and consumption growth: Mismanagement in demand and supply can lead to a threat. If a supplier is unable to provide storage facilities, consumption fluctuations can become problematic. In cases where a supplier has not considered urgent matters of sudden cold or gradual growing demand, the distribution could face difficulties due to lack of proper infrastructure.

Other physical threats: Earthquakes, pipeline failure, and terrorist attacks on facilities are among these threats. Some of these threats are taken seriously and some less seriously in Iran.

Pricing and marketing threats: These threaten the security of fuel exports. Pricing could threaten supply security by conflicts before and after contract. Energy security imposes costs on the supplier and the importer. Due to security issues and administrative allocation strategy, competitive pricing is a conflict between the authority and the private sector, making the market impossible.

Internal management threats: In addition to geopolitical issues which are compromised with the nature of the energy sector, national problems, such as economic crises and mass strikes, influence the security of supply.

Foreign and international political threats: Foreign countries which are stakeholders in energy trade in the region could be counted as threats by implication. In addition, actions implemented by other countries or unions, such as sanctions, could highly threaten safe and secure supply.

The objective function and constraints are as follows:

$$Min\ P_1d_1^+,\ P_2d_2^+,\ P_3d_3^+,\ P_3d_4^+,\ P_5d_5^+,\ P_6d_6^+, P_7d_7^+ \tag{9}$$

Subject to:

$$\sum_i \sum_j C_{ij} \cdot X_{ij} = Z_i \tag{10}$$

$$\sum_{k=1}^{i} \sum_{l=1}^{j} X_{kl} = \sum PR_k \tag{11}$$

$$d_k^+, d_k^- = 0; \quad \forall\, k \in i \tag{12}$$

$$X_{ij},\ d_k^+,\ d_k^- \geq 0; \quad \forall\, k \in i \tag{13}$$

In this paper, the developed model is a seven-objective and linear model. Therefore, goal programming and lexicographer methods have been used to solve the MODM models. It is worth noting that Lingo software is used to solve this problem.

3.4. Problem-Solving Approach with Lexicographer

One of the methods employed to solve goal programming problems when objectives are prioritized is lexicographers. In fact, the lexicographer uses a set of methods that an analyzer needs to get basic information from the decision-maker. The difference between lexicographers and the weighting method is receiving weights from the decision-maker. In the weighting method the weights taken from the decision-maker reveal the importance of the objective functions. However, in the lexicographer's method, the only order of priorities or preferences is determined by the decision-maker.

The lexicographer's approach is to select an objective function with the highest priority and the problem is optimized in the target space with the highest priority. Afterwards, if the found point is unique, it is considered the optimum. If multiple solutions are found, the point which satisfies the less prior objective functions as well.

This method consists of the following steps:

Step 1: Prioritize the desired goals;

Step 2: The goal has the highest priority from which the deviation is more important to the decision-maker, so in this step select the most important goal and minimize the related deviation; and

Step 3: The problem will be solved only for the highest prior function and its optimum will be searched. Two situations could happen at this stage:

Mode 1: If the solution is unique, the answer is found; and

Mode 2: If some points are found, lower prior functions determine which point is superior.

Steps above are iterated until reaching a unique answer.

4. Results and Discussion

In this section, priorities of allocating gas to different sectors of consumption as coefficients of proximity (rank) and relative proportional weight and priority of six different sectors of gas consumption are shown in Table 4. It should be noted that the relative weight of each sector is calculated by dividing the corresponding coefficient of affinity into the total coefficients of the proximity of all sectors of consumption. Regarding Table 4, the first rank of consumption is assigned to export and then injection to oil fields and petrochemical industries are in order, respectively.

Table 4. Rating, relative weight, and priority of the different sectors (P_i) of consumption.

Consumption Sectors	Rating	Relative Weight	Rank	Index of Sector
Household business	0.0249	0.006387	6	P_6
Industry	0.6071	0.155719	4	P_4
Petrochemical industries	0.7552	0.193706	3	P_3
Power plants	0.5884	0.150922	5	P_5
Injection to oil fields	0.9502	0.243722	2	P_2
Export	0.9729	0.249545	1	P_1

At this stage, the optimal allocation of gas resources to different sectors of consumption is implemented using the GP multi-purpose decision-making technique. Therefore, the solution for the objective function, the goal and system constraints of GP multi-purpose decision-making technique for 2018–2025 by using the relative weight of the various consumption sectors (presented in Table 4), as well as information extracted from the trend of the gas industry and energy balance sheet documents of Iran is to be found. The goal share of the various gas sectors, the amount of consumption of the various departments of consumption (based on the balance sheet of energy and hydrocarbons), the base gas volume, the allocable and predicted amount of each year is summarized in Figure 2, Tables 5 and 6.

Inputs of GP consists the goal portion (FC_{ij}) of consumption, price per cubic meter of natural gas in consumption areas, P_{ij}, base gas volume (Z_i), limited, allocated and predicted value, the volume of gas deficit (F_j), and minimum natural gas consumption of each consumption sector (AC_{ij}) in Figure 2, Tables 5–9, respectively.

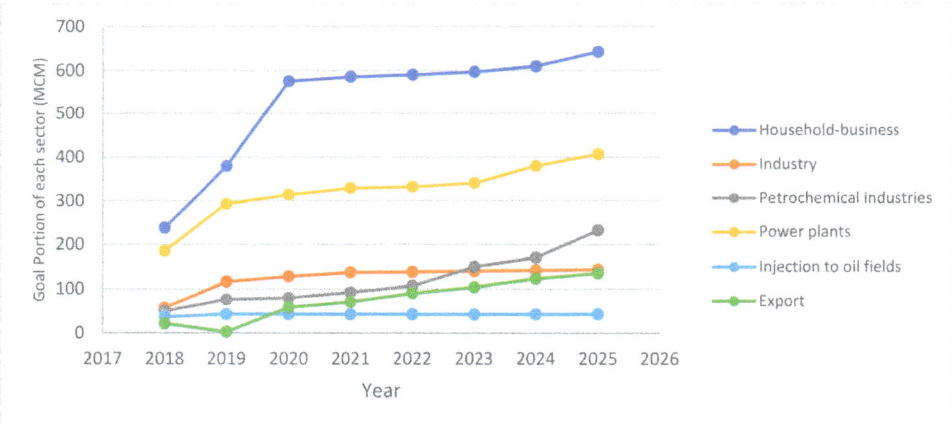

Figure 2. The goal portion (FC_{ij}) of various gas sectors (MCM: Million Cubic Meter).

Table 5. Price per cubic meter of natural gas in consumption areas P_{ij} (Rials).

Consumption Sectors	2018	2019	2020	2021	2022	2023	2024	2025
Household business	3864	3941	4021	4102	4184	4265	4347	4430
Industry	1320	1320	1320	1320	1320	1320	1320	1320
Petrochemical industries	3445	3445	3445	3445	3445	3445	3445	3445
Power plants	80	80	80	80	80	80	80	80
Injection to oil fields	130	130	130	130	130	130	130	130
Export	26,000	26,000	26,000	26,000	26,000	26,000	26,000	26,000

Table 6. Base gas volume (Z_i), limited, allocated and predicted value (MCM).

	2018	2019	2020	2021	2022	2023	2024	2025
The gas supply	588	709	809	818	825	527	824	807

Table 7. The volume of gas deficit (F_j) (MCM).

	2018	2019	2020	2021	2022	2023	2024	2025
The gas deficit	3.7	39.8	15.1	55.4	104.3	176.3	182.1	360.8

Table 8. Minimum natural gas consumption of each sector (AC_{ij}) (MCM).

Consumption Sectors	2018	2019	2020	2021	2022	2023	2024	2025
Household business	151.4	154	155.3	159	163	167.8	171.2	178.9
Industry	54.2	109	118	126.3	129.7	131.4	137	139.3
Petrochemical industries	50	75.3	80	91	104.5	143.2	151.8	231
Power plants	183	196.2	209.1	219.9	231.9	248	26.2	271.8
Injection to oil fields	22,500	22,500	22,500	22,500	22,500	22,500	22,500	22,500
Export	37.2	69	56.2	63.8	75.2	89	93.4	118.8

Table 9. The Gas surplus in different years (E_j) (MCM).

	2018	2019	2020	2021	2022	2023	2024	2025
The gas deficit	91.46	59.4	89.1	51	40.2	−66.7	−70.46	−247.68

Calculate Security Costs

Considering the equations, the cost of energy security in different sectors is calculated. The probability of occurrence of each of the threats in the gas security disorder is based on the recurrence of the expert's opinion as outlined in the following tables (Tables 10 and 11).

Table 10. The probability of occurrence of each of the threats (PR_{ij}) (Percentage).

Consumption Sectors	Investment	Technical	Demand and Consumption Growth	Physical	Price and Marketing	Internal Management	External Political
Household business	1%	1%	1%	0.3%	0.3%	0.4%	-
Industry	1%	1%	1%	1%	0.5%	1.5%	-
Petrochemical industries	1%	1%	0.5%	0.5%	2%	2%	-
Power plants	-	2%	-	1%	-	3%	-
Injection to oil fields	0.6%	2%	-	-	-	1.2%	-
Export	1%	2%	1%	3%	1%	1%	1%

Table 11. Price per cubic meter of gas per consumption unit (C_{ij}) (Rials).

Consumption Sectors	2018	2019	2020	2021	2022	2023	2024	2025
Household business	13,369.5	14,818	16,920	17,550	18,040	18,117	18,762	18,411
Industry	269.2	332.6	237.6	293	293.1	205.9	316.8	356.4
Petrochemical industries	72.3	217	24.1	337.6	578.7	1519.2	378.6	48.2
Power plants	10.4	19.6	209.6	217.2	216.8	198.8	187.2	268.8
Injection to oil fields	2340	11,180	3380	9620	35,880	32,760	66,560	22,620
Export	0.7	37.7	54.6	54.6	54.6	54.6	54.6	54.6

The goal programming model mentioned in the previous section after substituting numerical values associated with different input parameters is as follows:

$$Min\ P_6 d_1^+,\ P_4 d_2^+,\ P_3 d_3^+,\ P_5 d_4^+,\ P_2 d_5^+,\ P_1 d_6^+$$

Such that:

$$
\begin{array}{l}
3864X_{11} + 3941X_{12} + 4021X_{13} + 4102X_{14} + 4184X_{15} + 4265X_{16} + 4347X_{17} + 4430X_{18} + d_1^- - d_1^+ = 4,221,000 \\
1320X_{21} + 1320X_{22} + 1320X_{23} + 1320X_{24} + 1320X_{25} + 1320X_{26} + 1320X_{27} + 1320X_{28} + d_2^- - d_2^+ = 1,009,000 \\
3445X_{31} + 3445X_{32} + 3445X_{33} + 3445X_{34} + 3445X_{35} + 3445X_{36} + 3445X_{37} + 3445X_{38} + d_3^- - d_3^+ = 960,000 \\
80X_{41} + 80X_{42} + 80X_{43} + 80X_{44} + 80X_{45} + 80X_{46} + 80X_{47} + 80X_{48} + d_4^- - d_4^+ = 2,581,000 \\
130X_{51} + 130X_{52} + 130X_{53} + 130X_{54} + 130X_{55} + 130X_{56} + 130X_{57} + 130X_{58} + d_5^- - d_5^+ = 327,000 \\
26,000X_{61} + 26,000X_{62} + 26,000X_{63} + 26,000X_{64} + 26,000X_{65} + 26,000X_{66} + 26,000X_{67} + 26,000X_{68} + d_6^- - d_6^+ = 628,000
\end{array}
$$

$$
\begin{array}{l}
X_{11} + X_{21} + X_{31} + X_{41} + X_{51} + X_{61} \leq 588 \\
X_{12} + X_{22} + X_{32} + X_{42} + X_{52} + X_{62} \leq 709 \\
X_{13} + X_{23} + X_{33} + X_{43} + X_{53} + X_{63} \leq 809 \\
X_{14} + X_{24} + X_{34} + X_{44} + X_{54} + X_{64} \leq 818 \\
X_{15} + X_{25} + X_{35} + X_{45} + X_{55} + X_{65} \leq 825 \\
X_{16} + X_{26} + X_{36} + X_{46} + X_{56} + X_{66} \leq 827 \\
X_{17} + X_{27} + X_{37} + X_{47} + X_{57} + X_{67} \leq 824 \\
X_{18} + X_{28} + X_{38} + X_{48} + X_{58} + X_{68} \leq 807
\end{array}
$$

$$
\begin{array}{cccccc}
X_{11} \geq 151.4 & X_{21} \geq 54.2 & X_{31} \geq 50 & X_{41} \geq 183 & X_{51} \geq 22,500 & X_{61} \geq 37.2 \\
X_{12} \geq 154 & X_{22} \geq 109 & X_{32} \geq 75.3 & X_{42} \geq 196.2 & X_{52} \geq 22,500 & X_{62} \geq 69 \\
X_{13} \geq 155.3 & X_{23} \geq 118 & X_{33} \geq 80 & X_{42} \geq 209.1 & X_{53} \geq 22,500 & X_{63} \geq 56.2 \\
X_{14} \geq 159 & X_{24} \geq 126.3 & X_{34} \geq 91 & X_{44} \geq 219.9 & X_{54} \geq 22,500 & X_{64} \geq 63.8 \\
X_{15} \geq 163 & X_{25} \geq 129.7 & X_{35} \geq 104.5 & X_{45} \geq 231.9 & X_{55} \geq 22,500 & X_{65} \geq 75.2 \\
X_{16} \geq 167.8 & X_{26} \geq 131.4 & X_{36} \geq 143.2 & X_{46} \geq 248 & X_{56} \geq 22,500 & X_{66} \geq 89 \\
X_{17} \geq 171.2 & X_{27} \geq 137 & X_{37} \geq 151.8 & X_{47} \geq 262 & X_{57} \geq 22,500 & X_{67} \geq 93.4 \\
X_{18} \geq 178.9 & X_{28} \geq 139.3 & X_{38} \geq 231 & X_{48} \geq 271.8 & X_{58} \geq 22,500 & X_{68} \geq 118.8 \\
X_{11} \leq 237.9 & X_{21} \leq 57.6 & X_{31} \leq 50.3 & X_{41} \leq 185.6 & X_{51} \leq 22.2 & X_{61} \leq 38.1 \\
X_{12} \leq 380 & X_{22} \leq 117 & X_{32} \leq 76.2 & X_{42} \leq 293 & X_{52} \leq 43.5 & X_{62} \leq 33 \\
X_{13} \leq 576 & X_{23} \leq 129 & X_{33} \leq 80.1 & X_{43} \leq 314 & X_{53} \leq 43.5 & X_{63} \leq 58.8 \\
X_{14} \leq 586.2 & X_{24} \leq 138 & X_{34} \leq 92.5 & X_{44} \leq 329 & X_{54} \leq 43.5 & X_{64} \leq 71.2 \\
X_{15} \leq 591 & X_{25} \leq 139.2 & X_{35} \leq 107.3 & X_{45} \leq 332 & X_{55} \leq 43.5 & X_{65} \leq 91 \\
X_{16} \leq 597.6 & X_{26} \leq 141 & X_{36} \leq 150.6 & X_{46} \leq 341 & X_{56} \leq 43.5 & X_{66} \leq 105 \\
X_{17} \leq 610.1 & X_{27} \leq 142.8 & X_{37} \leq 171.2 & X_{47} \leq 380 & X_{57} \leq 43.5 & X_{67} \leq 124 \\
X_{18} \leq 642.5 & X_{28} \leq 144.5 & X_{38} \leq 232 & X_{48} \leq 407 & X_{58} \leq 43.5 & X_{68} \leq 136.2
\end{array}
$$

$$d_1^+ \cdot d_1^- = 0,\ d_2^+ \cdot d_2^- = 0,\ d_3^+ \cdot d_3^- = 0,\ d_4^+ \cdot d_4^- = 0,\ d_5^+ \cdot d_5^- = 0,\ d_6^+ \cdot d_6^- = 0$$

$$X_{ij},\ d_k^+,\ d_k^- \geq 0;\ k = 1,2,3,4,5,6$$

In the lexicographer method, the objective functions of the goal programming model are considered and solved separately to minimize the deviation from the specified goal according to their priority. Then, if it obtained a unique point, it is considered to be the optimal point. If multiple solutions are found in the space of multiple solutions, lower important functions are checked for the points.

According to the results obtained by solving the model with the first priority objective function (the first objective function), the obtained point is unique and the negative deviation of the first objective function from the specified cause is minimum and zero. The answer (Table 12 and Figure 3) is final and efficient and acceptable:

Table 12. Optimal allocation of gas to sectors based on cost minimization (MCM).

Consumption Sectors	2018	2019	2020	2021	2022	2023	2024	2025
Household business	151.4	154	155.3	159	163	167.8	171.2	46.1
Industry	54.2	109	118	126.3	129.7	131.4	137	139.3
Petrochemical industries	50	75.3	80.1	92.5	104.9	143.2	151.8	231
Power plants	185.6	293	314	328.9	294	248	262	271
Injection to oil fields	22.2	43.5	43.5	43.5	43.5	43.5	8.6	0
Export	38.1	33	58.8	67.7	89.8	93.1	93.4	118.8

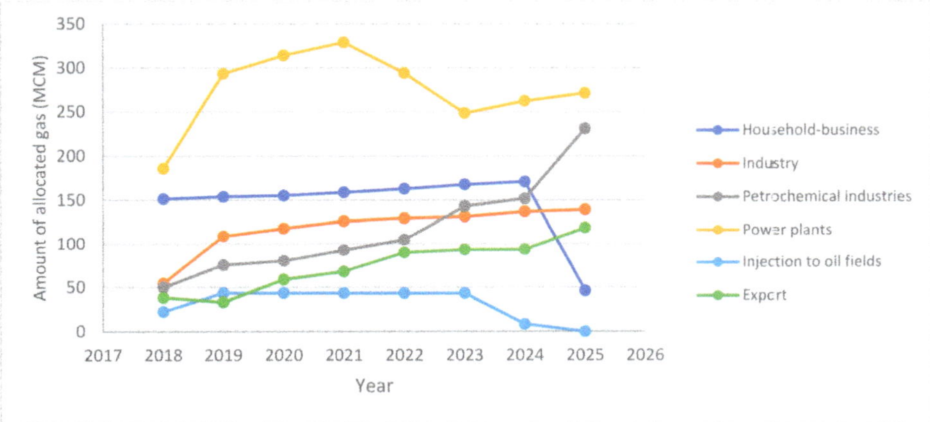

Figure 3. Optimal allocation of gas to sectors based on cost minimization.

Allocation of household surplus gas to other sectors is calculated as follows:

$$
\begin{aligned}
Min\ Z = 3864X_{11} \ & +3941X_{12} + 4021X_{13} + 4102X_{14} + 4184X_{15} + 4265X_{16} + 4347X_{17} + 4430X_{18} + 1320X_{21} \\
& +1320X_{22} + 1320X_{23} + 1320X_{24} + 1320X_{25} + 1320X_{26} + 1320X_{27} + 1320X_{28} + 3445X_{31} \\
& +3445X_{32} + 3445X_{33} + 3445X_{34} + 3445X_{35} + 3445X_{36} + 3445X_{37} + 3445X_{38} + 80X_{41} \\
& +80X_{42} + 80X_{43} + 80X_{44} + 80X_{45} + 80X_{46} + 80X_{47} + 80X_{48} + 130X_{51} + 130X_{52} + 130X_{53} \\
& +130X_{54} + 130X_{55} + 130X_{56} + 130X_{57} + 130X_{58} + 26,000X_{61} + 26,000X_{62} + 26,000X_{63} \\
& +26,000X_{64} + 26,000X_{65} + 26,000X_{66} + 26,000X_{67} + 26,000X_{68}
\end{aligned}
$$

$$
\begin{gathered}
X_{21} + X_{31} + X_{41} + X_{51} + X_{61} \leq 91.45 \\
X_{22} + X_{32} + X_{42} + X_{52} + X_{62} \leq 59.4 \\
X_{23} + X_{33} + X_{43} + X_{53} + X_{63} \leq 89.1 \\
X_{24} + X_{34} + X_{44} + X_{54} + X_{64} \leq 51 \\
X_{25} + X_{35} + X_{45} + X_{55} + X_{65} \leq 40.2 \\
X_{26} + X_{36} + X_{46} + X_{56} + X_{66} \leq 66.7 \\
X_{27} + X_{37} + X_{47} + X_{57} + X_{67} \leq 70.45 \\
X_{28} + X_{38} + X_{48} + X_{58} + X_{68} \leq 247.68 \\
X_{21} + X_{31} + X_{41} + X_{51} + X_{61} \geq 3.7 \\
X_{22} + X_{32} + X_{42} + X_{52} + X_{62} \geq 39.8 \\
X_{23} + X_{33} + X_{43} + X_{53} + X_{63} \geq 15.1 \\
X_{24} + X_{34} + X_{44} + X_{54} + X_{64} \geq 55.4 \\
X_{25} + X_{35} + X_{45} + X_{55} + X_{65} \geq 104.3 \\
X_{26} + X_{36} + X_{46} + X_{56} + X_{66} \geq 176.3 \\
X_{27} + X_{37} + X_{47} + X_{57} + X_{67} \geq 182.1 \\
X_{28} + X_{38} + X_{48} + X_{58} + X_{68} \geq 360.8 \\
X_{ij} \geq 0;
\end{gathered}
$$

According to the solution of the above formula, the optimal allocation of surplus gas to each of the sources of consumption is as follows (Table 13 and the Figure 4).

Table 13. Optimal allocation of surplus gas (MCM).

Consumption Sectors	2018	2019	2020	2021	2022	2023	2024	2025
Household business	3.864	3.941	4.021	4.102	4.184	4.265	4.347	4.430
Industry	1.24	1.24	1.24	1.25	1.24	1.32	1.24	1.24
Petrochemical industries	3.36	3.36	3.36	2.38	3.36	3.44	3.36	3.36
Power plants	0	0	0	0.8	0	0.8	0	0
Injection to oil fields	0.5	0.5	0.5	0	0.5	0.13	0.5	0.5
Export	26	26	26	26	26	26	26	66

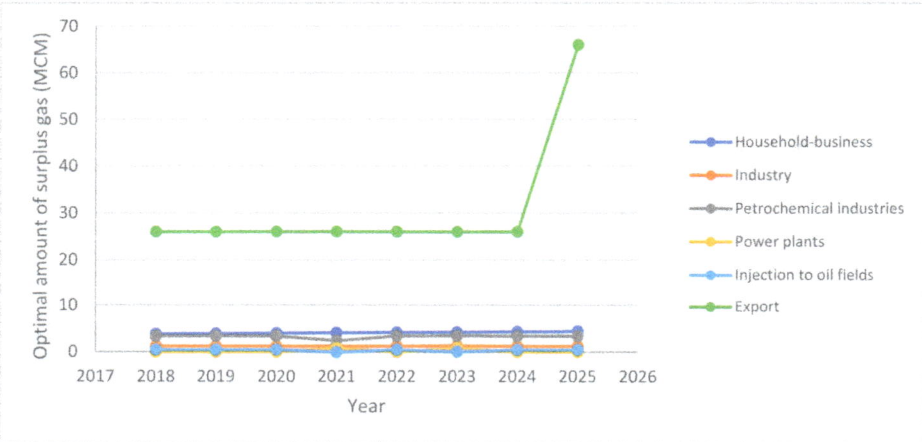

Figure 4. Optimal allocation of surplus gas.

Minimizing the cost of energy security is calculated as follows:

$$Min\ P_1 d_1^+,\ P_2 d_2^+,\ P_3 d_3^+,\ P_3 d_4^+,\ P_5 d_5^+,\ P_6 d_6^+$$

Such that:

$$\begin{aligned}13,369.5X_{11} &+14,818X_{12}+16,920X_{13}+17,550X_{14}+18,040X_{15}+18,117X_{16}+18,762X_{17}+18,411X_{18}+269.2X_{21}\\ &+332.6X_{22}+237.6X_{23}+293X_{24}+293.1X_{25}+205.9X_{26}+316.8X_{27}+356.4X_{28}+72.3X_{31}\\ &+217X_{32}+24.1X_{33}+337.6X_{34}+578.7X_{35}+1519.2X_{36}+378.6X_{37}+48.2X_{38}+100.4X_{41}\\ &+190.6X_{42}+209.6X_{43}+217.2X_{44}+216.8X_{45}+198.8X_{46}+187.2X_{47}+268.8X_{48}+2340X_{51}\\ &+11,180X_{52}+3380X_{53}+9620X_{54}+35,880X_{55}+32,760X_{56}+66,560X_{57}+22,620X_{58}+0.7X_{61}\\ &+37.7X_{62}+54.6X_{63}+54.6X_{64}+54.6X_{65}+54.6X_{66}+54.6X_{67}+54.6X_{68}<=327,712\end{aligned}$$

$$\begin{gathered}X_{11}+X_{12}+X_{13}+X_{14}+X_{15}+X_{16}+X_{17}+X_{18}<=4\\ X_{21}+X_{22}+X_{23}+X_{24}+X_{25}+X_{26}+X_{27}+X_{28}<=6\\ X_{31}+X_{32}+X_{33}+X_{34}+X_{35}+X_{36}+X_{37}+X_{38}<=7\\ X_{41}+X_{42}+X_{43}+X_{44}+X_{45}+X_{46}+X_{47}+X_{48}<=5\\ X_{51}+X_{52}+X_{53}+X_{54}+X_{55}+X_{56}+X_{57}+X_{58}<=2\\ X_{61}+X_{62}+X_{63}+X_{64}+X_{65}+X_{66}+X_{67}+X_{68}<=10\\ d_1^+\cdot d_1^-=0 d_2^+\cdot d_2^-=0\\ d_3^+\cdot d_3^-=0 d_4^+\cdot d_4^-=0\\ d_5^+\cdot d_5^-=0 d_6^+\cdot d_6^-=0\\ X_{ij},\ d_k^+,\ d_k^-\ \geq 0;\ k=1,2,3,4,5,6\end{gathered}$$

Considering the above problem in Lingo 17 software, the optimal allocation of energy security is as follows (Table 14 and Figure 5).

Table 14. Optimal allocation of energy security costs (Rials).

Consumption Sectors	2018	2019	2020	2021	2022	2023	2024	2025
Household business	0	1.23	1.23	1.23	0.29	0	0	0
Industry	0	1	1.23	1.23	1.23	1.23	0	0
Petrochemical industries	0	0.8	1.23	1.23	1.23	1.23	1.23	0
Power plants	0	1.23	1.23	1.23	1.23	0.61	0	0
Injection to oil fields	0	0.7	1.23	0	0	0	0	0
Export	1.23	1.23	0	0	0	0	0	7.5

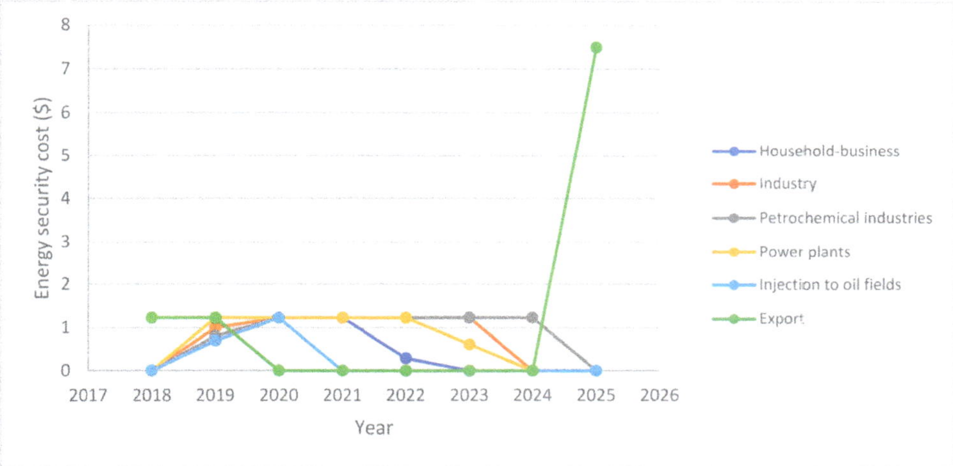

Figure 5. Optimal allocation of energy security costs.

According to Table 14, the optimal cost of energy security is calculated as follows:

$$
\begin{aligned}
(1 \times 332.6) + \; & (1.23 \times 237.6) + (1.23 \times 293) + (1.23 \times 293.1) + (1.23 \times 205.9) + (1.23 \times 14{,}818) + (1.23 \times 18{,}920) \\
& + (1.23 \times 17{,}550) + (0.29 \times 18{,}040) + (0.7 \times 11{,}180) + (1.23 \times 3380) + (1.23 \times 19.6) \\
& + (1.23 \times 209.6) + (1.23 \times 217.6) + (1.23 \times 216.8) + (0.61 \times 198.8) + (0.8 \times 217) + (1.23 \times 44.1) \\
& + (1.23 \times 337.6) + (1.23 \times 578.7) + (1.23 \times 1519.2) + (1.23 \times 378.6) + (1.23 \times 0.7) \\
& + (1.23 \times 37.7) + (7.5 \times 54.6) = 90{,}201.802
\end{aligned}
$$

Regarding Table 13 gas surplus was present throughout the gas system. It shows that the gas surplus is allocated to different consumption sectors. Table 14 shows the cost of energy security which is only calculated for the sectors which do not have a supply shortage. In other words, this cost is assigned to the various sectors according to the cost of gas production and the priority of the consumer sector as specified in Table 14.

5. Conclusions

In this paper, the optimal allocation of limited NG resources to different sectors of consumption (including household business, industry, petrochemical industries, power plants, injection to oil fields, and export) in Iran during 2018–2025 is studied. The multi-objective linear planning model has been used for modeling. Lexicography method is employed and the objective functions of the goal programming model are solved according to their assigned priority, in order to minimize the deviation from the specified goal. Energy security strategies differ in each country and region. Energy-importing countries often consider energy security as the security of energy supply, while energy-exporting countries prefer to refer to it as security of energy demand. It should be noted that energy security is crucial for decision-makers in terms of supply, moreover, it is highly influential on energy pricing as well political power. Two main objective functions are considered in this study: (1) maximizing energy security index (minimizing energy security cost), and (2) minimizing the relative weight of different consumption sectors from negative to a goal by using GP multi-objective decision-making technique. Therefore, the objective function and systematic constraints of multi-objective GP decision-making techniques are presented based on the relative weight of the various sectors and also on the basis of information extracted from the gas industry prospectus and energy and hydrocarbon balance sheets. The goal share of different natural gas sectors (based on the gas industry outlook in Iran), the base consumption of the various sectors (based on energy and hydrocarbon balances in Iran), and the volume of basic natural gas, the allocable and predicted limit per year are given in the tables. The results of this study are used for optimal allocation of natural gas to different consumption sectors as well as for future planning. Household business, power plants, petrochemical industries, industry, export, and injection to oil fields are the highest consuming sectors in 2025, respectively. Also based on cost minimization function, power plants, petrochemical industries, and industries in general are the more consuming sectors, respectively.

The results show that if Iran's gas production plans are successful and the level of gas production achieves the levels announced in the development plans, it needs to target a great level of gas exports to maximize social welfare. In other words, the gas exports should not be one of the priorities of Iran's policy when the production level is faced with constraint (which is confirmed by other studies over the past decade), and domestic consumptions and, in particular, injection into oil fields has higher security. But if policies of increasing production are successful and the level of Iran's production is increased with a level that is predicted in the sixth development plan, the mass exportation of gas will lead to increased social welfare. This shows the need for extensive planning investments in this area. One of the most important policy implications is the increasing transmission capacities of Iran. It is also worth mentioning that one of the reasons for the high level of energy consumption in the final sectors is the lack of attention to correct the pattern of consumption. Of course, in this regard the large subsidies received by the energy sector, especially the gas sector, is an important reason for it. Also, for the natural gas demand model, official prices of natural gas have been used, which also include subsidies, and should be considered in the analysis of the results. Therefore, although the increased volume of natural gas needs special attention in the next three decades and the necessary infrastructure for the extraction of gas resources in the area of final gas consumption must be developed, at the same time crucial strategies should be used that lead to improving consumption patterns.

Author Contributions: Conceptualization, I.M.-K. and S.G.; methodology, S.G. and I.M-K.; software, I.G.S.; validation, I.M.-K. and V.D., formal analysis, I.M.-K.; investigation, I.M.-K. and S.G.; resources, V.D.; writing—original draft preparation, S.G. and I.G.S.; writing—review and editing, I.G.S.; visualization, S.G. and V.D.; supervision, I.M.-K. All authors have read and agreed to the published version of the manuscript.

Funding: This research received no external funding.

Institutional Review Board Statement: Not applicable.

Informed Consent Statement: Not applicable.

Data Availability Statement: The data of this study is available from the authors upon request.

Conflicts of Interest: The authors declare no conflict of interest.

References

1. Economides, J.; Wood, A. The state of natural gas. *J. Nat. Gas Sci. Eng.* **2009**, *1*, 1–13. [CrossRef]
2. BP. Statistical Review of World Energy, London. Available online: https://www.bp.com (accessed on 3 October 2020).
3. Rowse, J. On Hyperbolic Discounting in Energy Models: An Application to Natural Gas Allocation in Canada. *Energy J.* **2008**, *29*, 135–158. [CrossRef]
4. Rowse, J. Allocation of Canadian Natural Gas to Domestic and Export Markets. *Can. J. Econ.* **1986**, *19*, 417–442. [CrossRef]
5. Khotanzad, A.; Elragal, H.; Lu, T.-L. Combination of artificial neural-network forecasters for prediction of natural gas consumption. *IEEE Trans. Neural Netw.* **2000**, *11*, 464–473. [CrossRef]
6. Liu, L.-M.; Lin, M.-W. Forecasting residential consumption of natural gas using monthly and quarterly time series. *Int. J. Forecast.* **1991**, *7*, 3–16. [CrossRef]
7. Jayaraman, R.; Colapinto, C.; Liuzzi, D.; La Torre, D. Planning sustainable development through a scenario-based stochastic goal programming model. *Oper. Res.* **2016**, *17*, 789–805. [CrossRef]
8. Jayaraman, R.; Colapinto, C.; La Torre, D.; Malik, T. A Weighted Goal Programming model for planning sustainable development applied to Gulf Cooperation Council Countries. *Appl. Energy* **2017**, *185*, 1931–1939. [CrossRef]
9. Lima, J.P.; Barroso, L.A.; Granville, S.; Pereira, M.V.F.; Fampa, M.H.C. Computing leastcore allocations for firm-energy rights: A Mixed Integer Programming procedure. In Proceedings of the 2016 IEEE Power and Energy Society General Meeting (PESGM), Boston, MA, USA, 17–21 July 2016; pp. 1–5.
10. Liu, Y.; Yu, S.; Zhu, Y.; Wang, D.; Liu, J. Modeling, planning, application and management of energy systems for isolated areas: A review. *Renew. Sustain. Energy Rev.* **2018**, *82*, 460–470. [CrossRef]
11. Council, W.E. *World Energy Scenarios Composing Energy Futures to 2050*; PSI: London, UK, 2013.
12. Salvi, B.; Subramanian, K.; Panwar, N. Alternative fuels for transportation vehicles: A technical review. *Renew. Sustain. Energy Rev.* **2013**, *25*, 404–419. [CrossRef]
13. Romm, J. The car and fuel of the future. *Energy Policy* **2006**, *34*, 2609–2614. [CrossRef]
14. Arslan, R.; Ulusoy, Y.; Tekin, Y.; Surmen, A. An evaluation of the alternative transport fuel policies for Turkey. *Energy Policy* **2010**, *38*, 3030–3037. [CrossRef]
15. Baptista, P.C.; Silva, C.M.; Farias, T.L.; Heywood, J.B. Energy and environmental impacts of alternative pathways for the Portuguese road transportation sector. *Energy Policy* **2012**, *51*, 802–815. [CrossRef]
16. Sehatpour, M.-H.; Kazemi, A.; Sehatpour, H.-E. Evaluation of alternative fuels for light-duty vehicles in Iran using a multi-criteria approach. *Renew. Sustain. Energy Rev.* **2017**, *72*, 295–310. [CrossRef]
17. Santisirisomboon, J.; Limmeechokchai, B.; Chungpaibulpatana, S. Impacts of biomass power generation and CO_2 taxation on electricity generation expansion planning and environmental emissions. *Energy Policy* **2001**, *29*, 975–985. [CrossRef]
18. Nakicenovic, N.; Riahi, K. *Model Runs with MESSAGE in the Context of the Further Development of the Kyoto-Protocol*; German Advisory Council on Global Change: Berlin, Germany, 2003.
19. Lund, H. *Advanced Energy Systems Analysis Computer Model Documentation Version 11.4*; Aalborg University: Aalborg, Denmark, 2014; pp. 1–188.
20. Mirjat, N.H.; Uqaili, M.A.; Harijan, K.; Das Valasai, G.; Shaikh, F.; Waris, M. A review of energy and power planning and policies of Pakistan. *Renew. Sustain. Energy Rev.* **2017**, *79*, 110–127. [CrossRef]
21. Strachan, N.; Kannan, R. Hybrid modelling of long-term carbon reduction scenarios for the UK. *Energy Econ.* **2008**, *30*, 2947–2963. [CrossRef]
22. Liu, Q.; Shi, M.; Jiang, K. New power generation technology options under the greenhouse gases mitigation scenario in China. *Energy Policy* **2009**, *37*, 2440–2449. [CrossRef]
23. Ball, M.; Wietschel, M.; Rentz, O. Integration of a hydrogen economy into the German energy system: An optimising modelling approach. *Int. J. Hydrog. Energy* **2007**, *32*, 1355–1368. [CrossRef]
24. Chiodi, A.; Deane, P.; Gargiulo, M.; Gallachóir, B.Ó. The Role of Bioenergy in Ireland's Low Carbon Future–Is it Sustainable? *J. Sustain. Dev. Energy Water Environ. Syst.* **2015**, *3*, 196–216. [CrossRef]
25. Tavakoli, M.; Shokridehaki, F.; Marzband, M.; Godina, R.; Pouresmaeil, E. A two stage hierarchical control approach for the optimal energy management in commercial building microgrids based on local wind power and PEVs. *Sustain. Cities Soc.* **2018**, *41*, 332–340. [CrossRef]
26. Valinejad, J.; Marzband, M.; Funsho Akorede, M.; D Elliott, I.; Godina, R.; Matias, J.; Pouresmaeil, E. Long-term decision on wind investment with considering different load ranges of power plant for sustainable electricity energy market. *Sustainability* **2018**, *10*, 3811. [CrossRef]
27. Jayaraman, R.; La Torre, D.; Malik, T.; Pearson, Y.E. Optimal work force allocation for energy, economic and en-vironmental sustainability in the United Arab Emirates: A goal programming approach. *Energy Procedia* **2015**, *75*, 2999–3006. [CrossRef]
28. Kumar, A.; Sah, B.; Singh, A.R.; Deng, Y.; He, X.; Kumar, P.; Bansal, R. A review of multi criteria decision making (MCDM) towards sustainable renewable energy development. *Renew. Sustain. Energy Rev.* **2017**, *69*, 596–609. [CrossRef]
29. Zografidou, E.; Petridis, K.; Petridis, N.E.; Arabatzis, G. A financial approach to renewable energy production in Greece using goal programming. *Renew. Energy* **2017**, *108*, 37–51. [CrossRef]
30. Kumar, D.; Rahman, Z.; Chan, F.T.S. A fuzzy AHP and fuzzy multi-objective linear programming model for order allocation in a sustainable supply chain: A case study. *Int. J. Comput. Integr. Manuf.* **2017**, *30*, 535–551. [CrossRef]

31. Zografidou, E.; Petridis, K.; Arabatzis, G.; Dey, P.K. Optimal design of the renewable energy map of Greece using weighted goal-programming and data envelopment analysis. *Comput. Oper. Res.* **2016**, *66*, 313–326. [CrossRef]
32. Sharma, D.; Jana, R.; Gaur, A. Fuzzy goal programming for agricultural land allocation problems. *Yugosl. J. Oper. Res.* **2007**, *17*, 31–42. [CrossRef]
33. Ren, C.F.; Li, R.H.; Zhang, L.D.; Guo, P. Multiobjective Stochastic Fractional Goal Programming Model for Water Resources Optimal Allocation among Industries. *J. Water Resour. Plan. Manag.* **2016**, *142*, 04016036. [CrossRef]
34. Flisberg, P.; Frisk, M.; Rönnqvist, M.; Guajardo, M. Potential savings and cost allocations for forest fuel transportation in Sweden: A country-wide study. *Energy* **2015**, *85*, 353–365. [CrossRef]
35. Mekonnen, D.; Bryan, E.; Alemu, T.; Ringler, C. Food versus fuel: Examining tradeoffs in the allocation of biomass energy sources to domestic and productive uses in Ethiopia. *Agric. Econ.* **2017**, *48*, 425–435. [CrossRef]
36. Chong, C.; Liu, P.; Ma, L.; Li, Z.; Ni, W.; Li, X.; Song, S. LMDI decomposition of energy consumption in Guangdong Province, China, based on an energy allocation diagram. *Energy* **2017**, *133*, 525–544. [CrossRef]
37. Atabaki, M.S.; Aryanpur, V. Multi-objective optimization for sustainable development of the power sector: An economic, environmental, and social analysis of Iran. *Energy* **2018**, *161*, 493–507. [CrossRef]
38. Adnan, N.; Nordin, S.M.; Rahman, I.; Amini, M.H. A market modeling review study on predicting Malaysian consumer behavior towards widespread adoption of PHEV/EV. *Environ. Sci. Pollut. Res.* **2017**, *24*, 17955–17975. [CrossRef]
39. Agrawal, V.; Shenoy, U.V. Unified conceptual approach to targeting and design of water and hydrogen networks. *AIChE J.* **2006**, *52*, 1071–1082. [CrossRef]
40. Ehsan, A.; Yang, Q. Optimal integration and planning of renewable distributed generation in the power distribution networks: A review of analytical techniques. *Appl. Energy* **2018**, *210*, 44–59. [CrossRef]
41. Vega-Garita, V.; Harsarapama, A.P.; Ramirez-Elizondo, L.; Bauer, P. Physical integration of PV-battery system: Advantages, challenges, and thermal model. In Proceedings of the 2016 IEEE International Energy Conference (ENERGYCON), Leuven, Belgium, 4–8 April 2016; pp. 1–6.
42. Calpa, M.; Castillo-Cagigal, M.; Matallanas, E.; Caamaño-Martín, E.; Gutiérrez, Á. Effects of Large-scale PV Self-consumption on the Aggregated Consumption. *Procedia Comput. Sci.* **2016**, *83*, 816–823. [CrossRef]
43. Ja'fari Samimi, A.; Dehghani, T. Application of Dynamic Model for Optimizing Iranian Gas Reserves. *Q. J. Econ. Res.* **2007**, *30*, 165–192.
44. Renani, M.; Sharifi, A.; Khosh Akhlagh, R.; Dinmohammadi, M. Determining the Priorities for Using Iranian Gas Reserves (Study Period 2006–2028). *Q. J. Econ. Res.* **2009**, *9*, 151–182.
45. Mohaghar, A.; Mehregan, M.; Abolhasani, G. Application of Decision-making Techniques to Prioritize Consumption and Optimal Allocation of Natural Gas with the Fuzzy Approach. *J. Energy Econ.* **2010**, *7*, 91–119.
46. Ng, D.W.K.; Lo, E.S.; Schober, R. Energy-Efficient Resource Allocation in OFDMA Systems with Hybrid Energy Harvesting Base Station. *IEEE Trans. Wirel. Commun.* **2013**, *12*, 3412–3427. [CrossRef]
47. Hutagalung, A.M. The Economic Value of Indonesia's Natural Gas: A Quantitative Assessment of Rhree Gas Policies. Ph.D. Thesis, University of Twente, Enschede, The Netherlands, 2014.
48. Orlov, A. An assessment of optimal gas pricing in Russia: A CGE approach. *Energy Econ.* **2015**, *49*, 492–506. [CrossRef]
49. Zhang, W.; Yang, J.; Zhang, Z.; Shackman, J.D. Natural gas price effects in China based on the CGE model. *J. Clean. Prod.* **2017**, *147*, 497–505. [CrossRef]
50. Daneshzand, F.; Amin-Naseri, M.R.; Elkamel, A.; Fowler, M.W. A System Dynamics Model for Analyzing Future Natural Gas Supply and Demand. *Ind. Eng. Chem. Res.* **2018**, *57*, 11061–11075. [CrossRef]
51. Daneshzand, F.; Amin-Naseri, M.R.; Asali, M.; Elkamel, A.; Fowler, M. A system dynamics model for optimal al-location of natural gas to various demand sectors. *Comput. Chem. Eng.* **2019**, *128*, 88–105. [CrossRef]
52. Alavi, S.E.; Lotfalipour, M.R.; Falahi, M.A.; Effati, S. The Optimal Allocation of Iran's Natural Gas. *Iran. Econ. Rev.* **2020**, *24*, 75–98.
53. Hashemi pour Rafsanjani, H.; rajabi Kalcateh, Y. Optimization of Allocating Gas to One of the Oil Field of Iran for Lifting by Gas Thesis. In *Sanctions and Export Deflection: Evidence from Iran*; Paris School of Economics: Paris, France, 2011.
54. Ali-Khani, A.; Rshidi Komeijani, A. Random Planning Model for Allocating the Gas Resources in Iran with the Ap-proach of Energy Security Cost. *Manag. Future Stud.* **2012**, *23*, 25–36.
55. Kazemi, A.; Shakouri, H.; Shakiba, S.; Hossein Zadeh, M. Selecting the Appropriate Model for Allocation of Energy in Iran by Using the Process of Hieratical Analysis. *Publ. Iran Energy* **2013**, *16*, 31–61.
56. Chedid, R.; Mezher, T.; Jarrouche, C. A fuzzy programming approach to energy resource allocation. *Int. J. Energy Res.* **1999**, *23*, 303–317. [CrossRef]
57. Borges, A.R.; Antunes, C.H. A fuzzy multiple objective decision support model for energy-economy planning. *Eur. J. Oper. Res.* **2003**, *145*, 304–316. [CrossRef]
58. Qardaji, W.; Yang, W.; Li, N. Understanding hierarchical methods for differentially private histograms. *Proc. VLDB Endow.* **2013**, *6*, 1954–1965. [CrossRef]
59. Maroufmashat, A.; Sattari, S. Estimation of natural gas optimum allocation to consuming sectors in year 2025 in Iran. *Energy Sources Part B Econ. Plan. Policy* **2016**, *11*, 587–596. [CrossRef]
60. Alikhani, R.; Azar, A. A hybrid fuzzy satisfying optimization model for sustainable gas resources allocation. *J. Clean. Prod.* **2015**, *107*, 353–365. [CrossRef]

61. Wang, T.; Lin, B. China's natural gas consumption and subsidies—From a sector perspective. *Energy Policy* **2014**, *65*, 541–551. [CrossRef]
62. Dagher, L. Natural gas demand at the utility level: An application of dynamic elasticities. *Energy Econ.* **2012**, *34*, 961–969. [CrossRef]
63. Shahbaz, M.; Arouri, M.; Teulon, F. Short- and long-run relationships between natural gas consumption and economic growth: Evidence from Pakistan. *Econ. Model.* **2014**, *41*, 219–226. [CrossRef]
64. Orlov, A. Distributional effects of higher natural gas prices in Russia. *Energy Policy* **2017**, *109*, 590–600. [CrossRef]
65. Dayo, F.B.; Adegbulugbe, A.O. Utilization of Nigerian natural gas resources: Potentials and opportunities. *Energy Policy* **1988**, *16*, 122–130. [CrossRef]
66. Boucher, J.; Smeers, Y. Alternative Models of Restructured Electricity Systems, Part 1: No Market Power. *Oper. Res.* **2001**, *49*, 821–838. [CrossRef]
67. Orlov, A. An assessment of proposed energy resource tax reform in Russia: A static general equilibrium analysis. *Energy Econ.* **2015**, *50*, 251–263. [CrossRef]
68. Zhu, X.; Yi, J.; Chen, S.; Jiang, H. A multi-objective algorithm for task scheduling and resource allocation in cloud-based disassembly. *J. Manuf. Syst.* **2016**, *41*, 239–255.
69. Gazijahani, F.S.; Salehi, J. Stochastic multi-objective framework for optimal dynamic planning of interconnected microgrids. *IET Renew. Power Gener.* **2017**, *11*, 1749–1759. [CrossRef]

sustainability

MDPI

Article

Harnessing Wind Energy Potential in ASEAN: Modelling and Policy Implications

Youngho Chang [1,*] and Han Phoumin [2]

1 School of Business, Singapore University of Social Sciences, Singapore 599494, Singapore
2 Economic Research Institute for ASEAN and East Asia (ERIA), Jakarta 10270, Indonesia; han.phoumin@eria.org
* Correspondence: yhchang@suss.edu.sg

Abstract: This study examines whether and how harnessing more wind energy can decrease the cost of meeting the demand for electricity and amount of carbon emissions in the Association for Southeast Asian Nations (ASEAN) region, using the ASEAN integrated electricity trade model. Three scenarios are considered: a counterfactual business-as-usual (BAU) scenario, which assumes no wind energy is used; an actual BAU scenario that uses the wind-generation capacity in 2018; and a REmap scenario, which employs the wind-generation capacity from the Renewable Energy Outlook for ASEAN. Simulation results suggest that dispatching more wind energy decreases the cost of meeting the demand for electricity and amount of carbon emissions. However, these emissions increase during the late years of the study period, as the no- or low-emitting energy-generation technologies are crowded out.

Keywords: wind energy; power trade; counterfactual scenario; ASEAN

JEL Classification: Q41; Q42

Citation: Chang, Y.; Phoumin, H. Harnessing Wind Energy Potential in ASEAN: Modelling and Policy Implications. *Sustainability* **2021**, *13*, 4279. https://doi.org/10.3390/su13084279

Academic Editor: Tomonobu Senjyu

Received: 14 January 2021
Accepted: 24 March 2021
Published: 12 April 2021

1. Introduction

Wind energy can be considered the most promising renewable source for generating electricity. Currently, about 5.30 percent of the world's electricity is generated by wind power; 1429.6 terawatt-hours (TWh), of the 27,004.7 TWh of electricity generated in 2019, came from wind energy [1].

Table 1 shows the amount of the electricity generated in 2019. Coal has the largest share, followed by natural gas and hydroelectric power.

Among the electricity generated from renewable energy sources, wind energy has the largest share. Table 2 shows the amount of electricity generated by renewable energy in 2019. Wind covered slightly more than 50 percent of electricity generated by renewable energy.

For electricity generated from renewable sources, hydropower is the mode most utilised in the Association of Southeast Asian Nations (ASEAN) region followed by geothermal energy and solid biofuels. Wind energy comprised a very small share of the renewable energy in the region [2]. Similarly, hydropower had most of the installed capacity of renewable energy in the ASEAN region, and the capacity of wind generation was quite low [2].

ASEAN member countries have massive wind energy potential, however [2]. Across the region, there are many suitable sites where the speed of wind is ideal for harnessing electricity. Harnessing energy from wind can help provide clean energy at affordable prices and reduce carbon emissions. Yet, the potential utilisation rates are not great due to the intermittency of electricity generated from wind, a relatively high levelized cost of electricity (LCOE), and high balance-of-system costs. Financing renewable energy projects, including wind farms, is also a key barrier to improve the utilisation rate of the renewable potential [3].

Table 1. Electricity Generation by Fuels for the World. (Terawatt-hours).

	Oil	Natural Gas	Coal	Nuclear	Hydro	Renewables	Others	Total
Electricity	825.3	6297.9	9824.1	2796.0	4222.2	2805.5	233.6	27,004.7
Share (%)	3.06	23.32	36.38	10.35	15.64	10.39	0.86	100.0

Note: 'Others' comprises sources not specified elsewhere. Source: [1].

Table 2. Renewable Electricity Generation in the World. (Terawatt-hours).

	Wind	Solar	Others	Total
Electricity	1429.6	724.1	651.8	2805.5
Share (%)	50.96	25.81	23.23	100.0

Note: Others include geothermal, biomass, and other sources of renewable energy not already itemised. Source: [1].

Among ASEAN countries, Viet Nam has good sources of wind energy. Its current share of wind energy in its power generation mix in 2020 was 1.7 percent, lower than that of solar energy (12.8 percent). The potential of offshore wind energy there is 261 gigawatts (GW) (fixed) and 214 GW (floating). Fourteen offshore wind projects have been proposed, which total 28 GW [4]. Indeed, Viet Nam aims to install 12 GW to 15 GW of onshore wind energy and 10 GW to 12 GW of offshore wind energy by 2030 [5].

Some obstacles exist for Viet Nam's wind energy projects, however, especially offshore in terms of environmental, social, and technical constraints. The offshore sites include protected areas or essential habitats that house vulnerable marine species, birds, and bats. In addition, those sites include oil-related activities, energy and communications infrastructure, and aquaculture. They are commercial fishing grounds, tourism spots, and have great historical and cultural significance. To be fully utilised, they also must also clear technical constraints such as marine traffic, air traffic, and military use [4].

Using a cross-border power trade model in ASEAN [6], this study aims to demonstrate that renewable energy resources, especially wind energy, can help ensure energy sustainability and climate change adaptation. As a basis of evaluation for how wind energy can contribute to meet the electricity demand in ASEAN, it constructs a counterfactual business-as-usual (BAU) scenario in which no wind energy is used. Following this, an actual BAU scenario is used, using 2018 as the starting year. Finally, this study adopts a REmap scenario against which the counterfactual and actual BAU scenarios are evaluated to see how much wind energy can help meet the demand for electricity and reduce carbon emissions. An International Renewable Energy Agency (IRENA) study is used to show how renewable energy can contribute to the energy landscape in the ASEAN region using 2025 as a target year [7].

The second section reviews prospects of harnessing wind energy and factors dragging this objective. The third section presents principles of harnessing wind energy that constitute the basis of the simulation model, and the fourth section discusses the methodology of this study, its key assumptions, and data. The fifth section discusses results of this study, and the sixth section presents policy implications derived from the study.

2. Harnessing Potential Wind Energy

2.1. Prospects

Huge potential exists for global wind power [8]. It can create more than 40 times the current worldwide consumption of electricity and more than 5 times the total global use of energy in all forms [9]. Wind energy can also bring non-energy benefits, as utilisation does not affect global temperature but does reduce carbon emissions and other air pollutants [10].

Some new technologies are currently exploring ways of harnessing energy from wind. A system that combines wind energy and hydropower in which the excess electricity generated from the wind farm is used to pump water from a lower tank to a higher level, which was installed in the island of Ikaria in Greece, appears to be feasible for low-cost electricity production [11]. Navarre, a Spanish region, has exhibited how even small towns can become a big player in wind energy [12]. Some have also made efforts to harness energy from high-altitude wind where the speed of wind is faster and, hence, renders higher potential [13]. Moreover, power generated from offshore wind can be delivered via synoptic-scale interconnection, which appears to solve the underutilisation of wind power due to the fluctuation of electricity generated [14].

2.2. Drag Factors

Harnessing energy from renewable sources can have some negative environmental consequences. Indeed, the United Kingdom's Sustainable Development Commission was criticized for its failure to minimise the negative environmental consequences of wind energy such as noise, visual intrusion in sensitive landscapes, and bird strikes. The fair balancing of the advantages and disadvantages of harnessing wind energy in specific situations should be evaluated and after which wind energy should be utilised [15]. Wind farms also have a poor reputation; for example, it was reported that 40,000 birds in a year ran into wind turbine blades in the United States [16]. The modern type of wind turbine whose height is 125 m is almost as high as the London Eye, whose height is 135 m. After 16 years of litigation, relentless opposition from industrialists, and financial and political setbacks made a plan to build a wind farm in Cape Cod, Massachusetts, fail. The wind farm could have provided clean energy to 200,000 homes on Cape Cod and would have helped develop wind farms to nearby regions [17]. Financial viability also affects the development of wind energy, as, for example, the credit crunch drastically affected wind-energy projects in the United States during the Global Financial Crisis in 2008 [18].

In addition, the large-scale deployment of wind turbines appears to reduce wind speed and, in turn, lower turbine efficiency. The reduced wind speed eventually leads to set low generation limits [19].

Wind energy, especially onshore wind, is a mature technology that has achieved a certain level of reliability. However, the reliability, or load factor, is affected negatively by the age of the wind turbines. In the United Kingdom, the normalised load factor declined from about 24 percent during peak (i.e., age 1 year) to 15 percent at age 10 years, and 11 percent at age 15 years. The normalised load factor for Danish wind farms showed a similar decline—from 22 percent at age 1 year to 18 percent at age 15 years. Offshore Danish wind farms exhibited huge declines in their normalised load factors—from 39 percent at their peak to 15 percent at age 10 years [20].

2.3. Positive Signs of Harnessing Wind Energy

Wind turbines mounted on buildings appear to be feasible for reducing carbon emissions by contributing significantly to energy requirements in buildings. The aggregate electricity generated from these wind turbines range from 1.7 to 5.0 TWh per year and reduce carbon emissions by from 0.75 million to 2.5 million tons per year [21]. An energy company, Royal Dutch Shell, and an operator of oil tankers, Maersk, are also attempting to use wind power to cut tankers' fuel bills. They try to install the oil tanker with two 'rotor sails' to propel the vessel, which is one of other ideas of supplying energy to the oil tanker such as solar-powered sails and kites [22].

3. Principles of Wind Energy

3.1. Wind Energy as Kinetic Energy

Wind energy is kinetic energy that is transformed from potential energy. Scottish physicist William Rankine stated in 1881 that, "the object is gaining the potential to move 'by the occurrence of such changes, actual energy disappears and is replaced by Potential or Latent Energy'" [23]. Taking the definition of 'work' as the force multiplied by the distance moved in the direction of the force, the amount of energy harnessed from wind is determined by the speed of the wind and volume of air moved. When air-flow passes a wind turbine at a given speed, a moving turbine constructs a hypothetical cylinder with the swept area as the length of the wind blade and the height as the speed of wind per second. The hypothetical cylinder captures air mass, which is kinetic energy, and is eventually transformed into electricity.

3.2. Kinetic Energy in a Wind Turbine: Calculation

Suppose a wind turbine with a diameter of 60 metres and a radius of 30 metres and the wind speed (v) of 9 metres per second.

- Swept area (A) is π x $r^2 = \pi \times 30^2$
- Wind speed (v) is 9 metres per second (9 m/s)
- Volume of the cylinder (V) is $v \times A = 9 \times \pi \times 30^2$ = 25,447 cubic metres per second
- Density of air (the mass per cubic metre) is 1.29 kg per cubic metre
- Mass of air arriving per second (m) is 1.29 × 25,447 = 32,827 kg per second
- The kinetic energy of a mass m moving with speed v is $\frac{1}{2}\ mv^2 = \frac{1}{2} \times 32{,}827 \times 9^2$ = 1,329,494 joules per second = 1.33 megawatts (MW).

The principles of kinetic energy suggest that the longer the wind blade and the faster the wind speed, the more energy will be transformed from kinetic energy to electric energy (i.e., electricity). The modern type of wind turbine has a capacity of 1.8 MW [23].

3.3. Economic Considerations of Wind Energy

The cost of wind energy can be calculated as follows. This calculation is based on the information given in Boyle [23]:

The cost per unit (g) is expressed in Equation (1):

$$g = \frac{(C \times R)}{E} + M \tag{1}$$

where:

g = the cost per unit of electricity generated;
C = the capital cost of the wind farm;
R = the capital recovery factor or the annual capital charge rate (expressed as a fraction);
E = the wind farm annual energy output;
M = the cost of operating and maintaining the wind farm annual output.

The required annual rate of return net of inflation (R) is expressed as:

$$R = [x/(1 - (1 + x))^{-n}] \tag{2}$$

where:

x = the required annual rate of return net of inflation;
n = the number of years over which the investment in the wind farm is to be recovered.

The annual energy output of the wind farm (E) is expressed as:

$$E = (hP_rF)T \tag{3}$$

where:

h = the number of hours in a year (8760);
P_r = the rated power of each wind turbine in kilowatts;
F = the net annual capacity factor of the turbines at the site;
T = the number of turbines.

The cost of operating and maintaining the wind farm annual output (M) is expressed as:

$$M = KC/E \tag{4}$$

where:

M = the operation and maintenance costs;
K = the factor representing the annual operating costs of a wind farm as a fraction of the total capital cost.

Generally, a wind turbine operates at only around 25 percent of turbine capacity due to inconsistent, imperfect wind. On better land-based wind sites, a capacity factor of 35 percent to 40 percent or more is achievable [23]. A wind turbine is quick to install, so it will be generating power before significant interest on capital. It is competitive with conventional power generation at sufficiently windy sites.

3.4. Unit or Levelised Costs of Wind Energy

A typical wind turbine has three parts: fiberglass blades, a standard gearbox, and a generator. Boyle [23] described the cost of a 600-kilowatt (kW) wind turbine in Denmark. Installation costs are $1800 to $2200 per kW, the turbine lasts about 20 years, the load factor is 25 percent, and the turbine generates 1,314,000 kilowatt-hours (kWh) per year. If a real discount or interest rate is assumed at 10 percent, the installation cost is $2000 per kW, or about $1,200,000. The unit or LCOE are $0.106 per kWh.

Table 3 presents the cost of generating electric power by various sources. The data are taken from estimated generation costs in the United States in 2017 for a comparison purpose.

Table 3. Estimated Cost of Generating Electric Power, 2017.

Type	Cost (2010 $ per Megawatt-Hour)
Gas (all types) *	66.1–127.9
Hydro	88.9
Wind	96.0
Coal (all types) *	97.7–138.8
Geothermal	98.2
Advanced Nuclear	111.4
Biomass	115.4
Solar Photovoltaic	152.7
Solar Thermal	242.0

* Includes carbon capture and sequestration. Source: [24].

Wind energy appears to be competitive with gas and coal. Moreover, the cost of electricity generated from wind is even lower than that of geothermal, although hydro is lower than wind. The cost competitiveness of wind in terms of power generation is also confirmed by the latest cost data provided by IRENA (Table 4).

Table 4. Weighted Average LCOE of Renewable Power Generation Technologies (kilowatt-hours).

	Biomass	Geothermal	Hydro	Solar Photovoltaic	Concentrated Solar Power	Offshore Wind	Onshore Wind
2010	0.076	0.049	0.037	0.378	0.346	0.161	0.086
2019	0.066	0.073	0.047	0.068	0.182	0.115	0.053

LCOE = levelized cost of energy. Notes: 1. The LCOE is the weighted average LCOE from utility-scale renewable power generation technologies from 2010 to 2019. 2. The fossil fuel LCOE range: = is $0.05–$0.18 per kilowatt-hour. Source: [25].

The LCOEs of geothermal and hydropower slightly increased in 2019 compared to 2010. The LCOEs of solar photovoltaic and concentrated solar power decreased immensely, while the LCOEs of offshore and onshore wind energy fell a small amount. Among various renewable power technologies, however, the LCOE of onshore wind energy is the second lowest after hydro. The LCOE of fossil fuels ranges from about $0.05 per kWh to about $0.18 per kWh [25]. Except for concentrated solar power and offshore wind energy, all other renewable power-generation technologies have become competitive with fossil fuel power-generation technologies. The cost-competitiveness of wind energy is confirmed further if the cost of carbon disposal and the price of carbon are added to the LCOE.

Boyle [23] presented a comparison of the costs of various sources of electricity generation at a 10 percent discount rate. The cost included capital payments, operation and maintenance, fuel, carbon disposal, and carbon price. Fifteen power-generation technologies were considered: combined-cycle gas turbine, conventional coal, combined-cycle gas turbine with carbon capture and storage, coal with carbon capture and storage, nuclear-pressurised water reactor, roof-mounted solar photovoltaic thin-film panels, large biomass non-combined heat and power, run of river, reservoir hydro, onshore wind, offshore wind, tidal barrage, tidal stream, floating, and geothermal. The five lowest-cost technologies were run of river, reservoir hydro, combined-cycle gas turbine, onshore wind energy, and nuclear-pressurised water reactor. The LCOE of onshore wind is still higher than combined-cycle gas turbine. If a carbon price is added or the costs of carbon disposal for the combined-cycle gas turbine are included, then onshore wind energy is competitive with these technologies.

4. Methodology, Assumptions, and Data

This study explores how harnessing wind energy in the ASEAN region can reduce the cost of meeting the electricity demand and estimates the amount of carbon emissions that can be reduced.

4.1. Methodology

This study adopts the ASEAN integrated electricity grid model developed by Chang and Li [6] and modifies wind energy-related information. The objective of the integrated power trade model is to minimise the cost of meeting demand for electricity in the ASEAN region from 2018 to 2040. Costs have four components: capital cost, operation cost, transmission cost, and carbon cost. As it has an integrated electricity market and grid, power trade (i.e., the import of electricity) is allowed for up to 30 percent of domestic demand. Detailed description of the model is presented in Appendix A.

4.2. Assumptions

To meet domestic demand and trade surplus electricity, this study made some key assumptions. First, the total installed capacity of power generation in the region is greater or equal to the total demand for electricity in the region. Second, the total output of electricity generation in each country is constrained by the load factor of the installed capacity of all types of electricity generation in the county. Third, the electricity supply of all countries in the region to a certain country should be greater than or equal to the demand for electricity in that country. Fourth, the total supply of electricity from one country to all countries (including the country itself) in the region must be smaller or equal to the total available supply capacity of that country at a given time.

4.3. Data

This study updates the initial capacity given in Chang and Li [6] using the data taken from ASEAN Centre for Energy (ACE) [26] and International Renewable Energy Agency (IRENA) [25]. Figure 1 shows the initial installed capacity in ASEAN by plant type in 2018.

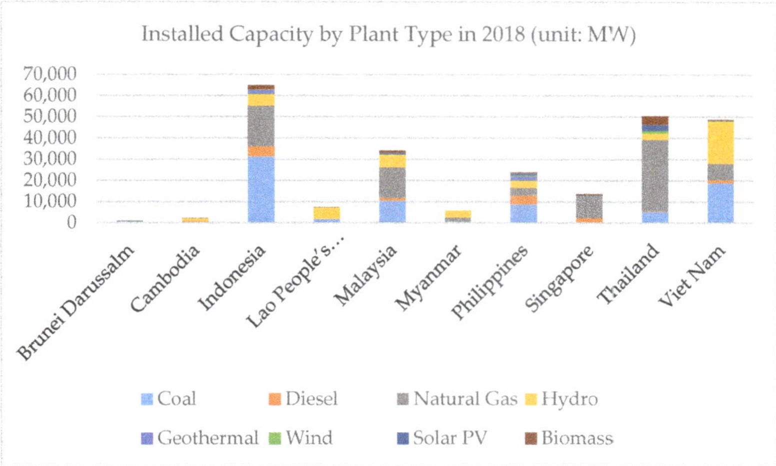

Figure 1. Installed Capacity by Plant Type in Association for Southeast Asian Nations (ASEAN), 2018. (Megawatts). PV = photovoltaic. Sources: [7] and [26].

4.4. Scenarios

This study establishes three scenarios: a counterfactual business-as-usual (BAU) scenario, an actual BAU scenario, and a REmap scenario. First, as the objective of this study is to estimate how much wind energy can help reduce the cost of meeting the electricity demand in the ASEAN region, it sets a counterfactual BAU scenario as a hypothetical base case. This assumes that no wind energy is used at all. In other words, there is no initial capacity of wind energy, and there is no added capacity of wind energy for the entire study period. This scenario presents the maximum possible contribution of wind energy to the cost of meeting the demand for electricity in the ASEAN region.

Second, an actual BAU scenario is set in 2018 in which the current initial capacity of wind energy is considered.

Third, a REmap scenario adopts the capacity of wind energy assumed in the REmap 2025 case in IRENA and ACE [7]. The REmap approach takes all available energy sources, including renewables, and considers energy supply and demand in power, heating, transport, and cooking. It aims to find a viable way of achieving the gap between the share of renewable energy under the reference case that is 17 percent and the target share of renewable energy for the region that is 23 percent. Full utilisation of potential wind energy is to be implemented in 2025 (Table 5).

Table 5. Expected Wind Capacity under REmap Scenario.

Country	Wind Capacity (Megawatts)	Remarks
Brunei Darussalam	0	
Cambodia	200	
Indonesia	2900	
Lao People's Democratic Republic	0	
Malaysia	100	
Myanmar	500	
Philippines	1100	

Table 5. *Cont.*

Country	Wind Capacity (Megawatts)	Remarks
Singapore	270	Offshore wind
Thailand	1800	
Viet Nam	5700	

Source: [7].

As stated previously, Viet Nam is expected to utilise its huge potential of wind energy and install the largest capacity of wind energy (5700 MW) among the 10 ASEAN countries. Indonesia is next at 2900 MW, and Thailand and the Philippines are in third and fourth with installed wind capacity of 1800 MW and 1100 MW, respectively.

5. Results, Discussions, and Policy Implications

5.1. No Wind Energy

The counterfactual BAU scenario presents the highest cost of meeting electricity demand in the ASEAN region and has the largest carbon emissions.

5.1.1. Cost of Electricity Generation in ASEAN Countries

When all capacities of wind energy are intentionally removed from the available technologies, three distinct trends emerge compared to the actual BAU case (Table 6). First, more low-cost technologies, such as hydropower, are used across many countries from 2026 to 2040. Second, renewable energy technologies, such as geothermal energy for Indonesia and the Philippines, are dispatched. Along with early utilisation of geothermal energy, more biofuel energy is utilised in Singapore. The Philippines appears to tap into biofuel energy as well. Third, more carbon-intensive and costly carbon-generation technologies, such as coal with carbon capture and storage and gas with carbon capture and storage, appear to be dispatched later in 2036 and 2040.

Table 6. Cost of Meeting Electricity Demand in the ASEAN Region ($ billion).

Scenarios	Cost	Difference
Counterfactual BAU	421.05	-
BAU	418.20	0.7%
REmap	409.36	2.8%

BAU = business as usual. Source: Authors.

When ASEAN countries utilise wind energy, however, the cost of meeting electricity demand in the region is lowered by about 0.7 percent. The share of wind energy, out of the total installed generation capacity in the ASEAN region, is about 0.8 percent. The cost of wind energy is almost the same as the share of installed generation capacity. Figure 2 presents the cost of meeting electricity demand in ASEAN countries.

The total cost of meeting the demand for electricity in the ASEAN region is $421.05 billion if no wind energy is utilised at all, i.e., the counterfactual BAU scenario. Under the BAU scenario in which the current level of wind energy is assumed, the total cost is $418.20 billion, about 0.7 percent lower than that of the counterfactual BAU scenario. The total cost of the counterfactual BAU scenario is $421.05 billion while that of the REmap scenario is $409.36 billion. The difference between the counterfactual scenario and the REmap scenario is 2.8 percent, which is more than three times the difference between the cost of the counterfactual scenario and BAU scenario, if the capacity of wind energy assumed under the REmap scenario of IRENA and ACE [7] is to be fully utilised from 2025.

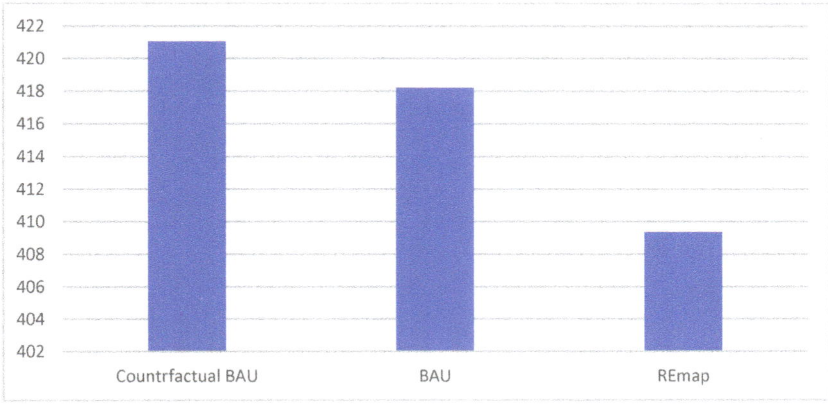

Figure 2. Total Cost of Meeting the Demand for Electricity in ASEAN ($ billion). BAU = business as usual. Source: Authors.

5.1.2. Carbon Emissions

The difference in carbon emissions between the counterfactual scenario and REmap scenario is interesting (Figure 3). The difference in the quantity ranges from 0.62 million tons in 2039 to 29.71 million tons in 2025, mostly because new capacity of wind energy is assumed to be installed in 2025. Excluding this, the next highest difference is achieved in 2028. The amount of carbon emissions under the counterfactual BAU scenario is slightly higher than the REmap scenario in 2038, probably due to the lower capacity of hydro, which is added in 2038.

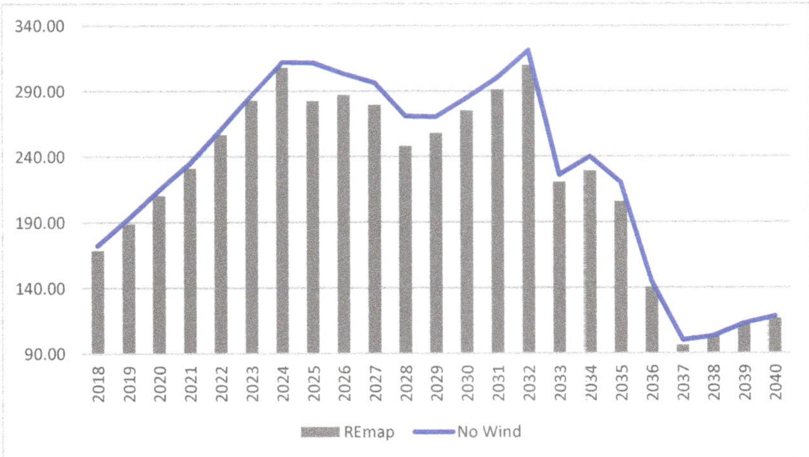

Figure 3. Trajectory of Carbon Emissions under Counterfactual BAU and Remap (million tons). BAU = business as usual. Source: Authors.

Thus, utilising more wind energy could reduce carbon emissions further. The simulation of the REmap scenario shows that a few countries in ASEAN, such as Brunei Darussalam, Malaysia, Singapore, and Thailand, appear to fully utilise their potential for wind energy. If other countries are able to harness their potential for wind energy, then the reduction in carbon emissions could be even larger.

5.2. Actual Business-as-Usual Scenario and REmap Scenario

A more realistic evaluation of how wind energy can reduce carbon emissions is shown by comparing the simulation results of the actual BAU scenario with those of the REmap scenario in which the full utilisation of potential for wind energy is expected to start from 2025. Figure 4 presents possible amount of carbon emissions reduced in the REmap scenario.

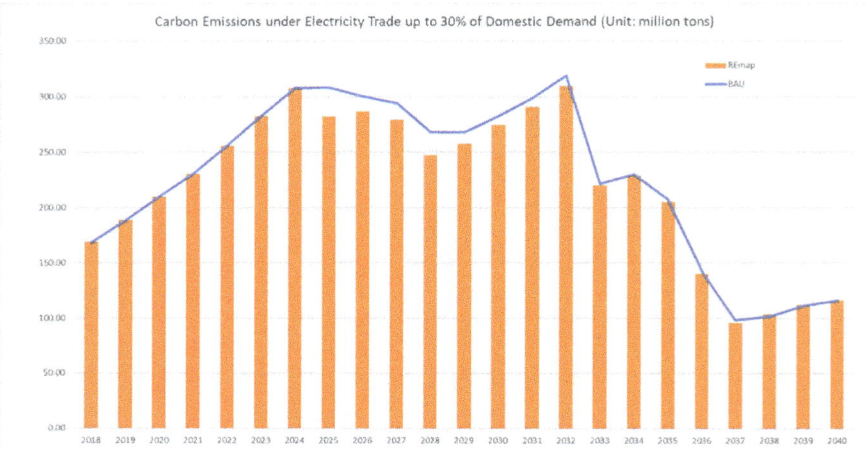

Figure 4. Reductions in Carbon Emissions under REmap Scenario (million tons). BAU = business as usual. Source: Authors.

The difference in the quantity of carbon emissions ranges from 1.44 million tons in 2034 to 26.22 million tons in 2025, mostly because new capacity of wind energy is assumed to be installed in 2025. Excluding this, the next highest difference is achieved in 2028. Carbon emissions under the REmap scenario appear to be higher than those under the actual BAU scenario during the last three years of the study period, caused by less hydro capacity during those years.

6. Conclusions and Policy Implications

ASEAN countries have good potential to harness wind energy, especially Viet Nam. Wind energy, however, is not commensurate with the degree of potential capacity. The intermittency of wind and high system costs are the main reasons for low development.

This study found that there would be 0.7 percent higher costs in meeting the demand for electricity in ASEAN countries if no wind energy was utilised. The costs of meeting the demand for electricity in ASEAN under the REmap scenario appear to be about 2.8 percent lower than that of the counterfactual scenario. As expected, the amount of carbon emissions from both the actual BAU scenario and the REmap scenario are lower than that of the counterfactual scenario, especially from 2025 when wind energy is extensively harnessed.

The trajectories of carbon emissions exhibit a visible gap between the counterfactual BAU scenario and REmap scenario from 2025 to 2032 and a lesser visible difference toward 2040. All three scenarios show that the level of carbon emissions would peak around the early 2030s when carbon-emitting power-generation technologies are more extensively dispatched to meet the increasing demand for electricity in the ASEAN region.

The REmap scenario shows that both the cost of meeting the demand for electricity and amount of carbon emissions decrease compared to the counterfactual BAU scenario and actual BAU scenario. However, the amount of carbon emissions appears to increase during later periods, as low- or no-carbon-emitting technology is crowded out. Considering the possible reverse in the trajectories of carbon emissions, whether the added capacity of wind energy will increase the amount of carbon emissions needs to be evaluated. If the reversal in the amount of carbon emissions appears to be the case, then such a case should not proceed.

This study draws a few policy implications from the findings presented above.

First, as shown in the REmap scenario, more wind capacity appears to accelerate the decreasing trend of carbon emissions. Wind energy should, thus, be promoted in ASEAN countries. As the cost of harnessing wind energy is expected to decrease further, more wind energy will lower the cost of meeting the electricity demand in ASEAN.

Second, the amount of carbon emissions could be larger when more wind capacity is dispatched, although the cost of meeting the demand for electricity will decrease. When a decision to add more wind capacity is made, a rigorous evaluation should proceed to determine whether the wind capacity will crowd out no- or low-carbon-emitting technologies, such as hydro, and eventually increase carbon emissions in the long term.

Third, harnessing more viable renewable energy power-generation technologies in the ASEAN region could decrease the level of carbon emissions. It is uncertain, however, if dispatching more of such technologies would decrease the costs of meeting the demand for electricity. ASEAN countries need to decrease the costs of renewable energy power-generation technologies, therefore, through more research and development.

Harnessing renewable energy power-generation technologies is not immune from damaging the environment and can have negative repercussions on the economy, as identified in Viet Nam's development of offshore wind energy. Thus, ASEAN must evaluate possible negative impacts of harnessing renewable energy on the environment and economy.

This study can be improved with more detailed data such as country-specific peak and offpeak demand. Such data can make the simulation study produce more realistic results. The results, in turn, will present more effective and relevant policy implications.

Author Contributions: Conceptualization, Y.C. and H.P.; methodology, Y.C. and H.P.; software, Y.C.; validation, Y.C. and H.P.; formal analysis, Y.C.; investigation, Y.C.; resources, Y.C.; data curation, Y.C.; writing—original draft preparation, Y.C.; writing—review and editing, Y.C. and H.P.; visualization, Y.C.; supervision, Y.C. and H.P.; project administration, Y.C. and H.P.; funding acquisition, H.P. All authors have read and agreed to the published version of the manuscript.

Funding: This research was partially funded by Economic Research Institute for ASEAN and East Asia (ERIA).

Institutional Review Board Statement: Not applicable.

Informed Consent Statement: Not applicable.

Data Availability Statement: Not applicable.

Conflicts of Interest: The authors declare no conflict of interest.

Appendix A

Appendix A.1. Model Description

This study adopts a dynamic linear programming framework in power generation first developed by Turvey and Anderson [27] and later adapted by Chang and Tay [28] and Chang and Li [6]. In the latest study by Chang and Li [6], significant extensions of the original models were made. A new country dimension was added to allow an international framework with cross-border electricity trade. The new model also added the cost of cross-border power transmission as well as transmission loss into account. Carbon emissions from power generation as well as the carbon cost of power generation were explicitly considered. The model was solved using General Algebraic Modelling System (GAMS).

This section documents how the capital expenditure (CAPEX) and operational expenditure (OPEX) of a certain type of power generation is represented in the model and how carbon emissions and cross-border transmission cost are explicitly represented in the model. Following this, it presents the objective function and various constraints to make the power trade model work.

Appendix A.2. CAPEX

The capital expenditure (CAPEX) of a certain type of power generation capacity at a certain point of time is modelled as follows. The total capital cost of a certain type of power generation capacity during the period of this study is expressed as $\sum_{i=1}^{I}\sum_{v=1}^{T}\sum_{m=1}^{M} c_{miv} * x_{miv}$. x_{miv} is the capacity of plant type m, vintage v, in country i. Vintage indicates the time a certain type of capacity is built and put into use. c_{miv} is the corresponding capital cost per unit of capacity of the power plant. For simulation purpose and consistency in presentation with the other cost terms, a time dimension to the equation besides the vintage dimension is added. This allows that the capital cost is amortized using a capital recovery factor.

Appendix A.3. OPEX

The operational expenditure (OPEX) of a certain type of power generation capacity at a certain point of time is modelled as follows. The operation cost of a certain type of power generation capacity in year t is expressed as $\text{Opex}(t) = \sum_{i=1}^{I}\sum_{j}^{J}\sum_{v=-V}^{t}\sum_{p=1}^{P}\sum_{m=1}^{M} F_{mitv} * u_{mijtvp} * \theta_{jp}$. u_{mijtvp} is the power output of plant m, vintage v, in year t, country i, block p on the load, and exported to country j. F_{mitv} is the corresponding operating cost that varies with v, and θ_{jp} is the time interval of load block p within each year in the destination country.

Appendix A.4. Carbon Emissions

Carbon emissions of different types/technologies of power generation capacity are modelled as follows and the cost of carbon emissions is explicitly considered. The amount of carbon emissions produced is expressed as $\sum_{m=1}^{M}\sum_{i=1}^{I}\sum_{j=1}^{J}\sum_{v=-V}^{T} u_{mijtvp} * \theta_{jp} * ce_m$, and the carbon cost in year t is $CC(t) = cp_t * \left(\sum_{m=1}^{M}\sum_{i=1}^{I}\sum_{j=1}^{J}\sum_{v=-V}^{T} u_{mijtvp} * \theta_{jp} * ce_m\right)$. ce_m is the carbon emissions per unit of power plant capacity of type j plant, and cp_t is the carbon price per unit of carbon emissions in year t.

Appendix A.5. Cross-Border Transmission Cost

The costs of cross-border transmission come in two forms—a tariff and transmission loss. The tariff is paid to recover the capital investment and operational cost of the grid line. The transmission loss could be significant if the distance of transmission is long and is explicitly considered in Equation (4). The tariff of transmission, $tp_{i,j}$, is the unit MWh transmission cost of power output from country i to country j. The total cost of cross-border power transmission in year t, using $tp_{i,j}$, is expressed as $TC(t) = \sum_{i=1}^{I}\sum_{j=1}^{J}\sum_{v=-V}^{T}\sum_{p=1}^{P} u_{mijtvp} * \theta_{jp} * tp_{i,j}$.

Appendix A.6. Objective Function

The objective of the power trade model is to minimize the total cost of electricity during the period of this study. The objective function is written as:

$$\text{obj} = \sum_{i=1}^{I}\sum_{v=1}^{T}\sum_{m=1}^{M} c_{miv} * x_{miv} + \sum_{t=1}^{T}\{\text{Opex}(t) + CC(t) + TC(t)\} \quad \text{(A1)}$$

Appendix A.7. Constraint Conditions

There are several constraints that are required to optimize the above objective function.

Equation (A2) shows a first set of constraints, which requires total power capacity to meet total power demand in the region. Q_{itp} is the power demand of country i in year t for load block p.

$$\sum_{i=1}^{I}\sum_{j=1}^{J}\sum_{m=1}^{M}\sum_{v=-V}^{t} u_{mijtvp} \geq \sum_{i=1}^{I} Q_{itp} \tag{A2}$$

The second one, shown in Equation (A3), states the constraint of load factor lf_{mi} of each installed capacity of power generation. kit_{mi} is the initial vintage capacity of type m power plant in country i.

$$u_{mijtvp} \leq lf_{mi} * (kit_{mi} + x_{miv}) \tag{A3}$$

The third constraint, shown in Equation (A4), says that power supply of all countries to a certain country must be greater than the country's power demand. $tl_{i,j}$ is the ratio of transmission loss in cross-border electricity trade between country i and country j.

$$\sum_{j=1}^{J}\sum_{m=1}^{M}\sum_{v=-V}^{t} u_{mijtvp} \cdot tl_{ij} \geq Q_{itp} \tag{A4}$$

The fourth constraint, shown in Equation (A5), states that total supply of power of one country to all countries (including itself) must be smaller than the summation of the country's available power capacity at the time.

$$\sum_{j=1}^{J} u_{mijtvp} \leq \sum_{m=1}^{M}\sum_{v=-V}^{t} lf_{mi} * (kit_{mi} + x_{miv}) \tag{A5}$$

The fifth constraint, shown in Equation (A6), is capacity reserve constraint. pr is the rate of reserve capacity as required by regulation. Moreover, $p = 1$ represents the peak load block.

$$\sum_{i}^{I}\sum_{m=1}^{M}\sum_{v=-V}^{t} lf_{mi} * (kit_{mi} + x_{miv}) \geq (1 + pr) * \sum_{i}^{I} Q_{it,p=1} \tag{A6}$$

Hydro-facilities have the so-called an energy factor constraint as shown in Equation (A7). ef_{mi} is the energy factor of plant type m in country i. Other facilities have $ef = 1$.

$$\sum_{p=1}^{P}\sum_{j=1}^{J} u_{mijtvp} \leq ef_{mi} * (kit_{mi} + x_{miv}) \tag{A7}$$

Lastly, development of power generation capacity faces resource availability constraint, which is shown in Equation (A8). $XMAX_{mi}$ is the type of resource constraint of plant type m in country i.

$$\sum_{v=1}^{T} x_{miv} \leq XMAX_{mi} \tag{A8}$$

References

1. BP. *BP Statistical Review of World Energy 2020*; BP: London, UK, 2020.
2. United Nations Economic and Social Commission for Asia and the Pacific (UNESCAP). *Statistical Perspectives Energy and Development in the ASEAN Region*; UNESCAP: Bangkok, Thailand, 2019.
3. Blazquez, J.; Galeotti, M.; Martin-Moreno, J.M. An Empirical Study on the Drivers of Financial Leverage of Spanish Wind Farms. In *OIES Papers, No. EL-40*; Oxford Institute for Energy Studies: Oxford, UK, 2020.
4. Ngo, T.N. Power System and Recent Trends in Wind Energy Development in Vietnam. Singapore International Energy Week: Singapore. Available online: https://static.agora-energiewende.de/fileadmin/user_upload/VIET_presentation_SIEW_2020_final.pdf (accessed on 29 October 2020).
5. Minh, H.; Teske, S.; Pescia, D.; Pujantoto, M. Options for Wind Power in Vietnam by 2030. 2020. Available online: https://hal-enpc.archives-ouvertes.fr/hal-02329698v2/document (accessed on 29 October 2020).

6. Chang, Y.; Li, Y. Power Generation and Cross-Border Grid Planning for the Integrated ASEAN Electricity Market: A Dynamic Linear Programming Model. *Energy Strategy Rev.* **2013**, *2*, 153–160. [CrossRef]
7. International Renewable Energy Agency (IRENA); ASEAN Centre for Energy (ACE). *Renewable Energy Outlook for ASEAN*; IRENA: Abu Dhabi, United Arab Emirates, 2019.
8. Marris, E. Snapshot: Global Wind Power. *Nature* **2008**, *454*, 264. [CrossRef]
9. Lu, X.; McElroy, M.B.; Kiviluoma, J. Global Potential for Wind-Generated Electricity. *Proc. Natl. Acad. Sci. USA* **2009**, *106*, 10933–10938. [CrossRef] [PubMed]
10. Keith, D.W.; DeCarolis, J.F.; Denkenberger, D.C.; Lenshow, D.H.; Malyshev, S.L.; Pacala, S.; Rasch, P.J. The Influence of Large-Scale Wind Power on Global Climate. *Proc. Natl. Acad. Sci. USA* **2004**, *101*, 16115–16120. [CrossRef] [PubMed]
11. Bakos, G.C. Feasibility Study of a Hybrid Wind/Hydro Power-System for Low-Cost Electricity Production. *Appl. Energy* **2002**, *72*, 599–608. [CrossRef]
12. Fairless, D. Energy-Go-Round: How Did a Little Spanish Province Become One of the World's Wind-Energy Giants? *Nature* **2007**, *447*, 1046–1048. [CrossRef] [PubMed]
13. Vance, E. High Hopes. *Nature* **2009**, *460*, 564–566. [CrossRef]
14. Kempton, W.; Pimenta, F.M.; Veron, D.E.; Colle, B.A. Electric Power from Offshore Wind via Synoptic-Scale Interconnection. *Proc. Natl. Acad. Sci. USA* **2010**, *107*, 7240–7245. [CrossRef] [PubMed]
15. Keay, M. Wind Power in the UK: Has the Sustainable Development Commission Got It Right. In *Oxford Energy Comment May 2005*; Oxford Institute for Energy Studies: Oxford, UK, 2005.
16. Marris, E.; Fairless, D. 'Wind Farms' Deadly Reputation Hard to Shift. *Nature* **2007**, *447*, 126. [CrossRef] [PubMed]
17. Seelye, K.Q. After 16 Years, Hopes for Cape Cod Wind Farm Float Away. *New York Times*, 19 December 2017.
18. Schiermeier, Q. Credit Crunch Threatens US Wind-Energy Projects. *Nature* **2008**, *455*, 572–573.
19. Miller, L.M.; Kleidon, A. Wind Speed Reductions by Large-Scale Wind Turbine Deployments Lower Turbine Efficiencies and Set Low Generation Limits. *Proc. Natl. Acad. Sci. USA* **2016**, *113*, 13570–13575. [CrossRef] [PubMed]
20. Hughes, G. *The Performance of Wind Farms in the United Kingdom and Denmark*; Renewable Energy Foundation: London, UK, 2012.
21. Dutton, A.G.; Halliday, J.A.; Blanch, M.J. *The Feasibility of Building-Mounted/Integrated Wind Turbines (BUWTs): Achieving Their Potential for Carbon Emission Reductions*; Final Report; Carbon Trust: London, UK, 2005.
22. Clark, P. Shipping Giants Trial Use of Wind-Power to Cut Tankers' Fuel Bills. *Financial Times*, 14 March 2017.
23. Boyle, G. *Renewable Energy: Power for a Sustainable Future*; Oxford University Press: Oxford, UK, 2012.
24. Dahl, C.A. *International Energy Markets: Understanding Pricing, Policies, and Profits*; PennWell: Tulsa, OK, USA, 2015.
25. International Renewable Energy Agency (IRENA). *Renewable Power Generation Costs in 2019*; IRENA: Abu Dhabi, United Arab Emirates, 2020.
26. ASEAN Centre for Energy (ACE). 2020. Available online: https://aseanenergy.org/ (accessed on 20 November 2020).
27. Turvey, R.; Anderson, D. *Electricity Economics: Essays and Case Studies*; The Johns Hopkins University Press: Baltimore, MD, USA, 1977.
28. Chang, Y.; Tay, T. Efficiency and Deregulation of the Electricity Market in Singapore. *Energy Policy* **2006**, *34*, 2498–2508. [CrossRef]

sustainability

MDPI

Article

Environmental Efficiency Analysis for Multi Plants Production Technologies

Mojtaba Ghiyasi [1,*] and Farhad Taghizadeh-Hesary [2,*]

1 Faculty of Industrial Engineering and Management Science, Shahrood University of Technology, Shahrood 3614773955, Iran
2 Social Science Research Institute, Tokai University, 4-1-1 Kitakaname, Hiratsuka-shi 259-1292, Kanagawa, Japan
* Correspondence: mog@shahoodut.ac.ir (M.G.); farhad@tsc.u-tokai.ac.jp (F.T.-H.)

Abstract: The current article extends the literature by proposing new models for estimating the classical and environmental performance of multi-plant firms. This yields some new indices for capturing the environmental performance vs. classical economic performance at the local and global level. The proposed approaches and indices were applied for the economic and environmental performance assessment of 46 power plants in Iran. The primary result emphasizes considering not only local environmental performance but also global performance to have a broad insight of environmental performance assessments. Moreover, we find only a few power plants with a resistant environmental performance at the global level. Proposed models in this article are general because they can be utilized in environmental analysis of any multiple plant production units.

Keywords: DEA; multi plant firms; environmental assessment; local-global performance

Citation: Ghiyasi, M.; Taghizadeh-Hesary, F. Environmental Efficiency Analysis for Multi Plants Production Technologies. *Sustainability* **2021**, *13*, 3989. https://doi.org/10.3390/su13073989

Academic Editor: Enrique Rosales-Asensio

Received: 24 February 2021
Accepted: 29 March 2021
Published: 2 April 2021

1. Introduction

Environmental issues are becoming more important as a consequence of growing pollution-generating technologies. Greenhouse gas emissions reduction is one of the main concerns of all societies in this century. A five percent reduction of greenhouse gas emissions on average was decided in the Kyoto Protocol for 2008–2012 compared with 1990. This reduction level was decided to be 50 percent on average in Copenhagen. Industries are one of the main sources of emissions production in all countries. The reduction path has been decided to be gradual, since cutting down the emissions may be possible because it they are a byproduct in industries. The key factor in emissions reduction is the performance and the efficiency of production technology. Data envelopment analysis (DEA) is a mathematical programming-based approach for efficiency analysis of a group of decision-making units (DMUs) proposed by [1]. In this paper, we propose new models for estimating the classical and environmental performance of multi-plant firms. Then we develop some new indices for capturing the environmental performance vs. classical economic performance at the local and global level. The proposed approach is utilized for the economic and environmental performance assessment of 46 power plants in Iran. The primary result emphasizes considering not only local environmental performance but also global performance to have a broad insight into environmental performance assessments. Primary results show that we have only a few power plants that are resistant to environmental performance at the country level when we use models with non-uniform scaling factors for desirable and undesirable outputs. This is due to the higher discrimination power of associated economic and environmental efficiency measures of these indices. Another important result is that the geographical location does not affect the environmental or economic performance. This finding encourages considering both local and global environmental performances to have a broad environmental performance that may be used for any type of local and global environmental planning by decision-makers. The rest of the paper is organized as follows.

Section 2 reviews the relative literature with the classical and environmental efficiency analysis of multi plants firms. Section 3 provides the primary models and material in the first subsection. In the second subsection, we develop environmental multi-plant DEA models dealing with undesirable outputs. The third subsection proposes a mixed, uniform, and non-uniform multi-plant DEA model for considering desirable and undesirable outputs simultaneously. Section 4 applies the proposed models for local and global classical and environmental efficiency analysis of Iranian power plants.

2. Literature Review

This method has extended for incorporating environmental issues into the efficiency analysis. Before that, the production function estimation operates for a single output case while DEA models can consider multiple outputs by [2]. A linear programming model was developed by [3] to analyze the efficiency of multi-plant firms in a DEA framework. An extension of the previous paper for the limited data was performed by [4]. They also used the multi-plant technology for efficiency analysis of multi-units [5]. Unlike the classical DEA models that were extended for dealing with pollution generating technology, the multi-plant DEA models cannot consider environmental issues. Analyzing the environmental performance of production units by DEA methods is growing, and this method has been intensively used for environmental efficiency analysis of different sectors in the last decades. A review for the application of DEA models in environmental and energy studies was done by [6]. A study for the UK's regional environmental efficiency using directional distance function DEA models was implemented by [7]. They investigate the link between regional environmental efficiency and economic growth and found a "U" shape form for the link mentioned above. An investigation of the environmental efficiency of transportation sectors in 30 Chinese provinces was done by [8] between 2003 and 2012. They found the transportation sectors to be inefficient in most provinces. Another study by [9] used the DEA model to determine greenhouse gas emissions and carbon sequestration in small-scale maize production in Niger State, Nigeria. An environmental DEA model capable of handling zero and negative data was proposed by [10] and used for US industrial sectors' environmental efficiency analysis. An investigation of the corporate suitability of US industrial sectors was performed by [11] via an environmental efficiency analysis. They emphasized the role of the proposed DEA environmental assessment for corporate leaders in identifying how to invest in technology innovation to reduce undesirable output. A study on the role of the Central Government's policy was doen by [12] in China. They performed provincial level environmental analysis and concluded that though the Central Government's environmental policies fail to solve the inner contradiction between economic and environmental systems. In another study, the economic and environmental performance of wastewater treatment plants was investigated by [13] to find how it is potentially possible to reduce greenhouse gas emissions in the Valencia region on the Mediterranean coast of Spain. Classical DEA models were used by [14] for efficiency analysis and ranking of Iranian power plants at the country level. In another study by [15], classical DEA models were utilized for the environmental efficiency of thermal and hydroelectric power plants in Iranian provinces. They found the average technical efficiency for the hydroelectric power plant in 2011 and 2010 are 62% and 53%, respectively, and 82% and 77%, for thermal power plants in 2011 and 2010, respectively. The technical and environmental efficiency of 16 selected thermal power plants in Iran was investigated by [16] during 2011–2015, using DEA. The technical efficiency of 26 thermal power plants in Iran was analyzed by [17] in the period of 2003–2008 using DEA and the Malmquist index. Iranian industrial sector emissions was studied by [18] using the input-output analysis during 1380–1390. They found that the production of final goods with the highest positive change is the most important factor affecting the increase of emissions. Another investigation was performed by [19] to measure the CO_2 emission levels of Iranian provinces. They considered and analyzed the impact of population, urbanization, energy intensity, and per capita income on environmental degradation. The interactions between the Iranian industries' productive activities, the intensity of energy consumption by these activities, and the resulting environmental impacts (specifically CO_2 emissions) were implemented by [20]. The current paper contributes to the literature in two ways. As a theoretical contribution, in this paper, we extend the multi-plant DEA model for dealing with undesirable outputs like pollution. We propose uniform and non-uniform multi-plant DEA models that consider the undesirable outputs. Moreover, other indices are also proposed for analyzing the local and global classical and environmental performance of production units. For the application side, we utilized the proposed model for the local and global classical and environmental efficiency analysis of 46 power plants in 21 provinces in Iran.

3. Materials and Methods

3.1. Classical Multi-Plant Firm Production Technology

Consider J multi-plant firms, numbered $j = 1, 2, \ldots, J$ and each firm has K_j plants numbered $k = 1, 2, \ldots, K_j$. Assume each plant consumes M inputs and produces N desirable outputs. Let x^j_{ik} be the i-the input $(1 \le i \le M)$, y^j_{rk} be the r-the desirable output $(1 \le r \le N)$ of plant k at firm j. The general production technology can be represented by the output correspondence of $P^j : \mathbb{R}^M_+ \to P^j(x) \subseteq \mathbb{R}^N_+$. Considering this setting and constant returns to scale for the production technology that consists of only the desirable output, we can consider the production set of $T^j_C = \{(x, y) \,|\, x \ge \sum_{k=1}^{K_j} \lambda^j_k x^j_k, y \le \sum_{k=1}^{K_j} \lambda^j_k y^j_k, \lambda \ge 0\}$ for j-th firm. The following linear programming model can be found based on this production set for assessing the performance of the plant "o" of j-th firm.

$$
\begin{aligned}
\varphi^j_{Lo} &= Max\varphi \\
&s.t \\
\sum_{k=1}^{K_j} \lambda^j_k x^j_{ik} &\le x^j_{io}, \quad i = 1, 2, \ldots, m(1) \\
\sum_{k=1}^{K_j} \lambda^j_k y^j_{rk} &\ge \varphi y^j_{ro}, \quad r = 1, 2, \ldots, s \\
\lambda^j_k &\ge 0, \quad k = 1, 2, \ldots, K_j
\end{aligned}
$$

If we consider all plants operating in all production spaces, we have a broader production system, and plants may face more competitive environments. This setting is seen in the global industry, and thus, for assessing the performance plant "o" considering all firms and associated plants, we may use the following linear programming:

$$
\begin{aligned}
\varphi^j_{Go} &= Max\varphi \\
&s.t \\
\sum_{j=1}^{J} \sum_{k=1}^{K_j} \lambda^j_k x^j_{ik} &\le x^j_{io}, \quad i = 1, 2, \ldots, m(2) \\
\sum_{j=1}^{J} \sum_{k=1}^{K_j} \lambda^j_k y^j_{rk} &\ge \varphi y^j_{ro}, \quad r = 1, 2, \ldots, s \\
\lambda^j_k &\ge 0, \quad k = 1, 2, \ldots, K_j, j = 1, 2, \ldots, J.
\end{aligned}
$$

Please note that in the classical efficiency analysis, we consider only inputs and desirable outputs; thus, other measures like emission, etc., that can be considered as undesirable outputs are not considered in the above models and associated measures. The following subsection deals with undesirable outputs in the production process of multi-plant firms.

3.2. Multi-Plant Firm Environmental Production Technology

Assume that each plant consumes not only M inputs and produces N desirable outputs but also P undesirable outputs. Let x^j_{ik} be the i-the input ($1 \le i \le M$), y^j_{rk} be the r-the desirable output ($1 \le r \le N$), and z^j_{hk} be the h-the undesirable output ($1 \le h \le P$) of plant k at firm j. We considered the general production technology that considers the undesirable outputs for the j-th plant by the output correspondence $P^j : \mathbb{R}^M_+ \to P^j(x) \subseteq \mathbb{R}^{N+P}_+$. For dealing with both desirable and undesirable outputs, we considered the strong disposability for desirable output and the weak disposability for undesirable outputs proposed by [21] as follows. Strong disposability says if $y \in P^j(x)$ then $y' \in P^j(x)$ for $y' \le y$ while weak disposability implies that $y \in P^j(x)$ then $\theta y \in P(x)$ for $0 \le \theta \le 1$. Considering y and z as desirable and undesirable outputs, we assumed that desirable outputs are strong disposal, and undesirable outputs are weak disposal in the context that if $(y, z) \in P^j(x)$, then $(y', z) \in P^j(x)$ for $y' \le y$.

Considering this and the constant returns to scale, we found the following output set for plant j:

$$T^j_{EL} = \{(x, y) \Big| x \ge \sum_{k=1}^{K_j} \lambda^j_k x^j_k, y \le \sum_{k=1}^{K_j} \lambda^j_k y^j_k, z = \sum_{k=1}^{K_j} \lambda^j_k z^j_k, \lambda \ge 0\}.$$

Considering this environmental production technology for the multi-plant firm, we may use the following linear programming model for assessing the performance of plant "o" in the j-the firm.

$$\begin{aligned} \varphi^j_{ELo} &= Max \varphi \\ &s.t \\ \sum_{k=1}^{K_j} \lambda^j_k x^j_{ik} &\le x^j_{io}, \quad i = 1, 2, \ldots, m (3) \\ \sum_{k=1}^{K_j} \lambda^j_k y^j_{rk} &\ge \varphi y^j_{ro}, \quad r = 1, 2, \ldots, s \\ \sum_{k=1}^{K_j} \lambda^j_k z^j_{hk} &= z^j_{ho}, \quad h = 1, 2, \ldots, P \\ \lambda^j_k &\ge 0, \quad k = 1, 2, \ldots, K_j \end{aligned}$$

In contrast with the classical efficiency analysis and associated efficiency measure that was dealt in the previous section, in the environmental efficiency analysis of multi-plant firms, we considered any undesirable output that may be produced as a byproduct of the desired output.

Theorem 1. *The environmental efficiency of an arbitrary plan in a firm is not greater than its classical efficiency measure.*

The above theorem says if a production unit is efficient, then it is not necessarily environmentally efficient. In order to have acceptable environmental performance, production units need to take care of associated environmental issues that may not be considered in the classical efficiency analysis (Proof of Theorem 1 is available in Appendix A).

If we consider all firms and owned plants, we face a more competitive environment, and then we can use the following model for environmental efficiency of plants "o". In fact, in the global environmental analysis, we considered all plants' environmental performance belonging to all firms.

$$\varphi^j_{EGo} = Max\varphi$$
$$s.t$$
$$\sum_{j=1}^{J}\sum_{k=1}^{K_j}\lambda^j_k x^j_{ik} \leq x^j_{io},\ i = 1,2,\ldots,m(4)$$
$$\sum_{j=1}^{J}\sum_{k=1}^{K_j}\lambda^j_k y^j_{rk} \geq \varphi y^j_{ro},\ r = 1,2,\ldots,s$$
$$\sum_{j=1}^{J}\sum_{k=1}^{K_j}\lambda^j_k z^j_{hk} = z^j_{ho},\ h = 1,2,\ldots,P$$
$$\lambda^j_k \geq 0,\ k = 1,2,\ldots,K_j, j = 1,2,\ldots,J$$

Theorem 2. *The classical efficiency of an arbitrary plant within its firm is not greater than its efficiency measure when considering all firms.*

Corollary 1. *The global classical environmental efficiency is greater than the local environmental efficiency, and we have the following relationship for the classical efficiency of* P^j_o *and its classical and environmental efficiency* $\varphi^j_{Go} \geq \varphi^j_{Lo} \geq \varphi^j_{ELo}$.

Proof of Corollary 1. This can be concluded while considering Theorems 1 and 2.

In order to measure the local-global efficiency measure of P^j_o we proposed Local-Globalindex $= \frac{\varphi^j_{Go}}{\varphi^j_{Lo}}$ and for estimating the classical-environmental efficiency measure of P^j_o we proposed Local-Environmental Index $= \frac{\varphi^j_{Lo}}{\varphi^j_{ELo}}$. Regarding Corollary 1, we saw that Local-Global Index $= \frac{\varphi^j_{Go}}{\varphi^j_{Lo}} \geq 1$. If this index is equal to unity, then it means that the evaluated plant could survive in the competitive global environment, since the local efficiency measure and the global efficiency measure are identical. Corollary 1 also concludes that Local-Environmental Efficiency $= \frac{\varphi^j_{Lo}}{\varphi^j_{ELo}} \geq 1$. This index determines whether a plant is environmentally friendly or not. Unity value shows that the efficiency measure does not depend on the technology's environmental structure and is environmentally friendly if we assess it in a classical or environmental production technology context. Suppose this index is greater than unity, then the environmental issue matters and can affect its performance P^j_o.

Theorem 3. *The global environmental efficiency of a plant is not greater than its local environmental efficiency.*

Theorem 4. *The global efficiency of a plant is not greater than its global environmental efficiency.*

Corollary 2. *The global classical environmental efficiency is not greater than the local environmental efficiency, and we have the following relationship for the classical efficiency of* P^j_o *and its classical and environmental efficiency:* $\varphi^j_{Go} \geq \varphi^j_{EGo} \geq \varphi^j_{ELo}$.

If we are interested in analyzing the global environmental performance, we can use the newly introduced index Global-Environmental Index $= \frac{\varphi^j_{Go}}{\varphi^j_{EGo}}$. Regarding Corollary 2, this index is also greater than or equal to unity $\frac{\varphi^j_{Go}}{\varphi^j_{EGo}} \geq 1$. If it is unity, then the environmental and classical efficiency of P^j_o are equal globally. Otherwise, the environmental issue matters. We could see that a production unit's technically well performance does not necessarily imply an acceptable environmental performance. In other words, if a production unit is economically efficient, then we may not necessarily conclude that it is environmentally efficient too. For analyzing the environmental performance of P^j_o the local and global production space, we introduced Environmetal Local-Global Index $= \frac{\varphi^j_{EGo}}{\varphi^j_{ELo}} \geq 1$. This index is also less than or equal to unity $\frac{\varphi^j_{EGo}}{\varphi^j_{ELo}} \geq 1$ and it indicates the situation of the environmental performance of P^j_o in the local and global production. If it is equal to unity, then the environmental performance of P^j_o is resistance globally. This means P^j_o has a similar environmental performance both locally and globally. However, if it is greater than unity, then we face a sub-optimal local performance for P^j_o.

3.3. *Joint Scaling of Desirable and Undesirable Outputs*

Proposed models in the previous subsection only consider the desirable output and look for possible expansion of this type of output. However, we may take care of both desirable and undesirable outputs simultaneously. In previous models, we sought a possible expansion of desirable output while keeping undesirable output. But, there might be a possibility of areduction of undesirable outputs that are not considered in the previous models. Therefore, we proposed the following mixed model taking both desirable and undesirable factors into consideration.

$$
\begin{aligned}
\varphi^j_{UELo} &= Max 1 + \varphi \\
& s.t \\
\sum_{k=1}^{K_j} \lambda^j_k x^j_{ik} &\leq x^j_{io}, \quad i = 1,2,\ldots,m (5) \\
\sum_{k=1}^{K_j} \lambda^j_k y^j_{rk} &\geq (1+\varphi) y^j_{ro}, \quad r = 1,2,\ldots,s \\
\sum_{k=1}^{K_j} \lambda^j_k z^j_{hk} &= (1-\varphi) z^j_{ho}, \quad h = 1,2,\ldots,P \\
\lambda^j_k &\geq 0, \ k = 1,2,\ldots,K_j
\end{aligned}
$$

The optimal value of the above model is less than or equal to zero. If it is zero, then the under-evaluation unit is efficient, otherwise it is inefficient.

Theorem 5. *If a plant is locally mixed efficient then its local efficiency score is greater than or equal to the mixed local efficiency score.*

An associated global model that simultaneous changes desirable and undesirable outputs can also be proposed by the following:

$$\varphi^{j}_{UEGo} = Max 1 + \varphi$$
$$s.t$$
$$\sum_{j=1}^{J}\sum_{k=1}^{K_j} \lambda^{j}_{k} x^{j}_{ik} \leq x^{j}_{io}, \quad i = 1, 2, \ldots, m (6)$$
$$\sum_{j=1}^{J}\sum_{k=1}^{K_j} \lambda^{j}_{k} y^{j}_{rk} \geq (1+\varphi) y^{j}_{ro}, \quad r = 1, 2, \ldots, s$$
$$\sum_{j=1}^{J}\sum_{k=1}^{K_j} \lambda^{j}_{k} z^{j}_{hk} = (1-\varphi) z^{j}_{ho} h = 1, 2, \ldots, P$$
$$\lambda^{j}_{k} \geq 0, \quad k = 1, 2, \ldots, K_j, j = 1, 2, \ldots, J.$$

Theorem 6. *If a plant is a locally mixed efficient plant, then its local efficiency score is greater than or equal to the mixed local efficiency score.*

Proof of Theorem 6. It is similar to the proof of Theorem 5.

The previous subsection's proposed indices can be updated by the new mixed uniform measure of the model (5) and model (6). However, we cannot compare the efficiency measures using mixed models and peer models in the previous subsection, since the structure production technologies are different. Therefore, associated indices may be meaningless. However, we can still compare the local and global environmental performance of production units by the new mixed index of

$$\text{Uniform Environmental Local-Global Index} = \frac{\varphi^{j}_{UEGo}}{\varphi^{j}_{UELo}}$$

Theorem 7. *The global mixed efficiency measure of a production unit is greater than or equal to its local mixed efficiency measure.*

Proof of Theorem 7. It is similar to the proof of Theorem 4.

Using Theorem 7, we then have

$$\text{Uniform Environmental Local-Global Index} = \frac{\varphi^{j}_{UEGo}}{\varphi^{j}_{UELo}} \geq 1.$$

The percentage of the desirable output expansion and the undesirable output reduction may not necessarily be equal; thus, we proposed the following model for the local and global environmental efficiency measurement of P^{j}_{o}.

$$\varphi^j_{NUELo} = Max1 + \varphi + \gamma$$
$$s.t$$
$$\sum_{k=1}^{K_j} \lambda^j_k x^j_{ik} \leq x^j_{io},\ i = 1,2,\ldots,m(7)$$
$$\sum_{k=1}^{K_j} \lambda^j_k y^j_{rk} \geq (1+\varphi) y^j_{ro},\ r = 1,2,\ldots,s$$
$$\sum_{k=1}^{K_j} \lambda^j_k z^j_{hk} = (1-\gamma) z^j_{ho},\ h = 1,2,\ldots,P$$
$$\lambda^j_k \geq 0,\ k = 1,2,\ldots,K_j \varphi^j_{NUEGo} = Max1 + \varphi + \gamma$$
$$s.t$$
$$\sum_{j=1}^{J}\sum_{k=1}^{K_j} \lambda^j_k x^j_{ik} \leq x^j_{io},\ i = 1,2,\ldots,m(8)$$
$$\sum_{j=1}^{J}\sum_{k=1}^{K_j} \lambda^j_k y^j_{rk} \geq (1+\varphi) y^j_{ro},\ r = 1,2,\ldots,s$$
$$\sum_{j=1}^{J}\sum_{k=1}^{K_j} \lambda^j_k z^j_{hk} = (1-\gamma) z^j_{ho},\ h = 1,2,\ldots,P$$
$$\lambda^j_k \geq 0,\ k = 1,2,\ldots,K_j, j = 1,2,\ldots,J$$

Theorem 8. *The global mixed efficiency measure of a production unit is greater than or equal to its local mixed efficiency measure.*

Proof of Theorem 8. It is similar to the proof of Theorem 7.

Using the model (7) and (8), we proposed the non-uniform environmental local-global index as follows:Non-uniform Environmetal Local-Global Index $= \frac{\varphi^j_{NUECo}}{\varphi^j_{NUELo}}$ and using Theorem 8, we could see that Non-uniform Environmetal Local-Global Index $= \frac{\varphi^j_{NUEGo}}{\varphi^j_{NUELo}} \geq 1$.

The local model of (7) and global model (8) consider the non-uniform scaling factor for desirable and undesirable output. In contrast, this factor is considered to be uniform in the local model (5) and global model (6). If we consider $\gamma = \varphi$ in the model (7) and model (8), then we get models (5) and (6), respectively. Therefore, we could consider model (5) and model (6) as a particular case of the model (7) and model (8), respectively.

Table 1 lists and summarizes the variables and parameters used in the current paper.

Table 1. Decision variables and parameters.

Symbol	Description
	Parameter
x^j_{ik}	The i-the input of plant k at firm j
y^j_{rk}	The r-the desirable output of plant k at firm j
z^j_{hk}	The h-the undesirable output of plant k at firm j
	Decision variable
φ	Poteintial output enlargement
λ^j_k	The intensity variable of plant k at firm j
	Other symbols
P^j	Output correspondence of firm j
T^j_C	Production technology of firm j
φ^j_{Lo}	Local efficiency measure of plant o at firm j
φ^j_{Go}	Global efficiency measure of plant o at firm j
T^j_{EL}	Environmental local production technology of firm j
φ^j_{LEo}	Local environmental efficiency measure of plant o at firm j
φ^j_{GEo}	Global environmental efficiency measure of plant o at firm j
φ^j_{ULEo}	Uniform local environmental efficiency measure of plant o at firm j
φ^j_{UGEo}	Uniform global environmental efficiency measure of plant o at firm j
φ^j_{NULEo}	Non-uniform local environmental efficiency measure of plant o at firm j
φ^j_{NUGEo}	Non-uniform global environmental efficiency measure of plant o at firm j

4. An Application for Local and Global Environmental Efficiency Analysis of Power Plants

This section applies the proposed approach for environmental efficiency analysis of 46 power plants in 21 provinces of Iran. Provinces were assumed as plants at the local level, and the country was assumed as firm at the global level. Total assets were assumed as inputs, electricity production was taken as the desirable output, and pollution was assumed as an undesirable output. Table 2 reports the descriptive statistics of the data. We are willing to share our unnamed data set and codes for those who wish to replicate the results of this research.

Table 2. Statistical summary of data.

Variable Type	Variable Name	Mean	Standard Error	Min	Max
Input	Total assets	760.6957	508.7195	42	2043
Output	Electricity production	754,238	664,627.8	2622.333	2,936,547
Output	Pollution	482,712	425,361.8	1,879,390	1678.293

In the first analysis, we assessed the classical local and global performance of all power plants. The results are reported in Table 3.

We could observe some important facts confirming the proposed methodology and theorem. Most of the production units were found to be efficient at the local level, considering both classical and environmental productions. However, this was not the case at the global level. The classical global efficiency measures were found to be higher than or at least equal to the classical local efficiency measures. We had the same observation when we considered environmental technologies. This is a rational observation since the global environment is a more competitive space, and an under-evaluation unit needs to compete with more rivals at the global level. The same observation appears when we compared the environmental performance vs. the classical performance. This is also an expectable observation since if a production unit is technically efficient and not necessarily environmentally efficient, it does not matter at the local or global level. The local and global analysis is performed at the province and country-level, respectively. This shows that when considering the efficiency status of production units classical or environmental behavior may not reveal the whole picture of the production behavior. Thus, decision-makers are highly encouraged to consider the production behavior of DMUs at both local and global levels. Considering both classical and environmental production technology at the local and global levels, we reported the proposed indices in Table 4. Using this report, we analyzed and tracked the classical and environmental performance of production units at the local and global levels.

Table 3. Classical and environmental efficiency measures at the local and global level.

Power Plant	Local Efficiency	Local Environmental Efficiency	Global Efficiency	Global Environmental Efficiency
P1	1	1	7.56080881	7.56080881
P2	1	1	8.0914273	8.0914273
P3	1.0548	1.0487	8.37519308	8 37519308
P4	1.4824	1.3254	15.3949782	15.3949782
P5	1	1	6.04726569	6.04726569
P6	1	1	5.92873151	5.92873151
P7	1.0025	1.0015	9.31173672	9.31173672
P8	1.0145	1.0112	222.98936	222.98936
P9	1	1	10.0435612	10.0435612
P10	1	1	1.01051666	1.01051666
P11	1	1	15.322035	15.322035
P12	1	1	5.60066461	5.60066461
P13	1	1	5.45235982	5.45235982
P14	1	1	22.0498452	22.0498452
P15	1	1	5.31678385	5.31678385
P16	1.0024	1	108.740439	108.740439
P17	1	1	6.63809458	6.63809458
P18	1	1	31.9781505	31.9781505
P19	1	1.0458	5.50775158	5.50775158
P20	1	1	7.57622863	7.57622863
P21	1	1	1	1
P22	1.0458	1.0415	7.2218846	7.2218846
P23	1	1	9.15474257	9.15474257
P24	1	1	10.0825616	10.0825616
P25	1	1	1.97976715	1
P26	1	1	1	1
P27	1	1	7.33669521	7.33669521
P28	1	1	6.2231639	6.2231639
P29	1	1	1.15865301	1.15865301
P30	1	1	6.78926893	6.78926893
P31	1	1	9.06070276	9.06070276
P32	1	1	1.76888648	1.76888648
P33	1	1	1	1
P34	1	1	3.15012452	3.15012452
P35	1	1	1.55292259	1.55292259
P36	1	1	4.25912395	4.25912395
P37	1	1	5.46666869	5.46666869
P38	1	1	2.79610979	2.79610979
P39	1	1	4.93834184	4.93834184
P40	1	1	3.59403477	3.59403477
P41	1	1	4.67331934	4.67331934
P42	1	1	2.49761552	2.49761552
P43	1	1	18.3410179	18.3410179
P44	1	1	4.00072678	4.00072678
P45	1	1	4.81755703	4.81755703
P46	1	1	7.67969877	7.67969877

Table 4. Local-Global classical and environmental efficiency indices using a single scaling factor.

Power Plant	Local-Global Index	Environmental Local-Global Index	Local Environmental Index	Global Environmental Index
P1	7.560809	7.560809	1	1
P2	8.091427	8.091427	1	1
P3	7.940077	7.986262	1.005817	1
P4	10.38517	11.61534	1.118455	1
P5	6.047266	6.047266	1	1
P6	5.928732	5.928732	1	1
P7	9.288515	9.29779	1.000999	1
P8	219.8022	220.5195	1.003263	1
P9	10.04356	10.04356	1	1
P10	1.010517	1.010517	1	1
P11	15.32203	15.32203	1	1
P12	5.600665	5.600665	1	1
P13	5.45236	5.45236	1	1
P14	22.04985	22.04985	1	1
P15	5.316784	5.316784	1	1
P16	108.4801	108.7404	1.0024	1
P17	6.638095	6.638095	1	1
P18	31.97815	31.97815	1	1
P19	5.507752	5.266544	0.956206	1
P20	7.576229	7.576229	1	1
P21	1	1	1.09558	1
P22	6.905608	6.934119	1.004129	1
P23	9.154743	9.154743	1	1
P24	10.08256	10.08256	1	1
P25	1.979767	1	1	1.979767
P26	1	1	1	1
P27	7.336695	7.336695	1	1
P28	6.223164	6.223164	1	1
P29	1.158653	1.158653	1	1
P30	6.789269	6.789269	1	1
P31	9.060703	9.060703	1	1
P32	1.768886	1.768886	1	1
P33	1	1	1	1
P34	3.150125	3.150125	1	1
P35	1.552923	1.552923	1	1
P36	4.259124	4.259124	1	1
P37	5.466669	5.466669	1	1
P38	2.79611	2.79611	1	1
P39	4.938342	4.938342	1	1
P40	3.594035	3.594035	1	1
P41	4.673319	4.673319	1	1
P42	2.497616	2.497616	1	1
P43	18.34102	18.34102	1	1
P44	4.000727	4.000727	1	1
P45	4.817557	4.817557	1	1
P46	7.679699	7.679699	1	1

The local-global index shows only a few power plants that have a unity measure. Thus, we had just these power plants that have a resistant performance at the global level. P21, P26, and P33 are technically efficient both in the local and global environment. This result provides valuable information in the process of target setting for decision-makers. These power plants that are efficient at both local and global levels may be used for target setting instead of those that are efficient only at the local or global level. Next, we analyzed the environmental local-global index and again found a few power plants with independent environmental performance, regardless of the local or global level. We observed for P25, P26, and P33 that their environmental local-global index is equal.

In contrast with the previous analysis, we observed more production units with a unity measure of the global environmental index. This shows that the environmental performance was better managed at the local level, and policymakers need more attention towards managing the global environmental issue. Such information may be used in the process of environmental target setting. Another interesting observation is the similarity of the local-global index in the classical and environmental space. We observed that these indices are almost similar (second and third column). This shows that power plants' technical and environmental performance have the same pattern when considering the local and global levels. Therefore, the geographical location does not affect the technical and environmental performance of power plants. In the next analysis, we looked at the case from a different angle by the local and global environmental proposed indices. We are interested in measuring the environmental effects at the local and global levels. These indices are reported in the fourth column and fifth column of Table 4. We observed that at the local levels, we have a few power plants with a greater than equal value. This shows that there is a potential for environmental improvement for those power plants. However, when we looked at the global index, we observed only one power plant with such a situation. More deep investigation revealed that this power plant is owned by a border province with an old generation technology that struggles with providing gas and has used fuel for electricity production in some situations. In order to consider the desirable and undesirable output simultaneously, we used models (5) and model (6) for the local and global performance assessment in the subsequent analysis. The results are reported in Table 5.

We found less efficient power plants when we used the joint model, not only in classical production but also in environmental production. This was also an expectable observation; when we considered just desirable or undesirable output separately, we had an easier job reaching the efficient frontier rather than when considering both desirable and undesirable outputs simultaneously. For the efficiency measure of power plants using a mixed model, that efficient power plants using this model were also efficient using model 3 and model 4 at the local and global level. This could be found by comparing the second and third columns of Table 5 and peers in Table 3. Note that we considered the scaling factor for both desirable and undesirable outputs in the mixed models of (5–6) while we had no scaling factor on undesirable output in the models (3–4). Using mixed environmental efficiency measures of power plants at the local and global level, we calculated the uniform environmental local-global index for all power plants reported in the sixth column of Table 5. We had only two power plants P26 and P34, which were resistant in the local and global environmental assessment. This emphasizes a more competitive space and a high potential for environmental improvement at the global level. Decision-makers and any environmental planning should consider this at the country level. Uniform factor analysis considers the simultaneous improvement of both desirable and desirable output; thus, we found more potential improvement, including desirable output enlargement and undesirable output reduction, in this analysis. However, the scaling factor of desirable and undesirable output may not be uniformly considered in the previous analysis. Thus, in the following analysis, we considered the non-uniform but joint scaling factor for the desirable and undesirable outputs. To this end, we used mixed models (7) and model (8) at the local environmental and global environmental levels, respectively. Table 6 lists the result of new efficiency measures and updated index regarding the local-global performance of power plants associated with new measures using models with non-uniform scaling factors.

The first observation was more discrimination power using models with the non-uniform scaling factor. Regarding the fact that the later model can be considered as a generalized model compared with models with a uniform scaling factor, we could expect this observation. More potential for environmental improvement was found using models with non-uniform scaling factors. We also saw that only the three power plants P21, P26, and P33 are environmentally resistant globally. Deeper analysis shows that these power plants are classically efficiency efficient in both the local and global space. More investigation reveals that these three power plants are classically efficient and environmentally efficient at both local and global levels. On the other hand, these are the only power plants that gained the unity value for all measures and associated indices. This fact shows that those power plants performing well in a classical and environmental manner can be the most favorable targets for other power plants at both local and global levels. However, the average local-global environmental index was about six, which emphasizes considering the local environmental performance and global environmental performance in any environmental planning at the local or country level. This emphasizes the classical and environmental efficiency at the local level and performs and indicates the global analysis of the classical and environmental efficiency status of production units.

Table 5. Local-Global classical and environmental efficiency indices using a uniform scaling factor.

Power Plant	Uniform Local Environmental Efficiency	Uniform Global Environmental Efficiency	Uniform Environmental Local-Global Index
P1	1	1.944752099	1.944752099
P2	1.0354	1.971346572	1.971346572
P3	1.0584	1.968863777	1.8774328
P4	1.3354	1.99330027	1.503923548
P5	1	1.914830522	1.914830522
P6	1	1.936111985	1.936111985
P7	1.0015	1.96905577	1.966106611
P8	1.0112	1.99961275	1.977465141
P9	1	1.978610344	1.978610344
P10	1	1.005230826	1.005230826
P11	1	1.993344478	1.993344478
P12	1	1.908453623	1.908453623
P13	1	1.905882958	1.905882958
P14	1	1.99133302	1.99133302
P15	1	1.904868025	1.904868025
P16	1.0017	1.998726843	1.998726843
P17	1	1.935417814	1.935417814
P18	1	1.996998422	1.996998422
P19	1.0574	1.889884246	1.807118231
P20	1	1.912816684	1.912816684
P21	1	1	1
P22	1	1.953804471	1.875952444
P23	1	1.969964547	1.969964547
P24	1	1.980108822	1.980108822
P25	1	1.519158388	1.519158388
P26	1	1	1
P27	1	1.937004632	1.937004632
P28	1	1.888059933	1.888059933
P29	1	1.129664166	1.129664166
P30	1	1.950313132	1.950313132
P31	1	1.969786473	1.969786473
P32	1	1.504709418	1.504709418
P33	1	1	1
P34	1	1.603532748	1.603532748
P35	1	1.27041928	1.27041928
P36	1	1.84189491	1.84189491
P37	1	1.912792441	1.912792441
P38	1	1.72454979	1.72454979
P39	1	1.786074705	1.786074705
P40	1	1.663260315	1.663260315
P41	1	1.756371902	1.756371902
P42	1	1.793855307	1.793855307
P43	1	1.989310537	1.989310537
P44	1	1.916055549	1.916055549
P45	1	1.839949652	1.839949652
P46	1	1.959037015	1.959037015

Table 6. Local-Global classical and environmental efficiency indices using a non-uniform scaling factor.

Power Plant	Non-Uniform Local Environmental Efficiency	Non-Uniform Global Environmental Efficiency	Non-Uniform Environmental Local-Global Index
P1	1	8.09142729	8.09142729
P2	1.43902307	8.37519308	5.82005474
P3	1.68100065	15.3949782	9.15822260
P4	1	6.21153570	6.21153570
P5	1	5.92873151	5.92873151
P6	1.21494955	9.37077363	7.71289112
P7	35.5037591	222.989359	6.28072535
P8	2.58316967	10.0933260	3.90734147
P9	1	1.01051666	1.01051666
P10	1.96369059	15.3220349	7.80267270
P11	1.00388961	5.77286465	5.75049743
P12	1	5.62912577	5.62912577
P13	1.48543471	22.0498452	14.8440352
P14	1	5.49180457	5.49180457
P15	10.1495961	108.740439	10.7137700
P16	1.39885265	6.71122533	4.7975642
P17	5.36906170	31.9781505	5.95600354
P18	1	5.72410083	5.72410083
P19	5.53539284	7.62722101	1.37790058
P20	1	1	1
P21	1	1	1
P22	1	7.221884598	7.221884598
P23	2.36310261	10.0839866	4.26726566
P24	2.02729392	2.08513153	1.02852946
P25	1	1	1
P26	1	7.46219151	7.46219151
P27	1	6.39031986	6.39031986
P28	1	1.70963893	1.70963893
P29	1	6.78926892	6.78926892
P30	1	9.06070276	9.06070276
P31	1	2.18082477	2.18082477
P32	1	1	1
P33	1	3.15012452	3.15012452
P34	1	1.55292259	1.55292259
P35	1	4.54565729	4.54565729
P36	1	5.60808255	5.60808255
P37	1	3.16396536	3.16396536
P38	1	4.93834183	4.93834183
P39	1	3.59403476	3.59403476
P40	1	4.67331934	4.67331934
P41	1	2.90542582	2.90542582
P42	1	18.3410178	18.3410178
P43	1	4.00072677	4.00072677
P44	1	4.81755702	4.81755702
P45	1	7.67969877	7.67969877
P46	1	8.09142729	8.09142729

5. Conclusions

The current paper proposes new models for environmental assessment and local-global analysis of pollution generating production units. In the application section, proposed models are used for Iranian power plants' local and global classical and environmental performance. However, the theoretical foundation and associated indices introduced in this paper can be used in any type of local-global analysis that is involved with environmental aspects. The proposed models put one step forward in contrast with classical efficiency analysis. It is highly recommended to utilize all developed indices for having a broad picture of classical and environmental performance in any performance analysis. One may perform well at the local level or may have an acceptable performance using classical models, but deeper analysis on the global level or considering environmental issues may provide better insight into the production. The current paper considers the production technology assuming convexity and constant returns to scale assumptions. Extending to other production technology types may not be a straightforward task, and we are still working on this. Investigating the production's scale effects is another important aim that can be achieved in a future research line.

Author Contributions: Conceptualization, M.G.; Data curation, M.G.; Formal analysis, F.T.-H.; Methodology, M.G.; Project administration, M.G.; Software, M.G.; Supervision, F.T.-H.; Validation, F.T.-H.; Writing—original draft, M.G.; Writing—review & editing, F.T.-H. All authors have read and agreed to the published version of the manuscript.

Funding: Farhad Taghizadeh-Hesary acknowledges the financial support of the JSPS KAKENHI (2019–2020) Grant-in-Aid for Young Scientists No. 19K13742 and Grant-in-Aid for Excellent Young Researcher of the Ministry of Education of Japan (MEXT).

Institutional Review Board Statement: Not applicable.

Informed Consent Statement: Not applicable.

Data Availability Statement: Data are gathered from province and country sources. In the country source, we used the annual report of the energy ministry, and in the local level, we used power plants data of provinces.

Conflicts of Interest: The authors declare no conflict of interest.

Appendix A

Proof of Theorem 1. Consider plant "*o*" of firm j denote it by P_o^j. The classical efficiency of this plant is gauged by the optimal value of model (1), that is, φ_{Lo}^j and its environmental efficiency measure is φ_{ELo}^j that is the optimal value of model (3). Let $(\overline{\lambda}^j, \varphi_{ELo}^j)$ be the optimal solution of model (3); thus, we have

$$\sum_{k=1}^{K_j} \overline{\lambda}_k^j x_{ik}^j \le x_{io}^j i = 1,2,\ldots,m$$

$$\sum_{k=1}^{K_j} \overline{\lambda}_k^j y_{rk}^j \ge \varphi_{ELo}^j y_{ro}^j r = 1,2,\ldots,s$$

$$\sum_{k=1}^{K_j} \overline{\lambda}_k^j z_{hk}^j = y_{ho}^j h = 1,2,\ldots,P$$

$$\overline{\lambda}_k^j \ge 0 k = 1,2,\ldots,K_j$$

Ignoring the third set of constraint from the above constraint set, we have

$$\sum_{k=1}^{K_j} \overline{\lambda}_k^j x_{ik}^j \le x_{io}^j i = 1,2,\ldots,m$$

$$\sum_{k=1}^{K_j} \overline{\lambda}_k^j y_{rk}^j \ge \varphi_{ELo}^j y_{ro}^j r = 1,2.\ldots,s$$

$\lambda_k^j \geq 0 k = 1, 2, \ldots, K_j$ and this means that $(\overline{\lambda}^j, \varphi_{ELo}^j)$ is a feasible solution for the model (1) that implies $\varphi_{Lo}^j \geq \varphi_{ELo}^j$, where φ_{Lo}^j is the optimal value of the classical model of (1), namely, the classical efficiency measure of the plant under evaluation.

Proof of Theorem 2. Consider P_o^j then model (1) finds the classical efficiency of this plant, that is, φ_{Lo}^j. Let $(\lambda^{j*}, \varphi_{Lo}^j) \in \mathbb{R}_+^{K_j+1}$ be the optimal solution of model (1), then if satisfies associated constraint set of

$$\sum_{k=1}^{K_j} \lambda_k^{j*} x_{ik}^j \leq x_{io}^j i = 1, 2, \ldots, m$$

$$\sum_{k=1}^{K_j} \lambda_k^{j*} y_{rk}^j \geq \varphi_{Lo}^j y_{ro}^j r = 1, 2, \ldots, s$$

$$\lambda_k^{j*} \geq 0 k = 1, 2, \ldots, K_j$$

Using this, we have a feasible solution for the model (2) that gauges the plant's efficiency under evaluation, considering all firm plants. Observe that $(\lambda^{j**}, \varphi_{Lo}^j) \in \mathbb{R}_+^{\sum_{j=}^{J} K_j+1}$ satisfies the following constraint set

$$\sum_{j=1}^{J} \sum_{k=1}^{K_j} \lambda_k^{j**} x_{ik}^j \leq x_{io}^j i = 1, 2, \ldots, m$$

$$\sum_{j=1}^{J} \sum_{k=1}^{K_j} \lambda_k^{j**} y_{rk}^j \geq \varphi_{Lo}^j y_{ro}^j r = 1, 2, \ldots, s$$

$$\lambda_k^{j**} \geq 0 k = 1, 2, \ldots, K_j, j = 1, 2, \ldots, J$$

where $\lambda_k^{j**} = \lambda_k^{j*}$ for the firm that owned plant "o" and $\lambda_k^{j**} = 0$ for other firms. This implies $\varphi_{Go}^j \geq \varphi_{Lo}^j$, that is, the classical efficiency of P_o^j within its firm is not greater than its efficiency measure when considering all firms.

Proof of Theorem 3. Mathematically, we can provide a similar argument to the proof of Theorem 2 to prove this theorem. However, we may also look at the problem from a production technology view. The production space for measuring the global environmental efficiency measure is larger than the production space for measuring the local environmental efficiency measure. This provides a broader production set when we consider the global production. Therefore we cannot expect lesser output efficiency measures in such production space compared with the local production space.

Proof of Theorem 4. Similar to the proof of Theorem 1, if we consider the model (4) that gauges the global environmental efficiency of P_o^j then its optimal solution satisfies the following set of constraints.

$$\sum_{j=1}^{J} \sum_{k=1}^{K_j} \lambda_k^{j*} x_{ik}^j \leq x_{io}^j i = 1, 2, \ldots, m$$

$$\sum_{j=1}^{J} \sum_{k=1}^{K_j} \lambda_k^{j*} y_{rk}^j \geq \varphi_{EGo}^{j*} y_{ro}^j r = 1, 2, \ldots, s$$

$$\sum_{j=1}^{J}\sum_{k=1}^{K_j}\lambda_k^{j*}z_{hk}^j = y_{ho}^j h = 1,2,\ldots,P$$

$$\lambda_k^{j*} \geq 0 k = 1,2,\ldots,K_j, j = 1,2,\ldots,J$$

where $(\lambda_k^{j*}, \varphi_{EGo}^{j*}), k = 1,2,\ldots,K_j, j = 1,2,\ldots,J$ is the optimal solution of model (4). If we consider the first and the second set of constraints in the above system of in-equality, then we reach the following

$$\sum_{j=1}^{J}\sum_{k=1}^{K_j}\lambda_k^{j*}x_{ik}^j \leq x_{io}^j i = 1,2,\ldots,m$$

$$\sum_{j=1}^{J}\sum_{k=1}^{K_j}\lambda_k^{j*}y_{rk}^j \geq \varphi_{EGo}^{j*}y_{ro}^j r = 1,2,\ldots,s$$

$$\sum_{j=1}^{J}\sum_{k=1}^{K_j}\lambda_k^{j*}z_{hk}^j = y_{ho}^j h = 1,2,\ldots,P$$

$$\lambda_k^{j*} \geq 0 k = 1,2,\ldots,K_j, j = 1,2,\ldots,J$$

This implies that $(\lambda_k^{j*}, \varphi_{EGo}^{j*})$ is a feasible solution of model (2) and therefore $\varphi_{Go}^{j*} \geq \theta_{EGo}^{j*}$, that is, the global efficiency of a plant is not greater than its global environmental efficiency.

Proof of Theorem 5. Assume P_o^j is efficient using the mixed model of (5), thus we have

$$\sum_{k=1}^{K_j}\lambda_k^{*j}x_{ik}^j \leq x_{io}^j i = 1,2,\ldots,m$$

$$\sum_{k=1}^{K_j}\lambda_k^{*j}y_{rk}^j \geq (1+\varphi^*)y_{ro}^j = y_{ro}^j r = 1,2,\ldots,s$$

$$\sum_{k=1}^{K_j}\lambda_k^{*j}z_{hk}^j = (1-\varphi^*)z_{ho}^j = z_{ho}^j h = 1,2,\ldots,P$$

$$\lambda_k^{*j} \geq 0 k = 1,2,\ldots,K_j$$

where, $(\lambda^{*j}, \varphi^*) = (\lambda^{*j}, \varphi_{UELo}^j - 1)$ is the optimal solution of mixed model (5). This implies $(\lambda^{*j}, \varphi^*) = (\lambda^{*j}, \varphi_{UELo}^j) = (\lambda^{*j}, 1)$ is a feasible solution of model (3), that is,

$$\sum_{k=1}^{K_j}\lambda_k^{*j}x_{ik}^j \leq x_{io}^j i = 1,2,\ldots,m$$

$$\sum_{k=1}^{K_j}\lambda_k^{*j}y_{rk}^j \geq \varphi_{UELo}^j y_{ro}^j = y_{ro}^j r = 1,2,\ldots,s$$

$$\sum_{k=1}^{K_j}\lambda_k^{*j}z_{hk}^j = z_{ho}^j h = 1,2,\ldots,P$$

$$\lambda_k^{*j} \geq 0 k = 1,2,\ldots,K_j$$

and this implies that $\varphi_{ELo}^j \geq \varphi_{UELo}^j = 1$.

References

1. Charnes, A.; Cooper, W.; Rhodes, E. Measuring the efficiency of decision making units. *Eur. J. Oper. Res.* **1978**, *2*, 429–444. [CrossRef]
2. Farrell, M.J. The measurement of productive efficiency. *J. R. Stat. Soc. Ser. A Gen.* **1957**, *120*, 253–290. [CrossRef]
3. Färe, R.; Primont, D. Efficiency measures for multiplant firms. *Oper. Res. Lett.* **1984**, *3*, 257–260. [CrossRef]
4. Färe, R.; Primont, D. Efficiency measures for multiplant firms with limited data. In *Measurement in Economics*; Springer: Berlin/Heidelberg, Germany, 1988; pp. 177–186.
5. Färe, R.; Primont, D. Measuring the efficiency of multi-unit banking: An activity analysis approach. *J. Bank. Financ.* **1993**, *17*, 539–544. [CrossRef]
6. Zhou, P.; Ang, B.W.; Poh, K.-L. A survey of data envelopment analysis in energy and environmental studies. *Eur. J. Oper. Res.* **2008**, *189*, 1–18. [CrossRef]
7. Halkos, G.E.; Tzeremes, N.G. A conditional directional distance function approach for measuring regional environmental efficiency: Evidence from UK regions. *Eur. J. Oper. Res.* **2013**, *227*, 182–189. [CrossRef]
8. Song, X.; Hao, Y.; Zhu, X. Analysis of the environmental efficiency of the Chinese transportation sector using an undesirable output slacks-based measure data envelopment analysis model. *Sustainability* **2015**, *7*, 9187–9206. [CrossRef]
9. Sadiq, M.; Singh, I.P.; Suleiman, A.; Isah, M.A.; Umar, S.M.; Maude, A.M.; Lawal, A.; Sallawu, H. Application of Data Envelopment Analysis (DEA) to determine GHG emission and carbon sequestration in small-scale maize production in Niger State, Nigeria. *Agric. Bionutr. Res.* **2015**, *1*, 1–19.
10. Sueyoshi, T.; Yuan, Y. Comparison among US industrial sectors by DEA environmental assessment: Equipped with analytical capability to handle zero or negative in production factors. *Energy Econ.* **2015**, *52*, 69–86. [CrossRef]
11. Wang, D.; Li, S.; Sueyoshi, T. DEA environmental assessment on US Industrial sectors: Investment for improvement in operational and environmental performance to attain corporate sustainability. *Energy Econ.* **2014**, *45*, 254–267. [CrossRef]
12. Xia, Q.; Li, M.; Wu, H.; Lu, Z. Does the central government's environmental policy work. Evidence from the provincial-level environment efficiency in China. *Sustainability* **2016**, *8*, 1241. [CrossRef]
13. Molinos-Senante, M.; Hernández-Sancho, F.; Mocholí-Arce, M.; Sala-Garrido, R. Economic and environmental performance of wastewater treatment plants: Potential reductions in greenhouse gases emissions. *Resour. Energy Econ.* **2014**, *38*, 125–140. [CrossRef]
14. Pourkazemi, M.; Heydari, K. Data envelopment analysis and its application in evaluating the efficiency of power plants in Iran. *Modarres Hum. Sci.* **2002**, *6*, 35–54.
15. Javaheri, A.; Ostadzad, H. Estimating Efficiency of Thermal and Hydroelectric Power Plants in Iranian Provinces. *Iran. J. Econ. Stud.* **2014**, *3*, 19–42.
16. Emami meibodi, A.; Amadeh, H.; Amini, F. Evaluation of Technical and Environmental Efficiency of Selected Thermal Power Plants of Iran. *Q. J. Energy Policy Plan. Res.* **2017**, *3*, 33–67.
17. Emami Meibodi, A.; Afghah, M.; Rahmani Sefati, M.H. Measuring Technical Efficiency and Productivity in Steam, Gas and Compound Cycle Power Plants. *Sci. J. Manag. Syst.* **2009**, *6*, 79–103.
18. Jafari Samimi, A.; Najari, F. Evaluating the Contribution of Factors Affecting on Pollution Changes in Iran's Industrial Sector: Structural Decomposition Approach in the Input-Output Method. *J. Environ. Sci. Stud.* **2019**, *3*, 831–839.
19. Falahi, F.; Hekmati Farid, S. Determinants of CO_2 Emissions in the Iranian Provinces (Panel Data Approach). *Iran. Energy Econ.* **2013**, *2*, 129–150. (In Persian)
20. Torabi, T.; Varesi, M. Studying the Environmental Pollution of Industries in Iran using an Input-Output Approach (Special Case of CO_2). *J. Environ. Sci. Technol.* **2009**, *11*, 77–92. (In Persian)
21. Färe, R.; Grosskopf, S.; Lovell, C.K.; Pasurka, C. Multilateral productivity comparisons when some outputs are undesirable: A nonparametric approach. *Rev. Econ. Stat.* **1989**, *71*, 90–98. [CrossRef]

sustainability

Article

ASEAN's Energy Transition towards Cleaner Energy System: Energy Modelling Scenarios and Policy Implications

Han Phoumin [1,*], Fukunari Kimura [1,2] and Jun Arima [1,3]

[1] Economic Research Institute for ASEAN and East Asia (ERIA), Think Tank, Jakarta 10270, Indonesia; vzf02302@nifty.ne.jp (F.K.); junarima@g.ecc.u-tokyo.ac.jp (J.A.)
[2] Faculty of Economics, Keio University, Tokyo 108-8345, Japan
[3] Graduate School of Public Policy, Tokyo University, Tokyo 113-0033, Japan
* Correspondence: han.phoumin@eria.org

Abstract: The Association of Southeast Asian Nations (ASEAN) faces tremendous challenges regarding the future energy landscape and how the energy transition will embrace a new architecture—including sound policies and technologies to ensure energy access together with affordability, energy security, and energy sustainability. Given the high share of fossil fuels in ASEAN's current energy mix (oil, coal, and natural gas comprise almost 80%), the clean use of fossil fuels through the deployment of clean technologies is indispensable for decarbonizing ASEAN's emissions. The future energy landscape of ASEAN will rely on today's actions, policies, and investments to change the fossil fuel-based energy system towards a cleaner energy system, but any decisions and energy policy measures to be rolled out during the energy transition need to be weighed against potentially higher energy costs, affordability issues, and energy security risks. This paper employs energy modelling scenarios to seek plausible policy options for ASEAN to achieve more emissions reductions as well as energy savings, and to assess the extent to which the composition of the energy mix will be changed under various energy policy scenarios. The results imply policy recommendations for accelerating the share of renewables, adopting clean technologies and the clean use of fossil fuels, and investing in climate-resilient energy quality infrastructure.

Keywords: business as usual (BAU); Alternative Policy Scenarios (APSs); energy transition; renewables; clean technologies; fossil fuels; and resiliency

Citation: Phoumin, H.; Kimura, F.; Arima, J. ASEAN's Energy Transition towards Cleaner Energy System: Energy Modelling Scenarios and Policy Implications. *Sustainability* **2021**, *13*, 2819. https://doi.org/10.3390/su13052819

Academic Editor: George Xydis

Received: 28 January 2021
Accepted: 1 March 2021
Published: 5 March 2021

1. Introduction

The world has been struggling with the coronavirus disease (COVID-19) pandemic since March 2020, which has damaged the world economy—including the Association of Southeast Asian Nations (ASEAN). The global economy is being pushed into a recession by the COVID-19 pandemic due to preventive and containment measures such as country lockdowns, travel restrictions, and slow or even negative growth in many sectors such as tourism, retail, and industry. The magnitude of the economic impacts is hard to predict as it depends on the success of the pandemic containment efforts around the world. The International Monetary Fund (IMF) projected the world economy and the ASEAN 5 (Indonesia, Malaysia, the Philippines, Singapore, and Thailand) to contract sharply by −4.9% and −2.5% respectively in 2020, much worse than during the 2008–2009 financial crisis (IMF, 2020) [1]. Such an economic downturn is contracting energy demand and energy-related carbon dioxide (CO_2) emissions around the globe, but this crisis is seen as temporary and both energy demand and CO_2 emissions will bounce back once the economy starts to recover. Global energy demand increased 10 times from 1999 to 2019, and keeps increasing (IEA, 2017) [2]. The gravity of energy demand has shifted to Asia, and emerging economies account for half of global growth in gas demand. Many of the Organisation for Economic Co-operation and Development (OECD) countries will see energy demand peak, while some countries will experience negative growth due to energy efficiency and other factors such as population growth and industrial structures. However, ASEAN will be the opposite, as it will need more energy to steer its economic growth [3].

To reconfirm the present situation of ASEAN's high reliance on fossil fuel, this study also estimates the energy demand and supply in ASEAN. It found that ASEAN will see strong growth in fossil fuel demand to steer economic growth from 2017 (The energy modelling uses 2017 for the baseline information as it is the most up-to-date baseline data in the ASEAN Member States (AMS)) to 2050 in which fossil fuels (oil, coal, and gas) had the dominant share in the primary energy mix in 2017, at 78.0%, while their combined share is projected to increase to 81.7% in 2050 (Tables A1–A8). Oil will be the largest energy source in the primary energy mix in 2050, at 39.6%, down from 36.9% in 2017. Coal was the second largest energy source after oil in 2017, at 21.6%, and is projected to have a 22.4% share in 2050. Natural gas is projected to have the second largest share of the primary energy mix in 2050, at 24.7%, overtaking coal. Thus, this is a real concern as to whether ASEAN's energy transition will be achieved or not within the context of Climate Change timeframe in which the net zero emission should start from the mid of this century?

Elsewhere, especially the OECD, moves away from fossil fuel dependence to a system based on cleaner energy through a higher share of renewables, but ASEAN will continue to rely on the fossil fuel, and it tries to find ways on how to use fossil fuels more cleanly in an energy transition. For instance, coal use has been drastically reduced in the OECD and more developed countries due to the role of gas, renewables, and advanced technologies. However, as the most abundant and reliable energy resource in ASEAN, coal use will continue to be the second largest energy source in power generation after gas in the foreseeable future, to meet fast-growing electricity demand [4]. The increase in coal use for power generation in ASEAN countries will lead to the widespread construction of coal-fired power plants, which will result in increased greenhouse gas (GHG) and CO_2 emissions if the best available clean coal technology (CCT) is not employed (Phoumin, 2015) [5].

Meanwhile, the climate narrative which has prevailed since the Conference of the Parties (COP) 21 in 2015 and is likely to continue at the upcoming COP 26, promotes the banning of public coal financing throughout the world, through financial instruments and influence over multilateral development banks and OECD member countries. Based on the Greenhouse Gas Emissions Data (United States Environmental Protection Agency, 2020) [6], emissions from fossil fuel combustion and industrial processes contributed about 78% of the increase in GHG emissions from 1970 to 2011. China, the United States (US), Europe, and India are the largest emitters, contributing 30%, 15%, 9%, and 6% of global GHG emissions, respectively [6]. With substantial new generation capacity required to generate power, unabated coal-fired power generation plants are increasingly being constructed in developing Asia. Therefore, these trends reflect the urgent need to address the environmental sustainability of powering emerging Asia's economic development.

The Economic Research Institute for ASEAN and East Asia (ERIA) held a round discussion on "ASEAN's Future Energy Landscape" on 10 September 2020 [7], whose leaders in ASEAN expressed that managing the energy transition in ASEAN will need to consider the presence of fossil fuels (coal, oil, and natural gas) in the short- and medium-term energy system. At that round table discussion [7], ASEAN's leaders and experts expressed that it will be crucial to explore ways in which to use fossil fuels in an environmentally sustainable manner to act as a bridge to a carbon-free energy future, rather than simply ruling out them completely. For successful implementation of the energy transition and climate change policy objectives, policymakers also indicated the need to balance the other equally important policy objectives of energy security, energy access, and affordability.

For climate friendly, ASEAN's position to shift towards a cleaner energy system will have fundamental impacts on environmental sustainability and emission reduction. The evidence has been shown elsewhere in New York's power generation mix in which the drastic reduction of emissions came from the shift of fossil fuel dependence to increasing share of onshore and offshore wind power (Isik, M.; Kaplan, P.O., 2021) [8]. Here at ASEAN, the pace at which ASEAN Member States (AMS) have adopted national power development plans and policies has created a drastic change in the energy system, as more renewables have penetrated the electrical grid. For instance, the recent development of accelerating pace of share of solar in power mix in Cambodia and Vietnam is achievably remarkable surprises for ASEAN [9]. However, one of the greatest challenges in all countries in ASEAN of increasing the share of variable renewable energy (e.g., wind and solar) in the power mix is the high cost of upgrading and integrating the systems that need more investment in grids, the internet of things, technological know-how, and quality energy infrastructure [10]. In a recent virtual conference on Asia—Carbon Capture, Utilization, and Storage (CCUS)—organised by ERIA on 18 February 2021 [11], experts in Asia generally expressed that ASEAN would need to create an energy bridging from the current fossil based energy system to a cleaner energy system that will need to consider the role of cleaner use of fossil fuels through innovative technologies such as clean coal technologies and CCUS, the technology that can remove carbon dioxide from flue gas and atmosphere, followed by recycling carbon dioxide for utilization and further determining safe and permanent storage options. In this way, the CCUS can reduce CO_2 and GHG emissions. Therefore, urgent steps need to be taken to decarbonise the energy sector through pathways to a low-carbon economy which require the rapid deployment of the clean use of fossil fuel technologies, renewable energy development, and a doubling of energy efficiency, given that the energy sector accounts for two-thirds of global GHG emissions.

As ASEAN as well as developing countries around the world are embarking on energy transition, it is very crucial to ensure everyone is not left behind or become victims of the energy transition as they may be denied access to energy. In this regard, McCauley, D. and Heffron, R. (2018) [12], called on researchers to explore the multiple implications of the transition to a post-carbon society through the application of their proposed new triumvirate of tenets (distributional, procedural, and restorative), as this just transition framework enables researchers to more explicitly reflect upon the intersectionality of environment, climate, and energy, assess justice issues from a truly interdisciplinary perspective and ultimately contribute to meaningful long-term solutions. Details of Energy Justice Metric has been developed to facilitate the decision-makings to formulate energy policy with respect to the difficulties faced in balancing between the competing aims of economics, politics and the environment which form the trilemma of energy policy (Heffron, R, et.al., 2015) [13].

Thus, the paramount of climate change awareness and policy towards energy security and affordability will need to be flexible in the context of ASEAN, considering the role of fossil fuels in an energy transition. In the minds of ASEAN's leaders as expressed through various conferences as mentioned above, meeting the growing energy demand would need appropriate energy policies and cooperation that can help facilitating energy-related infrastructure investments. These common energy challenges need to be addressed through concerted efforts—including collective measures and actions—to rapidly deploy energy efficiency and energy savings, highly efficient and low-emissions coal-fired power plant technology, and nuclear safety; and to double the share of renewable energy in the overall energy mix for inclusive and sustainable development.

This study aims to analyses the potential impacts of proposed additional energy-saving goals, action plans, and policies in ASEAN on energy consumption, by fuel, sector, and greenhouse gas (GHG) emissions as to whether under various policy commitment and targets will ASEAN be able to achieve climate neutrality by 2050? In addition, the study results will support ASEAN's energy policy directions to achieve the followings:

i. Improve the efficiency and environmental performance of fossil fuel use.
ii. Reduce dependence on conventional fuels through intensified EEC programs; increased share of hydropower; and expansion of renewable energy systems, biofuel production and/or utilization, and, for interested parties, civilian nuclear power.
iii. Mitigate GHG emissions through effective policies and measures to help abate global climate change.

Therefore, this study will explore the best energy mix under various Alternative Policy Scenarios (APSs) and the associated emissions. Under the APSs, key considerations are realistic assumptions in terms of technologies, resource endowment, energy efficiency, and system integration challenges, when the power generation mix has a higher share of intermittent renewables such as wind and solar energy. The study employs the energy outlook by each country and ASEAN under the ERIA's Energy Outlook and Energy Potentials in ASEAN and East Asia. However, for the purpose of ASEAN's scope of study, only aggregated ASEAN' data is shown. The paper is organized as follows. Section 2 reviews the literature, Section 3 discusses the research methodology, Section 4 describes the results and discussion, and Section 5 is the conclusion and policy implications.

2. Literature Review

2.1. Global Commitment to Emissions Reduction (COP 21)

The Paris Agreement, negotiated at the Paris Climate Conference (COP 21), is the first universal legally binding global climate change agreement, adopted by the majority of leaders on 22 April 2016. It aims to limit the average temperature rise to well below 2 °C above pre-industrial levels (baseline: 1850–1900) and to pursue efforts to limit the temperature increase to 1.5 °C above pre-industrial levels, recognizing that this would significantly reduce the risks and impacts of climate change (EU, 2020) [14].

Bridging the gap from current policies and actions to climate neutrality by the end of this century is very challenging. The world will need to reduce emissions by 7.6% per year from 2020 to 2030 to limit global warming to 1.5 °C. If we do nothing, temperatures are expected to rise 3.2 °C above pre-industrial levels by the end of century—posing a serious threat to our living environment (UNEP, 2019) [15]. If emissions cuts are delayed, it will become very difficult to meet the limit of a global temperature rise of well below 1.5 °C by 2100. UNEP (2019) [15], stated that delaying emissions cuts until 2025 would steepen the need to cut emissions to 15.5% per year, which would be extremely difficult to achieve, especially for the developing world. As parties to the Paris Agreement, countries have submitted comprehensive national climate action plans known as Nationally Determined Contributions (NDCs). Some countries have not yet finalized their NDCs, but have carried out preparatory work known as Intended Nationally Determined Contributions (INDCs).

About 78% of all global emissions come from G20 nations, requiring their strong commitment to long-term zero emissions targets by 2100. Amongst the G20 nations, China, the US, the European Union (EU) 28, (The EU 28 refers to the 28 countries which were members of the EU until 31 January 2020 when the United Kingdom left the group (Austria, Belgium, Bulgaria, Croatia, Cyprus, Czech Republic, Denmark, Estonia, Finland, France, Germany, Greece, Hungary, Ireland, Italy, Latvia, Lithuania, Luxembourg, Malta, Netherlands, Poland, Portugal, Romania, Slovenia, Slovakia, Spain, Sweden, and United Kingdom).) and India contributed more than 55% of the total emissions over the last decade (UNEP, 2019) [15]. Thus, the speed of emissions reduction is very concerning, and full decarbonization of the energy sector may go beyond renewables and energy efficiency. The carbon sinks will rely on the clean use of fossil fuels with carbon capture, utilization, and storage (CCUS). Developing countries may face difficulties in achieving emissions reduction targets without international support, such as technologies for the clean use of fossil fuels and the other climate abatement initiatives. However, their emissions contribution remains small compared with that of the G20 nations. Developing nations can contribute more in terms of the conservation of natural resources such as forestry and the management of improved agricultural practices.

The fifth session of the UN Environment Assembly (UNEA-5), provides leadership, catalyzes, intergovernmental actions on the environment, and contributes to the implementation of the UN 2030 Agenda for Sustainable Development and its Sustainable Development Goals (SDGs) (UNEP, 2021) [16]. UNEA-5 will also provide an opportunity for Member States and Stakeholders to take ambitious steps towards building back better and greener by ensuring that investments in economic recovery after the COVID-19 pandemic contribute to sustainable development.

Amongst other policy measures to reduce the GHE, energy efficiency is very crucial as it is also known as the hidden fuel—fuel that can be translated to be an energy resource to a nation as it will become energy available for economic activities and supply to the greater population. Energy efficiency can offer relatively cheaper and quicker solutions to help ASEAN embark on the road to economic recovery through the provision of available and sustainable energy for economic activities and reduce emissions ASHRAE (2020) [17]. Energy efficiency can be applied to all sectors including power generation, commercial and residential building, and transportation. In this regard, guidance offered by the ISO standards (ISO 50001 for energy management, ISO for 14065 for greenhouse gas validation, and verification or ISO 14001 for environmental management and others) will help countries in the process of implementing changes within the ASEAN energy system transition.

2.2. ASEAN and EU Energy Policy Directions

The ASEAN Plan of Action for Energy Cooperation (APAEC) is the high-level energy cooperation framework within ASEAN where leaders in ASEAN adopted the first APAEC at the 17th ASEAN Ministers on Energy Meeting (AMEM) held on 3 July 1999 in Bangkok, Thailand, in which ASEAN's leaders had agreed to implement a framework for cooperation in energy to enable ASEAN countries to obtain security of supply, whereby ASEAN Member States (AMS) work towards lessening dependence on imported oil and towards accelerating the development of indigenous energy sources and energy conservation (ACE, 2015) [18]. APAEC is implemented in two phases: Phase I covers the period 2016–2020 and consists of short to medium-term measures to enhance energy security cooperation and to take further steps towards connectivity and integration. Phase II covers the period 2021–2025 which is developed based on the progress of Phase I implementation.

Phase 2 of the ASEAN Plan of Action for Energy Cooperation (APAEC), which is under preparation for endorsement by the ASEAN Ministers on Energy Meeting in 2020, will set key energy policy targets and will have energy policy implications for energy infrastructure related investment in the region (ASEAN Centre for Energy, 2020) [19]. Key targets include the revision of the new energy efficiency and conservation target from a 30% reduction in energy intensity by 2025 (based on 2005 levels) to more ambitious levels—a new target of 35–40% reduction is likely—and will involve the expansion of energy efficiency and conservation measures to transport and industries. It will also establish a new sub-target for the share of renewables in installed power capacity, which will complement the existing target of a 23% share of renewables in the total primary energy supply (TPES) by 2025. APAEC Phase 2 will also include policy measures to pursue smart grids and renewable energy grid integration; and measures to address emerging and alternative technologies such as hydrogen, energy storage, bioenergy, nuclear energy, and CCUS. APAEC Phase 2 will maintain the focus on energy connectivity and market integration, but will add a sub-theme on the energy transition and energy resilience on how the region will need to have a strategy to deal with fossil fuels and new technologies.

The ASEAN region has wide economic development gaps in terms of gross domestic product (GDP), population growth, energy use, and technologies. However, each country is committed to addressing the common climate change issue. Countries share their commitments through various policies such as energy intensity targets or through targets for the share of renewables in the energy mix. Nevertheless, emerging countries face energy access and affordability issues, while promoting renewables and other clean energy technologies remains expensive. Although solar and wind module costs have dropped drastically, the system cost remains expensive when applied in developing countries. Making these clean and green technologies available to developing countries in ASEAN will require policy attention, including regulations and financing mechanisms, with support from developed countries.

The EU aims to be climate neutral by 2050 (EU, 2020) [14]. Amongst other targets, the 2030 climate and energy framework includes EU-wide targets and policy objectives for 2021–2030. The key targets for 2030 include (i) at least 40.0% cuts in GHG emissions from 1990 levels, (ii) at least a 32.0% share for renewable energy, and (iii) at least a 32.5% improvement in energy efficiency. For GHG emissions, a cut of at least 40.0% below 1990 levels is targeted by 2030. This will enable the EU to move towards a climate-neutral economy and implement its commitments under the Paris Agreement. For renewables, the binding renewable energy target for the EU for 2030 is at least 32.0% of final energy consumption, including a review clause by 2023 for an upward revision of the target. For energy efficiency, a headline target of at least 32.5% is to be achieved collectively by the EU in 2030, with an upward revision clause by 2023. To help achieve these targets, a transparent and dynamic governance process will help deliver on the 2030 climate and energy targets in an efficient and coherent manner. The EU has adopted integrated monitoring and reporting rules to ensure progress towards its 2030 climate and energy targets and its international commitments under the Paris Agreement.

2.3. INDCs' Commitments and Targets submitted by ASEAN Member States to the United Nations Framework Convention on Climate Change (UNFCCC)

COP 21 was a very successful conference, at which leaders around the globe showed their solidarity in fighting global climate change. Countries laid out targets or programs aimed at reducing CO_2 emissions. Some countries have clear policies and targets, while others have no targets—especially developing countries. In the AMS, the key commitments are varied, reflecting each country's socio-economic and environmental situation. The following paragraphs summarize the key commitments of AMS for mitigating climate change (Kimura and Phoumin, 2018) [20].

Cambodia proposes a GHG mitigation contribution for 2020–2030 (UNFCCC, 2015) [21], conditional on the availability of support from the international community. Cambodia is expected to contribute a maximum reduction of 3100 gigagrams of carbon dioxide equivalent ($GgCO_2eq$) by 2030 compared with 2010 baseline emissions of 11,600 $GgCO_2eq$. The Lao People's Democratic Republic (Lao PDR) is a highly climate-vulnerable country whose GHG emissions were only 51,000 $GgCO_2eq$ in 2000—negligible compared with total global emissions. The Lao PDR has ambitious plans to reduce its GHG emissions through increased carbon stock by expanding forest cover to 70% of the country's land area by 2020. The Lao PDR electricity grid draws on renewable resources for almost 100% of output, and the government has laid the foundations for implementing a renewable energy strategy that aims to increase the share of small-scale renewable energy to 30% of total energy consumption by 2030.

Viet Nam's intended unconditional contribution (Developing countries announced two sets of mitigation targets to be reached under the Paris Agreement. The low target or unconditional target can be reached without outside support. However, the conditional target can be reached only with outside support.) to GHG emissions reduction efforts during 2021–2030 is to reduce its GHG emissions by 8% in 2030 compared with the BAU scenario, in which the emissions intensity per unit of GDP will decline by 20% from 2010 levels and forest coverage will increase by 45%. Under its conditional contribution, Viet Nam intends to cut emissions by 25% from 2010 levels if international support is received through bilateral and multilateral cooperation (UNFCCC, 2015) [21]. Further, the emissions intensity target per unit of GDP will be reduced by 30% from 2010 levels. Thailand expects its GHG emissions to reach 555 million tones of carbon equivalent ($MtCO_2e$) by 2030 in the BAU case, with 76.8% mainly from the energy and transport sectors. According to Thailand's INDC, the country intends to reduce GHG emissions by 20% of the BAU emissions in 2030. This means that Thailand's amount of GHG emissions reduction should be 111 $MtCO_2e$ in 2020.

From 2016 to 2030, Myanmar aims to increase the share of renewables in rural electrification to 30%, increase hydropower capacity to 9.4 gigawatts, and distribute about 260,000 energy-efficient cooking stoves to rural areas (UNFCCC, 2015) [21]. For energy efficiency, Myanmar aims to achieve 20% electricity-saving potential of the forecast electricity consumption by 2030. Under the INDC framework, Brunei Darussalam targets reducing its energy consumption by 63% by 2035 against the BAU scenario. Furthermore, the country aims to achieve a 10% share of renewable energy in power generation by 2035. With regards to the transport sector, the target is to reduce CO_2 emissions by 40% from morning peak-hour vehicle use by 2035 compared with the BAU scenario. Another target in its INDC is to enhance the stocks of carbon sinks by increasing the current 41–55% of the country's total forest area in 2016.

Indonesia's INDC specifies conditional and unconditional mitigation targets. It intends to reduce 29% of its emissions against the BAU scenario by 2030 in the unconditional scenario. If there is additional international support, Indonesia intends to reduce an additional 12% of the emissions. The intended contributions cover five sectors: Energy (including transport); industrial processes and product use; agriculture; land use, land use change, and forestry; and waste. The amount of emissions under the 29% and 41% reduction targets would be 0.848 $GtCO_2eq$ and 1.119 $GtCO_2eq$, respectively. Malaysia intends to reduce its GHG emissions intensity of GDP by 45% by 2030 relative to the emissions intensity of GDP in 2005 (UNFCCC, 2015) [21]. This consists of 35% on an unconditional basis and a further 10% conditional upon receipt of climate finance, technology transfer, and capacity building from developed countries.

The Philippines targets a GHG emissions reduction of 70% by 2030 relative to its BAU scenario of 2000–2030. The mitigation contribution is conditioned on the extent of financial resources—including technology development and transfer—and capacity building that will be made available to the Philippines (Kimura and Phoumin, 2018) [20]. Singapore pledged in 2009 to reduce carbon emissions unconditionally from 7–11% lower than its BAU level by 2020. It committed to a further 16% reduction by 2020 after the COP 21 in Paris on 12 December 2015.

3. Methodology and Scenario Assumptions

This study aims to address the research questions on what are the projected impacts on greenhouse gas emissions of different potential energy mixes for ASEAN countries, and which energy mix should the region pursue? Thus, this study employs energy models of ASEAN countries using the Long-range Energy Alternatives Planning (LEAP) system software, an accounting system used to develop projections of energy balance tables based on final energy consumption and energy input/output in the transformation sector. LEAP is an energy modelling software that can conduct a variety of analyses of energy systems including Demand Analysis, Transformation Analysis, Resource Analysis, and Environmental Analysis. The data structures in a LEAP are organized using a hierarchical tree. The types of data entered at each branch depend on the type of branch, its position in the tree (for example whether it is a Demand or Transformation branch). At the heart of LEAP is the concept of scenario analysis. Scenarios are self-consistent story-lines of how a future energy system might evolve over time in a particular demographic and socio-economic setting and under a particular set of policy conditions. Using LEAP, scenarios can be built and then compared to assess their energy requirements, social costs and benefits and environmental impacts. All scenarios start from a common base year. We can use scenarios to ask an unlimited number of "what if" questions, such as: What if more efficient appliances are introduced, what if different electric generation capacity expansion plans are pursued, what if indigenous reserves of oil and gas are discovered, what if renewable energy technologies are introduced, etc.

In LEAP, the energy consumption is calculated as the product of an activity level and an annual energy intensity (energy use per unit of activity). Overall activities are defined as the products of the individual activities entered along a complete branch of the Demand tree. Total energy consumption is thus calculated by the equation:

$$Energy\ Consumption = (Activity\ level) \times (Energy\ intensity). \quad (1)$$

In general, Equation (1) can apply only when energy intensity data is available. However, in most case, energy intensity by activity is not available. Thus, regressions are used to predict the future energy consumption by activity and sector.

In this study, data availability varies in the 10 ASEAN Member States (AMS). It is very challenging to collect long-term historical data in countries such as Cambodia, the Lao PDR, and Myanmar. Further, there are many missing data points in the historical data that need to be estimated. The LEAP application is very useful in dealing with such minimal data, and it allows expert judgement on how the future growth of demand in each fuel should be estimated. If good historical data are available, linear forecasting is used to forecast future values based on a time series of historical data. The new values are predicted using linear regression, assuming a linear trend (y = mx + c) where the Y term corresponds to the variable to be forecast and the X term is years. Multiple regressions are used to predict the future growth of energy demand by sector, such as transport, industry, and the commercial and residential sectors. Whenever the energy intensity data is not available, energy demand equations are applied and could be summarized in the equations below:

$$\sum_{i=1}^{10} TFEC_{ijkt} = TFEC_{1jkt} + TFEC_{1jkt} + \cdots + TFEC_{10jkt} \tag{2}$$

where index (i) represents the country in ASEAN; index (j) is the energy consumption by sector; index (k) is the energy type; and index (t) is time. Thus, the Total Final Energy Consumption in ASEAN is the aggregated summation of all energy consumption in 10 countries of ASEAN. The TFEC of each country is derived from regression estimates of energy consumption by sector (j) and by energy type (k) at time (t) as follows:

$$TFEC_{ijkt} = \propto_0 + \beta_1 X_{ijkt} + \beta_2 TFEC_{ijk(t-1)} + \varepsilon_{ijkt} \tag{3}$$

where:

$TFEC_{ijkt}$ is the Total Final Energy Consumption of country (i), by sector (j), by energy type (k), at time (t). $TFEC_{ijk(t-1)}$ is the lag variable. The variables $\beta_1 X_{ijt}$ are the independent variables including per capita gross domestic product, energy relative price, population growth, car ownership, and floor area. The summation of all energy demand by sector forms the Total Final Energy Consumption (TFEC). From here, the LEAP will further generate the Total Primary Energy Supply (TPES) by certain assumption of the losses in the transformation sector. Due to many equations run by each country in ASEAN, this study omits the results of the equation, and it shows only the results of the demand generated by LEAP.

In this modelling work using the LEAP application, the baseline for the 10 AMS was 2017—the latest available baseline data. For future energy demand, the projected demand growth is based on government policies, population and economic growth, and other key variable such as energy prices, using the International Energy Agency (IEA) world energy model (IEA, 2019) [22]. The BAU case is future predicted energy demand based on the government's current energy policies. However, the APSs are somewhat different to the BAU case in terms of policy changes and targets, as they have a greater share of renewables, including possible nuclear uptake if the government's alternative policies include nuclear as an energy option and more efficient power generation and energy efficiency in the final energy consumption.

Key variables and assumptions used in the model include the average annual growth rate of the population and the GDP, and energy efficiency and renewable targets (Figure 1 and Table 1).

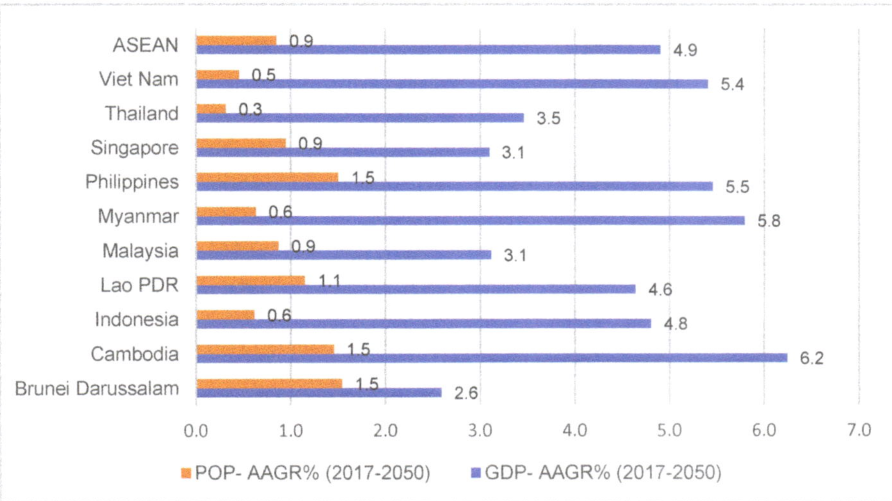

Figure 1. Average Annual Growth Rate of GDP (%) and Population in AMS, 2017–2050. AMS = ASEAN Member States, ASEAN = Association of Southeast Asian Nations, GDP-AAGR = average annual growth rate of GDP, POP-AAGR = average annual growth rate of the population. Source: Authors' calculations.

Table 1. Other Assumptions of Energy Saving Targets under the APS by AMS.

Country	Assumptions
Brunei Darussalam	Electricity: 35% reduction target by 2050
Cambodia	Specific fuel efficiency target by 2050 included (coal, oil, gas, biomass industry, 10%; electricity efficiency target, 20%)
Indonesia	Sectoral target by 2050 (commercial and residential, 10%; transport, 20%; bioethanol blending increase to 15% from 3–7% in 2010)
Lao PDR	Biodiesel: 20% blend from 1–5% in 2010; utilization of biofuels equivalent to 10% of road transport fuels
Malaysia	16% electricity saving by 2050 in industry, commercial, and residential sectors; 16% oil saving in final consumption by 2050; replacement of 5% of diesel in road transport with biodiesel
Myanmar	Target saving by 2050 included (transport and residential by 20%; industry, commercial, and others by 10%); replacement of 8% of transport diesel with biodiesel
Philippines	20% saving of oil and electricity by 2050; displacement of 20% of diesel and gasoline with biofuels by 2025
Thailand	Energy efficiency targets by 2050 included (transport, 70%; residential, 10%; commercial, 40%; and industry, 20% reduction of final energy demand); biofuels to displace 12.2% of transport energy demand
Viet Nam	20% reduction for all sectors; 10% ethanol blend in gasoline for road transport

AMS = ASEAN Member State, APS = alternative policy scenario, ASEAN = Association of Southeast Asian Nations, Lao PDR = Lao People's Democratic Republic. Source: Kimura and Phoumin (2019) [23].

This study generates and compares four possible energy mixes and their effects on greenhouse gas emissions over time. We break these pathways into four types, each with their own assumptions, including the APS, APS_RE, APS_EI, and APS_EmT. The APS refers to Alternative Policy Scenario and assumes that states will increase more efficient final energy consumption, more efficient power generation, a higher share of renewables, and the introduction of nuclear power plants, based on each AMS government policy. The assumptions used in the APS are described in the table below. The APS_RE is the APS with a higher share of renewable targets at the ASEAN level. In the APS_RE, the targets are increases of 23%, 30%, and 50% in the share of renewables in the primary energy supply by 2025, 2030, and 2050, respectively, from 2005 levels. The increase in the renewable share is expected from solar, wind, geothermal, and hydro. As hydro and geothermal energy are limited by resources, the maximum share is set based on the resource endowment. The APS_EI is the APS using energy intensity reduction targets of 30%, 40%, and 50% from 2005 levels by 2025, 2030, and 2050, respectively. A greater reduction in energy intensity means that the energy consumption per unit of GDP becomes more efficient as a result of the application of energy efficiency, technological development, or any economic structural transformation of the economies shifting from energy-intensive sectors such as industry to less energy-intensive sectors such as services. The APS_EmT is the APS using emission reduction targets of 40% and 80% from the BAU scenario by 2030 and 2050, respectively. This is the top-down policy target in which the energy mix composition needs to be changed towards cleaner energy to meet such targets. This will have many policy implications if the AMS wish to reduce emissions by as much as half from the BAU scenario by 2050.

4. Results and Analyses

The results of various energy supply and demand scenarios in ASEAN are in Tables A1–A8. ASEAN's energy system is predicted to be more efficient because energy intensity is expected to drop from the baseline in the future scenarios. However, the energy system will largely depend on fossil fuel consumption. The results from the energy model predicted that all ASEAN's emissions in the future scenarios will remain high because fossil fuel remains the dominant share in the future energy mix. Fossil fuel consumption—coal, oil, or natural gas—is associated with emissions, although natural gas has less emissions than coal and oil. It is also important to note that the trend of natural gas use in the energy transition is very promising, as its share has grown quickly in the primary energy mix as well as in power generation. Thus, ASEAN's energy transition will need to consider cleaner use of fossil fuels through clean technologies and a gradually increasing share of renewables and clean energy. Any policy changes to meet the emissions reduction in ASEAN need to be cautioned about high energy costs, energy access, affordability, and energy security risks. Below are the key results from the study.

More efficient use of energy. ASEAN's primary energy supply grows at an annual average rate of 3.1% from 2017 to 2050 under the BAU scenario, reaching 1823 million tones of oil equivalent (Mtoe) in 2050 from 639 Mtoe in 2017 (Figure 2). However, under the APS of ambitious emissions reduction targets (APS_EmT), the primary energy supply is predicted to reduce by 21% and 44% from the BAU in 2030 and 2050, respectively (Tables A1 and A2). ASEAN as a group achieves a significant reduction in energy intensity of 30.3% in the BAU case (a drop of energy intensity from 228 in 2017 to 154 in 2050). However, the scenario of emissions reduction targets (APS_EmT) could achieve a reduction of 60% in energy intensity in 2050 from the BAU scenario (a drop of energy intensity from 228 in 2017 to 86 in 2050) (Figure 3).

Reliance on fossil fuel consumption. The results from the energy demand and supply modelling under various policy scenarios draw attention to the high reliance on fossil fuel use in ASEAN's energy system. The total combined share of fossil fuels (oil, gas, and coal) in the primary energy supply was 78% in 2017; and they are predicted to have an 87%, 82%, and 80% share in 2050 under the BAU, APS, and APS with emission reduction targets (APS_EmT) scenarios, respectively (Figures 4 and 5).

Oil remains the dominant fuel in the primary energy supply, with a share of 37% in 2017. The share of oil is projected to be 42%, 41%, and 38% in the BAU scenario, APS, and APS_EmT in 2050, respectively (Figures 6 and 7). Oil is mainly used in the transport and industrial sectors in the final energy demand. The share of oil in the final energy demand was 45% in 2017, and its share grows to 51%, 50%, and 49% in 2050 for the BAU scenario, APS, and APS_EmT, respectively. This indicates that ASEAN as a group will rely heavily on oil consumption for the foreseeable future. For most countries in ASEAN, the growing oil import dependency will need to be safeguarded by resilient infrastructure and mechanisms such as oil stockpiling (either government stock or inventory stock by the oil importing companies). Most countries in ASEAN have a stock requirement of 15–50 days, varying from country to country. However, the stock requirement for OECD members will need to be at least 90 days of net oil imports to meet the emergency oil stock holding requirement in case of supply disruption (IEA, 2020) [24].

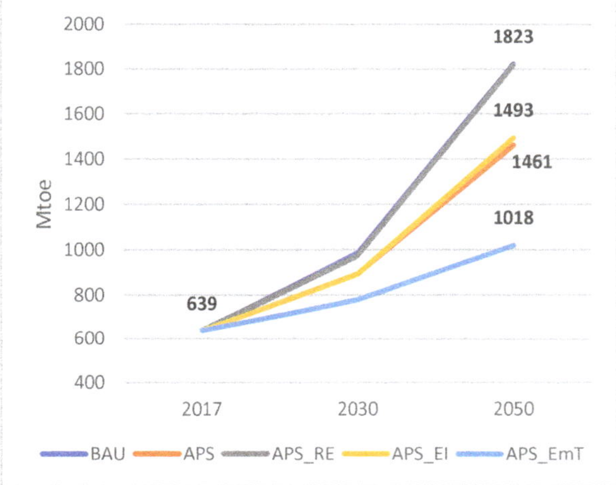

Figure 2. Primary Energy Supply (TPES) in ASEAN. APS = alternative policy scenario, APS_EI = alternative policy scenario with energy intensity targets, APS_EmT = alternative policy scenario with emission reduction targets, APS_RE = alternative policy scenario with renewable targets, ASEAN = Association of Southeast Asian Nations, BAU = business as usual, Mtoe = million tones of oil equivalent, TPES = total primary energy supply. Source: Authors' calculations.

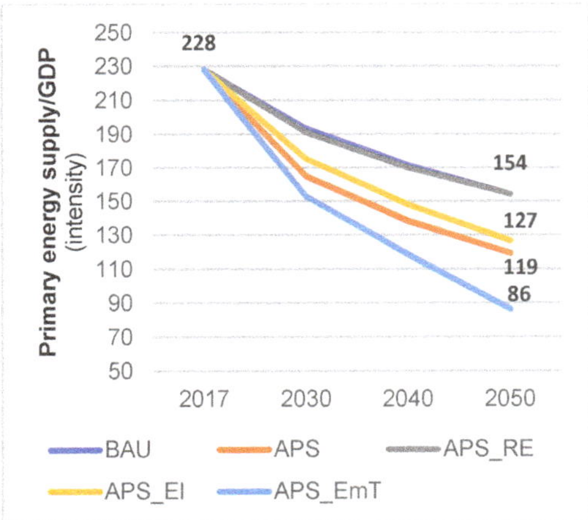

Figure 3. Energy Intensity in ASEAN. APS = alternative policy scenario, APS_EI = alternative policy scenario with energy intensity targets, APS_EmT = alternative policy scenario with emission reduction targets, APS_RE = alternative policy scenario with renewable targets, ASEAN = Association of Southeast Asian Nations, BAU = business as usual, GDP = gross domestic product. Source: Authors' calculations.

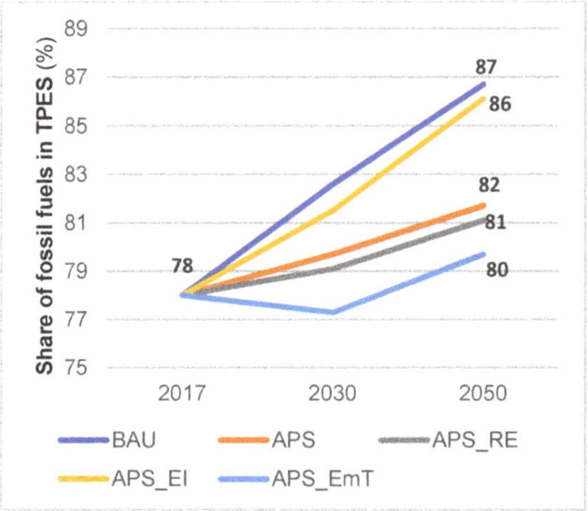

Figure 4. Share of Fossil Fuels (Coal, Oil, Gas) in the TPES. APS = alternative policy scenario, APS_EI = alternative policy scenario with energy intensity targets, APS_EmT = alternative policy scenario with emission reduction targets, APS_RE = alternative policy scenario with renewable targets, BAU = business as usual, TPES = total primary energy supply. Source: Authors' calculations.

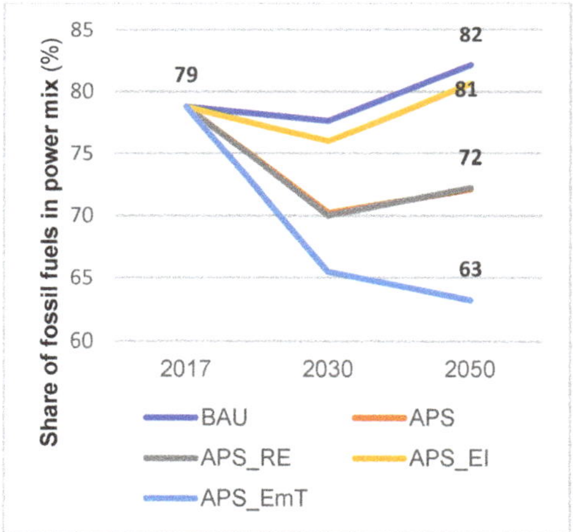

Figure 5. Share of Fossil Fuels in the Power Mix. APS = alternative policy scenario, APS_EI = alternative policy scenario with energy intensity targets, APS_EmT = alternative policy scenario with emission reduction targets, APS_RE = alternative policy scenario with renewable targets, BAU = business as usual. Source: Authors' calculations.

The share of coal in the primary energy supply was 22% in 2017; and it is predicted to be 23%, 17%, and 14% in the BAU scenario, APS, and APS_EmT in 2050, respectively. Coal has the second largest share in power generation, at 37% in 2017; and it is predicted to be 36%, 27%, and 19% in the BAU scenario, APS, and APS_EmT in 2050, respectively. Under the APS of emission reduction targets (APS_EmT), the share of coal is projected to drop significantly for both the primary energy supply as well as the share in the power generation mix (Figures 8 and 9).

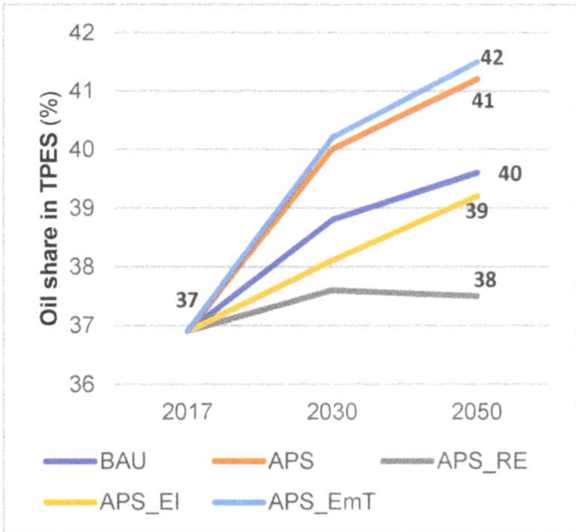

Figure 6. Oil Share in TPES in ASEAN. APS = alternative policy scenario, APS_EI = alternative policy scenario with energy intensity targets, APS_EmT = alternative policy scenario with emission reduction targets, APS_RE = alternative policy scenario with renewable targets, ASEAN = Association of Southeast Asian Nations, BAU = business as usual, TPES = total primary energy supply. Source: Authors' calculations.

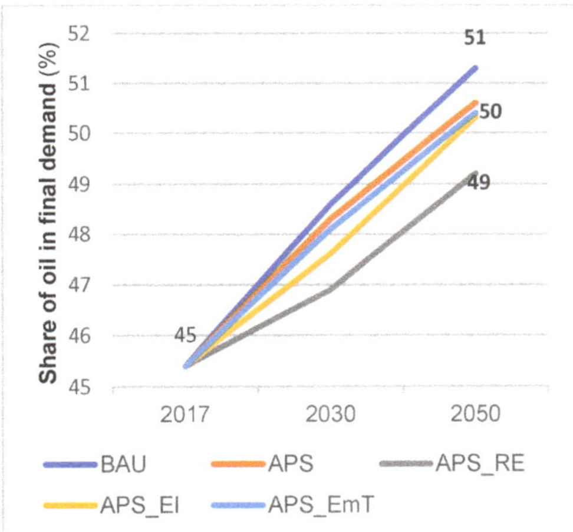

Figure 7. Oil Share in Final Demand in ASEAN. APS = alternative policy scenario, APS_EI = alternative policy scenario with energy intensity targets, APS_EmT = alternative policy scenario with emission reduction targets, APS_RE = alternative policy scenario with renewable targets, ASEAN = Association of Southeast Asian Nations, BAU = business as usual. Source: Authors' calculations.

Although ASEAN relies heavily on fossil fuels (oil, coal, and gas), some AMS have shifted drastically to use more gas in power generation and other final uses, such as the industrial and transportation sectors. ASEAN as a group had a 20% share of gas in the primary supply in 2017, but its share in the primary energy supply is projected to increase to 25% and 23% in 2050 for the BAU case and APS, respectively. Remarkably, the share of gas, at 40% in 2017, was a dominant fuel in the power generation mix; and it is projected to increase to 46%, 45%, and 44% in 2050 for the BAU case, APS, and APS_EmT, respectively (Figures 10 and 11).

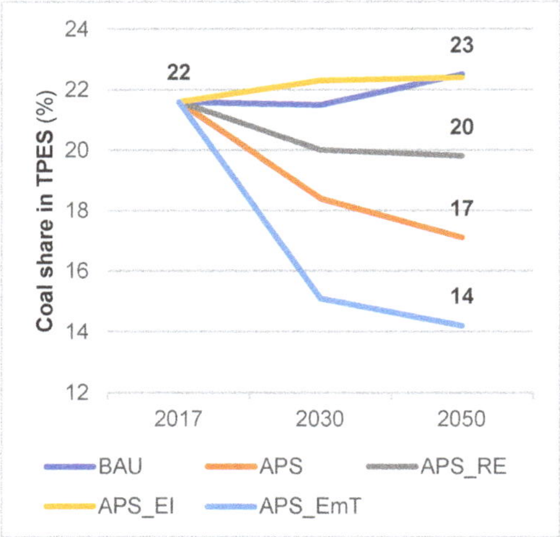

Figure 8. Coal Share in TPES in ASEAN. APS = alternative policy scenario, APS_EI = alternative policy scenario with energy intensity targets, APS_EmT = alternative policy scenario with emission reduction targets, APS_RE = alternative policy scenario with renewable targets, ASEAN = Association of Southeast Asian Nations, BAU = business as usual, TPES = total primary energy supply. Source: Authors' calculations.

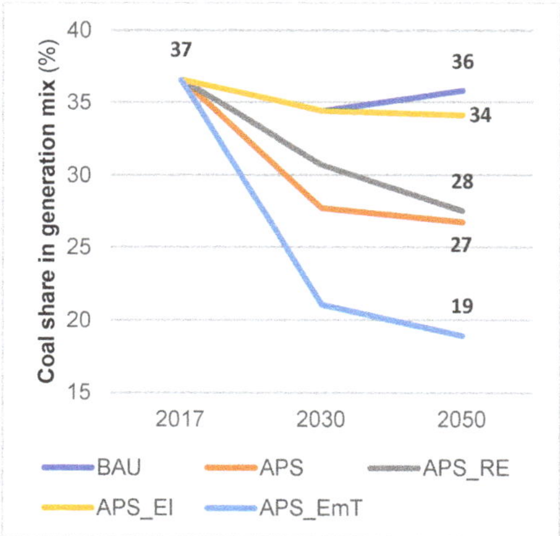

Figure 9. Coal Share in Generation Mix in ASEAN. APS = alternative policy scenario, APS_EI = alternative policy scenario with energy intensity targets, APS_EmT = alternative policy scenario with emission reduction targets, APS_RE = alternative policy scenario with renewable targets, ASEAN = Association of Southeast Asian Nations, BAU = business as usual. Source: Authors' calculations.

Increasing but not sufficient share of renewables. The share of renewables (hydropower, geothermal, biomass, wind, and solar) in the power mix was 21% in 2017. Its share is projected to increase to 36%, 28%, and 27% in the APS_EmT, APS_RE, and APS in 2050 (Figure 12). The share of renewables is projected to be higher in 2030 than 2050 because hydropower and geothermal resources are limited. However, the share of wind and solar is projected to increase from 2% in 2017 to 18%, 12%, and 11% in 2050 under the APS_EmT, APS_RE, and APS, respectively (Figure 13).

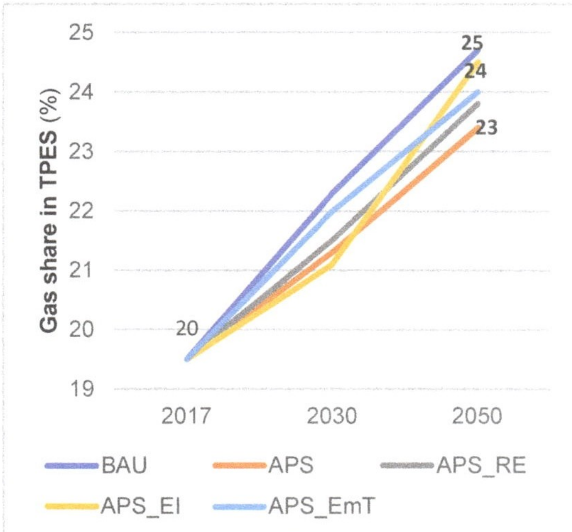

Figure 10. Gas Share in TPES in ASEAN. APS = alternative policy scenario, APS_EI = alternative policy scenario with energy intensity targets, APS_EmT = alternative policy scenario with emission reduction targets, APS_RE = alternative policy scenario with renewable targets, ASEAN = Association of Southeast Asian Nations, BAU = business as usual, TPES = total primary energy supply. Source: Authors' calculations.

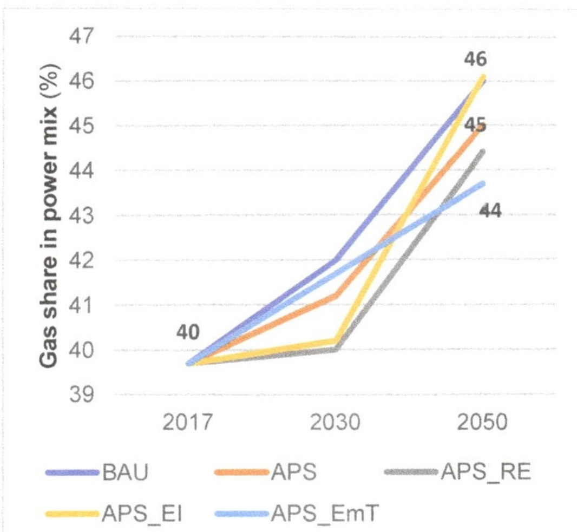

Figure 11. Gas Share in Generation Mix. APS = alternative policy scenario, APS_EI = alternative policy scenario with energy intensity targets, APS_EmT = alternative policy scenario with emission reduction targets, APS_RE = alternative policy scenario with renewable targets, ASEAN = Association of Southeast Asian Nations, BAU = business as usual. Source: Authors' calculations.

Although renewables are key to achieving emissions reductions, their share in the energy mix is not high enough to decarbonize emissions to meet the climate target of reducing emissions to net zero from 2050 until the turn of this century (Figures 14 and 15).

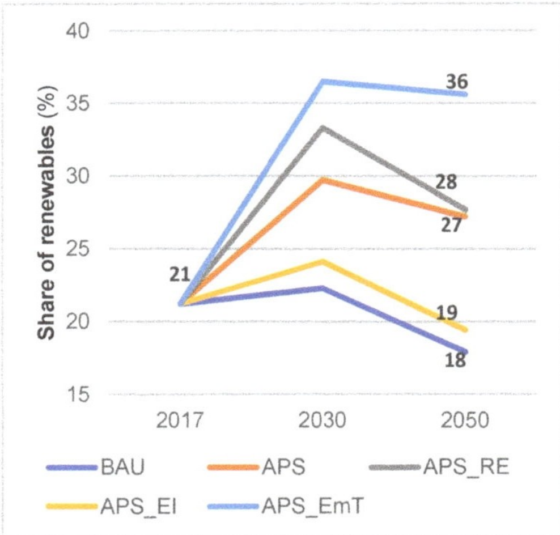

Figure 12. Renewables Share in Power Mix. APS = alternative policy scenario, APS_EI = alternative policy scenario with energy intensity targets, APS_EmT = alternative policy scenario with emission reduction targets, APS_RE = alternative policy scenario with renewable targets, ASEAN = Association of Southeast Asian Nations, BAU = business as usual. Source: Authors' calculations.

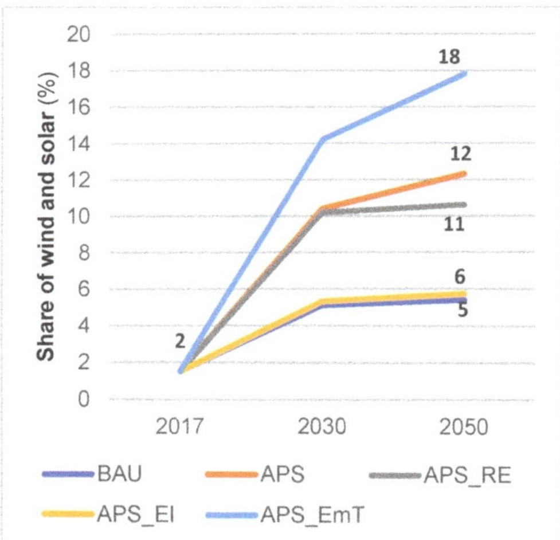

Figure 13. Share of Wind and Solar in Power Mix. APS = alternative policy scenario, APS_EI = alternative policy scenario with energy intensity targets, APS_EmT = alternative policy scenario with emission reduction targets, APS_RE = alternative policy scenario with renewable targets, ASEAN = Association of Southeast Asian Nations, BAU = business as usual. Source: Authors' calculations.

Achieving the APS_EmT is very unlikely because this scenario assumes the most efficient technologies and the highest share of renewables to achieve emissions reduction targets. Although the emissions reduction target was set at 80% from the BAU scenario to the APS_EmT, given the plausible challenges of integrating wind and solar in ASEAN's system, only 55% could be achieved for all combined types of renewables. Thus, the remaining emissions coming from fossil fuels will need to be decarbonized through CCUS technologies or the growth of natural carbon stock.

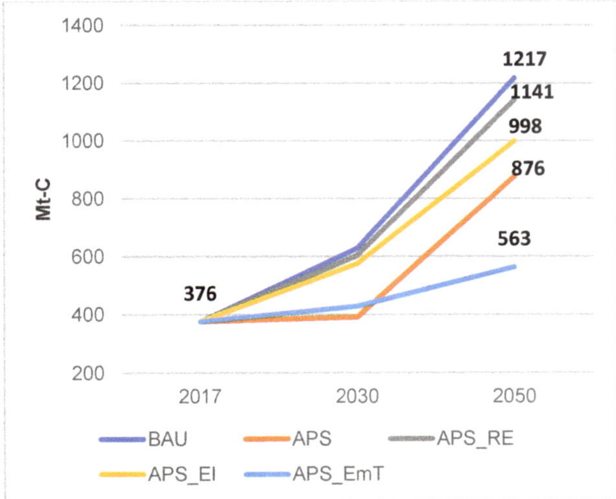

Figure 14. Emission Reduction in Various Scenarios. APS = alternative policy scenario, APS_EI = alternative policy scenario with energy intensity targets, APS_EmT = alternative policy scenario with emission reduction targets, APS_RE = alternative policy scenario with renewable targets, ASEAN = Association of Southeast Asian Nations, BAU = business as usual. Mt-C = million tonnes of carbon. Source: Authors' calculations.

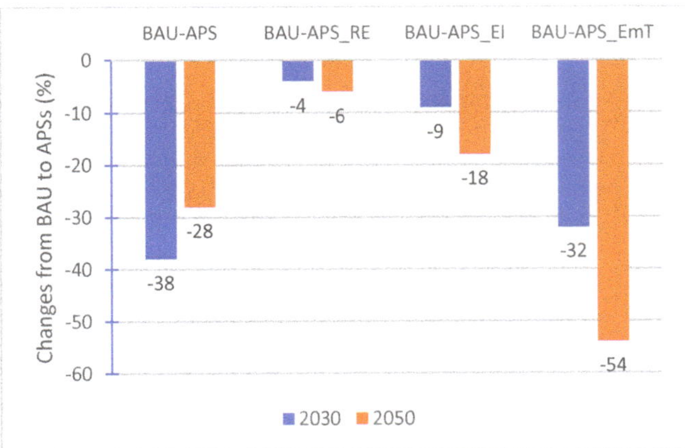

Figure 15. Emission Reduction in the Power Mix. APS = alternative policy scenario, APS_EI = alternative policy scenario with energy intensity targets, APS_EmT = alternative policy scenario with emission reduction targets, APS_RE = alternative policy scenario with renewable targets, ASEAN = Association of Southeast Asian Nations, BAU = business as usual. Source: Authors' calculations.

ASEAN's emissions keep increasing in the foreseeable scenarios. ASEAN as a group will see emissions doubling or tripling from 2017 to 2050, varying from the BAU case to the APSs. In the BAU scenario, emissions could reach 1217 million tonnes of carbon (Mt-C), almost triple the baseline level of 376 Mt-C in 2017. However, emissions could also be lower, at 876 Mt-C for the APS and 563 Mt-C for the APS_EmT (Figure 14). To limit the global temperature rise to 1.5 °C by 2100, emissions will need to be slashed by 45% from 2010 levels by 2030, then reach net zero emissions by 2050 (The Climate Reality Project, 2018) [25]. Thus, ASEAN as a group will miss this target and it will make it more difficult to cut emissions by 2050.

Required investment in power generation. Figure 16 is the estimated required investment for solar and wind energy. For the Small Modular Reactors (SMR) is not included in this estimation because ASEAN does not have plan to introduce nuclear power plant soon, although they keep this option open for the future. Thus, only solar and wind are estimated.

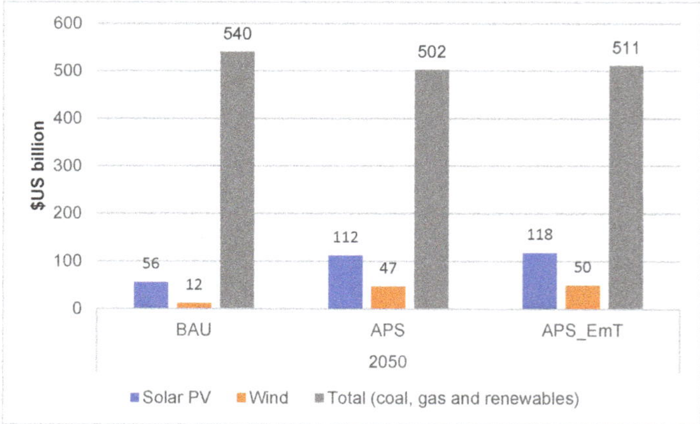

Figure 16. Required Investment for Variable Renewable Energy (Solar and Wind) by 2050. APS = alternative policy scenario, APS_EI = alternative policy scenario with energy intensity targets, APS_EmT = alternative policy scenario with emission reduction targets, APS_RE = alternative policy scenario with renewable targets, ASEAN = Association of Southeast Asian Nations, BAU = business as usual. Source: Authors' calculations.

Accelerating the share of variable renewables, such as solar and wind, in ASEAN's power mix will require \$56 billion–\$118 billion from the BAU scenario to the APSs in the case of solar photovoltaic and \$12 billion–\$50 billion in the case of wind, in 2050 (Figure 16). The total investment in the power generation of additional capacity will be \$540 billion in the BAU scenario and \$511 billion in the APSs—reflecting the reduced investment in fossil fuels and the increase in renewables, which will have less capital costs, driven by technological development, expected in 2050.

5. Implications of the Scenario Results

In 2020, fossil fuels (oil, coal, and natural gas) have the largest share of ASEAN's primary energy mix, at 78%. They are expected to continue to have a dominant share in the BAU scenario in 2050, at 86%, but could drop slightly to an 82% and 80% share under the APS and APS emission reduction target (APS_EmT) respectively in 2050, when considering more efficient power generation, an increasing share of renewables, and energy efficiency measures (Tables A1–A8). Although oil has the largest share in the primary energy mix, natural gas and coal are the dominant energy sources in the power generation mix, at 37% and 44% respectively in 2017; and their share is projected to be 46% and 36% respectively in 2050.

Need for cleaner use of fossil fuels and clean technologies. The composition of the future energy system depends on the current actions, policies, and future policy changes. However, all decisions need to be weighed against potentially higher energy costs, affordability, and energy security risks. Coal consumption has dropped globally in recent years, but Southeast Asia has seen the opposite trend—coal consumption has been concentrated in power generation although its share of the primary energy supply remains the same from the BAU scenario to the APS, while the actual quantity of coal consumption is predicted to increase significantly from 143 Mtoe in 2017 to 251 Mtoe in 2050. The relatively high level of coal consumption in ASEAN could be attributable to affordability and energy security issues. As coal will be the second most dominant source of energy for power generation, there is a real concern that many ASEAN countries cannot afford clean technologies such as CCT (advanced ultra-supercritical (A-USC) or ultra-supercritical (USC) technology) due to the higher up-front cost of these technologies compared with conventional high-emissions coal power plants (subcritical technology). At the same time, ASEAN as a bloc has lower emissions standards for coal-fired power plants than advanced countries such as Germany, Japan, and the Republic of Korea, where CCT is mandatory (Figure 17) [26]. This means that ASEAN countries have relatively high allowable emissions in terms of sulphur oxides (SOx), nitrogen oxides (NOx), and particulate matter (PM).

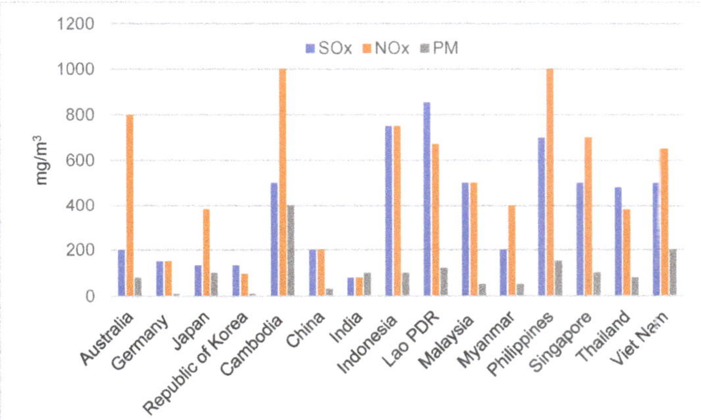

Figure 17. Emissions Standards for Newly Constructed Coal-Fired Power Plants in Selected Countries. Lao PDR = Lao People Democratic Republic, mg/m^3 = milligram per cubic metre, NOx—nitrogen oxides, PM = particulate matter, SOx—sulphur oxides. Source: Motokura et al. (2017) [26].

For ASEAN, the reliance on fossil fuel will be expected in the foreseeable future by 2050, even though ASEAN member states will invest in new infrastructure for cleaner-burning fossil fuels such as Clean Coal Technologies or Carbon Capture, Utilization and Storage (CCUS). However, the introduction of cleaner burning fossil fuels will help mitigate the pollutants nitrogen oxides, particulate matter, and sulfur oxides which are the major source contributing to direct human health such as respiratory system and cancer. Further, it also can help to cut down the carbon dioxide emissions if compared with traditional and inefficient coal-fired power generation.

Promoting natural gas uses in ASEAN's energy transition. Natural gas has a significant role to play in ASEAN's transition to a cleaner energy system. ASEAN as a group is forecast to continue to be a net natural gas exporter until 2030, but the situation will change due to declining domestic natural gas production and increasing domestic energy demand in ASEAN (Kobayashi and Phoumin, 2018) [27]. Demand for liquefied natural gas (LNG) in ASEAN is driven by increasing demand from the power generation and industrial sectors. Most AMS will see rising LNG imports in the foreseeable future because of sustained growth in electricity demand, the public preference for a cleaner fuel, and depleting domestic production. Prospects for the use of natural gas in ASEAN are optimistic, and demand is likely to increase 3.5 times in the BAU case (from 129 Mtoe in 2027 to 450 Mtoe in 2050)—depending on the future stability of gas and LNG market prices, and whether ASEAN and East Asia can create a competitive gas/LNG market in the future, with potential supply of gas/LNG from Australia, US, and other sources. Thus, ASEAN is expected to be a key market for future gas demand, so investment in gas infrastructure (such as gas pipelines and LNG receiving terminals) is crucial to support the increasing demand for gas in ASEAN.

ASEAN's scaling up renewable share and adoption of smart grid. Energy sustainability in ASEAN and around the globe requires an increased share of renewables in the energy mix to decarbonize emissions. Currently, ASEAN's power generation mix is dominated by coal, gas, and hydropower (Tables A1–A8). Intermittent renewables (solar and wind) comprise the most abundant energy resources in ASEAN, but have contributed negligible amounts (1.4% in 2017, 2.4% in 2020, and 10% and 12% in 2050 for the APS) to the power mix. Many ASEAN grid operators hold misperceptions about intermittent renewable energy. Although the production cost of renewable energy has dropped dramatically in recent years, its share in the power generation mix remains small. The misperceptions about renewable energy stem from its variable and intermittent nature, which adds costs to grid systems as it requires back-up capacity from conventional gas power plants. Technically, wind and solar power output varies depending on the strength of the wind or the amount of sunshine. However, this risk of variable energy output can be minimized if power systems are integrated within countries and within the ASEAN region. The aggregation of output from solar and wind from different geographical locations has a balancing effect on the variability (NREL, 2020) [28]. However, the ASEAN Power Grid is making slow progress and the integrated ASEAN power market may remain unrealized due to several reasons, such as regulatory and technical harmonization issues between the ASEAN Power Grid and utilities.

Challenges of power system integration in ASEAN. In the recent development of the power mix in ASEAN, some countries have accelerated the increase in the share of solar in the power mix without properly considering the poor gird infrastructure and power system integration challenges. As a result, electricity from solar has been curtailed. It is important to note that the shift from fossil fuels towards renewables in the energy transition will involve costs and investments for all energy-related infrastructure, which will hugely affect energy affordability. For AMS that can afford significant investments in renewable energies, an important concern is the need for electricity storage and smart grids to support higher renewable energy penetration levels in the electricity sector. Smart grid technologies are already making significant contributions to electricity grids in some developed countries of the OECD. However, these technologies are undergoing continual refinement and hence are vulnerable to potential technical and non-technical risks. Renewable energy growth will thus be constrained by infrastructure development as well as by the evolution of technology, including the capacity to assess and predict the availability of renewable energy sources (Kimura, Pacudan, and Phoumin, 2017) [29]. These capacities of smart grids offer additional benefits, notably the promise of higher reliability and overall electricity system efficiency.

Long-term emissions reduction and COVID-19. Due to the drastic decline in energy consumption, daily global emissions dropped by 17% in the first quarter of 2020 compared with 2019 levels (Le Quéré et al., 2020) [30]. However, an economic recovery could see the levels of CO_2 emissions bouncing back very quickly. Indeed, global data from late May 2020 show an all-time high for CO_2 levels, as countries started to reopen their economies. The sudden drop in current emissions has nothing to do with low-carbon energy policy measures—it is just the impact of the pandemic slowing down all economic activities. It is also understandable that the energy structure cannot be changed overnight, given its large dependence on fossil fuels. The results have shown that ASEAN emissions will be 1217 Mt-C in the BAU and 565 Mt-C to 876 Mt-C in the APSs, in which they are supposed to fall to zero emissions if the rise in temperature is to keep within 1.5 °C by the end of this century. This means that ASEAN will not be able to achieve the emissions reduction targets. This necessitates a serious review of the commitment in the NDCs or INDCs to limit the emissions to half by 2030 and reach net zero emissions by 2050. It also points to the urgent need for carbon sink technologies such as CCUS.

ASEAN's energy transition from a system based on fossil fuels to a system based on cleaner energy use will rely on investment in quality infrastructure—including renewable and cleaner use of fossil fuels, and CCUS—to reduce global GHG emissions and avoid the most serious impacts of climate change. Clean technologies and CCUS are the obvious choice to reduce fossil fuel emissions in ASEAN, while accelerating the use of renewables and the application of energy efficiency in all sectors.

Need for quality energy infrastructure and investment. To satisfy the growing energy demand in ASEAN, huge energy-related infrastructure investment is necessary between now and 2050. This study estimates that about $500 billion–$550 billion will be necessary in the power generation sector, of which combined variable renewables (wind and solar) will require $68 billion–$168 billion from the BAU scenario to the APSs, respectively. More broadly, the IEA (2017) [2], projected that $2.1 trillion will be required for oil, gas, coal, and power supply infrastructure in ASEAN. More than 60% of investment goes to the power sector, with transmission and distribution accounting for more than half of the total necessary investment. Globally, the Ministry of Finance of Japan (2019) estimated that the infrastructure investment gap is estimated to be $15 trillion from now until 2040. Asia alone will have a $4.6 trillion investment gap from now until 2040 (Ministry of Finance, Japan, 2019) [31]. the huge potential for energy infrastructure related investment will need to be guided by appropriate policies to promote quality infrastructure and resilience in ASEAN for growth and sustainability. Thus, ASEAN will need to prepare an array of policies suited to specific conditions to facilitate investment opportunities.

6. Conclusions and Policy Implications

The results of various scenarios have shown that ASEAN's current and future energy mix relies greatly on fossil fuels. The current share of fossil fuels is almost 80% in the primary energy supply and its future share is projected to be 87% under the BAU scenario and 78% under the APS. ASEAN's emissions will remain very high in all APS scenarios. To limit the temperature rise to 2° Celsius, emissions will need to fall to half by 2030 and reach net zero emissions by 2050 from 2010 levels. Thus, the clean use of fossil fuels through clean technologies and CCUS will be the only technological options to decarbonize emissions from fossil fuel use. In the energy transition, natural gas should be promoted as a transitional fuel in ASEAN, given the abundant supply from Australia. Renewables, energy efficiency, and green hydrogen (Green hydrogen refers to the hydrogen production from renewable electricity.) should be accelerated—along with the adoption of clean ecotechnologies—in the medium to long term in ASEAN's future energy system. Policies to manage ASEAN's energy transition need to be weighed against potentially higher energy costs, affordability, and energy security risks. Oil is the dominant energy source in the transport sector, while natural gas and coal are the dominant energy sources for power generation in ASEAN. The higher share of natural gas in ASEAN's power mix is a step in the right direction in promoting natural gas use in the energy transition towards a cleaner energy system.

In many ASEAN countries, coal use in power generation has been locked into the foreseeable future energy mix, as current and future coal-fired power generation generally involves 20- to 35-year power purchasing agreements with state-owned utilities to provide electricity. Thus, ignoring coal use in ASEAN means ignoring the reality and emissions of coal use. Considering the clean use of coal as part of ASEAN's energy transition is crucial to address the priorities of energy affordability and climate change. The deployment of CCT is urgent in the ASEAN region. Although ASEAN's energy targets have been set to include more renewables, ASEAN faces challenges in implementing such targets because renewables remain expensive in terms of the system integration cost to achieve high penetration in the grid system. Smart grids using the internet of things will provide a new green investment infrastructure which allows more penetration of renewables, but significant investment is required such as hard grids, internet of things technologies and applications, data management, and human resources.

A cleaner energy system in ASEAN relies on today's actions, policies, and investments to accelerate a higher share of renewables, the adoption of clean technologies and clean use of fossil fuels, and investment in climate-resilient energy quality infrastructure. The need for variable renewable investment in the power mix is estimated to be $118 billion in the APSs. Finally, willingness to pay is crucial if ASEAN is to leapfrog from its current energy system towards more efficient and clean technologies and a higher share of renewables in the energy mix.

Below are the key policy implications from the study:

- AMS will require assistance from developed countries to support the deployment of clean coal technologies, so that some developing countries in ASEAN will be able to afford clean coal technologies (e.g., USC or A-USC) to remove pollutants and increase the efficiency of power plants.
- The current climate narrative and policy approach of banning coal use should be reviewed to assist emerging Asia to afford CCTs, if alternative energy options are not available or feasible for emerging Asia in the medium term to meet energy demand. Treating CCTs as technology solutions in the energy transition will be a win–win solution for the world in terms of mitigating emissions and for Asia in sustaining energy accessibility and affordability.
- Emerging Asia will rely on whatever CCTs are available in the market at an affordable price. The up-front cost of such USC or A-USC technology is higher than that of supercritical (SC) and sub-critical (C) technology. Thus, it is necessary to lower such costs through policies such as attractive financing loan schemes for USC technologies, or a strong political institution to deliver public financing for CCTs to emerging Asia.
- A policy framework should clearly state the corporate social responsibilities of developed and developing nations, respectively, by highlighting the near- and long-term policy measures towards the coal industry and coal-fired power generation. As emissions in ASEAN are expected to rise until 2050, carbon recycling technologies will be necessary. In this regard, the world needs to accelerate the research, development, and deployment of CCUS for commercialization in the near future.
- There is a need to accelerate smart grid infrastructure development and investment, and energy cooperation from developed countries to share the experience of energy system integration, to achieve a higher share of renewables in the power system.
- ASEAN should promote natural gas use in the energy transition, as it creates only half the emissions that coal produces. Thus, investment in natural gas infrastructure will be crucial to increase natural gas use in ASEAN.
- ASEAN should accelerate the penetration of renewables, while increasing the adoption of clean technologies and the deployment of CCUS in the foreseeable future.
- ASEAN's leaders should consider the gradual removal of blanket fossil fuel subsidies, but should replace them with subsidies targeted at vulnerable groups to help meet their basic energy needs and support their well-being.
- Other energy policy measures should consider the potential higher energy costs, energy affordability and accessibility, and energy security risks. Regular surveys to assess people's willingness to pay for energy costs will be key in planning policy measures/reforms.

Author Contributions: Conceptualization, H.P., F.K. and J.A.; Data curation, H.P.; Formal analysis, H.P.; Investigation, H.P.; Methodology, H.P.; Software, H.P.; Supervision, F.K. and J.A.; Writing—original draft, H.P. All authors have read and agreed to the published version of the manuscript.

Funding: This research received no external funding.

Institutional Review Board Statement: Not applicable.

Informed Consent Statement: Not applicable.

Data Availability Statement: Not applicable.

Conflicts of Interest: The authors declare no conflict of interest.

Appendix A

Table A1. Estimates of Primary Energy Supply and Percentage Changes from BAU to APSs, 2030 (Mtoe).

Item	2017	2030								
	Baseline	**BAU**	**APS**	**% Change (BAU vs. APS)**	**APS_RE**	**% Change (BAU vs. APS_RE)**	**APS_EI**	**% Change (BAU vs. APS_EI)**	**APS_EmT**	**% Change (BAU vs. APS_EmT)**
Coal	143	220	164	−25	195	−12	199	−10	118	−46
Oil	228	374	357	−5	366	−2	340	−9	314	−16
Natural gas	119	214	190	−11	209	−2	188	−12	172	−20
Nuclear	0	0	0	0	0	0	0	0	0	0
Hydro	16	24	24	0	25	7	23	−4	24	0
Geothermal	20	32	32	1	34	6	30	−5	32	2
Biomass	105	102	102	1	113	11	97	−5	97	−5
Solar, wind, ocean	1	6	12	90	12	81	6	−7	10	62
Biofuels	7	12	11	−7	18	48	10	−17	13	5
Electricity	−1	2	0	−108	2	19	0	−104	0	−106
Total	**639**	**986**	**893**	**−9**	**974**	**−1**	**893**	**−9**	**780**	**−21**

APS = alternative policy scenario, APS_EI = alternative policy scenario with energy intensity targets, APS_EmT = alternative policy scenario with emission reduction targets, APS_RE = alternative policy scenario with renewable targets, BAU = business as usual, Mtoe = million tonnes of oil equivalent. Source: Authors' calculations.

Table A2. Estimates of Primary Energy Supply and Percentage Changes from BAU to APSs, 2050 (Mtoe).

Item	2017	2050								
	Baseline	BAU	APS	% Change (BAU vs. APS)	APS_RE	% Change (BAU vs. APS_RE)	APS_EI	% Change (BAU vs. APS_EI)	APS_EmT	% Change (BAU vs. APS_EmT)
Coal	143	409	251	−39	360	−12	335	−18	145	−65
Oil	228	721	602	−17	681	−6	586	−19	423	−41
Natural gas	119	450	342	−24	432	−4	366	−19	245	−46
Nuclear	0	0	6	557	0	0	0	0	7	718
Hydro	16	31	30	−3	35	16	30	−3	28	−8
Geothermal	20	63	74	17	101	61	51	−19	41	−35
Biomass	105	99	104	4	127	28	87	−13	91	−8
Solar, wind, ocean	1	14	25	80	24	71	12	−16	24	72
Biofuels	7	28	23	−20	50	76	21	−26	13	−56
Electricity	−1	7	6	−12	6	−10	6	−7	3	−59
Total	**639**	**1823**	**1461**	**−20**	**1817**	**0**	**1493**	**−18**	**1018**	**−44**

APS = alternative policy scenario, APS_EI = alternative policy scenario with energy intensity targets, APS_EmT = alternative policy scenario with emission reduction targets, APS_RE = alternative policy scenario with renewable targets, BAU = business as usual, Mtoe = million tonnes of oil equivalent. Source: Author's calculations.

Table A3. Estimates of Final Energy Consumption and Percentage Changes from BAU to APSs, 2030 (Mtoe).

Item	2017	2030								
	Baseline	BAU	APS	% Change (BAU vs. APS)	APS_RE	% Change (BAU vs. APS_RE)	APS_EI	% Change (BAU vs. APS_EI)	APS_EmT	% Change (BAU vs. APS_EmT)
Industry	148	248	227	−8	241	−3	220	−11	199	−20
Transportation	129	231	201	−13	231	0	201	−13	184	−20
Others	141	190	177	−7	189	−1	176	−8	158	−17
Non-energy	62	80	80	0	66	−18	66	−18	66	−18
Total	**480**	**750**	**686**	**−9**	**727**	**−3**	**663**	**−12**	**607**	**−19**

APS = alternative policy scenario, APS_EI = alternative policy scenario with energy intensity targets, APS_EmT = alternative policy scenario with emission reduction targets, APS_RE = alternative policy scenario with renewable targets, BAU = business as usual, Mtoe = million tonnes of oil equivalent. Source: Authors' calculations.

Table A4. Estimates of Final Energy Consumption and Percentage Changes from BAU to APSs, 2050 (Mtoe).

Item	2017	2050								
	Baseline	**BAU**	**APS**	**% Change (BAU vs. APS)**	**APS_RE**	**% Change (BAU vs. APS_RE)**	**APS_EI**	**% Change (BAU vs. APS_EI)**	**APS_EmT**	**% Change (BAU vs. APS_EmT)**
Industry	148	453	386	−15	448	−1	381	−16	250	−45
Transportation	129	483	374	−23	486	1	376	−22	246	−49
Others	141	294	253	−14	294	0	253	−14	190	−36
Non-energy	62	126	126	0	109	−13	109	−13	109	−13
Total	**480**	**1356**	**1139**	**−16**	**1337**	**−1**	**1119**	**−17**	**794**	**−41**

APS = alternative policy scenario, APS_EI = alternative policy scenario with energy intensity targets, APS_EmT = alternative policy scenario with emission reduction targets, APS_RE = alternative policy scenario with renewable targets, BAU = business as usual, Mtoe = million tonnes of oil equivalent. Source: Authors' calculations.

Table A5. Estimates of Power Generation Mix and Percentage Changes from BAU to APSs, 2030 (TWh).

Item	2017	2030								
	Baseline	**BAU**	**APS**	**% Change (BAU vs. APS)**	**APS_RE**	**% Change (BAU vs. APS_RE)**	**APS_EI**	**% Change (BAU vs. APS_EI)**	**APS_EmT**	**% Change (BAU vs. APS_EmT)**
Coal	381	608	449	−26	582	−4	552	−9	298	−51
Oil	26	23	21	−9	21	−8	22	−5	10	−57
Natural gas	414	743	669	−10	660	−11	645	−13	591	−20
Nuclear	0	0	0	0	0	0	0	0	0	0
Hydro	183	267	276	4	397	49	267	0	278	4
Geothermal	23	37	37	1	39	7	35	−5	38	2
Others	14	91	169	86	193	113	86	−5	201	122
Total	**1041**	**1768**	**1622**	**−8**	**1892**	**7**	**1607**	**−9**	**1416**	**−20**

APS = alternative policy scenario, APS_EI = alternative policy scenario with energy intensity targets, APS_EmT = alternative policy scenario with emission reduction targets, APS_RE = alternative policy scenario with renewable targets, BAU = business as usual, TWh= terawatt-hour. Source: Authors' calculations.

Table A6. Estimates of Power Generation Mix and Percentage Changes from BAU to APSs, 2050 (TWh).

Item	2017	2050								
	Baseline	BAU	APS	% Change (BAU vs. APS)	APS_RE	% Change (BAU vs. APS_RE)	APS_EI	% Change (BAU vs. APS_EI)	APS_EmT	% Change (BAU vs. APS_EmT)
Coal	381	1232	772	−37	1054	−14	1005	−18	398	−68
Oil	26	12	12	1	12	0	11	−3	12	0
Natural gas	414	1582	1303	−18	1700	7	1359	−14	919	−42
Nuclear	0	0	21	2137	0	0	0	0	28	2757
Hydro	183	356	344	−3	537	51	346	−3	326	−8
Geothermal	23	73	86	17	118	61	59	−19	47	−35
Others	14	185	356	93	406	120	167	−10	376	104
Total	**1041**	**3439**	**2895**	**−16**	**3827**	**11**	**2948**	**−14**	**2105**	**−39**

APS = alternative policy scenario, APS_EI = alternative policy scenario with energy intensity targets, APS_EmT = alternative policy scenario with emission reduction targets, APS_RE = alternative policy scenario with renewable targets, BAU = business as usual, TWh = terawatt-hour. Source: Authors' calculations.

Table A7. Estimates of CO_2 Emissions and Percentage Changes from BAU to APSs, 2030 (Mt-C).

Item	2017	2030								
	Baseline	BAU	APS	% Change (BAU vs. APS)	APS_RE	% Change (BAU vs. APS_RE)	APS_EI	% Change (BAU vs. APS_EI)	APS_EmT	% Change (BAU vs. APS_EmT
Coal	147	227	144	−37	197	−13	197	−13	122	−46
Oil	138	249	147	−41	258	3	238	−5	202	−19
Natural gas	91	152	100	−34	148	−3	139	−9	105	−31
Total	**376**	**628**	**391**	**−38**	**603**	**−4**	**574**	**−9**	**429**	**−32**

APS = alternative policy scenario, APS_EI = alternative policy scenario with energy intensity targets, APS_EmT = alternative policy scenario with emission reduction targets, APS_RE = alternative policy scenario with renewable targets, BAU = business as usual, Mt-C = million tonnes of carbon equivalent. Source: Authors' calculations.

Table A8. Estimates of CO_2 Emissions and Percentage Changes from BAU to APSs, 2050 (Mt-C).

Item	2017	2050								
	Baseline	**BAU**	**APS**	**% Change (BAU vs. APS)**	**APS_RE**	**% Change (BAU vs. APS_RE)**	**APS_EI**	**% Change (BAU vs. APS_EI)**	**APS_EmT**	**% Change (BAU vs. APS_EmT**
Coal	147	432	264	−39	360	−17	317	−27	151	−65
Oil	138	503	395	−21	507	1	437	−13	280	−44
Natural gas	91	281	216	−23	275	−2	244	−13	132	−53
Total	**376**	**1217**	**876**	**−28**	**1141**	**−6**	**998**	**−18**	**563**	**−54**

APS = alternative policy scenario, APS_EI = alternative policy scenario with energy intensity targets, APS_EmT = alternative policy scenario with emission reduction targets, APS_RE = alternative policy scenario with renewable targets, BAU = business as usual, Mt-C = million tonnes of carbon equivalent. Source: Authors' calculations.

References

1. IMF. World Economic Outlook Reports. Available online: https://www.imf.org/en/publications/weo (accessed on 30 June 2020).
2. IEA. *World Energy Outlook 2017*; International Energy Agency: Paris, France, 2017; Available online: https://www.iea.org/reports/world-energy-outlook-2017 (accessed on 20 September 2020).
3. Phoumin, H.; Kimura, F.; Arima, J. Potential Renewable Hydrogen from Curtailed Electricity to Decarbonize ASEAN's Emissions: Policy Implications. *Sustainability* **2020**, *12*, 10560. [CrossRef]
4. Ali, H.; Phoumin, H.; Weller, S.R.; Suryadi, B. Cost–Benefit Analysis of HELE and Subcritical Coal-Fired Electricity Generation Technologies in Southeast Asia. *Sustainability* **2021**, *13*, 1591. [CrossRef]
5. Phoumin, H. Enabling Clean-Coal Technologies in Emerging Asia, Working Paper for the National Bureau of Asian Research 2015 Pacific Energy Summit, Beijing, 27–29 May 2015. Available online: https://www.nbr.org/publication/enabling-clean-coal-technologies-in-emerging-asia/ (accessed on 20 July 2020).
6. United States Environmental Protection Agency. Global Greenhouse Gas Emissions Data. 2020. Available online: http://www.epa.gov/climatechange/ghgemissions/global.html (accessed on 15 February 2020).
7. Economic Research Institute for ASEAN and East Asia (ERIA). *Round Table Discussion on ASEAN's Future Energy Landscape*; ERIA: Jakarta, Indonesia, 2020.
8. Isik, M.; Kaplan, P.O. Understanding Technology, Fuel, Market and Policy Drivers for New York State's Power Sector Transformation. *Sustainability* **2021**, *13*, 265. [CrossRef] [PubMed]
9. Phoumin, H.; Mikkal, E.H.; Nikos, T.; Courtney, W. Powering Southeast Asia: Meeting the Region's Electricity Needs. NBR Special Report #89. Available online: https://www.nbr.org/publication/powering-southeast-asia-meeting-the-regions-electricity-needs/ (accessed on 18 December 2020).
10. Nepal, R.; Phoumin, H.; Khatri, A. Green Technological Development and Deployment in the Association of Southeast Asian Economies (ASEAN)—At Crossroads or Roundabout? *Sustainability* **2021**, *13*, 758. [CrossRef]
11. Economic Research Institute for ASEAN and East Asia (ERIA). Asia-CCUS Network Conference. In Proceedings of the Virtual Conference on Asia-CCUS, Jakarta, Indonesia, 10 February 2020.
12. McCauley, D.; Heffron, R. Just Transition: Integrating climate, energy and environmental justice. *Energy Policy* **2018**, *119*, 1–7. [CrossRef]
13. Raphael, J.H.; McCauley, D.; Benjamin, K.S. Resolving society's energy trilemma through the Energy Justice Metric. *Energy Policy* **2015**, *87*, 168–176.
14. EU. EU Action: 2050 Long-Term Strategy. 2020. Available online: https://ec.europa.eu/clima/policies/strategies/2050 (accessed on 27 April 2020).
15. UNEP. *Emissions Gap Report 2019*; United Nations Environment Programme: Nairobi, Kenya, 2019; Available online: https://www.unenvironment.org/resources/emissions-gap-report-2019 (accessed on 26 September 2020).
16. UN Environment Assembly (UNEP). The Fifth Session of the UN Environment Assembly (UNEA-5). 2021. Available online: https://www.unep.org/environmentassembly/unea5 (accessed on 21 February 2021).
17. ASHRAE. The ASHRAE Digital: Lighthouse and Industry 4.0. 2020. Available online: https://online.flipbuilder.com/Mashrae.bog/ynzh/ (accessed on 21 February 2021).
18. ASEAN Centre for Energy (ACE). *ASEAN Plan of Action for Energy Cooperation 2016–2025: Phase I: 2016–2020*; ACE: Jakarta, Indonesia, 2015.
19. ASEAN Centre for Energy. The ASEAN Member States Are on Their Way to Implement APAEC Phase II, News Release. 2020. Available online: https://aseanenergy.org/the-asean-member-states-are-on-their-way-to-implement-apaec-phase-ii/ (accessed on 25 September 2020).
20. Kimura, S.; Phoumin, H. (Eds.) *Technical Improvement Report on Energy Outlook and Energy Saving Potential in East Asia*; ERIA Research Project Report 2016, No. 8; Economic Research Institute for ASEAN and East Asia: Jakarta, Indonesia, 2018; Available online: https://www.eria.org/research/technical-improvement-report-on-energy-outlook-and-energy-saving-potential-in-east-asia/ (accessed on 30 June 2020).
21. UNFCCC. *Intended Nationally Determined Contributions: Submissions*; United Nations Framework Convention on Climate Change: New York, NY, USA, 2015; Available online: https://www4.unfccc.int/sites/submissions/indc/Submission%20Pages/submissions.aspx (accessed on 14 April 2020).
22. IEA. *World Energy Model*; International Energy Agency: Paris, France, 2019; Available online: https://www.iea.org/reports/world-energy-model/macro-drivers (accessed on 17 September 2020).
23. Kimura, S.; Phoumin, H. (Eds.) *Energy Outlook and Energy Saving Potential in East Asia 2019*; Economic Research Institute for ASEAN and East Asia: Jakarta, Indonesia, 2019.
24. IEA. Oil Stocks of IEA Countries. 15 September 2020. Available online: https://www.iea.org/articles/oil-stocks-of-iea-countries (accessed on 21 September 2020).
25. The Climate Reality Project. 2030 or Bust: 5 Key Takeaways from the IPCC Report, Blog. 18 October 2018. Available online: https://www.climaterealityproject.org/blog/2030-or-bust-5-key-takeaways-ipcc-report (accessed on 30 September 2020).

26. Motokura, M.; Lee, J.; Kutani, I.; Phoumin, H. (Eds.) *Improving Emission Regulation for Coal-fired Power Plants in ASEAN*; ERIA Research Project Report 2016, No. 2; ERIA: Jakarta, Indonesia, 2017; Available online: https://www.eria.org/RPR_FY2016_02.pdf (accessed on 10 June 2020).
27. Kobayashi, Y.; Phoumin, H. (Eds.) *Natural Gas Master Plan for Myanmar*; ERIA Research Project Report 2017, No. 17; Economic Research Institute for ASEAN and East Asia: Jakarta, Indonesia, 2018; Available online: https://www.eria.org/publications/natural-gas-master-plan-for-myanmar/ (accessed on 27 July 2020).
28. NREL. *Wind and Solar on the Power Grid: Myths and Misperceptions*; National Renewable Energy Laboratory: Golden, CO, USA, 2020. Available online: https://www.nrel.gov/docs/fy15osti/63045.pdf (accessed on 1 July 2020).
29. Kimura, S.; Pacudan, R.; Phoumin, H. (Eds.) *Development of the Eco Town Model in the ASEAN Region through Adoption of Energy-Efficient Building Technologies, Sustainable Transport, and Smart Grids*; ERIA Research Project Report 2015, No. 20; Economic Research Institute for ASEAN and East Asia: Jakarta, Indonesia, 2017; Available online: https://www.eria.org/RPR_FY2015_No.20.pdf (accessed on 29 September 2020).
30. Le Quéré, C.; Jackson, R.B.; Jones, M.W.; Smith, A.J.P.; Abernethy, S.; Andrew, R.M.; De-Gol, A.J.; Willis, D.R.; Shan, Y.; Canadell, J.G.; et al. Temporary reduction in daily global CO2 emissions during the COVID-19 forced confinement. *Nat. Clim. Chang.* **2020**, *10*, 647–653. [CrossRef]
31. Ministry of Finance, Japan. Quality Infrastructure Investment G20 Deliverables under Japanese Presidency, Presentation, The Fifth Tokyo Fiscal Forum, Tokyo, 20–21 November 2019. Available online: https://www.mof.go.jp/pri/research/seminar/fy2019/tff2019_s1_01.pdf (accessed on 29 September 2020).

sustainability

MDPI

Article

Quantitative Assessment of Energy Supply Security: Korea Case Study

Herie Park [1] and Sungwoo Bae [2,*]

1 Division of Electrical and Biomedical Engineering, Hanyang University, Seoul 04763, Korea; bakery@hanyang.ac.kr
2 Department of Electrical Engineering, Hanyang University, Seoul 04763, Korea
* Correspondence: swbae@hanyang.ac.kr; Tel.: +82-2-2220-2309

Abstract: Ensuring energy supply security has become one of the most important purposes for many countries. To make the strategies for ensuring the energy supply security of a country, it is essential to quantitatively assess the security. This paper aims to present a methodology to evaluate the energy supply security of a country by using different indices of energy dependence and energy diversity, which have been raised as two main paradigms of energy supply security. This study also proposes two indices reflecting the correlation between a country's energy diversity and energy import dependence to evaluate its energy supply security based on easily accessible data. The presented methodology and indices were applied to the evaluation of the primary energy supply security of Korea from 1991 to 2018. The results show that a country highly dependent on energy imports is not evaluated as secure enough in terms of energy supply even if it obtains higher energy diversity. This finding supports the importance of the correlation of energy dependence and energy diversity of a country to ensure its energy supply security. This approach could be further adapted to other countries and help them to make their energy policy and strategies.

Keywords: energy supply security; energy dependence; energy diversity; energy policy

Citation: Park, H.; Bae, S. Quantitative Assessment of Energy Supply Security: Korea Case Study. *Sustainability* **2021**, *13*, 1854. https://doi.org/10.3390/su13041854

Academic Editors: Farhad Taghizadeh-Hesary
Received: 16 December 2020
Accepted: 1 February 2021
Published: 8 February 2021

1. Introduction

The increase in population and social growth has accelerated energy consumption, bringing concerns over energy security [1]. Most countries have set their energy policies and strategies to ensure energy security for sustainable development and growth [2–4]. However, because of the geographical inequality of energy resources and resource depletion, many countries have difficulty in lowering their dependence on energy imports and ensuring their energy supply security. Moreover, energy supply security is exposed by exterior risks, such as energy price fluctuations, supply disruption, and geopolitical uncertainty.

To reduce the risk of energy supply security related to energy import dependence, Månsson et al. [5] listed solutions such as improving diversity, financial portfolios, and reliable supply and transit routes. Chalvatzis and Ioannidis [6] also focused on energy dependence and diversity as two main paradigms of energy supply security and insisted that energy dependence has raised the importance of diversity in energy security paradigms. They categorized the primary energy import dependence of different countries and analyzed their primary energy diversity to provide reference benchmarks among countries. Matsumoto et al. [7] also classified the overall improvement of primary energy diversity in EU countries using a clustering method.

Moreover, countries that are highly dependent on energy imports such as Singapore, Taiwan, Japan, and Korea have also put great effort into diversifying energy resources to reduce their energy security vulnerability. These countries have perceived the importance of energy diversification and formulated policy concepts in the direction of diversifying energy resources [8,9]. Chuang and Ma [10] evaluated the diversity of primary energy in the energy supply structure of Taiwan through the most commonly used diversification indices: the Shannon-Wiener Index (*SWI*) and the Herfindahl-Hirschman Index (*HHI*). In another case study, Matsumoto conducted a socioeconomic study of the primary energy security performance of Japan considering the diversity of primary energy based on the *SWI* [11]. These research works provided quantifiable indices and helped to compare the given values to that of other countries.

However, as the *SWI* and *HHI* only deal with the number and proportion of energy resources, there is no information on the origin of energy resources. The results have thus led to demonstrate the diversity of already available energy resources in the county. To express the diversity and other factors related to energy resources simultaneously, a few researchers have proposed a diversity index weighted by the dependence on energy suppliers and their political stability [5,7,10,11], the diversity and reliability index [10], and the diversity and reliability variance index [10] based on the *SWI* and the *HHI*. However, these indices mostly require more specific data that are not easily accessible, nor available to the public. In addition, there has been a lack of study on the direct relationship between energy dependence and diversity.

This study thus suggests modified indices of energy diversity based on the *SWI* and *HHI*, which are weighted by import dependence-related parameters, as supplementary data to assess the energy supply security of a country. As these indices basically stem from the preliminarily obtained import dependence and primary energy diversity, the results are relatively easily obtained and compared with the reference values. The suggested indices are applied to evaluate the energy supply security of any country. In order to validate the effectiveness of the indices, this study adapted the indices to the case of Korea, a country highly dependent on energy import and making an effort to ensure their energy supply security, and compared the results with that of the conventional metrics.

The main contributions of this study are summarized as follows:

- This study suggests modified indices of energy diversity considering a weighted factor stemming from energy import dependence. Most previous studies have analyzed energy dependence and diversity separately. These studies seem to lack a fundamental factor indicating the import dependence of the target energy source. Therefore, the modified indices of energy diversity weighted by import dependence-related parameters are proposed in this study. These indices reflect the correlation between a country's energy diversity and energy import dependence to evaluate its energy supply security.
- This study performs a comprehensive case analysis of the energy supply security in Korea and introduces the energy transition of the country based on the proposed indices as well as conventional metrics of energy dependence and diversity, using a relevant historical dataset from 1991 to 2018. The proposed approach could be adapted in other countries to evaluate their energy supply security regarding energy diversity and energy import dependence as well as help them to make their energy policy and strategies.

The remainder of this paper is organized as follows: Section 2 introduces a literature review on global, regional, and national analysis of energy security. Section 3 presents the analysis methodology of energy supply security, including indices of energy dependence and energy diversity, proposed energy diversity indices, and data for evaluating the energy supply security of Korea. Section 4 describes the results and discussion on the case study of Korea. Section 5 provides the main conclusions and suggestions.

2. Literature Review

2.1. Global Level Analysis of Energy Supply Security

The main themes of energy supply security have changed slightly as the interests of society, technology trends, and geopolitical and economic situations change. The definition of energy supply security has been represented by a number of factors such as availability [12,13], infrastructure [13], energy price [12,14,15], environment [15–17], efficiency [15,18,19], social effects [17], and governance [19]. These definitions have even varied depending on the target application domains and their detailed subjects. While "reliable supplies of energy at reasonable prices to support the economy and industry" has been used broadly in the literature as the definition of energy supply security [14], other researchers have extended the definition to social welfare and socio-cultural effects [15,17]. Therefore, researchers, policymakers, and stakeholders have faced difficulty in establishing a global definition of energy security and its measuring method.

A number of studies have tried to integrate these concepts of energy supply security. Winzer [20] listed the definition of energy supply security by reviewing papers published in different domains in order to map the conceptual boundaries of energy security. Kruyt et al. [19] classified energy security indicators incorporating regionalization, globalization, economic efficiency, and environmental acceptability. Based on these factors, several frameworks for defining and methodologies for evaluating energy security have been suggested at the international, regional, and national levels. The analyses at the international level have provided the position, direction, and strengths and weaknesses of energy security in different countries [9,21–23]. These investigations have recommended policy prescriptions based on the evaluation results of energy security and have served as a reference for energy security among countries with differing conditions.

2.2. Regional and National Level Analysis of Energy Supply Security

From a more specific and comparative perspective, many studies have been conducted at regional and national levels [13,24–30]. Le et al. [24] reported the different levels of energy insecurity in Asia, focusing on 12 variables such as CO_2 emissions, CO_2 intensity, energy imports, energy intensity, energy use, fossil fuel energy consumption, and renewable energy consumption. Erahman et al. [25] presented the energy situation of Indonesia and 14 indicators for measuring energy security in the five dimensions of availability, affordability, accessibility, acceptability, and efficiency. Glynn et al. [26] provided a method to assess energy security considering sovereignty, infrastructural robustness, and market resilience for a future Irish energy system. Geng and Ji [27] presented China's energy supply security and its evolutionary characteristics by analyzing external availability, affordability, technologies, efficiency, and resource reserves. Narula et al. [28] introduced India's multi-dimensional sustainable energy security using 23 metrics considering availability, affordability, efficiency, and acceptability. Zhao and Yang [13] introduced a regional energy security assessment focusing on integrated systems with renewables in China. Malik et al. [29] analyzed Pakistan's energy security under the 4-A framework, namely availability, applicability, acceptability, and affordability over the six years. Gopal et al. [30] examined the energy security of the Indian electricity sector over a decade. They suggested 11 indicators representing economic, environmental, and social dimensions for their analysis. These researchers have attempted to determine commonly-used indicators assessing the energy supply security of each country and identified its unique characteristics for addressing the appropriate energy security of the target countries.

2.3. Energy Supply Security Analysis: Korea Case Study

Several research works on the evaluation of the energy supply security of Korea have also been conducted during the past decade. Their purpose is to understand the position and characteristics of energy supply security in Korea and to suggest energy policy prescriptions. Due to insufficient natural resources alongside economic and geopolitical uncertainty, the accessibility and availability of energy have been concerns for Korea. The top priority in energy supply security has thus been to avoid any disruption of the energy supply [9]. Jun et al. [23] presented the energy security costs for energy sources contained within the Korean electricity market. The minimum cost of ensuring energy supply security was measured as a supply disruption in the case of electricity generation and an energy portfolio with the lowest cost is suggested. Ryu et al. [31] investigated three scenarios with different levels of carbon emission, energy security, and electricity generation costs in Korea. Through these scenarios, the appropriate portions of specific energy resources for electricity generation were given for executing national energy policy strategies. Ahn et al. [32] also suggested the optimal allocation of energy resources for electricity generation by scenario-based analysis for sustainable development and lessened external risks to energy security in Korea. These studies have mostly concentrated on the energy mix for electricity as a form of final energy. Although there have been many attempts to analyze secure electricity generation in the country, there is a lack of research that investigates the primary energy supply. As energy resources have recently become more diverse than ever before with the development of energy conversion technologies and applications, this study includes various resources, including fossil fuels, nuclear, and renewable energy.

3. Methodology

To evaluate energy supply security, different analytical approaches and indicators, such as energy resource reserves [19], cost of energy security [23], energy dependence [6], energy diversity [6,7,10,11,19], and energy market concentration [33] have been presented in the literature. Amongst them, energy dependence and diversity have been regarded as the two main paradigms for evaluating energy supply security [6]. This study applied these two indicators and proposed modified indices of energy diversity to quantitatively assess the energy supply security of Korea, a country highly dependent on energy imports.

3.1. Energy Dependence

3.1.1. Energy Import Ratio

Most countries are engaged in the import and export of resources and products. The energy import ratio can be used as a simple index to estimate the energy dependence of a country. It describes the ratio between energy imports and the total imports of the country established as follows:

$$D_1 = \frac{EI_T}{AI_T} = \frac{\sum_{i=1}^{N} EI_i}{AI_T} \tag{1}$$

where D_1 is the energy import ratio, AI_T represents the total of all imports, and EI_T is the primary energy import total obtained by the sum of each energy import total. The value of i denotes the index of the types of energy sources, N is the number of the types of energy sources, and EI_i is the amount of the *ith* primary energy import.

3.1.2. Energy Import Dependence

Energy import dependence is one of the most used metrics for assessing energy supply security. It describes the percentage of total energy imports over the total energy supply of a country. Fossil fuels, such as oil, coal, gas, and uranium, are the energy sources that this index evaluates across different regions and countries [6,19]. It is expressed as follows:

$$D_2(\%) = \frac{EI_T}{ES_T} \times 100(\%) = \frac{\sum_{i=1}^{N} EI_i}{\sum_{i=1}^{N} ES_i} \times 100(\%) \tag{2}$$

where D_2 is the energy import dependence, and ES_T and ES_i represent the total amounts of all types of primary energy supply and the ith primary energy supply, respectively. In this term, EI_T and ES_T use the same monetary or physical unit.

3.2. Energy Diversity Indices

As one of the types of indices interrelated with energy dependence, the energy supply diversity indices have been quantified by several analytical approaches [34]. Stirling presented research focusing on diversity in the energy sector and suggested three important elements: variety, balance, and disparity [35,36]. As the disparity of energy options is not easily quantified, indices without consideration of disparity are widely applied in the measurement of diversity [6,10,34]. Hill identified an entire family of possible quantitative assessments of diversity in terms of proportional abundances of the species in a sample [37]. As the Hill index family, the *SWI* and *HHI* are the most commonly used indices in the energy sector. These two indices are applied in this study to evaluate energy diversity, especially the balance and variety of energy options in Korea.

3.2.1. The Shannon-Wiener Index (SWI)

The *SWI* is expressed by accounting for the share of each primary energy source as follows:

$$SWI = -\sum_{i=1}^{N} p_i \ln p_i \tag{3}$$

$$p_i = \frac{ES_i}{ES_T} \tag{4}$$

where p_i is the share of primary energy supply by the ith energy source ES_i in the total primary energy supply ES_T. The smallest value of the *SWI* is zero in the case where there is only one primary energy supply option in the country. It means the sole option takes 100% of the total primary energy supply. Then, the share of this energy source p_i becomes one. Consequently, ln (1) equals zero, and the *SWI* thus becomes zero. By contrast, the theoretical maximum value of the *SWI* may be obtained when the shares of primary energy sources are even, which means that p_i equals $1/N$ and that the number of types of energy sources becomes larger. If the share of each option is uneven, the value of the *SWI* is less than its theoretical maximum value, SWI_{Max}.

$$SWI_{Max} = -N\frac{1}{N}\ln\frac{1}{N} = \ln N \tag{5}$$

3.2.2. The Herfindahl-Hirschman Index (HHI)

The *HHI* shows the concentration of the individual share of energy options. It is expressed by the sum of the square of the proportion of each energy source as follows:

$$HHI = \sum_{i=1}^{N} P_i^2 = \sum_{i=1}^{N} (p_i \times 100)^2 \tag{6}$$

where P_i is the percentage of the primary energy supply by the ith energy source ES_i in the total primary energy supply ES_T. As this index uses the square of the percentage of each energy source, the index highlights quantitatively bigger energy resources rather than smaller ones in whole options. This means that the *HHI* emphasizes abundant energy resources. Opposite to the *SWI*, the lower the value of the *HHI*, the higher the diversity. The theoretical minimum value of *HHI* is expressed as follows:

$$HHI_{\min} = N\left(\frac{1}{N} \times 100\right)^2 = \frac{1}{N} \times 10000 \tag{7}$$

3.3. Energy Diversity Weighted by Energy Import Dependence

The presented diversity indices, such as the *SWI* and *HHI*, only depend on the quantity and the proportion of each energy resource, considering the proportion of each energy resource is effective when energy resources are less dependent on other countries in the energy supply security of a country. However, its effectiveness could be reduced in a country highly dependent on energy imports. This country could not be evaluated as a secure country in terms of energy supply security, even though the country obtains a higher value of the *SWI* and a lower value of the *HHI*. In other words, these indices seem to lack a fundamental factor indicating the dependence of the target energy source on imports. Therefore, this study proposed modified indices of energy diversity by weighting the import dependence-related parameters. The main concept of these indices is to imply energy import dependence into the diversity indices. Weighted factors stemming from the import dependence of each energy type are multiplied into the energy diversity indices, the SWI and HHI. The proposed indices are expressed below:

$$D\_SWI = -\sum_{i=1}^{N}\left(1 - \frac{EI_i}{ES_i}\right)p_i \ln p_i \quad (8)$$

$$D\_HHI = \sum_{i=1}^{N}\left(1 + \frac{EI_i}{ES_i}\right)P_i^2 \quad (9)$$

where the D_SWI and the D_HHI demonstrate the modified SWI and HHI, which imply energy import dependence. In the case of the D_SWI, $(1 - EI_i/ES_i)$ is the inverse weight obtained by the share of ith energy source imports in supply. Its range is from 0 to 1 and reflects how much the ith energy type is self-sufficiently supplied to the country. The theoretical maximum value of the D_SWI may be obtained when the number of types of energy sources is higher and that the quantity of each source is even. In addition, all the considered energy sources should be self-sufficient which means $(1 - EI_i/ES_i)$ equals 1. By contrast, the D_SWI becomes smaller when the share of energy imports increases ($EI_i/ES_i > 0$), even though the value of the SWI is the SWI_{Max}. In the case of D_HHI, however, its theoretical minimum value could be obtained when the share of ith energy source imports in supply EI_i/ES_i is zero and the HHI reaches the HHI_{min} simultaneously. As the weighting is a process to assign a relatively important factor in each index [38,39], these proposed indices can be effective, especially to evaluate the primary energy supply security of the countries highly dependent on energy imports such as Singapore, Taiwan, Japan, and Korea.

3.4. Data for Evaluation

In this study, time series data were obtained from a dataset from the Korea Energy Economics Institute [40]. The assessment was conducted with 10 primary energy sources, including anthracite coal, bituminous coal, petroleum for energy use, LPG, petroleum for non-energy use, LNG, general and micro-hydro, pumped hydro, nuclear, and renewables for the period from 1991 to 2018. The data were used for measuring four indices which are the import ratio of energy, energy import dependence, energy diversity represented by the SWI and HHI, and the mixed indices of energy import dependence and energy diversity.

4. Results and Discussion

4.1. Energy Dependence

Korea's energy dependence was first evaluated by quantitatively measuring the energy import ratio and energy import dependence. Figure 1 shows the trend of Korea's energy import ratio, as expressed in Equation (1). The units of EI_T and AI_T are both USD. Korea's energy imports account for over 14% of total imports during the period between 1991 and 2018. From 1991 to 2012, the energy import ratio increased globally by 35.6% with fluctuations. From 2013 to 2016, the ratio decreased significantly by 19.9%. Then the value rebounded, respectively, to 22.9% and 27.3% in 2017, 2018. Slight increases and decreases in the import ratio appeared in the periods 1999–2000, 2004–2005, 2007–2008, 2008–2009, 2010–2011, 2014–2016, and 2016–2018.

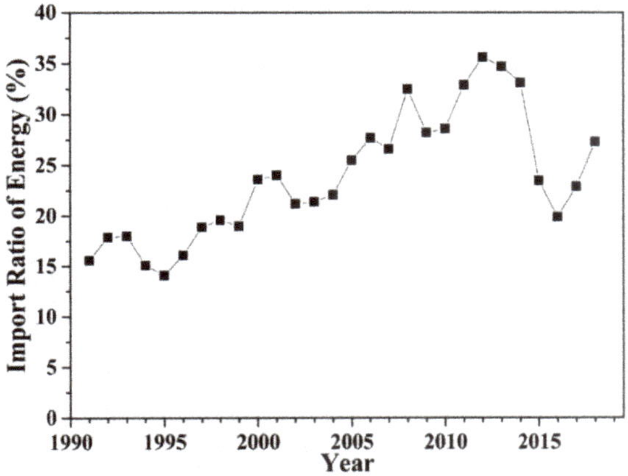

Figure 1. Energy import ratio.

It is worth noting that these trends match well enough to the oil price fluctuations in Korea as shown in Figure 2. As oil imports are a large share of Korea's total energy imports, it is reasonable that the energy import ratio in monetary terms is significantly influenced by the oil price. The ratio drops with a decrease in oil prices after 2013, even though the amount of oil imports increases in this period [41]. As this situation may not be stable, Korea needs a buffer system against the surge of oil prices to ensure its energy supply security and a stable domestic economy.

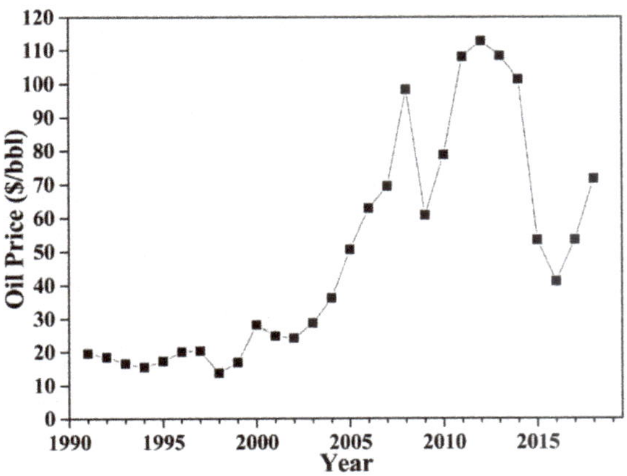

Figure 2. Oil price trend in Korea.

Under these conditions, the oil stock levels of net imports in Korea have consistently been recorded above 180 days since September 2009. This is double the value of the IEA 90-day requirements. According to historical data, the oil stock levels of Korea hit 309 days in March 2016. Figure 3 shows the oil stock levels in days of net imports given by the International Energy Agency [42].

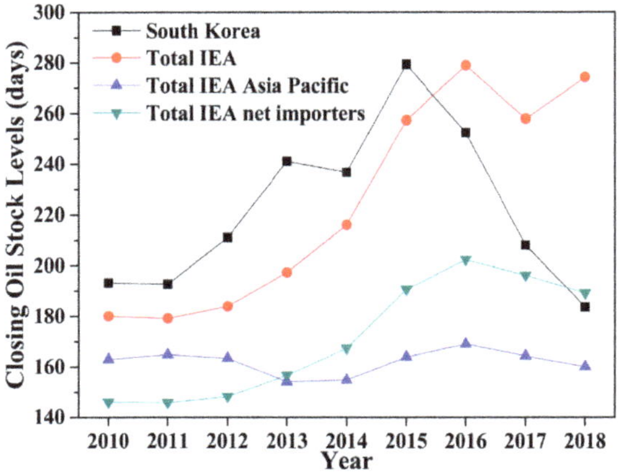

Figure 3. Closing oil stock levels of net imports.

Compared with the average value of total IEA countries, including net exporters such as Canada, Denmark, and Norway, Korea reserved more oil than other IEA countries until 2015. This indicator shows Korea's awareness of the importance of both the accessibility and availability of energy. However, the oil stock levels of Korea decreased by 184 days in 2018. This value is lower than that of the average value of total IEA net importers in the same period.

Figure 4 illustrates Korea's energy import dependence. A physical term of primary energy imports rather than a monetary term was used to remove the influence of oil price on the dependence as shown in Figure 1. The value of energy import dependence has been greater than 90% since 1991. From 1991 to 1997, energy dependence increased from 91.20% to 97.79% with the economic development of the country. After the liquidity crisis in 1997, the dependence moderately decreased because of the drop in domestic consumption. After 2012, the value decreased more visibly to 94.5%.

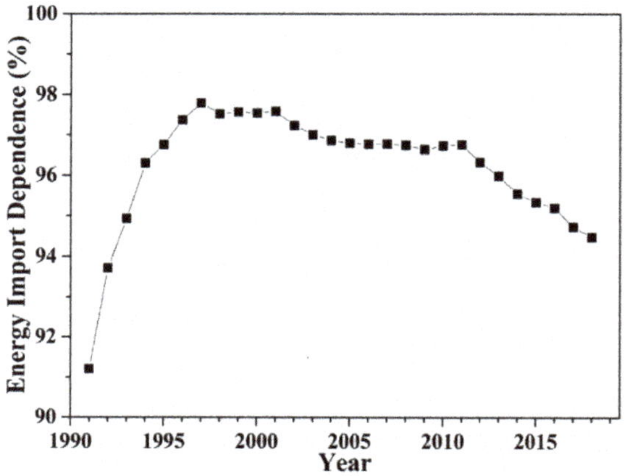

Figure 4. Korea's energy import dependence.

Amongst the primary energy sources supplied to Korea, almost all fuels have been imported from foreign countries. For example, the entire quantities of bituminous coal, crude oil, and uranium are imported from outside of the country. Energy resources such as anthracite coal and LNG are partially produced in Korea, but the quantity is very small. In the case of anthracite coal, its import dependence was approximately 10% in the early 1990s. However, because of anthracite coal reserve limits, the import dependence gradually increases, reaching 89.19% in 2018. Besides, although the Korea Gas Corporation produced LNG, the amount of LNG is less than 2% of the total supply in the country. To lower energy import dependence, the proportion of energy produced in Korea should be increased. This does not only mean the use of reserved fossil fuel but also the gradual increase of renewable energy such as hydro energy, solar energy, wind energy, and bioenergy, among others. In line with this, Korea is planning a scenario 2030, aiming to achieve 20% of the total amount of electricity generation from renewables by 2030 [43,44].

4.2. Energy Diversity

The energy diversity of Korea was quantitatively evaluated by using the *SWI* and the *HHI*. The considered primary energy sources were anthracite, bituminous coal, petroleum for energy use, LPG, petroleum for non-energy use, LNG, general hydro, pumped hydro, nuclear, and renewables. There are ten primary energy sources (n = 10). Figure 5 shows the Korean Total Primary Energy Supply (TPES) and its composition during the period between 1991 and 2018 [40]. Except in the years 1998 and 2009 when the economic crises occurred, primary energy supply increases continuously, reaching 307.5 million tons of oil equivalent (Mtoe) in 2018, and energy shares are more balanced.

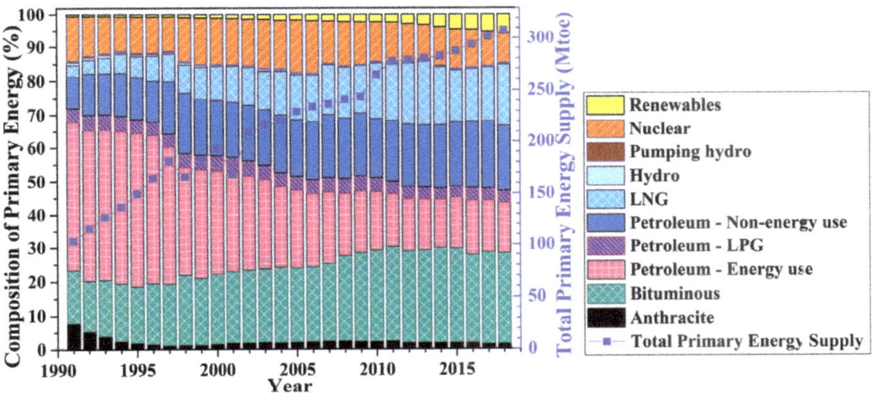

Figure 5. Korean total primary energy supply and its composition.

Figure 6and Figure 7 show the overall energy diversity of Korea represented by the *SWI* and *HHI* during the period between 1991 and 2018. From the calculation results of the *SWI*, the *SWI* gradually decreases from 1991 to 1995, as shown in Figure 6. The lowering of this value during this period is due to the more unbalanced composition of primary energy. Korean economic growth was promoted in the 1990s before the economic crisis occurred in 1997. During this period, energy consumption, especially driven by oil consumption, increases. Simultaneously, the consumption of imported bituminous coal increases, whereas the anthracite coal reserve reduces. Hence, the energy supply structure is concentrated in petroleum, and bituminous coal accounts for 74.7% in 1995. In consequence, the minimum value of the *SWI* is obtained as 1.629 in 1995. The lowest value means the lowest energy diversity because of the unevenness of energy shares of the year. Later, the value follows an increasing trend with small fluctuations from 1996 to 2018. In this period, there is a significant increase in the *SWI* in 1998. This globally increasing trend is explained first by the decrease in oil consumption due to the economic crisis in Korea. Second, there was a structural transition of energy supply from crude oil to nuclear energy, LNG, and renewables. The share of nuclear energy increases by 125%, whereas the portion of petroleum for energy use decreases by 79% in 1998. Third, the national energy policy has promoted the use of other types of primary energy rather than oil. After 2004, the *SWI* reaches a value higher than 81% of the theoretical maximum value of the *SWI*, which is about 2.302 as obtained by Equation (5). In 2016, the highest diversity is obtained as 1.877, whereas the value drops again in 2017 to 1.862. Thereafter, the value reaches 1.864 in 2018.

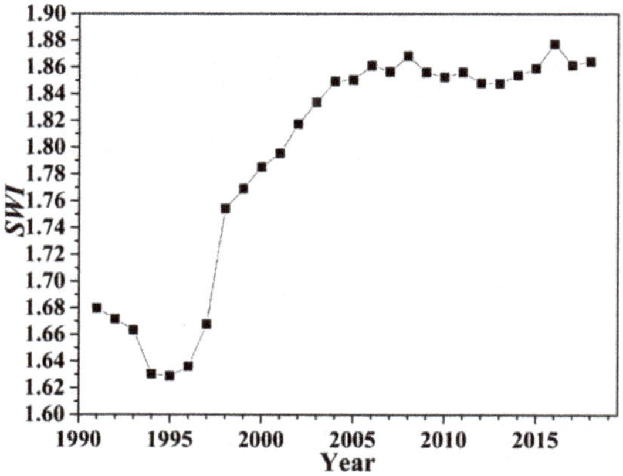

Figure 6. The *SWI* (Shannon-Wiener Index) of Korea from 1991 to 2018.

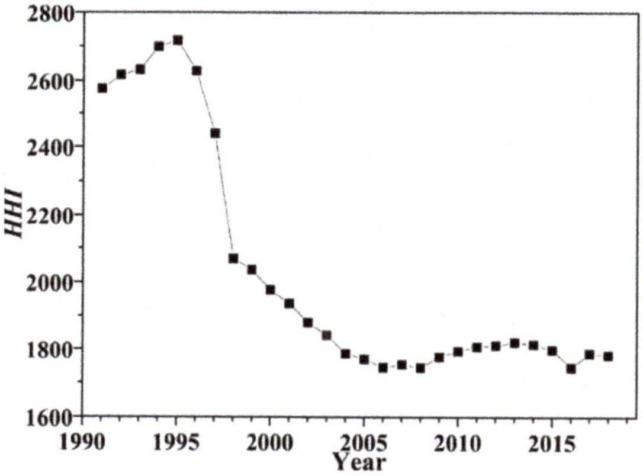

Figure 7. The *HHI* (Herfindahl-Hirschman Index) of Korea from 1991 to 2018.

The evolution of the *HHI* is shown in Figure 7. The *HHI* seems to be a scaled reflection of the *SWI* in the x-axis. This trend is explained by the fact that these two indicators oppositely represent energy diversity. As mentioned, the higher value of the *SWI* and the lower value of the *HHI* indicate the higher diversity. As inversely following the trend of the *SWI*, the *HHI* increases by 2710 in 1995 and decreases by 1736 in 2016. The value rebounded by 1787 in 2017 and decreased again by 1781 in 2018. These similar diversity trends between the *SWI* and *HHI* are shown from 1991 to 2010 and from 2013 to 2018.

However, the diversity trends obtained by the *SWI* and *HHI* for 2010–2011 and 2011–2012 are unmatched. In 2011, the diversity shown by the *SWI* increases, whereas the diversity given by the *HHI* decreases. Similarly, the diversity obtained by the *SWI* drops, whereas the diversity calculated by the *HHI* increases in 2012. These unmatched diversity trends, of which phenomena were similarly presented by Chuang and Ma [10], were led by the relative abundance perspective. The primary energy option which takes a smaller quantity is emphasized in the *SWI*, and the more abundant is underlined in the *HHI*. In 2011, the shares of rare resources and abundant resources increase simultaneously except for petroleum for energy use, LPG, and nuclear energy. This implies higher diversity through the increase in the shares of rare resources in the *SWI* and the lower diversity induced by the increase in the shares of abundant resources in the *HHI*. On the contrary, the shares of rare resources such as anthracite, LPG, petroleum for non-energy use, and general hydro decreases, whereas that of abundant resources like bituminous coal, petroleum for energy use, and nuclear energy increases in 2012. This shows the decreased diversity led by rare resources in the *SWI* and increased diversity caused by abundant resources in the *HHI*.

With the fixed number of the types of energy sources in this study, the proportion of each energy source only affects the *SWI* and *HHI*. The share of each energy resource is an effective factor to evaluate energy security when energy resources are less dependent on other countries. However, its effectiveness could not be ensured in a country highly dependent on energy imports such as Korea. It means that these indices overlook how safely the energy resources are supplied to the country. The energy import dependence-related index should thus be considered for the more accurate evaluation of energy supply security and policy suggestions in more detail.

4.3. Energy Diversity Weighted by Energy Import Dependence

The energy diversity of Korea was quantitatively evaluated by using the proposed energy dependence-related indices, the *D_SWI* and *D_HHI*. Figure 8 describes the *D_SWI*. The theoretical maximum value of the *D_SWI* is approximately 2.302. As expected, a different trend is shown in the *D_SWI* compared with that of the *SWI*. The maximum value of the *D_SWI* is found in 1991 when the import dependence reaches the lowest level. Then, the *D_SWI* drops continuously by 0.122 until 1997 when the country strongly relies on energy imports. The global trend of the *D_SWI* from 1998 to 2010 tends to recover with fluctuation. Then, the value continuously increases from 2011 and finally reaches 0.181 in 2018. This increasing trend is linked to the country's energy dependence that has decreased since 2011 as shown in Figure 4.

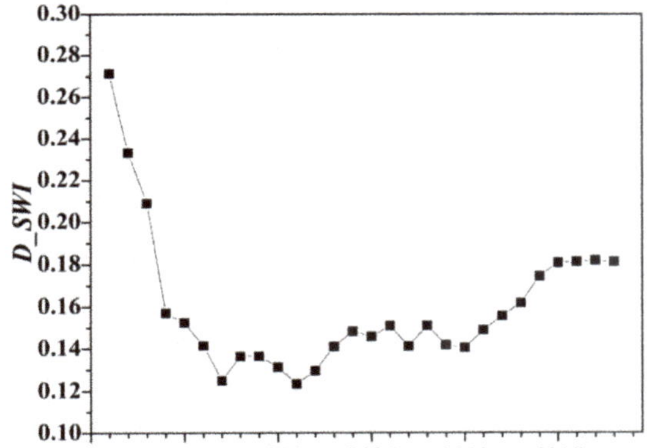

Figure 8. The *D_SWI* of Korea from 1991 to 2018.

However, this diversity index does not improve significantly compared with the past. The latest obtained value is even lower than that of the early 1990s (1991–1993) when the import dependence of the country was lower than 95%. This result shows the important role of the use of indigenous energy resources to improve energy supply security evaluated by the given index. This result implies the need to use indigenous energy resources and the challenges to promote self-sufficient energy such as anthracite coal and LNG. In addition, R&D on energy-efficient systems and energy systems driven by renewable energy should be supported to improve the energy supply security of the country in the long term.

Figure 9 illustrates the *D_HHI*. The theoretical minimum value of the *D_HHI* is 1000 when the energy import dependence is the minimum, which is zero. However, because the import dependence of Korea is far from zero, rather close to one, the *D_HHI* is nearly doubled by the *HHI*. This value follows the trend of *HHI*. As the share of oil imports has been over 40% for the period from 1991 to 1997, the effect of oil import on the *D_HHI* as well as on the *HHI* is comparatively high, whereas its effect on the *SWI* and the *D_SWI* is reduced. This means that the *D_HHI* and *HHI* still retain the importance of abundant resources despite the high weight of energy import dependence. It indicates that avoiding any abundant energy resources could lower the *D_HHI* more effectively than lowering their import dependence.

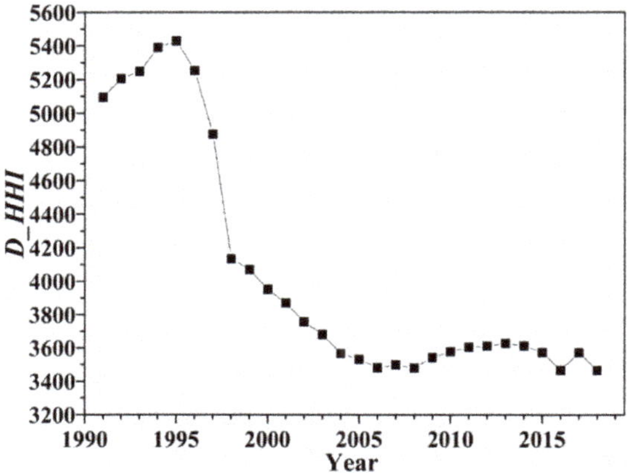

Figure 9. The *D_HHI* of Korea from 1991 to 2018.

In summary, the *SWI* and *HHI* showed a similar trend of energy diversity of the country as referring to other research on energy supply security [6,7,10,11,45]. In more detail, the *SWI* focuses on rare energy resources, while the *HHI* highlights the abundance of any energy resources. However, these two indices overlooked the energy import dependence of the target country. As introducing the energy import dependence as a weighting factor in the *D_SWI* and *D_HHI*, the correlation between the country's energy diversity and energy import dependence could be accounted for in the evaluation of energy supply security. Based on the import ratio of each energy source, the *SWI* is scaled from 0 to 1 for the *D_SWI*, whereas the *HHI* is multiplied by the value from 1 to 2 for the *D_HHI*.

Moreover, the overall evaluation of the energy supply security of Korea infers that the country should attempt to balance the use of different types of primary energy resources because they cannot liberalize its energy imports. To reduce the concentration of abundant import resources such as oil and coal, and increase the use of rare resources, electrification in the energy sector and a higher proportion of renewable energy in electricity generation could be alternative solutions to improve its energy supply security regarding the scenario 2030. Green financing and investment in renewable energy projects are recommended as practical solutions to promote a higher proportion of renewable energy and sustainable technologies, including hydrogen and electric vehicles [46,47]. Besides, the energy policy focusing on demand-side management should strongly be applied to the country for effective and efficient use of limited resources [24].

5. Conclusions

This study presented a methodology to evaluate the energy supply security of a country by using different indices of energy dependence and energy diversity. The considered indicators were energy import ratio and energy import dependence as the indices of energy dependence and the *SWI* and *HHI* as those of energy diversity. This study also proposed two indices, the *D_SWI* and *D_HHI*, that reflect the correlation between a country's energy diversity and energy import dependence.

The presented methodology and indices were applied to quantitatively assess Korea's energy supply security. Korea's energy imports were over 14% of total imports during the period between 1991 and 2018. The energy import dependence increased from 91.20% to 97.79% with the economic development of the country for the period from 1991 to 1997. After the liquidity crisis in 1997, the dependence gradually decreased and reached 94.49% in 2018. The energy diversity of Korea in terms of the *SWI* and *HHI* was also measured. The primary energy sources considered were anthracite, bituminous, petroleum, LPG, LNG, general hydro, pumped hydro, nuclear, and renewables. From the measurement of energy diversity, it was found that the energy transition from petroleum to nuclear and renewable energy contributed to increasing the diversity of primary energy.

From the calculation results of the *SWI* and *HHI*, similar diversity trends were obtained in the given periods from 1991 to 2010 and from 2013 to 2017. Diversity gradually decreased because of the concentration of oil consumption until 1995. Since then, diversity has globally improved with fluctuations. In addition, unmatched trends between the *SWI* and *HHI* appeared for 2010–2011 and 2011–2012 because of the emphasis of the *SWI* on rare resources and the *HHI* on abundant resources in the shares of energy.

However, a country highly dependent on energy imports cannot be evaluated as a secure country in terms of energy supply security, even if it obtains higher energy diversity. As Korea is highly dependent on energy imports because of its high energy consumption and insufficient energy resources, the analysis of the energy supply security of Korea should consider the types and the proportion of available energy resources and their import dependence. By focusing on the correlation between Korea's energy diversity and import dependence, this study also suggested modified indices of energy diversity based on the *SWI* and *HHI*, which are weighted by import dependence-related parameters. As a result, the *D_SWI* showed lower values of energy diversity, even in recent years, because of higher energy import dependence. The *D_HHI* almost doubled compared with the value of the *HHI* because of the higher energy import dependence of the country and followed the trend of the *HHI* which emphasizes abundant resources. This result implies that the correlation of energy dependence and energy diversity of a country should also be considered as an important factor to evaluate energy supply security and to make energy policy strategies.

In focusing on the correlation between energy diversity and import dependence, the national energy policy should promote renewable energy because its use could mitigate the environmental effects and energy dependence. It might permit more diversified energy options in the transition of electricity-driven transport and industry. In practice, the automotive industry and R&D of electric vehicles and renewable hydrogen vehicles could be boosted from several perspectives. Policy-supported R&D investment and infrastructure in renewable energy and energy-efficient technologies with demand-side management should be implemented to stabilize the energy supply security of the country in the long term. Comparative study on energy dependence and diversity at the international level should be conducted to further analyze the characteristics of energy security of each country and classify them to suggest referable policies as well.

Author Contributions: All the authors have been involved in the preparation of the manuscript. All the authors have read and agreed to the published version of the manuscript.

Funding: This research was funded by Basic Science Research Program through the National Research Foundation of Korea (NRF) funded by the Ministry of Education, grant numbers 2017R1D1A3B03035693 and 2020R1I1A1A01073797. This work was also supported by the Korea Institute of Energy Technology Evaluation and Planning (KETEP) grant funded by the Korea government (MOTIE), grant number 20192010107050.

Institutional Review Board Statement: Not applicable.

Informed Consent Statement: Not applicable.

Data Availability Statement: Not applicable.

Conflicts of Interest: The authors declare no conflict of interest.

References

1. Kafle, S.; Parajuli, R.; Bhattarai, S.; Euh, S.H.; Kim, D.H. A review on energy systems and GHG emissions reduction plan and policy of the Republic of Korea: Past, present, and future. *Renew. Sustain. Energy Rev.* **2017**, *73*, 1123–1130. [CrossRef]
2. Proskuryakova, L. Updating energy security and environmental policy: Energy security theories revisited. *J. Environ. Manag.* **2018**, *223*, 203–214. [CrossRef] [PubMed]
3. Vivoda, V. LNG import diversification and energy security in Asia. *Energy Policy* **2019**, *129*, 967–974. [CrossRef]
4. Song, Y.; Zhang, M.; Sun, R. Using a new aggregated indicator to evaluate China's energy security. *Energy Policy* **2019**, *132*, 167–174. [CrossRef]
5. Månsson, A.; Johansson, B.; Nilsson, L.J. Assessing energy security: An overview of commonly used methodologies *Energy* **2014**, *73*, 1–14. [CrossRef]
6. Chalvatzis, K.J.; Ioannidis, A. Energy supply security in the EU: Benchmarking diversity and dependence of primary energy. *Appl. Energy* **2017**, *207*, 465–476. [CrossRef]
7. Matsumoto, K.; Doumpos, M.; Andriosopoulos, K. Historical energy security performance in EU countries. *Renew. Sustain. Energy Rev.* **2018**, *82*, 1737–1748. [CrossRef]
8. Li, Y.; Shi, X.; Yao, L. Evaluating energy security of resource-poor economies: A modified principle component analysis approach. *Energy Econ.* **2016**, *58*, 211–221. [CrossRef]
9. Yao, L.; Shi, X.; Andrews-Speed, P. Conceptualization of energy security in resource-poor economies: The role of the nature of economy. *Energy Policy* **2018**, *114*, 394–402. [CrossRef]
10. Chuang, M.C.; Ma, H.W. Energy security and improvements in the function of diversity indices—Taiwan energy supply structure case study. *Renew. Sustain. Energy Rev.* **2013**, *24*, 9–20. [CrossRef]
11. Matsumoto, K.; Shiraki, H. Energy security performance in Japan under different socioeconomic and energy conditions. *Renew. Sustain. Energy Rev.* **2018**, *90*, 391–401. [CrossRef]
12. Asia Pacific Energy Research Centre. *Quest for Energy Security in the 21st Century: Resources and Constraints;* Asia Pacific Energy Research Centre: Tokyo, Japan, 2007.
13. Zhao, Z.; Yang, H. Regional Security Assessment of Integrated Energy Systems with Renewables in China: A Grid-Connected Perspective. *Sustainability* **2020**, *12*, 10299. [CrossRef]
14. Yergin, D. Energy Security in the 1990s. *Foreign Aff.* **1988**, *67*, 110. [CrossRef]
15. Dorian, J.P.; Franssen, H.T.; Simbeck, D.R. Global challenges in energy. *Energy Policy* **2006**, *34*, 1984–1991. [CrossRef]
16. Hughes, L. A generic framework for the description and analysis of energy security in an energy system. *Energy Policy* **2012**, *42*, 221–231. [CrossRef]
17. Vivoda, V. Evaluating energy security in the Asia-Pacific region: A novel methodological approach. *Energy Policy* **2010**, *38*, 5258–5263. [CrossRef]
18. Martchamadol, J.; Kumar, S. Thailand's energy security indicators. *Renew. Sustain. Energy Rev.* **2012**, *16*, 6103–6122. [CrossRef]
19. Kruyt, B.; Van Vuuren, D.; De Vries, H.; Groenenberg, H. Indicators for energy security. *Energy Policy* **2009**, *37*, 2166–2181. [CrossRef]
20. Winzer, C. Conceptualizing energy security. *Energy Policy* **2012**, *46*, 36–48. [CrossRef]
21. Sovacool, B.K.; Mukherjee, I. Conceptualizing and measuring energy security: A synthesized approach. *Energy* **2011**, *36*, 5343–5355. [CrossRef]
22. Wang, Q.; Zhou, K. A framework for evaluating global national energy security. *Appl. Energy* **2017**, *188*, 19–31. [CrossRef]
23. Jun, E.; Kim, W.; Chang, S.H. The analysis of security cost for different energy sources. *Appl. Energy* **2009**, *86*, 1894–1901. [CrossRef]
24. Le, T.-H.; Chang, Y.; Taghizadeh-Hesary, F.; Yoshino, N. Energy insecurity in Asia: A multi-dimensional analysis. *Econ. Model.* **2019**, *83*, 84–95. [CrossRef]
25. Erahman, Q.F.; Purwanto, W.W.; Sudibandriyo, M.; Hidayatno, A. An assessment of Indonesia's energy security index and comparison with seventy countries. *Energy* **2016**, *111*, 364–376. [CrossRef]
26. Glynn, J.; Chiodi, A.; Ó Gallachóir, B. Energy security assessment methods: Quantifying the security co-benefits of decarbonising the Irish Energy System. *Energy Strategy Rev.* **2017**, *15*, 72–88. [CrossRef]
27. Geng, J.-B.; Ji, Q. Multi-perspective analysis of China's energy supply security. *Energy* **2014**, *64*, 541–550. [CrossRef]
28. Narula, K.; Reddy, B.S.; Pachauri, S. Sustainable Energy Security for India: An assessment of energy demand sub-system. *Appl. Energy* **2017**, *186*, 126–139. [CrossRef]
29. Malik, S.; Qasim, M.; Saeed, H.; Chang, Y.; Taghizadeh-Hesary, F. Energy security in Pakistan: Perspectives and policy implications from a quantitative analysis. *Energy Policy* **2020**, *144*, 111552. [CrossRef]
30. Sarangi, G.K.; Mishra, A.; Chang, Y.; Taghizadeh-Hesary, F. Indian electricity sector, energy security and sustainability: An empirical assessment. *Energy Policy* **2019**, *135*, 110964. [CrossRef]
31. Ryu, H.; Dorjragchaa, S.; Kim, Y.; Kim, K. Electricity-generation mix considering energy security and carbon emission mitigation: Case of Korea and Mongolia. *Energy* **2014**, *64*, 1071–1079. [CrossRef]
32. Ahn, J.; Woo, J.; Lee, J. Optimal allocation of energy sources for sustainable development in South Korea: Focus on the electric power generation industry. *Energy Policy* **2015**, *78*, 78–90. [CrossRef]

33. Jansen, J.C.; Seebregts, A.J. Long-term energy services security: What is it and how can it be measured and valued? *Energy Policy* **2010**, *38*, 1654–1664. [CrossRef]
34. Bazilian, M.; Roques, F. *Analytical Methods for Energy Diversity & Security*; Elsevier: Amsterdam, The Netherlands, 2008.
35. Stirling, A. Diversity and ignorance in electricity supply investment: Addressing the solution rather than the problem. *Energy Policy* **1994**, *22*, 195–216. [CrossRef]
36. Stirling, A. *On the Economics and Analysis of Diversity*; Science Policy Research Unit, Centre for the Study of Evolution, University of Sussex: Brighton, UK, 1998; pp. 1–134.
37. Hill, M.O. Diversity and Evenness: A Unifying Notation and Its Consequences. *Ecology* **1973**, *54*, 427–432. [CrossRef]
38. Rowley, H.V.; Peters, G.M.; Lundie, S.; Moore, S.J. Aggregating sustainability indicators: Beyond the weighted sum. *J. Environ. Manag.* **2012**, *111*, 24–33. [CrossRef]
39. Gasser, P. A review on energy security indices to compare country performances. *Energy Policy* **2020**, *139*, 111339. [CrossRef]
40. Korea Energy Economics Institute. *KEEI Yearbook of Energy Statistics 2019*; Korea Energy Economics Institute: Ulsan, Korea, 2020.
41. OECD Organisation for Economic Co-Operation and Development, OECD Data Crude Oil Import Prices. Available online: https://data.oecd.org/energy/crude-oil-import-prices.htm (accessed on 7 February 2021).
42. IEA International Energy Agency, Closing Oil Stock Levels in Days of Net Imports. Available online: http://www.iea.org/netimports/ (accessed on 7 February 2021).
43. Kim, S.; Lee, H.; Kim, H.; Jang, D.-H.; Kim, H.-J.; Hur, J.; Cho, Y.-S.; Hur, K. Improvement in policy and proactive interconnection procedure for renewable energy expansion in South Korea. *Renew. Sust. Energ. Rev.* **2018**, *98*, 150–162. [CrossRef]
44. MOTIE Ministry of Education, Science & Technology's (MOTIE). *MOTIE Notice #2017-611, The 8th Basic Plan for Long-Term Electricity Supply and Demand (2017–2031)*; Ministry of Trade, Industry and Energy: Sejong-si, Korea, 2017.
45. Li, M.; Li, L.; Strielkowski, W. The Impact of Urbanization and Industrialization on Energy Security: A Case Study of China. *Energies* **2019**, *12*, 2194. [CrossRef]
46. Taghizadeh-Hesary, F.; Yoshino, N. Sustainable Solutions for Green Financing and Investment in Renewable Energy Projects. *Energies* **2020**, *13*, 788. [CrossRef]
47. He, W.; Abbas, Q.; Abbas, Q.; Mohsin, M.; Hanif, I.; Vo, X.V.; Taghizadeh-Hesary, F. Integration of renewable hydrogen in light-duty vehicle: Nexus between energy security and low carbon emission resources. *Int. J. Hydrog. Energy* **2020**, *45*, 27958–27968. [CrossRef]

sustainability

MDPI

Article

Cost–Benefit Analysis of HELE and Subcritical Coal-Fired Electricity Generation Technologies in Southeast Asia

Hassan Ali [1,*], Han Phoumin [2], Steven R. Weller [3] and Beni Suryadi [4]

1 School of Engineering Technology and Industrial Trades, College of the North Atlantic Qatar, Doha 24449, Qatar
2 Economic Research Institute for ASEAN and East Asia (ERIA), Jakarta 10270, Indonesia; han.phoumin@eria.org
3 School of Electrical Engineering and Computing, The University of Newcastle, Callaghan, NSW 2308, Australia; steven.weller@newcastle.edu.au
4 ASEAN Centre for Energy (ACE), Jakarta 12950, Indonesia; benisuryadi@aseanenergy.org
* Correspondence: hassan.ali@cna-qatar.edu.qa; Tel.: +974-4495-2198

Abstract: A large potential exists in the Southeast Asia region for deployment of high-efficiency, low-emission (HELE) electricity generation technologies. A cost–benefit analysis of HELE technologies compared to the less efficient subcritical electricity generation plants is thus carried out to find a persuasive scenario supporting quicker transition from subcritical stations towards HELE technologies in the region. A levelized cost of electricity (LCOE) analysis is carried out for both technologies under four potential policy scenarios. Scenario 1 does not take into consideration any carbon pricing or costs associated with the desulphurization (deSOx) and denitrification (deNOx) facilities. Scenario 2 (Scenario 3) incorporates carbon pricing (costs associated with the deSOx and deNOx facilities), and Scenario 4 includes both carbon pricing and costs associated with the deSOx and deNOx facilities. Under each scenario, a sensitivity analysis is performed to evaluate the uncertainty affecting the future coal prices. This study demonstrates that HELE technologies are competitive against the subcritical plants under all four scenarios and both the technologies derive benefit from lifetime extensions and low coal prices. It is revealed that future deployments of HELE technologies can be best expedited by factoring in carbon pricing in LCOE costs of coal-fired power plants under Scenario 2.

Keywords: high-efficiency; low-emission; carbon dioxide emissions; carbon pricing; subcritical; desulphurization; denitrification; cost–benefit analysis; levelized cost of electricity

Citation: Ali, H.; Phoumin, H.; Weller, S.R.; Suryadi, B. Cost–Benefit Analysis of HELE and Subcritical Coal-Fired Electricity Generation Technologies in Southeast Asia. *Sustainability* **2021**, *13*, 1591. https://doi.org/10.3390/su13031591

Received: 4 December 2020
Accepted: 25 January 2021
Published: 2 February 2021

1. Introduction

Currently, coal-fired electricity generation plants with a total capacity of about 1700 GW account for over 41% of the electricity generation worldwide [1]. Coal-fired electricity generation is responsible for over 28% of global carbon dioxide (CO_2) emissions [2], and scientific studies suggest that CO_2 emissions are responsible for global warming and associated devastating public health and environmental impacts.

As the pressure to act against global warming is increasing, several coal-using countries have been working on their national plans to kick in global efforts to reduce CO_2 emissions from their electricity generation sectors through development and deployment of high-efficiency, low-emission (HELE) coal-fired and renewable energy (RE) power generation technologies. HELE technologies utilize higher temperatures and pressures, compared to less-efficient subcritical technologies [3,4]. HELE electricity generating plants include supercritical (SC), ultra-supercritical (USC), advanced ultra-supercritical (A-USC), integrated gasification combined cycle (IGCC) and integrated gasification fuel cell (IGFC) technologies developed to increase the efficiency of coal-fired electricity generation plants, and thus reducing CO_2 and other greenhouse gas (GHG) and non-GHG emissions. HELE units emit

25–33% less CO_2 than the global average CO_2 emissions from existing electricity generation fleets and up to 40% less than the oldest technologies [4]. Table 1 shows the efficiency ratings; CO_2 intensity factors; and fuel consumption values for subcritical, SC, USC and A-USC power plants.

Table 1. HELE technologies: Low heating value (LHV)-based efficiency improvements, intensity factors and fuel consumption [5].

	Efficiency Rate (% Net LHV Basis)	CO_2 Intensity	Coal Consumption	Steam Temperature
A-USC	45–50%	670–740 g CO_2/kWh	290–320 g/kWh	700 °C
USC	Up to 45%	740–800 g CO_2/kWh	320–340 g/kWh	600 °C
SC	Up to 42%	800–880 g CO_2/kWh	340–380 g/kWh	Approx. 550–600 °C
Subcritical	Up to 38%	≥880 g CO_2/kWh	≥380 g/kWh	<550 °C

Every 1% improvement in the efficiency of coal-fired electricity generation plants results in a 2–3% reduction in CO_2 emissions [6]. In this regard, since the year 2000, HELE power plants have already reduced global CO_2 emissions by over 1 billion tons [7]. HELE technology is a vital first step to the carbon capture and storage (CCS). The International Energy Agency (IEA) Energy Technology Perspective (ETP) 2012 2 °C Scenario (2DS) indicates that to limit the average rise in global temperature to 2 °C, it is necessary to cut more than half of the energy sector-related CO_2 emissions by 2050 (compared to 2009) [3]. Combined with CCS, HELE technologies are expected to cut global average CO_2 emissions from coal-fired plants by as much as 90% to attain the 2DS by 2050 [5].

Southeast Asia consists of ten countries of the Association of Southeast Asian Nations (ASEAN): Brunei Darussalam, Indonesia, Cambodia, Lao People's Democratic Republic (PDR), Singapore, Malaysia, Philippines, Myanmar, Vietnam, and Thailand. ASEAN member countries have set a broad set of policies to fast-track the development of renewable electricity generation capacity. The region is aiming to generate 23% of its primary energy from RE sources by 2025, compared to 9.4% in 2014 [8]. While this upward trend towards RE is continuing, the vast availability of coal reserves in the region and its lower cost has made coal the largest and preferred source for electricity generation. The IEA forecasts that installed coal-fired electricity generation capacity will increase to around 160 GW by 2040, making a large contribution to growth in generation capacity of the region [9]. Additionally, coal-fired generation will overtake natural gas by 2040 to become the largest source of power capacity. Furthermore, the IEA confirms that low emission coal will be the generation of choice in the region and will provide 40% of electricity generation by 2040. There is a regional understanding among ASEAN nations that growing use of coal will necessitate a HELE technology energy pathway supported with renewables.

The levelized cost of electricity (LCOE) is often cited as a convenient summary measure of the overall competitiveness of different electricity generating technologies [10]. To influence the type of technology that project developers select in ASEAN countries, several LCOE studies focusing on RE technologies have been conducted [11–20]. Veldhuis et al. in [11], focused their study on LCOE of off-grid photovoltaic (PV) systems required to electrify Indonesian rural households and showed that off-grid PV systems are 19% cheaper as compared to electricity generation with diesel generator sets in most rural parts of Indonesia. In [12], Talavera et al. conducted a worldwide economic feasibility analysis of High Concentrator PV (HCPV) systems through the LCOE estimation. Blum et el. in [13], investigated the LCOE of isolated renewable hybrid mini-grid systems in Indonesia. In [16], Januar compared the economic viability of a 20 MW solar thermal and PV power plant in Rogkop, Indonesia, using the LCOE approach. Lau et al. in [17], presented a detailed analysis of PV grid parity based on the calculation of PV LCOE for the residential sector in Malaysia. In [18], the Asian Development Bank (ADB) conducted an LCOE analysis for REs for the greater Mekong sub-region (GMS) countries: Cambodia, Lao PDR, Myanmar,

Thailand, and Viet Nam. Huber et al. in [19] concluded that the most economical options for electricity generation in the ASEAN region are hydro, biomass and geothermal. The study in [20] analyses LCOE of selected RE technologies in several ASEAN countries, and advises necessary policies to reach a significant competitive edge for those selected RE technologies.

The study conducted by Phuangpornpitak and Kumar [21] provides an in-depth analysis of the renewable hybrid mini-grid systems with solar PV, wind, battery and diesel that have been installed in the national parks of Thailand. In [22], Keeley and Managi have assessed the economic viability of renewable hybrid mini-grid systems in Indonesia.

There is also some limited work on LCOE focused at coal technologies and comparison of coal technologies with other electricity generation technologies for the Southeast Asian region. A cost–benefit analysis of US, SC and subcritical plants is carried out by the Economic Research Institute for ASEAN and East Asia (ERIA) in [23]. The ERIA's study confirms that USC is generally competitive against SC and subcritical plants. Also, the World Coal Association (WCA) and ASEAN Centre for Energy (ACE) report (herein called WCA report) suggests that various coal-fueled electricity generation technologies are the lowest LCOE option available for mass deployment in Southeast Asia [4].

The ASEAN governments are promoting HELE technologies as a key step towards CO_2 mitigation. The ASEAN countries are thus making a transition from less efficient subcritical stations towards HELE coal-fueled facilities. Current research suggests that almost half of coal stations under construction or in development are expected to make use of advanced HELE coal-fired technology. The analysis also indicates that 23% of coal capacity currently under construction or in development is SC, while a further 29% of proposed projects have not finalized the technology choice [4]. HELE coal-fired technologies are more expensive to build than the subcritical technologies due to more expensive materials, complex boilers and precise control systems. High cost is a main restriction element for large-scale deployment of HELE technologies. By contrast, subcritical electricity generating plants have been traditionally preferred due to their lower upfront costs and shorter lead times. It is therefore highly likely that project developers end up accepting lower efficiency and poorer emission rates from subcritical coal-fired technology. On the other hand, to decarbonize the electricity sector by 2050 under 2DS, electricity generation from subcritical coal plants needs to be completely phased out by 2050 and following 2020, the more efficient CCS fitted HELE coal-fired plants are to be employed. The IEA thus recommends the implementation of national energy plans and policies to rapidly phase out construction and deployment of subcritical coal-fired plants [24]. Though the ASEAN is making a transition away from less efficient subcritical stations towards HELE coal-fueled facilities, current deployment progress is slow and subcritical units are still being deployed. Scope thus exists for necessary policy support to expedite transition from less efficient subcritical units to HELE units. The work in this paper is therefore aimed at demonstrating economic feasibility of HELE against subcritical, and finding a policy scenario that will result in the decline of subcritical coal-fired electricity generation even more rapidly, and shift new project investments towards being in favor of HELE technologies. This study is deemed important for ASEAN governments as a point of reference to formulate necessary policies and emission standards that expedite transition to HELE technologies, as well as to improve energy efficiency and reduce emissions.

This study is novel in the sense that we have included A-USC in this study, and under each scenario, a sensitivity analysis is performed to evaluate the uncertainty affecting the future coal prices on coal plants of a 20- and 25-year lifespan. Carbon pricing is an important policy tool to promote more energy-efficient, low-carbon technologies that emit less CO_2 emissions [25]. Due to rising concerns of air pollution from coal power stations, implementation of air pollution control technologies and regulations are crucial for sustainable development [26]. The study thus seeks to answer the question as to which one of these approaches can best help to expedite deployments of HELE technologies in

Southeast Asia or whether a mix of these approaches should be implemented. Based on the analysis of results, relevant policy recommendations are also discussed.

2. Methodology

2.1. LCOE

The LCOE represents the lifetime average cost of electricity as a constant unit price (in USD per megawatt-hour (USD/MWh) for a specific electricity generation project; it is a commonly used metric to assess overall competitiveness of different electricity generation projects. This mainstream technique is therefore used in our study.

The LCOE is calculated by dividing the project's overall expected lifetime costs (including construction, fuel, financing, maintenance, insurance, taxes and incentives) with the project's lifetime expected power output (MWh) [4,27–29]. As the value of the dollar today does not have the same economic value as the dollar in future, to properly add costs that occur at different points in time, they are converted into "present value" terms through the use of "discounting". The present values of all expense are thus divided by the present value of electricity generation to compute the LCOE as:

$$LCOE = \frac{\sum_{t=1}^{N} \frac{[I_t + M_t + F_t]}{(1+r)^t}}{\sum_{t=1}^{N} \frac{E_t}{(1+r)^t}}, \quad (1)$$

where:

I_t = Capital expenditure in the year t associated with the construction of the plant;
M_t = Non-fuel operating and maintenance costs in year t;
F_t = Fuel price expenditures in the year t;
E_t = Net electricity production in MWh in the year t;
N = Economic lifetime in years;
t = Year of lifetime $(1, 2, \ldots, N)$; and
r = Discount rate or interest rate.

If the net output of the plant is constant over the life of the plant, and if the operating, maintenance and fuel costs are also constant, Equation (1) can be reduced to:

$$LCOE = \frac{CAPEX \times FCF + O\&M_{fixed}}{CF \times 8760} + O\&M_{variable} + \Pi_{fuel} \times HR, \quad (2)$$

where:

- FCF is the fixed charge factor. The factor turns capital costs into a uniform annual amount and is given by:

$$FCF = \frac{r(1+r)^N}{(1+r)^N - 1}. \quad (3)$$

- $CAPEX$ is the capital expenditure (USD/MW). There are no publicly available CAPEX data sets for ASEAN countries. For our analysis, these figures are therefore replaced with engineering, procurement and construction (EPC) costs, in which other costs may incur additionally, such as land cost, cost of any additional emission controls and other financing costs;
- $O\&M_{fixed}$ is the fixed operation and maintenance (O&M) cost (USD/MW);
- CF is the capacity factor. It is a fraction between 0 and 1 representing the total generation of a plant as proportion to its nameplate capacity;
- 8760 is the number of hours in a year;
- $O\&M_{variable}$ is variable O&M cost (USD/MW);
- Π_{fuel} is the fuel price (USD/GJ (USD/MMBtu)); and
- HR is the heat rate (GJ/MWh (MMBtu/MWh)).

In addition to emitting (contributing to climate change), coal-fired power plants are a major CO_2 source of air pollution tied to heart and lung diseases. The toxic pollutants arising from coal power plants include sulphur oxides (SOx), nitrogen oxides (NOx), as well as mercury (Hg) and particulate matter (PM). Studies confirm that these emissions severely impact human health [30]. Our analysis suggests that the correct interpretation of LCOE results of coal-fired plants are blurred by the fact that a cost–benefit analysis does not reflect costs on society, such as CO_2, SOx, NOx, etc. Since HELE power plants pollute less SOx, NOx and CO_2 into the atmosphere than subcritical designs, their emission abatement, denitrification (deNOx) and desulphurization (deSOx) facilities and climate costs are expected to be less as compared to the subcritical plants of the same capacity. From the perception of global and ASEAN action on climate change, there is a clear imperative to make coal power generation sustainable by shifting incremental coal generation capacity under carbon pricing to make coal power generation more sustainable. Additionally, to improve public acceptance of the coal plants in the ASEAN region, there is a need to raise emission standards for coal plants in the region to the equivalent levels of the Organization of Economic Co-operation and Development (OECD) countries [26]. Therefore, we examine the role of carbon pricing and emission control technologies in transition to HELE technologies under four potential policy scenarios in Southeast Asia.

The cost of coal-fired electricity generation is heavily contingent on coal price. Since the Asian benchmark of thermal coal prices has been growing, based on (2), sensitivity of LCOE generation values is thus analyzed to evaluate the impact of rising coal prices in Southeast Asia on subcritical, SC, USC and A-USC coal-fired units with life spans of 20 and 25 years, under each scenario.

2.2. *Scenarios Description*

2.2.1. Scenario 1 (Base Scenario)

This scenario represents the continuous trend of the electricity sector's development from the past, and thus assumes no future for carbon pricing and no controls over NOx and SOx emissions in Southeast Asia. The associated carbon costs and NOx and SOx emission reduction costs are thus not accounted, and an LCOE analysis is simply based on base plant EPC, O&M, fuel costs and financing costs.

2.2.2. Scenario 2 (Climate Change Mitigation Scenario)

Carbon pricing is a tangible and cost-effective way of reducing risks, costs and GHG emissions. It provides a mechanism to account for the environmental, social and economic costs of climate change. In some Southeast Asian countries, a carbon pricing mechanism is fully implemented. Other Southeast Asian countries have planned for carbon pricing and have started to include analyses on the impact of carbon pricing in the electricity mix, while the remaining countries have not yet considered carbon pricing. Carbon pricing approaches in the ASEAN are "country specific", and there is no uniform carbon price across the region due to wide income disparity and poverty in the region. Among the ASEAN member states, Myanmar is the poorest economy. The Energy Master Plan (EMP) of Myanmar (2016) considers sensitivity analyses, based on the inclusion of carbon prices of USD 10/ton and USD 15/ton, for the development of an optimum power strategy for the country under a least-cost plan [31]. This scenario represents a future situation whereby a carbon price for coal power plants is fully implemented in the ASEAN region. Under this scenario, we thus assume a low-end carbon price of USD 10/ton in the ASEAN region to include poor economies, as well as for the achievement of low emissions to help limit global mean temperature under the 2DS. It is expected that the adoption of a carbon price in Southeast Asia under this scenario will accelerate the deployment of HELE coal-fired power plants.

2.2.3. Scenario 3 (Pollution Control Scenario)

The ERIA study [26] suggests that minimizing the emission of air pollutants in ASEAN countries is a pre-condition for the future use of coal power plants and for moving gradually to meet the current emission standards for coal plants of the OECD countries. In this regard, this scenario is considered to reflect a future situation where a legislation could take the form of ASEAN agreements to limit SOx and NOx emissions, linked through a uniform emissions standard mechanism in the ASEAN region. This scenario thus adds the cost of deSOx and deNOx facilities to the respective coal-fired plants in our analysis It is expected that strict pollution control technology requirements/adoption could add heavy financial costs to subcritical plants and thus help phase out generation from subcritical coal-fired electricity generation plants. The approach is expected to accelerate the deployment of HELE plants (all of which reduce NOx and SOx emissions in the ASEAN region).

2.2.4. Scenario 4 (Climate Change Mitigation and Pollution Control Scenario)

This potential policy scenario is a mix of Scenarios 2 and 3, and represents a situation whereby carbon price and strict emission controls are imposed on coal power plants in the ASEAN region. This scenario encourages both climate change mitigation and air pollution emission reduction efforts. Under this scenario, the costs of deSOx and NOx facilities and carbon pricing are thus integrated in the overall costs of coal-fired plants.

2.3. *General Assumptions*

ASEAN's preference for coal is to continue in the future as it remains the most economic source of long-term base-load generation. Collectively, large scale operational coal-fired electricity generation plants around the globe are key contributors to total emissions. We thus consider large coal plants that are connected to the grid to provide base load. High-efficiency coal-fired plants are typically large and fall in the range 600 MW to 1000 MW. Since 1000 MW plants are usually grid-connected to provide base load, the cost–benefit analysis was thus targeted at 1000 MW-capacity coal-fired plants. Base-load power plants typically have annual capacity factors (CFs) that exceed 75 percent, but are usually more likely to be 90–98% [32]. All coal plants were modelled with an assumed CF of 80%. The LCOE calculation is usually performed assuming ideal conditions; a CF of 80% is thus a moderate choice. Based on plant capacity and utilization rate, total annual generation was thus 7008 gigawatt-hours (GWh).

Within a project capital structure, a project may receive equity investment from a private equity firm or group of investors, with an insurance wrap from a development financial institute (DFI). The coal-fired plant life cycle is about 25–30 years, however, investors are likely looking for faster return/payback based on 20–25-year cash flow projections. Therefore, for this reason, the return cash flow is analyzed for 20 and 25 years of the expected lifetimes for each coal technology.

The efficiency figures listed in Table 1 are based on the low heating value (LHV) of the fuel and net output (LHV, net). Coal-fired station efficiencies based on the high heating value (HHV) are generally around 2% to 3% lower than those based on LHV efficiencies. We thus added three percentage points to the higher end LHV-based efficiencies in Table 1 to get HHV-based efficiencies for different coal-fired plants and associated heat rates (See Table 2).

Table 2. High heating value (HHV)-based coal-fired power plant efficiencies and heat rates.

	Efficiency Rate (% Net HHV Basis)	Heat Rate of Fuel (Btu/kWh) (HHV Basis)
A-USC	47%	7259.57 (Btu/kWh)
USC	42%	8123.81 (Btu/kWh)
SC	39%	8748.72 (Btu/kWh)
Subcritical	35%	9748.57 (Btu/kWh)

Coal has a calorific value of 4000 kcal/kg and emissions (adjusted from the Intergovernmental Panel on Climate Change (IPCC) default emission factors) of 1.43 kg-/kg-coal. The kWh generated from CO_2 per kg of coal was computed by dividing the coal heat content (in Btu per kg) with HR (in Btu per kWh). Coal requirements to generate one kWh of electricity (in kg-coal/kWh) were multiplied by the emission factor to obtain levelized kg CO_2 emissions per kWh.

The general assumptions for electricity generation plant specifications and coal composition are summarized in Table 3.

Table 3. General assumptions for cost benefit analysis.

		Values	Remarks
Plant	Capacity	1000 MW	For cash flow purposes
	Operation	20, 25 years	
	Operation rate	80%	
	Thermal efficiencies	47%(A-USC), 42% (USC), 39% (SC), 35%(subcritical)	HHV based values. A 3% decrease in thermal efficiency is assumed.
	Annual generation	7008 GWh	
Coal specifications	Heating value	4000 kcal/Kg or equivalently 1008.656 Btu/Kg	
	CO_2 emissions	1.43 kg-CO_2/kg coal	Based on IPCC 2006 default emission for stationary combustion in the energy sector [33].

2.4. Cost Assumptions and Methodologies

For the analysis in this paper, LCOE consists of base plant costs, deSOx and deNOx costs, financing costs and emission costs. Base plant costs are divided into EPC, O&M and fuel costs. Similarly, deSOx and deNOx costs consist of EPC, O&M and additional fuel costs (see Table 4).

Table 4. Levelized cost of electricity (LCOE) breakdown costs.

		Factors
LCOE	Base plant	EPC O&M Fuel cost
	deSOx deNOx	EPC O&M Additional fuel cost
	Financing	Internal Rate of Return (IRR)
	CO_2	Carbon

The cost assumption of EPC is adopted from [23]. The EPC cost consists of generator, turbine, boiler and auxiliary machine costs, construction costs and other management costs. The standard assumption is that all coal technologies pay equivalent connection costs and land costs. These costs are thus not taken into consideration.

For the 25-year life cycle of the plant, the SC and subcritical capital costs are discounted from USC capital costs (USD 1931 million per 1000 MW), based on a cost index from [34]. Subcritical plant capital costs are indexed at 100, while SC and USC are indexed at 106.5 and 108.5, respectively. Based on these indexes, capital costs for SC are estimated at USD 1897 billion, and capital costs for subcritical are estimated at USD 1786 million. Likewise, the A-USC capital cost is an escalating cost index factor of 107.5. Therefore, the EPC costs of different types of coal combustion technologies are: A-USC at USD 2100 million, USC at USD 1931 million, SC at USD 1897, and subcritical at USD 1786. Additionally, for the cost assumption for the 20-year life cycle of the plant, the cost estimates for different types of coal-fired power generation technologies are: A-USC at USD 2625 million, USC at USD 2413.75 million, SC at USD 2371.25 million, and subcritical at USD 2232.5 million.

Base plant O&M costs are calculated by dividing non-fuel O&M costs by annual generation (7008 GWh). The annual costs for this analysis are calculated by applying O&M cost differences from [35] to the annual O&M costs for USC from [34]. Annual O&M costs are thus estimated at: USD 0.6/kWh for A-USC, USD 0.7/kWh for USC, USD 0.72/kWh for SC and USD 0.75/kWh for subcritical power plants.

Thermal coal prices grew since the second half of 2016 due to robust Chinese demand and supply tightness at several production sites [36]. For example, free on board (FOB) Kalimantan 4200 kcal/kg gross as received (GAR) coal price rose 34% since the start of 2017 to USD 49.60/mt in January 2018. Coal prices in April 2020 dipped to their lowest level since 2010 due to the COVID-19 pandemic. Import prices of the majority of ASEAN coal reserves have seen a strong momentum since mid-November 2020 and have been trending upward due to steady economic recovery and thus high demand of coal by key consumers in China, Japan, India and South Korea. Coal demand by importing countries is expected to rise in the post-COVID-19 era due to increased economic activities and, consequently, coal prices will evolve to new high prices in the future. To assess the impact of high coal prices on LCOE values, a bandwidth is thus analyzed to reflect the situation. The annual average fuel price assumptions used in this study were: USD 50/ton, USD 80/ton, and USD 100/ton. For the breakdown of these costs and calculation methodologies, the reader is advised to refer to [23].

3. Results

Figures 1–4 show the LCOE sensitivity analysis results in USD cents/kWh for Scenario 1 through 4, respectively. The results of Scenario 1 in Figure 1 suggest that HELE plants are competitive against subcritical plants without coal pricing and deSOx and deNOx costs. A comparison of the results of Scenario 1 through Scenario 4 in Figures 1–4, respectively, reveals that as carbon price and deSOx and deNOx costs are included, LCOEs increase. However, HELE plants retain their competitive edge over the subcritical plants. The study suggests that Scenario 1 offers the best economic case for HELE plants due to the lowest LCOE values for HELE plants, followed by the Scenario 3, Scenario 2 and Scenario 4.

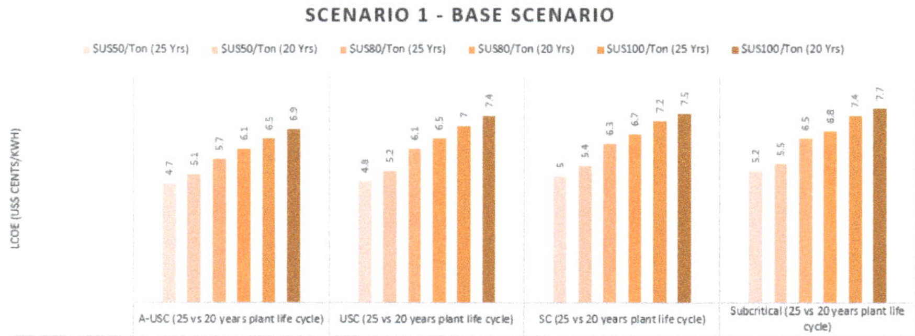

	A-USC (25 vs 20 years plant life cycle)	USC (25 vs 20 years plant life cycle)	SC (25 vs 20 years plant life cycle)	Subcritical (25 vs 20 years plant life cycle)
$US50/Ton (25 Yrs)	4.7	4.8	5	5.2
$US50/Ton (20 Yrs)	5.1	5.2	5.4	5.5
$US80/Ton (25 Yrs)	5.7	6.1	6.3	6.5
$US80/Ton (20 Yrs)	6.1	6.5	6.7	6.8
$US100/Ton (25 Yrs)	6.5	7	7.2	7.4
$US100/Ton (20 Yrs)	6.9	7.4	7.5	7.7

Figure 1. Scenario 1: Sensitivity Analysis of LCOE for different coal prices and economic life span of the subcritical and high-efficiency, low-emission (HELE) plants.

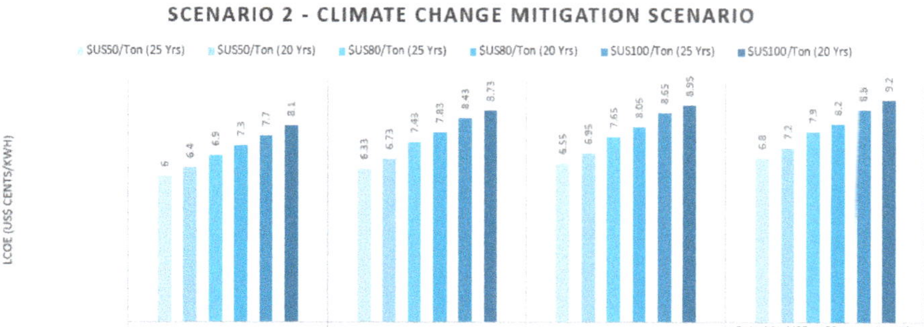

	A-USC (25 vs 20 years plant life cycle)	USC (25 vs 20 years plant life cycle)	SC (25 vs 20 years plant life cycle)	Subcritical (25 vs 20 years plant life cycle)
$US50/Ton (25 Yrs)	6	6.33	6.55	6.8
$US50/Ton (20 Yrs)	6.4	6.73	6.95	7.2
$US80/Ton (25 Yrs)	6.9	7.43	7.65	7.9
$US80/Ton (20 Yrs)	7.3	7.83	8.05	8.2
$US100/Ton (25 Yrs)	7.7	8.43	8.65	8.8
$US100/Ton (20 Yrs)	8.1	8.73	8.95	9.2

Figure 2. Scenario 2: Sensitivity Analysis of LCOE for different coal prices and economic life span of the subcritical and HELE plants.

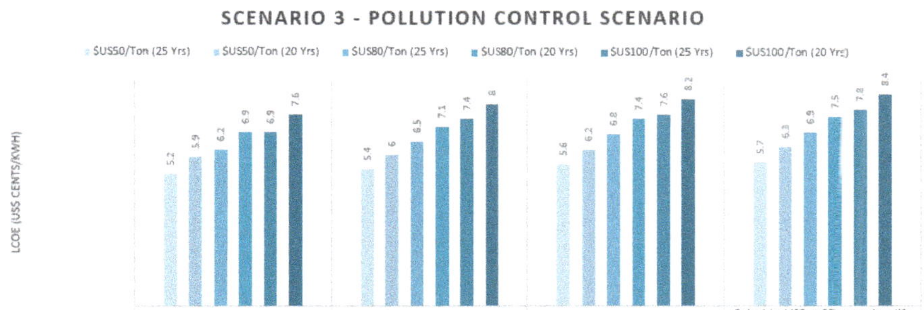

	A-USC (25 vs 20 years plant life cycle)	USC (25 vs 20 years plant life cycle)	SC (25 vs 20 years plant life cycle)	Subcritical (25 vs 20 years plant life cycle)
$US50/Ton (25 Yrs)	5.2	5.4	5.6	5.7
$US50/Ton (20 Yrs)	5.9	6	6.2	6.3
$US80/Ton (25 Yrs)	6.2	6.5	6.8	6.9
$US80/Ton (20 Yrs)	6.9	7.1	7.4	7.5
$US100/Ton (25 Yrs)	6.9	7.4	7.6	7.8
$US100/Ton (20 Yrs)	7.6	8	8.2	8.4

Figure 3. Scenario 3: Sensitivity Analysis of LCOE for different coal prices and economic life span of the subcritical and HELE plants.

In all scenarios, for different coal prices and operating life spans of 20 and 25 years, both coal technologies derive benefit from lifetime extensions and low coal prices. However, HELE coal technologies derive more benefit due to lower LCOE values compared to the subcritical technology. It is immediately apparent that, in all scenarios, A-USC offers the best economic value, followed by USC and SC.

The lower LCOEs of HELE against the subcritical technology are necessary to shift investment decisions in favor of HELE and thus expedite deployment of HELE plants in the region. We thus evaluated the difference of LCOE values between HELE and subcritical technologies for each scenario using the LCOE difference metric $\Delta LCOE = |LCOE_{HELE} - LCOE_{Subcritical}|$. These differences are displayed in Figures 5–8, for Scenario 1 through 4, respectively. A close comparison of results in Figures 5–8 suggests that Scenario 2 is the best scenario causing the highest difference in LCOE values of subcritical and HELE plants. A similar observation reveals that Scenario 4 is the second-best Scenario, followed by Scenario 3. A closer analysis of results reveals that Scenarios 2 through 4 allow a shift of economics in favor of HELE technologies at the price of increased LCOE values of coal plants. However, notice that since Scenario 2 LCOE values are lower than the Scenario 4 LCOE values, Scenario 2 thus emerges as the best driver scenario to displace subcritical plants. Notice that Scenario 3 yields low LCOE prices as compared to the Scenario 4. However, as compared to Scenario 3, the difference in LCOE values of HELE and subcritical technologies in Scenario 4 is highly attractive to shift economics strongly in favor of HELE plants. Scenario 4 thus emerges as the second-best driver scenario. A similar comparison between LCOE values of coal plants, and LCOE difference in Scenario 1 and Scenario 3 suggests that Scenario 3 is the third-best Scenario.

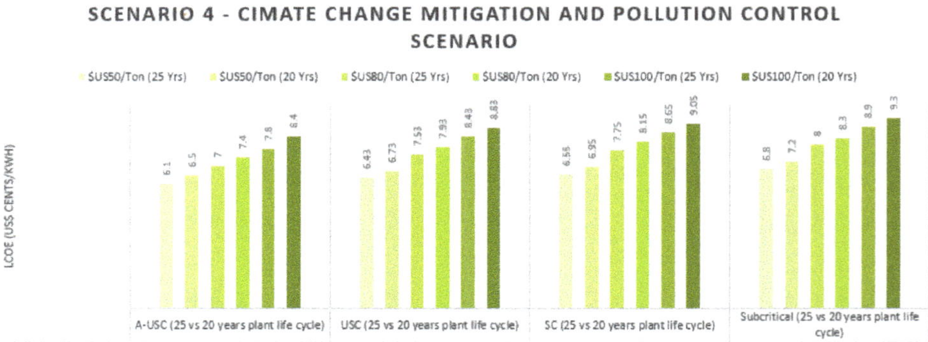

	A-USC (25 vs 20 years plant life cycle)	USC (25 vs 20 years plant life cycle)	SC (25 vs 20 years plant life cycle)	Subcritical (25 vs 20 years plant life cycle)
$US50/Ton (25 Yrs)	6.1	6.43	6.55	6.8
$US50/Ton (20 Yrs)	6.5	6.73	6.95	7.2
$US80/Ton (25 Yrs)	7	7.53	7.75	8
$US80/Ton (20 Yrs)	7.4	7.93	8.15	8.3
$US100/Ton (25 Yrs)	7.8	8.43	8.65	8.9
$US100/Ton (20 Yrs)	8.4	8.83	9.05	9.3

Figure 4. Scenario 4: Sensitivity Analysis of LCOE for different coal prices and economic life span of the subcritical and HELE plants.

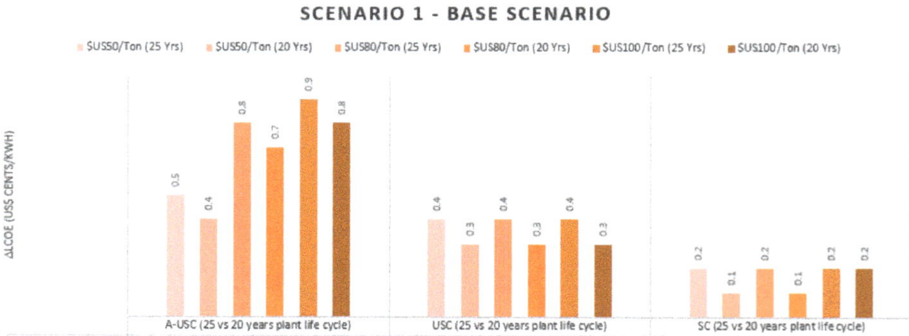

	A-USC (25 vs 20 years plant life cycle)	USC (25 vs 20 years plant life cycle)	SC (25 vs 20 years plant life cycle)
$US50/Ton (25 Yrs)	0.5	0.4	0.2
$US50/Ton (20 Yrs)	0.4	0.3	0.1
$US80/Ton (25 Yrs)	0.8	0.4	0.2
$US80/Ton (20 Yrs)	0.7	0.3	0.1
$US100/Ton (25 Yrs)	0.9	0.4	0.2
$US100/Ton (20 Yrs)	0.8	0.3	0.2

Figure 5. Scenario 1: LCOE differences between HELE and subcritical technologies for different coal prices and economic life span of the plant.

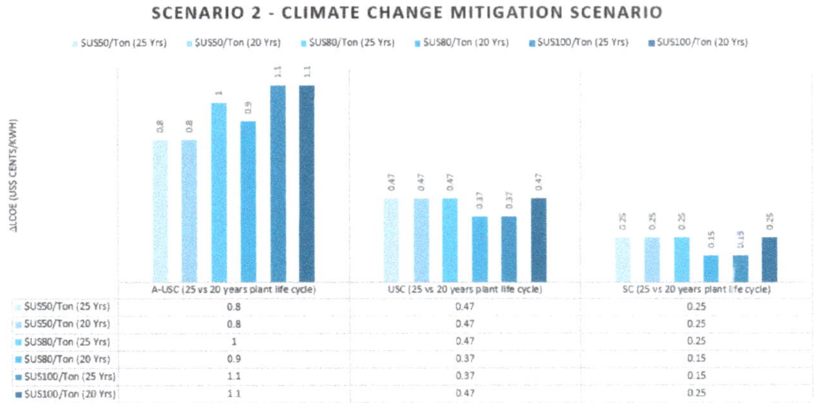

Figure 6. Scenario 2: LCOE differences between HELE and subcritical technologies for different coal prices and economic life span of the plant.

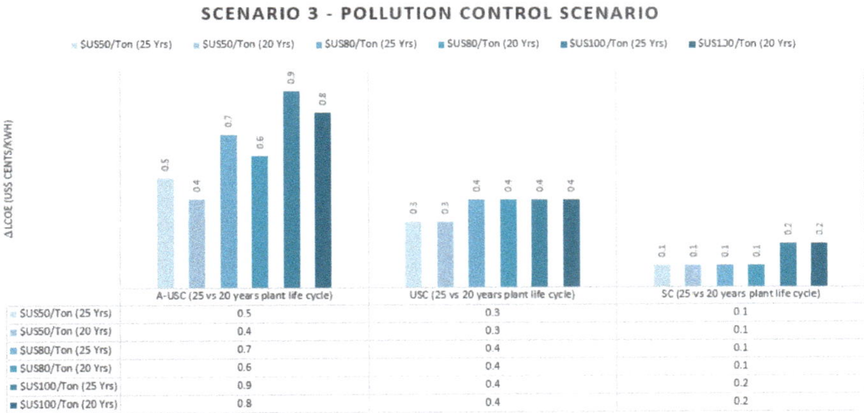

Figure 7. Scenario 3: LCOE differences between HELE and subcritical technologies for different coal prices and economic life span of the plant.

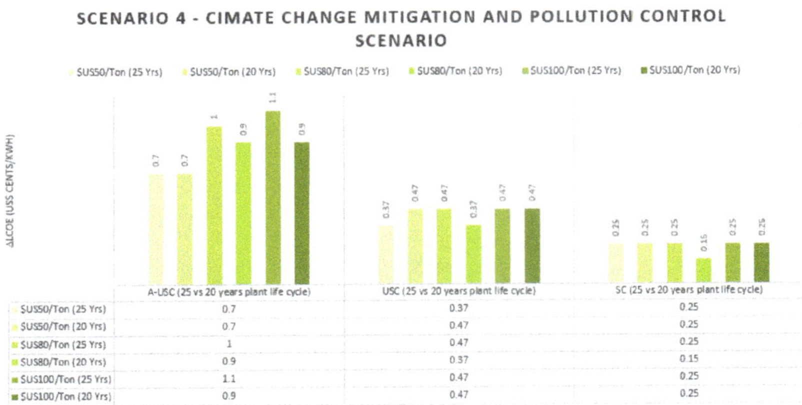

Figure 8. Scenario 4: LCOE differences between HELE and subcritical technologies for different coal prices and economic life span of the plant.

4. Discussion, Policy Implications and Future Directions

The WCA report [4] confirms that advanced coal technologies have a slightly higher LCOE compared to subcritical coal due to the initial higher capital costs. In contrast, results obtained in our work suggest that HELE technologies are economically competitive against subcritical plants. Since the WCA study does not cover details of LCOE calculations, methodology, data, and assumptions, the LCOE results for coal technologies in our work are not directly comparable to the WCA's LCOE results for coal technologies. In our work, a sensitivity analysis was carried out to evaluate the impact of different coal prices for 1000 MW-capacity coal-fired plants with life spans of 20 and 25 years and relies on IEA-listed thermal efficiencies for both technologies under different potential scenarios from ERIA's study in [23]. Therefore, our work in this paper is also not comparable with the ERIA study in [23]. Both the WCA and ERIA studies do not include an economic feasibility study of A-USC for the Southeast Asia region. In contrast, in our work, A-USC emerges as the most economically attractive choice, followed by USC and SC for the region.

The ASEAN member states will continue to rely on coal use to meet the growing electricity demand in the foreseeable future due to the energy supply security, as well as the competitive LCOE from coal plants compared to the other power generation technologies. Our study suggests that HELE coal-fired power plants are economically competitive against subcritical plants. This competitiveness of HELE technologies is associated with levelized avoided costs associated with high efficiency (and thus fuel savings) and low emissions.

To meet the 2DS targets, policies and associated measures are needed to address both the long-term and short-term challenges linked with electricity generation from coal-fired plants. Following are the policy implications of our study:

- In the short term, the implementation of an efficient and impactful harmonized carbon pricing policy for coal-fired plants in all ASEAN countries is necessary to support the first-best driver, Scenario 2, to displace the subcritical plants and shift investments to emerging HELE opportunities in the ASEAN market. This would yield clean coal technology (CCT) for Southeast Asia and bring many benefits for the environment and people of the region.
- ASEAN countries have relatively lower emission standards of SOx, NOx and particulates when compared with OECD countries [23,37]. It is therefore very important to regulate continuation of coal through stringent environmental and emission standards to pave the way for HELE technologies. A long-term carbon policy coupled with emissions standards and effective enforcement is thus needed to support the second-best driver, Scenario 4, to shift the balance in favor of HELE plants. However, since the inclusion of a carbon price and raising emission standards causes a further rise in LCOE values, the ASEAN countries need to better understand how this move will affect the regional economic developments before it becomes an effective policy tool.
- In general, the clean use of fossil fuel will need to be accelerated in the policy agenda in ASEAN. Therefore, policy reform to accommodate clean fuels and technologies is urgently needed to ensure that clean use of fossil fuel will play a significant role for energy transition from a fossil fuel-based energy system towards a clean energy system where renewables and clean fuels play a major role in the future energy mix.

The cost of solar and wind technologies is also expected to drop in the future. In situations with strict emission control for coal-fired electricity generating plants (with carbon pricing and strict emission standards in place), switching from coal to these renewables would thus be expected. Nevertheless, the intermittent nature and low load factors associated with wind and PV technologies will likely limit their effectiveness in the region. In our future work, we aim to extend our cost–benefit analysis study by including solar and wind sources of energy generation in the ASEAN region. Financing costs also account for a considerable share of LCOE and the competitiveness of technology [38]. In recent years, multilateral development banks have adopted more restrictive finance policies for coal electricity generating plants to reduce emissions [4]. Exploring the impact of variations in financing costs on the feasibility of HELE plants in the Southeast Asia region would be another interesting research direction.

5. Conclusions

Across Southeast Asia, there is a vital need to deploy HELE technologies, rather than employing less-efficient subcritical technology. Deployment of HELE technologies is progressing in Southeast Asia, but the overall rate of deployment falls short of achieving the 2DS. ASEAN should therefore make increased efforts to eliminate generation from subcritical plants and increase generation from HELE plants to meet the 2Ds targets. In this study, it is revealed that the pollution control scenario (i.e., implementation of a carbon pricing policy) surpasses the other scenarios in displacing subcritical plants sooner to pave the way for HELE technologies.

The study also reveals that:

- The climate change mitigation and pollution control scenario (i.e., a mix of carbon price and emission control technologies) is the second-best driver scenario (at the cost of increased LCOE prices as compared to Scenario 2);
- Reduced coal prices and increased life spans benefit both HELE and subcritical coal-fired power plants;
- HELE coal-fired power plants are economically competitive against subcritical plants;
- A-USC coal-fired power plants are the most economically attractive choice for deployment in Southeast Asia, followed by USC and SC plants.

The conclusion is that HELE plants are economically competitive against the subcritical plants, and in the short run, the Southeast Asian economies should focus on devising and implementing carbon pricing to support quicker deployment of HELE and displacement of subcritical technologies. Ultimately, in the long-run, a strong carbon price signal will be needed with strict emission standards to enable HELE transition.

While this study focuses specifically on ASEAN countries, its broader lessons are applicable for global deployment of HELE coal plants.

Author Contributions: Conceptualization, H.A.; Methodology, H.A., H.P. and S.R.W.; Software, H.P. and H.A.; Validation, H.A., H.P. and B.S.; Formal Analysis, H.A., H.P., S.R.W. and B.S.; Investigation, H.A.; Resources, H.A., H.P. and B.S.; Data Curation, H.A. and H.P.; Writing—Original Draft Preparation, H.A.; Writing—Review & Editing, H.A., H.P., S.R.W. and B.S.; Visualization, H.A. and H.P.; Supervision, H.A.; Project Administration, H.A. and H.P.; Funding Acquisition, H.A. and H.P. All authors have read and agreed to the published version of the manuscript.

Funding: Financial support from ERIA, and UON Singapore (UONS) (under Grant number UONS_SRG_1602) is gratefully acknowledged.

Data Availability Statement: New data generated is shared through this article. All other sources of data are cited throughout the paper.

Conflicts of Interest: The authors declare no conflict of interest.

Abbreviations

ACE	ASEAN centre for energy
ASEAN	Association of southeast Asian nations
A-USC	Advanced ultra-supercritical
Btu	British thermal unit
CAPEX	Capital expenditure
CCS	Carbon capture and storage
CCT	Clean coal technology
CF	Capacity factor
CO_2	Carbon dioxide
deNOx	Denitrification

deSOx	Desulphurization
EPC	Engineering, procurement and construction
ERIA	Economic Research Institute for ASEAN and East Asia
ETP	Energy technology perspectives
FCR	Fixed charge rate
GAR	Gross as received
gce/kWh	Grams coal equivalent per kWh
GHG	Greenhouse gas
g/kWh	Grams per kilowatt hour
GW	Gigawatt
GWh	Gigawatt hours
HELE	High-efficiency low-emissions
Hg	Mercury
HHV	High heating value
IEA	International energy agency
IGCC	Integrated gasification combined cycle
IGFC	Integrated gasification fuel cell
IPCC	Intergovernmental panel on climate change
kcal/kg	Kilocalorie per kilogram
kg-CO_2/kg-coal	Kg of CO_2 per Kg of coal
kg-coal/kWh	Kg of coal per kilowatt hour
kW	Kilowatt
kWh	Kilowatt hour
LCOE	Levelized cost of electricity
LHV	Low heating value
MMBtu	One million Btus
mt	Metric ton
MW	Megawatt
NOx	Nitrogen oxides
NO_2	Nitrogen dioxide
O&M	Operation and maintenance
$O\&M_{fixed}$	Fixed O&M charges
$O\&M_{variable}$	Variable O&M charges
OECD	Organization for economic co-operation and development
SC	Supercritical
SO_2	Sulphur dioxide
SOx	Sulphur oxides
ton/MWh	Ton per megawatt hour
TWh	Terawatt-hour
USC	Ultra-supercritical
US$	US dollars
2DS	2 °C scenario

References

1. Burnard, K.; Jiang, J.; Li, B.; Brunet, G.; Bauer, F. *Emissions Reduction through Upgrade of Coal-Fired Power Plants. Learning from Chinese Experience*; Partner Country Series; IEA: Paris, France, 2014.
2. Reddy, T.A.; Kreider, J.F.; Curtiss, P.S.; Rabl, A. *Heating and Cooling of Buildings: Principles and Practice of Energy Efficient Design*, 3rd ed.; CRC Press: Boca Raton, FL, USA, 2016.
3. International Energy Agency. *Energy Technology Perspectives 2012*; IEA Publications: Paris, France, 2010; Available online: https://webstore.iea.org/energy-technology-perspectives-2012 (accessed on 1 December 2020).
4. World Coal Association; ASEAN Centre for Energy. *ASEAN's Energy Equation: The Role of Low emission coal in Driving a Sustainable Energy Future*; A Joint Publication of WCA: London, UK; ACE: Jakarta, Indonesia, 2017; Available online: https://www.worldcoal.org/file_validate.php?file=ASEAN%27s%20energy%20equation_final_0.pdf (accessed on 23 October 2020).
5. International Energy Agency. *Technology Roadmap: High-Efficiency, Low-Emissions Coal-Fired Power Generation*; IEA Publications: Paris, France, 2012; Available online: https://www.iea.org/reports/technology-roadmap-high-efficiency-low-emissions-coal-fired-power-generation (accessed on 1 December 2020).
6. Sporton, B. Global Platform for Accelerating Coal Efficiency. *Cornerstone* **2015**, *3*, 15–18.

7. International Energy Agency. *Power Generation from Coal: Measuring and Reporting Efficiency Performance and CO_2 Emissions*; IEA Publications: Paris, France, 2010; Available online: https://webstore.iea.org/power-generation-from-coal-2010 (accessed on 1 December 2020).
8. IRENA & ACE. *Renewable Energy Outlook for ASEAN: A REmap Analysis*; International Renewable Energy Agency (IRENA): Abu Dhabi, United Arab Emerites; ASEAN Centre for Energy (ACE): Jakarta, Indonesia, 2016; Available online: https://www.irena.org/publications/2016/Oct/Renewable-Energy-Outlook-for-ASEAN (accessed on 15 January 2020).
9. International Energy Agency. *Southeast Asia Energy Outlook 2017*; Word Energy Outlook Special Report; IEA Publications: Paris, France, 2017; Available online: https://webstore.iea.org/weo-2017-special-report-southeast-asia-energy-outlook (accessed on 26 October 2018).
10. US Energy Information Administration. *Levelized Cost and Levelized Avoided Cost of New Generation Resources in the Annual Energy Outlook 2018*; US EIA: Washington, DC, USA, 2018; Available online: http://large.stanford.edu/courses/2018/ph241/asperger2/docs/eia-mar18.pdf (accessed on 10 November 2020).
11. Veldhuis, A.J.; Reinders, A.H.M.E. Reviewing the potential and cost-effectiveness of off-grid PV systems in Indonesia on a provincial level. *Renew. Sustain. Energy Rev.* **2015**, *52*, 757–769. [CrossRef]
12. Talavera, D.L.; Ferrer-Rodríguez, J.P.; Pérez-Higueras, P.; Terrados, J.; Fernández, E.F. A worldwide assessment of levelized cost of electricity of HCPV systems. *Energy Convers. Manag.* **2016**, *127*, 679–692. [CrossRef]
13. Blum, N.U.; Wakeling, R.S.; Schmidt, T.S. Rural electrification through village grids—Assessing the cost competitiveness of isolated renewable energy technologies in Indonesia. *Renew. Sustain. Energy Rev.* **2013**, *22*, 482–496. [CrossRef]
14. Holland, C.W.; Derbyshire, W. Electrifying 20 million households: An Indonesian rural electrification strategy. In Proceedings of the Electricity Engineers Association (EEA) Conference and Exhibition, Christchurch, New Zealand, 19–20 June 2009; pp. 19–20.
15. Abraham, K.; Phadke, H.; Riaz, M.; Strahl, J. REwiRE: Rural Electrification with Renewable Energy. International Impact Investing Challenge 2012. Available online: https://en.calameo.com/read/00142264778c385f5a5d1 (accessed on 16 January 2021).
16. Januar, R. Comparative Analysis of 20-MW Solar Thermal and PV Power Plant in Rongkop, Indonesia Using LCOE Simulation Method. *J. Clean Energy Technol.* **2017**, *5*, 383–387. [CrossRef]
17. Lau, C.Y.; Gan, C.K.; Tan, P.H. Evaluation of Photovoltaic Levelized Cost of Energy for PV Grid Parity Analysis in Malaysia. *Int. J. Renew. Energy Resour.* **2014**, *4*, 28–34.
18. Asian Development Bank. *Rural Renewable Energy Initiative in the Greater Mekong Subregion*; ADB: Manila, Philippines, 2015; Available online: https://www.adb.org/sites/default/files/publication/161898/renewable-energy-developments-gms.pdf (accessed on 1 December 2020).
19. Huber, M.; Roger, A.; Hamacher, T. Optimizing long-term investments for a sustainable development of the ASEAN power system. *Energy* **2015**, *88*, 180–193. [CrossRef]
20. ASEAN Centre for Energy. *Levelized Cost of Electricity of Selected Renewable Technologies in the ASEAN Member States*; ACE, ASEAN Renewable Energy Sub-Network, Eds.; ACE: Jakarta, Indonesia, 2016; Available online: https://agep.aseanenergy.org/wp-content/uploads/2018/04/Levelised-Cost-of-Electricity-of-Selected-Renewable-Technologies-in-ASEAN-Member-States.pdf (accessed on 23 September 2019).
21. Phuangpornpitak, N.; Kumar, S. PV hybrid systems for rural electrification in Thailand. *Renew. Sustain. Energy Rev.* **2007**, *11*, 1530–1543. [CrossRef]
22. Kelley, A.R.; Managi, S. The importance of financial cost for renewable energy projects: Economic viability assessment of renewable hybrid mini-grid systems in Indonesia. *Green Financ.* **2019**, *1*, 139–155. Available online: http://www.aimspress.com/article/10.3934/GF.2019.2.139 (accessed on 21 January 2021). [CrossRef]
23. ERIA. *A study on the Strategic Usage of Coal in the EAS Region: A Technical Potential Map and Update of the First-Year Study*; Otaka, Y., Phoumin, H., Eds.; Economic Research Institute for ASEAN and East Asia: Jakarta, Indonesia, 2015; Available online: http://www.eria.org/RPR-FY2014-36.pdf (accessed on 10 November 2020).
24. International Energy Agency. *Tracking Clean Energy Progress 2017—Energy Technology Perspectives 2017 Excerpt Informing Energy Sector Transformations*; IEA Publications: Paris, France, 2017; Available online: https://webstore.iea.org/energy-technology-perspectives-2017 (accessed on 1 December 2020).
25. Aldy, J.E.; Stavins, R. *The Promise and Problems of Pricing Carbon: Theory and Experience*; NBER Working Paper No. 17569; National Bureau of Economic Research (NBER), November 2011; Available online: https://www.nber.org/system/files/working_papers/w17569/w17569.pdf (accessed on 15 January 2021).
26. ERIA. *Improving Emission Regulation for Coal-fired Power Plants in ASEAN*; Motokura, M., Lee, J., Kutani, I., Phoumin, H., Eds.; ERIA Research Project FY2016: Jakarta, Indonesia, 2 August 2017; Available online: https://www.eria.org/research/improving-emission-regulation-for-coal-fired-power-plants-in-asean/ (accessed on 15 January 2021).
27. Holmes, S. *How Clean are Australia's 'Clean Coal' Power Stations*? Renew Economy. 2017. Available online: http://reneweconomy.com.au/clean-australias-clean-coal-power-stations-14224/ (accessed on 1 December 2020).
28. Tran, T.T.D.; Smith, A.D. Incorporating performance-based global sensitivity and uncertainty analysis into LCOE calculations for emerging renewable energy technologies. *Appl. Energy* **2018**, *216*, 157–171. [CrossRef]
29. Rhodes, J.D.; King, C.; Gulen, G.; Olmstead, S.M.; Dyer, J.S.; Hebner, R.E.; Beach, F.C.; Edgar, T.F.; Webber, M.E. A geographically resolved method to estimate levelized power plant costs with environmental externalities. *Energy Policy* **2017**, *102*, 491–499. [CrossRef]

30. Baig, K.S.; Yousaf, M. Coal Fired Power Plants: Emission Problems and Controlling Techniques. *J. Earth Sci. Clim. Chang.* **2017**. [CrossRef]
31. Emmerton, M.; Thorncraft, S.; Oksanen, S.; Soe, U.M.; Hlaing, K.K.; Thein, Y.Y.; Khin, U.M. *Republic of the Union of Myanmar: Institutional Strengthening of National Energy Management Committee in Energy Policy and Planning*, Financed by the Japan Fund for Poverty Reduction and the Technical Assistance Special Fund, Technical Assistance Consultant's Report, Project Number: 46389-001 December 2015. Available online: https://www.adb.org/sites/default/files/project-documents/46389/46389-001-tacr-en_17.pdf (accessed on 15 January 2021).
32. Hynes, J. How to Compare Power Generation Choices, Renewable Energy World, Issue 1. 2009. Available online: https://www.renewableenergyworld.com/2009/10/29/how-to-compare-power-generation-choices/ (accessed on 15 January 2021).
33. Intergovernmental Panel on Climate Change. *2006 IPCC Guidelines for National Greenhouse Gas Inventories*; IPCC Secretariat: Geneva, Switzerland, 2006.
34. Japan International Cooperation Agency. *Project for Promotion of Clean Coal Technology in Indonesia*; JICA: Tokyo, Japan, 2014.
35. United States Environmental Protection Agency. *New Coal-Fired Power Plant Performance and Cost Estimates*; US EPA: Washington, DC, USA, 2014.
36. Indonesian Coal Prices Seen Rising in Near to Medium Term: Sources. Available online: https://www.platts.cn/latest-news/coal/singapore/indonesian-coal-prices-seen-rising-in-near-to-26880890 (accessed on 1 December 2020).
37. Phoumin, H. Benefits of Increasing Emission Standards for Coal-fired Power Plants in ASEAN beyond Environment and Health, Sep 2017. Available online: https://www.worldcoal.org/benefits-increasing-emission-standards-coal-fired-power-plants-asean-beyond-environment-and-health (accessed on 1 December 2020).
38. Kost, C.; Shammugam, S.; Julch, V.; Huyen-Tran, N.; Scheigl, T. *Levelized Cost of Electricity Renewable Energy Technologies*, March 2018. Fraunhofer-Institut für Solare Energiesysteme. Available online: https://www.ise.fraunhofer.de/content/dam/ise/en/documents/publications/studies/EN2018_Fraunhofer-ISE_LCOE_Renewable_Energy_Technologies.pdf (accessed on 11 January 2021).

sustainability

MDPI

Review

Green Technological Development and Deployment in the Association of Southeast Asian Economies (ASEAN)—At Crossroads or Roundabout?

Rabindra Nepal [1,*], Han Phoumin [2] and Abiral Khatri [3]

1 Faculty of Business and Law, School of Business, University of Wollongong, Wollongong, NSW 2522, Australia
2 Economic Research Institute for ASEAN an East Asia (ERIA), Senayan Jakarta Pusat 10270, Indonesia; han.phoumin@eria.org
3 Department of Business-Society Management, Rotterdam School of Management, Erasmus University Rotterdam, 3062 PA Rotterdam, The Netherlands; khatri@rsm.nl
* Correspondence: rnepal@uow.edu.au

Abstract: Southeast Asia faces one of the fastest growths in electricity demand in the world, driven by increasing incomes, urbanization and industrialization. Development and deployment of green energy technologies offer a natural conduit to meet the growing electricity needs of the Association of Southeast Asian Economies (ASEAN) region while also serving as a viable strategy to adapt to climate change. The aim of this study is to formulate the policy lessons for the ASEAN economies and governments in facilitating the development and deployment of green technologies and alternatives energy options based on a specific case review of the ASEAN. The ASEAN economic region is prioritizing sustainable economic growth while minimizing the regional impacts of climate change through decarbonization. The study undertakes a case-specific analysis in reviewing green energy deployment in the context of green growth and energy transition using secondary data sources and discusses the current status and future options of renewable energy development in the ASEAN. We find that carbon capture and storage (CCS) technologies will allow the ASEAN to continue to use fossil fuels while achieving sustainable economic growth as coal demand increases in the region. The deployment of CCS technologies will also act as an enabler of hydrogen energy as a green energy solution in the region in the longer term. Boosting public acceptance to nuclear energy, implementing energy efficiency improvement policies and eliminating fossil fuels consumption subsidies are feasible short-term and medium-term policies. Increasing both the public and private sector energy investments and development of CCS technologies in the longer term are necessary complementary policies to maximize the benefits of greater deployment of renewable energy sources in the region and combat climate change.

Keywords: green technology; sustainability; climate change; Southeast Asia; energy policy

Citation: Nepal, R.; Phoumin, H.; Khatri, A. Green Technological Development and Deployment in the Association of Southeast Asian Economies (ASEAN)—At Crossroads or Roundabout?. *Sustainability* **2021**, *13*, 758. https://doi.org/10.3390/su13020758

Received: 26 November 2020
Accepted: 12 January 2021
Published: 14 January 2021

1. Introduction

Sustainable development is about achieving a more sustainable global future so that future societies face fewer challenges arising from resource scarcity and accumulating atmospheric pollutants. Sustainable development is a powerful development concept as it integrates the economic, societal and environmental aspects and the interrelationships among the energy, environment and societal concerns. Developing sustainably ensures that the availability of critical resources such as energy, water and food is available to both present and future generations and emphasizes mitigating the scope of the environmental problems across both geographic and generational boundaries [1]. However, the transition towards sustainability is still at an early stage in developing economic regions while economies around the world have been struggling to balance their economic growth priorities without deteriorating the natural resources. Developing economies and regions are expected to be hit hard by climate change due to their more limited adaptation capabilities even though they have produced relatively small amounts of greenhouse gases. The COVID-19 pandemic brings further uncertainty in adapting sustainability reforms to combat climate change given the economic downturn and border closures in many regions affecting resource mobility. Nonetheless, the role and importance of green technology as a climate change adaptation vehicle has always been crucial in providing a new perspective on sustainable development.

Developing economic regions like Southeast Asia currently face paramount challenges as well as opportunities in matching its increasing energy demand due to rising incomes, industrialization and urbanization with sustainable energy supply considering the transition to a lower-carbon economy. Greenhouse gas emissions have been rising rapidly at an average annual rate of 5% in recent decades among major Southeast Asian economies such as Indonesia, Malaysia Philippines, Thailand and Vietnam [2]. The region is poised to become a net energy importer of fossil fuels such as oil due to growing population, industrialization and urbanization despite the slowdown in economic growth [3]. The total population in the Association of Southeast Asian Economies (ASEAN) region will increase to 715 million by 2025 with the economy growing by more than 5% per year and therefore explaining a rapid rise in energy demand of at least 4% annually [4]. The overall growth in energy demand of more than 80% since 2000 has been met by a doubling in fossil fuel use engendering severe energy security concerns such as rising import dependence and environmental concerns due to an increase in energy-related carbon dioxide (CO_2) emissions [3]. For instance, the share of this geographic region to global emissions increased to 4% in 2018 (3% in 2010), while the number of deaths linked to outdoor and household air pollution in Southeast Asia is expected to spike to more than 650,000 a year by 2040, which is an increase from around 450,000 deaths in 2018 [5]. The average temperature in ASEAN is also rising by 0.1 to 0.3 degree Celsius per decade in the last 50 years and is projected to increase by 2–4 degrees Celsius by the end of the 21st century [6]. The electricity demand in the region is growing at an average of 6% and remains among the fastest in the world, while the region's demand for electricity is projected to double by 2040 [3]. In 2016, the ASEAN economies set a target of 23% to have its primary energy supply secured from renewable sources by 2025 [4]. However, it is also likely that the overall energy demand will grow by almost 50%, while the power generation will double by 2025 [3]. Some countries in the region will have to at least double their share of renewable energy every year, but a doubling of renewable capacity alone may not be enough to combat climate change. The rising energy demand and related CO_2 emissions in the ASEAN, therefore, implicates the heightened need for transitioning towards the development and deployment of greener energy sources as a necessary climate change adaptation strategy in the region.

There is also an ongoing discourse in the ASEAN to devise policy strategies to mitigate and adapt to climate change threats and to balance the trade-offs between economic development and environmental sustainability given the region's heavy reliance on fossil fuels. For instance, the power generation mix of the ASEAN is dominated by fossil fuels, which accounted for almost 80% in 2017 and are expected to account for 82% in 2050 in the absence of the region not transitioning to cleaner energy systems [7]. Therefore, the need to develop, deploy and adopt green technologies is imminent for Southeast Asia to address the twin challenges of rising energy demand and increasing emissions in ensuring energy sustainability as well as to mitigate the adverse impacts of climate change. The ASEAN economies should increase their share of renewables in the energy mix to 70% by 2040 to meet their sustainable development goals [8]. However, the progress towards the adoption of green technology such as renewables in Southeast Asia is slower than the anticipated potential across many Southeast Asian economies. Renewable energy only meets around 15% of demand with the rapid increase in hydropower and modern use of bioenergy in heating and transport [9]. Solar and wind power being the most abundant energy resources in the ASEAN only contribute negligibly to the power mix [7]. The large potential of sustainable use of modern bioenergy remains untapped in the region, although electricity from hydropower production has quadrupled since 2000. The Southeast Asian economies are yet to perform globally in renewable energy deployment due to various challenges despite having huge potentials for sustainable energy sources [10]. However, policymakers across Southeast Asia are intensifying their efforts in achieving a common goal of a secure, sustainable and affordable energy sector even though the region is diverse and dynamic [8]. The diversity in energy mix in the region also offers a viable opportunity to accelerate regional physical interconnections of power grids and make greater use of the resource and demand complementariness [11]. Boosting regional power grids in the ASEAN has also been a well-advocated energy policy agenda in the past [12].

The objective of this study is to analyze and review the energy–economy–environment in the ASEAN from an energy sustainability perspective in the context of green energy development and deployment as a powerful climate change adaption policy in the region. In doing so, the study recognizes the inevitable economy–environment tradeoffs between regional economic growth and adverse climate change impacts as a policy tool for policymakers to emphasize. The study assumes that successful energy transition in the ASEAN is only possible by increasing the share of renewables and clean energy and fuels. Therefore, based on our impartial and unbiased analysis using secondary data sources, we propose that the policymakers need to formulate and implement proper policies that are of short-term, medium-term and long-term nature. These policy proposals include scaling of renewable energy deployment, focusing on energy efficiency improvements, discouraging the use of fossil fuels by undertaking energy pricing reforms and embracing hydrogen, carbon capture, utilization and storage technologies. However, significantly accelerating the deployment of renewable energy in the region is challenged by greater levels of investment and financing requirements, which the region needs to overcome alongside managing the energy governance and financial risks.

Our study is also one of the limited studies focused on the deployment and advancement of green energy technologies as a necessary climate change adaption strategy in the context of sustainable development in the ASEAN. The findings of our study are significant to the ASEAN governments and policymakers in crafting a sustainable energy policy through the development and deployment of alternative greener energy options apart from the conventional renewable sources in adapting to adverse impacts of climate change in the region. We undertake a case review of the ASEAN region as a whole as a case-specific analysis based on a secondary data as a suitable approach to examine policy problems that do not easily lend themselves to rigorous quantitative analysis or that cannot be analyzed using econometrics due to the unavailability of disaggregated data [13].

The remainder of the paper is structured as follows. Section 2 is a brief literature review on the deployment of conventional renewable energy in the context of green growth and energy transition and also portrays the current status of conventional renewable energy deployment in the ASEAN. Section 3 discusses green energy innovation and alternative green energy options for the ASEAN in adapting to climate change. The three major proposed policy recommendations are discussed in Section 4. Section 5 concludes the paper.

2. Current Status of Renewable Energy Deployment in the ASEAN

Historically, the development and deployment of conventional green technologies have mostly been carried out in the context of national green growth strategies and energy transition. This is because economies around the world have set challenging targets to reduce greenhouse gas emissions and renewable energy technologies that offer decarbonization pathways and possibilities. Many developing countries have started to adapt strategies related to green growth into their economic development agenda. The gap between the technologies and the capabilities has been addressed on the basis of literature on technology transfer as well as industry insights across the technology lifecycle. Heuristics have been shown to be crucial when it comes to deployment of clean technology from a policymaker perspective, while technology-specific green growth strategies also depend on countries with different income levels as well as their requirements [14]. However, several challenges in the energy sector have been identified to meet the renewable energy targets that were set by ASEAN economic community between 2010–2015 [15]. For instance, there has been little progress in terms of regional cooperation in critical areas like physical infrastructure, while various financial constraints, regulatory differences and technological gaps are challenging energy security.

Governments in Indonesia, the Philippines and Singapore have focused on the institutionalization of sustainable energy by aligning with the domestic systems of renewable energy based on an analysis of their sustainable energy policies between 2000 and 2016 [16]. Li et al. [17] in 2020 found a tremendous mismatch between resources and energy demand despite the huge potential for renewable energy in the research about the energy transition in East and Southeast Asia. In another article of 2019 by Lee et al. [18], the impact of fossil fuel aerosols on air quality in Southeast Asia under varying hypothetical fuel consumption scenarios were examined. The results showed that the replacement of coal by natural gas in the power generation and industry sectors would lead to a 25% reduction in sulphate in the Southeast Asia region. Substituting biofuels by natural gas in the residential sectors would lead to a 42% reduction in black carbon concentration. The importance of clean air in the Southeast Asian region is also articulated under the findings of the International Energy Agency (IEA) [8] where energy-related air pollution death will rise to 650,000 by 2040 from an estimated 450,000 in 2018. Taghizadeh-Hesary and Rasoulinezhad [19] determined how energy transition patterns depend in 45 Asian economies when classified as per their income levels for the period of 1993–2018 using the generalized method of moments (GMM) estimation approach. The results showed that an increase in population slows the energy transition process across all income levels with economic growth, generating a positive relationship with the energy transition, while CO_2 emissions negatively influence energy transition.

There has also been increasing debate about the effects of energy transition in the form of broader global biodiversity threats. Studies have shown that renewable energy facilities can be land-intensive as well as impact conservation areas. Rehbein et al. (2020) [20] assessed that the extent of current and likely future development of renewable energy related to onshore wind, hydropower and solar photo-voltaic generation intervenes with the protected biodiversity areas. It was found that lack of proactive measures like non-removal of subsidies of fossil fuels and barriers to regional market integration by the ASEAN governments will act as barriers to the region in achieving the ambitious target of 23% renewables in the primary energy mix by 2025 [9].

The Table 1 below shows that studies focused on the deployment and advancement of green energy technologies in the context of green growth and energy transition are limited for Southeast Asia. Our review study attempts to extend this strand of literature by reviewing the specific case of the ASEAN economic region in search of viable climate change adaptation strategies. Existing studies have mostly focused on the technical and environmental aspects of conventional renewable energy with much disregard for innovative fuels like hydrogen and carbon capture and storage. Our research will contribute to the crafting of sustainable energy policy by the ASEAN governments through the deployment of innovative green technologies in fighting climate change as well as achieving sustainable economic growth in the region.

The ASEAN region is expected to achieve an accelerated economic growth over the next decade and experience 50% rise in energy demand. Importantly, the region is targeted to source 23% of its primary energy from renewable sources [4]. The global economic and energy indicators show that ASEAN region is becoming a net importer of fossil fuels given the rapidly growing economies in the region and increasing population size. Oil is likely to continue to dominate the road and transport demand in ASEAN. Similarly, coal demand will increase, driven by strong policy settings by countries to meet the economic growth targets. IEA estimated that more industrial consumers than the power plants drive the demand of natural gas although the increase in imports of oil is making sources such as Liquefied Natural Gas (LNG) less price competitive in the ASEAN.

Table 1. Literature Summary.

Year.	Authors	Data	Methodology	Main Findings
2019	Jalton Garces Taguibao	2000–2016 policy documents.	Examination of policy discourse of Southeast Asia	Governments in Southeast Asia tend to reduce the cost of electricity to make supply accessible and reliable.
2019	Hsiang He Lee, Oussama Iraqui and Chien Wang	Primary air pollutants and greenhouse gases from 2000 to 2008.	Weather Research and Forecasting model coupled with the modal aerosol dynamics model for Europe and the Secondary Organic Aerosol Model (SORGAM)	Southeast Asia's sulfate could be decreased by 25% if coal could be replaced by natural gas.
2019	J. Rehbein, J. Warson, J. Lane et al.	Key Biodiversity Areas, Protected Area and Wilderness Area Datasets on renewable sources.	Secondary research	Negative effect of energy transition on biodiversity.
2020	Farhad Taghizadeh-Hesary and Ehsan Rasoulinezhad	Energy transition (renewable energy consumption/non-renewable energy consumption) as a dependent variable, and official exchange rate, GDP and CO_2 emissions as independent variables	Panel Generalized Method of Moments (GMM) approach	An increase in population slows the energy transition process across all economies at different income levels

About 120 million people (around 10% of ASEAN's overall population) still do not have access to electricity in Southeast Asia, and the rural areas face a critical challenge in accessing the power [21]. There are about 45 million people in the region who rely on biomass as a fuel for cooking [8]. There is a tremendous potential for renewable energy, but renewable energy only accounts for 15% of the energy demand. On one hand, hydropower has increased by four times since 2000 along with the increase in the suse of bioenergy in heating and transport [3]. On the other hand, the share of solar photovoltaics and wind is small, although the costs have been declining in recent years. An efficient market-based energy efficiency framework could strengthen in their deployment but such framework is missing.

Southeast Asia's overall energy demand is also expected to grow by 60% by 2040 based on a stated policies scenario developed by IEA in 2019. The projection assumes that the size of the economy will double over the period and the majority of the population will be concentrated in the urban areas experiencing an increase of 120 million [22]. A structural economic shift towards less energy-intensive manufacturing and services sectors is expected along with greater energy efficiency improvements, which will lower the rate of energy demand compared to previous decades, although it will represent 12% of global energy rise by 2040. The oil demand will exceed 9 million barrels per day (mb/d) by 2040, which is currently at 6.5 mb/day [3].

However, Southeast Asian nations also have a geographic advantage in terms of natural resource endowments to produce renewable energy. For example, Indonesia and the Philippines have substantial geothermal energy potential, while Vietnam, Cambodia, Laos and Myanmar have large-scale hydropower potentials. Similarly, most areas in these countries have at least 12 h of sunshine on average and are suitable for solar electrification. The global renewable energy generation capacity stood at 2179 GW by the end of 2017, with the hydro sector holding the largest share with an installed capacity of 1271 GW. In the same period, Asia alone accounted for 64% of the global share in new renewable capacity. The majority of the growth in installed renewable capacities is driven by the new installations of solar and wind energy in the ASEAN (around 85% of all new renewable capacity installed). Thailand was one of the distinguishable countries in the ASEAN region that had the second-highest share in bioenergy capacity (430 MW). Indonesia topped the list in expanding the geothermal energy capacity by 306 MW and is soon approaching a 2 GW total geothermal capacity [23]. Similarly, Malaysia is the third-largest producer of photovoltaic cells in the world while solar is responsible for sourcing the majority (47% in 2017) of the grid-connected renewable power generation in Malaysia [24].

Likewise, Laos has around 80% of its primary energy demand sourced through renewable energy, and the country has realized its potential. Biomass energy, which comes from forestry and agricultural waste, comprises 68% of its energy and is used for household cooking and small-scale rural production. The remaining 12% is sourced from the hydropower sector. The commercial methods to utilize biomass in Laos includes direct combustion, gasification, anaerobic digestion and hybrid system [25]. Laos also took advantage of the 300 days of annual sunlight, which enabled its economy to equip 13,000 rural homes with solar panels. In Indonesia, the government has taken the initiative to build a large-scale floating power plant, which has opened doors for 60 other reservoirs across the country [26]. The country has huge potential for wind, and a 100-hectare wind farm was opened in South Sulawesi that has a capacity to power around 70,000 households [27].

The Philippines has the largest potential for wind energy in Southeast Asia, although a significant proportion of the population does not have access to electricity, compelling them to use alternate methods for cooking and lighting. Green start-ups have played a major role in the Philippines by benefitting from the natural energy resource endowments. A Filipino startup named Sustainable Alternative Lighting came up with a saltwater solution-powered lamp product that retains power for up to eight hours. Furthermore, the disposable component of the lamp lasts for 6 months and is also not expensive to replace. Around 51% people use firewood or charcoal in the Philippines [28].

Figure 1 shows that the region's energy demand is expected to rise to 60% by 2040. This expected rise in energy demand is far lower compared with the energy demand of the previous decades. The renewable share in power generation is expected to rise from 24% in 2020 to 30% by 2040. However, the expected rate of progress in renewable energy is still short of levels reached by other emerging economies such as China and India under the stated policies scenario. The hydropower sector that accounts almost 80% of the renewable share is the cornerstone of ASEAN's energy portfolio and the rise of wind and solar energy as well as biofuels, and bioenergy from waste products is likely to deliver promising growth. Furthermore, innovation in hydrogen carbon technologies could positively change the energy landscape of ASEAN [7].

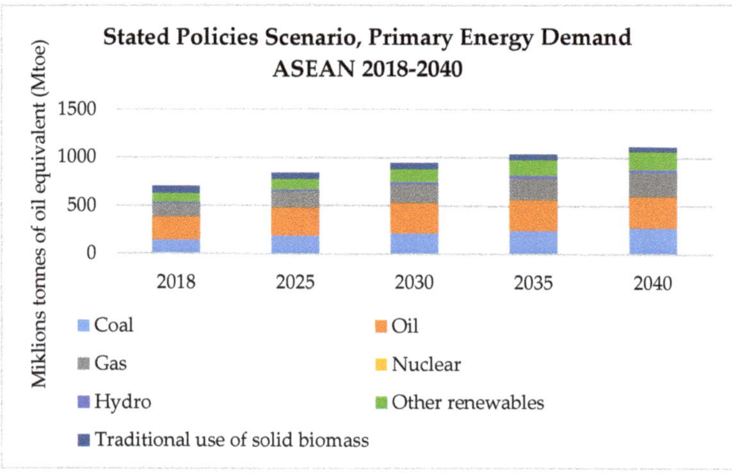

Figure 1. Primary Energy Demand in the Association of Southeast Asian Economies (ASEAN) (2018–2040). Source: Adapted from IEA [3].

Figure 2 portrays the economic and energy outlook of the Southeast Asian region in the global context. While the region is experiencing one of the highest increases in electricity demand at an average of 6% per year, the numbers of power systems in the region face major financial backlash. The use of overall energy demand cannot be undermined either, as the overall energy demand has grown by more than 80%, with fossil fuel use doubling. The renewable energy capacity potential in Southeast Asia is significant enough and is continuously growing as observed in Figure 2. Nevertheless, only 15% of the region's energy demand is met at present from renewable energy sources, although the scope for renewable energy in the region is tremendous. The falling costs of solar photovoltaics and wind could be encouraging news for supporting their deployment, especially for the small economies in the region such as Myanmar, Cambodia, Vietnam and Laos.

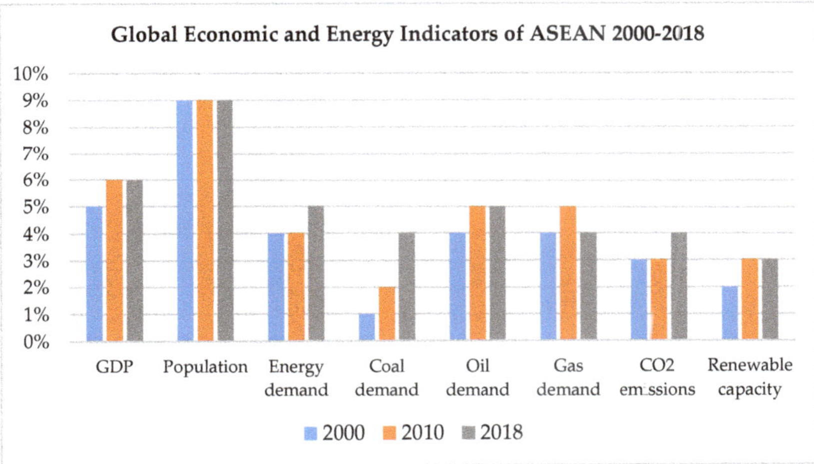

Figure 2. Economic and Energy Indicators in the ASEAN (2000–2018). Source: Adapted from IEA [3].

Hydropower output has also quadrupled in Southeast Asia since 2000 [3]. The costs for solar photovoltaics (PV) have been falling over time, but the share in total energy remains small. IEA [3] data also show that there has been a shift towards low energy-intensive manufacturing and services given the projected rate of energy demand growth is lower than the past two decades, which holds 12% share of the projected rise in global energy use to 2040. Figure 3 below also shows that space cooling is set to drive the growth in household electricity demand in the ASEAN by 2040.

Achieving a clean energy future in the ASEAN requires electrifying the transport sectors by deploying green technologies like electric vehicles. However, the congested roads and lack of proper infrastructure make it difficult to scale up and replace oil consumption. The rise of middle income and the increasing demand for household space cooling has increased the energy use for air conditioners in ASEAN by 7.5 times in the past 30 years. Indonesia, which is the most populated country in ASEAN, only has about 10% of its households with air conditioning, and less than 20% of households in the whole ASEAN region have air conditioning. However, these numbers are likely to keep growing, and an additional 200GW of capacity needs to be added by ASEAN countries by 2040, which will increase demand by 30% [29]. There are opportunities, at the same time, to increase energy efficiency policies, which could in turn enhance efforts to improve the building and equipment efficiency. Policymakers must understand that hydrogen is one of many alternatives available to fossil fuels, given the high significance in energy storage, longer-distance driving and faster filling [7].

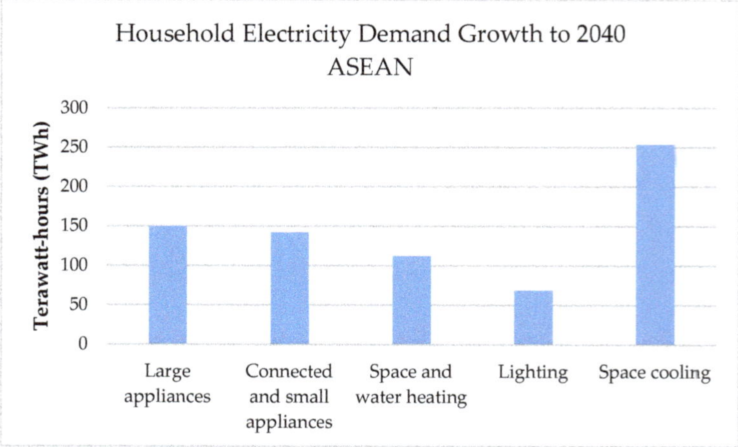

Figure 3. Household electricity demand growth in the ASEAN (by appliance source). Source: Adapted from IEA [3].

3. Green Innovation and Alternative Energy Options in the ASEAN

Meeting the energy-related sustainable development goals (SDGs) in the ASEAN requires deploying multiple technologies and policy approaches in the energy sector. As there are no silver bullets, international experiences of energy transitions can offer valuable guidance and insights in the development and deployment of green energy technologies in the ASEAN, considering that fossil fuels have dominated the planet for centuries and will continue to do so. While the removal of carbon from the atmosphere is urgent, innovative energy solutions should be adapted considering the environmental, technological and economical aspects. Policymakers need to have a practical orientation towards the frameworks that are being developed internationally towards the deployment of green technologies so that the energy transition becomes smooth. A report by IRENA showed that Southeast Asia has the highest share of jobs in renewable energy (83%), whereas it is the lowest in terms of energy efficiency jobs (only 7%) [30]. Renewable energy technology varies significantly across the member states in ASEAN, although there has been some significant progress made in renewable energy development.

3.1. Nuclear Energy

Nuclear power systems are comparatively clean and a reliable source of energy with the potential to contribute towards hydrogen economy. Many countries in Southeast Asia have also expressed increasing interest in nuclear energy given the economic benefit as well as low-carbon emissions for electricity supply [31]. In addition to renewables, the technological advancement of nuclear reactors is already considered to have the possibilities of transforming the clean energy sector in Southeast Asia [32]. The substantial possibility for cost-effective, efficient and large-scale hydrogen production utilizing heat derived from nuclear power station already exists. For example, the US Department of Energy introduced the Advanced High-Temperature Reactor (AHTR) technology build for hydrogen production with high-temperature water electrolysis or thermochemical cycles [33]. Several studies on the thermochemical cycle have delivered thermal-to-hydrogen energy efficiencies of, e.g., 50% for the adiabatic UT-3 cycle and 52% for the sulphur-iodine cycle [34]. Hence, economically sound and technologically superior hydrogen production capacities could be sourced from nuclear energy. Nuclear energy sector has also gained favor from international organizations like Intergovernmental Panel on Climate Change (IPCC) as an important energy option to attain the "zero emissions". However, there have been accidents like the Fukushima nuclear incident in Japan that have changed the political environment and perceptions towards nuclear energy. The commitment to mitigating greenhouse gas (GHG) was revised in Japan as a response. Although Japan committed to a 25% reduction in emissions from 1990 levels by 2020, it only decreased by 3.8% from 2005 levels, translating to a 3.1% increase in GHG from 1990 levels [35]. However, the nuclear reactors in Japan also restarted their operations since 2015 despite the lack of public acceptance.

An increase in global nuclear power production of about 80% is required by 2040 to achieve the sustainability target in which 85% of the global electricity needs to come from clean sources by 2040 compared with the existing 36%. The use of nuclear power has reduced carbon dioxide (CO_2) emissions by over 60 gigatons, which is equivalent to two years' worth of global energy-related emissions [3]. Hence, it would be much harder to achieve a sustainable energy system without a proper nuclear investment. Furthermore, nuclear plants also help to keep the power grids stable by limiting the seasonal fluctuation impact from other renewables and reduce dependence on imported fuels, which has been prevalent in major ASEAN countries. However, public acceptance and trust needs to be garnered by informing the public about the importance of the energy source as a viable energy technology to address societal needs.

3.2. Carbon Capture and Storage (CCS)

Achieving long-term economic growth in the ASEAN will involve the continued use of fossil fuels. Increasing demand for coal is expected to cause a rise in emissions around 66% by 2040 [8]. How can the ASEAN region continue to use fossil fuels to accelerate economic growth without hurting the environment? CCS offers a viable pathway to use cheaper energy sources such as fossil fuels while minimizing their environmental impacts, as the technology can prevent around 90% of CO_2 from entering the atmosphere by capturing the emissions produced from fossil-based electricity generation and uses. CCS technology also enables producing clean hydrogen from fossil fuels as the emitted carbon gets captured and geologically stored. Almost all of the world's hydrogen is sourced from gas, coal and the production of clean hydrogen using CCS technology and can be more cost-effective than producing clean hydrogen from renewables using electrolytes [7]. If combined with renewable biomass, CCS allows carbon dioxide to be taken out of the atmosphere and is carbon-negative.

Southeast Asia provides good opportunities for harnessing the CCS technology as the region has plentiful geological storage resources. Countries like Indonesia, Vietnam, Philippines and Thailand have 54 gigatons of storage capacity [36], which reflects that the region has sufficient capacity to conceal carbon dioxide. However, countries in the ASEAN region are developing CCS at different speeds. For instance. CCS technologies have gained much attention in Singapore across both the public and private sectors since 2017. Indonesia is also considering the development of large gas projects with high CO_2 concentrations, even though there is a need to further codify the CCS legal framework Malaysia, on the other hand, is focusing on the CCS development in the power and oil/gas sectors by undertaking capacity development and storage assessments alongside running legal and regulatory workshops. The Asian Development Bank (ADB) has also been promoting carbon capture, utilization and storage (CCUS) in Asia since 2009 [37]. The economic analysis by Asian Development Bank (ADB) showed that natural gas processing and power plants are the best capture source, as it is the lowest-cost option for carbon capture and storage in Southeast Asia [38]. However, the development and deployment of CCS in the ASEAN region need to overcome significant challenges such as generating investments, attracting climate financing, regional and international collaboration as well as establishing regulatory frameworks for CO_2 storage. An effective stakeholder engagement, especially through a smooth public dialogue could enhance carbon capture and storage development, which could increase the commercial viability.

3.3. *Hydrogen Energy*

Hydrogen is the most abundant chemical element available in the atmosphere and can be a viable source to electrify homes, transport and industry. Hydrogen is being pursued as a potential form of clean energy given its wide usage in areas such as ammonia production, petrochemical and oil refining industries and many others. Currently, around 95% of hydrogen is produced from coal and gas, also called "grey hydrogen", and a small portion is produced with carbon capture, sequestration and storage (CCS), called "blue hydrogen". Less than 5% of total hydrogen production is produced from renewables, also known as "green hydrogen" [7]. Green hydrogen obtained through the electrolysis of water could be a non-polluting alternative for energy. Green hydrogen could be adopted in sectors such as transport, power generation, construction buildings and energy storage as it can make a remarkable contribution to clean energy transitions. Hydrogen has the characteristics of being light, storable and energy-dense, and no direct emissions of greenhouse gases makes its an important part of a clean and secure energy future. A study has found out that the electricity demand would reach 3600 TWh, surpassing the total annual electricity generation of the whole European Union if all the current hydrogen production is to be transformed from green sources [39].

Hydrogen fuel has huge potential to combat climate change by facilitating the transition to low-carbon energy sources despite their existing low share in the global energy consumption. An increase in scope for renewable energy and continuous decrease in the costs demanded for innovative green technologies of which storage facilities developed through hydrogen is likely. Furthermore, research has shown that blending of hydrogen with natural gas could provide a smooth transition from the current hydrocarbon-based economy to a hydrogen carbon economy [40]. In a long-term transition toward a clean and sustainable energy future, hydrogen provides a flexible option and a more distributed energy system that ensures a clean and sustainable hydrogen future [41]. The system brought about by a hydrogen economy could provide an easy transition towards a renewable-based future for many countries in ASEAN, which demands infrastructure and high energy demands.

The cost of hydrogen will also decline by over 50% by 2040 if adopted across all sectors, making it as competitive as the price of gasoline [42].The current cost of supplying renewable is about five times higher than gas, but the cost will come down with an investment in hydrogen supply chains. Green hydrogen will serve as a catalyst to address the integration challenges hindered for wind and solar as the world is shifting towards a green economy. By 2023, many hydrogen projects in Organisation for Economic Cooperation and Development (OECD) countries are expected to be launched which includes major pipelines for distribution to end-users and electrolyzers [43]. Island countries, especially in the ASEAN region, will benefit substantially as hydrogen will accelerate the storage as a clean energy carrier [7].

However, the ASEAN region has not yet included hydrogen in its policy agenda in many countries as an alternative fuel. Nevertheless, policy measures are likely to be addressed on emerging and alternative technologies, as hydrogen and energy storage by ASEAN Plan of Action for Energy Cooperation (APAEC) Phase 2 is under preparation for endorsement at ASEAN Ministers on Energy Meeting. The OECD's action plan to increase the share of hydrogen in the energy mix could indeed be fulfilled with the support from APAEC. The energy leaders in ASEAN could also develop a clear strategy on ways to promote hydrogen use in transportation and power sectors not limited to refining, fertilizer and petrochemical industries. Countries such as Singapore, Malaysia, Thailand, Indonesia and the Philippines could learn lessons from OECD countries, China and European community to guide the investment in research and development for hydrogen produced from both renewables and non-renewables.

Southeast Asian countries can learn from neighboring economies such as China, which has already accelerated hydrogen investment support to local industries whereby about US$2 billion is being injected. Similarly, Japan has been promoting the global adoption of hydrogen for vehicles, power plants and other usages. Brunei in the ASEAN region has also taken a lead in the supply chain of hydrogen as it has supplied the liquefied hydrogen to Japan since 2019. However, more energy is consumed by the liquefied hydrogen as it needs a temperature of minus 253 degrees Celsius in order to transform the cooled gas into a liquid form [7].

Japan has been pioneering the renewable hydrogen economy in which the production of hydrogen through the reformatting process of renewable electricity such as solar and nuclear is likely to bring a breakthrough in decarbonizing emissions. Japan also became the first country in East Asia to adopt a basic hydrogen strategy, which will ensure that the production will reach cost parity with gasoline fuel and power generation in the long term. The society's willingness to pay is also a major factor despite the efforts by governments and private sectors to adopt hydrogen practices. South Korea is another country that has set a target for hydrogen usage with 10% of total energy consumption by 2030 and 30% by 2040 in order to power selected cities and towns [7]. The South Korean government has also made an announcement to create three hydrogen cities by 2022 where hydrogen will be used for major urban functions such as cooling, electricity, heating and transportation.

New research efforts are also underway with regards to investigating new methods for chemical-based liquid hydrogen carriers. A study [44] in 2013 introduced a methodology to quantitatively analyze the energy system by looking into the relationship of green car technology and greenhouse gas reduction in the regions of South Korea. The research suggested that technology such as decarbonization should be enhanced through the production of hydrogen to replace the existing fossil fuel sources in the foreseeable future. A recent study [7] also supports the promising future role of renewable hydrogen in energy transition to decarbonize ASEAN's emissions based on the examination of the potential scalability of renewable hydrogen production from curtailed electricity.

4. Policy Recommendations

The development and deployment of green technologies are viable and necessary in Southeast Asia to address the critical issues of climate change and adaptation in the context of increasing energy demands. The development and deployment of green energy technologies will improve environmental quality, human welfare and overall help developing economies to achieve sustainable development goals. ASEAN as a regional multinational organization has a pivotal role to play to not only fulfill its global commitments of the United Nations Climate Change Conference (COP 21) but also to facilitate cross-sectoral partnerships for sustainable economic development. This is important to achieving the ASEAN Community Vision 2025, which aims to sustain the momentum of regional integration [45].

There seems to be lack of adequate experience and expertise in some ASEAN member states such as Vietnam, Malaysia and Indonesia when it comes to the evaluation of risks of renewable energy investments, which has translated into lack of financial support and public capital immobility towards renewable energy investment. The cost of deploying renewable energy sector has been continuously falling, which has increased prospects for accelerated investment shifting away investor's choices from fossil fuels towards renewables. Green technologies such as hydropower, geothermal and hydrogen carbon have become substantially competitive. For instance, the costs of solar PV, concentrating solar, onshore wind and offshore wind have fallen respectively by 82%, 47%, 39% and 29% between 2010 and 2019 [46].

Around 56% of capacity additions for utility-scale renewable power achieved lower electricity costs in 2019 than the cheapest new coal plant. The annual potential savings were projected around $23 billion if 500 GW of existing coal were to be replaced by solar wind [47]. This global trend is an indication for the policymakers in ASEAN also to emphasize alternative green energy options and exploit the huge benefits they bring. Technologies to reduce emissions from the power sector such as carbon capture, utilization and storage are essential, and efficiency must be achieved in sectors such as vast cooling and road transport. The gasification of biomass and solar–thermal technology create alternatives in producing hydrogen from renewable energy sources. Similarly, the surplus wind electricity is also used for hydrogen production as a means for storing energy, which reduces the risk from the curtailment of solar and wind power [7].

4.1. Transitioning towards Hydrogen Carbon Economy

The ASEAN countries could emphasize on an efficient interplay between energy, environment and economy in moving towards a hydrogen carbon economy. Hydrogen has major implications in various sectors such as transport. Countries like India have welcomed foreign investment in fuel cell vehicles and hydrogen transportation infrastructure, which have already started in some pilot cities. Similarly, in Japan, the Tokyo Metropolitan government has increased the number of hydrogen buses to 100 in 2020 [48]. As for the ASEAN region, the Sarawak Local Government in Malaysia is starting to operate hydrogen buses soon. Singapore is also collaborating with companies from Japan to explore the development of hydrogen as a new clean fuel to decarbonize emissions.

Hydrogen production mostly comes from natural gas as it consists of 70 million tones, which is around three quarters of the annual global or 6% of natural gas use. Coal also contributes equally, as countries like China have a major stake while only some of thei production of hydrogen comes from oil and electricity [40]. It can be observed from Figure 4 below that the support investments for hydrogen technologies have increased recently in many countries with around 50 targets, mandates and several policy incentives especially focused on the transport sector. Hydrogen production mostly comes from natural gas, as it consists of 70 million tonnes, which is around three quarters of the annual global or 6% of natural gas use.

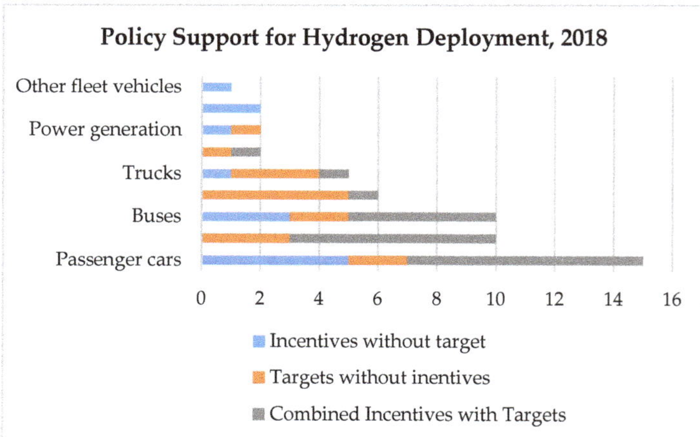

Figure 4. Support policies for hydrogen development. Source: [40].

There is not a one-size-fits-all solution when it comes to hydrogen policy. The production of both "blue" and "green" hydrogen include several opportunities and risks to the countries following the respective approaches, even though there are options available to deploy hydrogen products from both fossil fuels and low-carbon sources such as renewable electricity. On one hand, hydrogen based on fossil fuels may enable rapid scale-up in short term. However, there are minimal environmental benefits, and it requires carbon capture in the long term. On the other hand, the substantial application of hydrogen in big sectors such as transport and chemicals can bring efficiency to the energy system. This could bring numerous opportunities to exploit energy resources that are currently underutilized. The ASEAN government should align their ambition and approach for the use of hydrogen by considering the international practices as well as the market scope where it can be widely applied.

There still remains a considerable gap towards realizing its potential despite the wide spectrum of opportunities entailed by hydrogen with its industry application. An action-oriented plan and vision is required both for the near future to make hydrogen feasible for that future as the support for clean energy transition is growing among policymakers in the ASEAN. An intelligible policy is essential to meet the long-term hydrogen goals as there are various risks associated in investments, which could be detrimental to many stakeholders given the complexity of hydrogen value chains. A standard regulation is required across the ASEAN countries to mitigate uncertainties and co-ordination problems. The IEA has stated four key value chains as an opportunity in the coming decade to accelerate the speed of hydrogen deployment focusing on different regions of the world. The ASEAN is focused as part of the fourth value chain as a part of Asia Pacific, along with Middle East, North Africa and Europe, which are "the first shipping routes", in order to kick-start international hydrogen trade for the ultimate global low-carbon market [49].

Hydrogen can be directly produced from increasingly demanded coal in the ASEAN with near-zero greenhouse gas emissions as carbon capture and storage technology becomes available. However, the development and deployment of certain green technologies like the carbon capture, utilization and storage requires appropriate institutional and policy set up as a prerequisite. There are raw materials widely used in infrastructure such as in construction, aerospace and automotive sectors, whereby traditional materials are replaced by carbon-based materials such as carbon composites and manufactured graphite. These materials can largely absorb the enormous amount of carbon products as countries like Canada, Japan and the US have already constructed and developed bridges with such mechanisms. One major advantage of carbon-composites in comparison to traditional materials such as steel is that it does not erode and is five times stronger than the mainstream heavy construction equipment [33]. There could be a significant decrease in CO_2 emissions, which in turn would discontinue the cement-manufacturing plants by replacing concrete with carbon materials. There has been good progress made in terms of using carbon-based products as additives for substituting cements.

4.2. Adapting Green Energy Financing for Green Deployment

Finance is the engine of development of renewable energy projects, and financing of investments that provide environmental benefits through new financial instruments such as green bonds, green banks, carbon market instruments, fiscal policy, green central banking, fintech, community-based green funds is necessary to achieve the SDGs [50]. The ASEAN and the Southeast Asia governments should adopt these targeted funding channels, also known as green energy financing, for the greater deployment of green technologies in the region. A geographical mismatch between resource endowments and demand centers provides incentive for regional integration of power grids in order to bridge the gap but requires capital-intensive investments in physical interconnectors. Therefore, the hindrance of renewable energy development does not only include technological capacity but also access to finance in the ASEAN [51]. It is difficult for policymakers in determining the ways to make the transition towards a green economy from the existing coal generation in the absence of financing access, when generally, financial institutions show more interest in fossil fuel projects rather than in green projects. The cross-sector policy framework can enable integrative financing and development of renewable energy fostering energy efficiency and replacement of fossil fuels.

The Southeast Asian region has played a significant role under the agenda of "one community for sustainable energy" with publicly financed initiatives such as ASEAN Power Grid (APG) interconnection, Trans-ASEAN Natural Gas Pipeline (TAGP), energy efficiency, renewables, and regional level policy and planning [15]. All these initiatives require costly investments in capital expenditures and are risky which the private sector is not willing to bear and therefore requires appropriate public financing as evidenced. The breakthroughs in technology in the renewable sector can provide a resilient model forward on a low-carbon energy system by easing access to finance and overcoming financial barriers in the deployment of renewable energy. The stronger regional framework on green projects financing can serve as an extensive development plan and ensure a sustainable energy transition roadmap moving forward. Both regional coordination and cooperation with a strong political will from all the countries in the region will be vital for an integrated financial framework development in supporting infrastructure projects in the ASEAN.

The belt and road initiative introduced by China also has some major implications to the South East Asian economies such as promoting infrastructure projects through Chinese government financing in the region that relates with water resources and transboundary rivers. However, several positive and negative impacts may arise, creating political issues on the social and environmental front through these publicly financed projects [52]. Therefore, concerns can be raised while deploying green technology projects, especially when international collaborations take place. A regional governing institution focused on energy and use of market-based instruments can provide a platform for strengthening energy dialogues and facilitating the mobilization of green technologies to boost the energy infrastructure by attracting financing. Furthermore, the role of the private sector is also equally important and will not only ensure civic engagement but also support the leveraging of public funds. The policymakers around ASEAN have been increasingly trying hard to ensure reliable and affordable sustainable energy solutions. It is equally important to focus on efficiency while developing investment infrastructure for fuel and power supply.

Commitment for funding from both public and private sectors is crucial. For example, many public sources have played important roles in financing thermal power plant projects and large-scale renewables such as hydropower, while most wind and solar PV projects have relied on private finance supported by policy incentives. Civic engagement and initiatives from investors and companies also play an equally vital role. A finding of the South Korean government showed how government aid and other public finance is deployed. Figure 5 shows the amount of public and private investments in the ASEAN power generation by different sources. Public investments are dominant in the region, with significant involvement in electricity generation from coal and gas. The challenge for policymakers is to entice more private investments into the renewable power sector in the ASEAN.

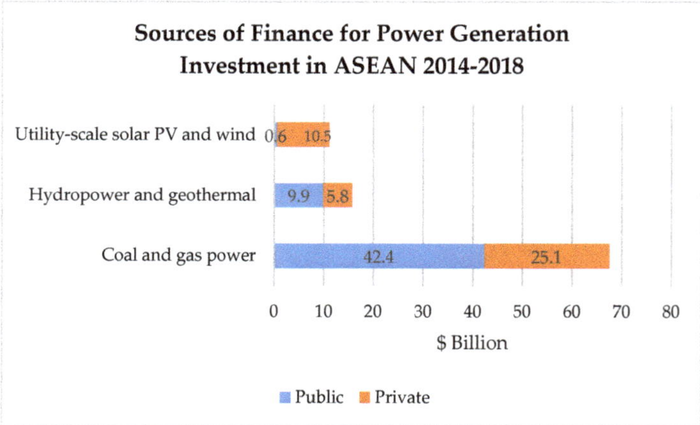

Figure 5. Financing sources. Source: Adapted from IEA [8].

Figure 5 above also indicates that more investments should be channeled towards sustainable energy, and the deployment of renewables should be scaled up although notable progress has been made towards disincentivizing the consumption of fossil fuel. Figure 6 below shows the share of financial instruments on total renewable energy investments in the ASEAN between 2009–2016. Loan and concessional loans remain the two most popular financing instruments, which highlights the need to increase the access to credits for the development and deployment of green energy in the ASEAN.

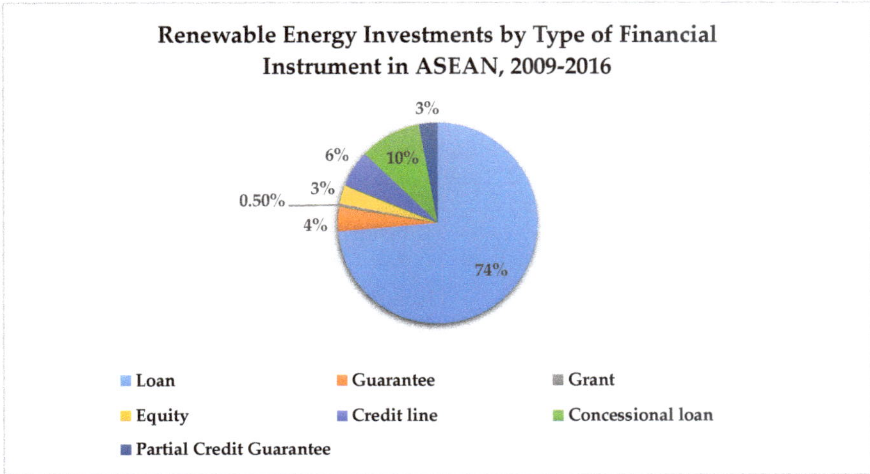

Figure 6. Renewable energy investments Type. Source: Adapted from IRENA [53].

4.3. Managing Energy, Environmental and Financial Risks

Green projects are associated with risks pertaining to being new technologies and their relatively lower rate of return. The rapid rise of energy demand in Southeast Asia is poised to bring several risks to the region from an energy financing perspective. The region is forecasted by IEA to register a net deficit in energy trade of $300 billion per year due to increasing imports of oil by 2040 [3]. The government budget will likely remain tightened as the increase in subsidies for renewable energy continues while also distorting the market-based energy prices. Setting energy prices based on market signals by reducing fossil fuel consumption subsidies will attract more sustainable energy consumption and investments in the ASEAN. While the progress is notable in eliminating fossil-fuel subsidies; the process still remains incomplete. The current dependence on the import of oil is 65% and is expected to rise to 80% in 2040, and it therefore remains a serious energy security concern for the region [18]. The high-carbon-intensive power sector in Southeast Asia, especially due to the rise in coal demand, is expected to amplify environmental risks through increased CO_2 emissions to almost 2.4 gigatons by 2040 [18]. This will negatively impact the environmental quality, adding to already existing poor urban air quality and congested transportation infrastructure.

The governments of ASEAN need to address the energy security risks by taking into account the financial, environmental and social viability of the projects. To achieve this, various frameworks could be developed in the process of procurement and contracting mechanisms in the renewable sector. The support to the financial system and enhancement of sustainability utilities could also strengthen the market. The challenge of limited infrastructure particularly in the Philippines and Indonesia, which are archipelagic in nature, has obstructed effective renewable energy deployment as they have fragmented electricity transmission grids. Similarly, the lack of regulatory frameworks on green technology development and deployment brings major challenges too. Countries like Brunei do not have a specific policy framework in place to regulate the development of renewable energy, although it has been reported to be in progress [54]. There was major infrastructure devastation in Laos PDR due to a lack of coordination, creating human risk as the failure of an auxiliary dam washed out 13 villages, affecting around 11,000 people [55]. The high-risk nature of hydropower dam construction should not be underestimated despite the huge potentials for hydropower with an unrealized power potential of 22.3 GW in the region.

Vietnam is another major player in the hydropower sector and has an estimated 16.68GW capacity, but the lessons from Laos have allowed the country to focus on less intrusive sources of renewable energy. The revised master plan of Vietnam has not focused on the development of large-scale hydropower as a renewable source of energy but promotes increasing the capacity to 21.6 GW in 2020, which is approximately 27.8 GW by 2030 with small and multipurpose projects [56]. Vietnam has a heavy reliance on coal-fired power, as in 2020 alone, the country's capacity stood at 49.3%. The coal's share in Vietnam is expected to reach 53.2% by 2030 as the development project is demanding more energy despite the efforts by the government's revised master plan to reduce reliance on coal. Given the cheaper cost associated with renewables and wind and solar, sources from coal could be shifted and reduce the current import of coal, which is around 30 million tones in Vietnam [57].

The proper coordination among the government agencies and the private sector is crucial towards prioritizing renewable energy policies into implementation. Awareness among the public about the benefits of using green technologies can boost energy efficiency as well environmental conservation initiatives. Multilateral power trading agreements will be crucial along with the expansion of cross-border transmission, which can lower the building and operating costs of ASEAN region's power systems. Countries such as Lao PDR export 67% of electricity generated from hydropower, which is almost 30% of all its total exports, with main buyers being in ASEAN countries itself such as Thailand, Vietnam and Cambodia [53]. The regional integration could facilitate the growing demand for energy by deploying green technologies such as wind and solar PV and most importantly the application of hydrogen carbon-based instruments.

5. Conclusions

The purpose of this study was to formulate the scenario-based policy lessons and framework in the case of ASEAN economies in facilitating the development and deployment of green technologies and alternative energy options. In doing so, the study reviewed the literature around green energy deployment in the context of green growth and energy transition and discussed the current status of renewable energy development in the ASEAN. Alternative energy options such as nuclear and hydrogen energy prospects were discussed, while the study proposed hydrogen fuel as a way forward in meeting the energy and environmental objectives in the ASEAN. Some of the underlying research questions that this study aimed to shed light on and expose as urgent areas of future research include: (i) Why is the deployment of renewable energy low in the ASEAN, and what frameworks are needed to better support their wider deployment of green technologies like wind and solar in Southeast Asia in the context of energy transition? (ii) What are the policy and institutional frameworks required to implement innovative green technologies such as carbon capture, utilization and storage in the region as demand for fossil fuels, in particular, coal escalates? (iii) What is the scope for energy efficiency improvements in the region within the context of the push towards greener technology development and deployment? (iv) How can cross-sectoral partnerships between the governments, businesses, and NGOs in the ASEAN help to mitigate the threats of climate change collectively? However, it was beyond the scope and not the aim of this paper to comprehensively provide answers to each of these individual questions.

The study concludes that carbon capture, utilization and storage (CCUS) will be a vital technology in the ASEAN to reduce emissions from the power sector and from industry while allowing the use of fossil fuels to achieve economic growth. The study proposes that transitioning to a hydrogen carbon economy, adapting to green energy finance for development and managing financial risks in promoting green energy development are necessary and urgent in the ASEAN region to adapt to climate change. The decreasing costs for renewable electricity, especially from solar PV and wind, seems to support the production of electrolytic hydrogen, making it a low-cost supply technology option for hydrogen. Similarly, the increasing pressure from international agreements such as COP21 will demand countries to deploy alternative fuel pathways in their energy mix.

The International Monetary Fund forecasted the global economy to grow negatively at 4.9% in 2020, which will demand that policymakers come up with major economic stimulus packages to combat the COVID-19 crisis [58]. Investment in clean energy with technological solutions will not only be an ideal option from an environmental standpoint but will also fulfil the unemployment gap by creating green-technology-related jobs while spurring economic growth and is perceived to be vital in emerging regions like ASEAN. In addition, the falling costs of renewables can also provide policymakers a perspective to revisit the energy policy planning documents and have a long-term vision about green technology deployment. Batteries, hydrogen and carbon capture are viable technologies as they have the potential to be deployed in mass scale, which could help in achieving global clean energy transition. According to a recent analysis done by IEA, governments are believed to be driving 70% of global energy investments [59]. A proper coordination and leadership from the ASEAN governments to engage multiple stakeholders is important to achieve climate change goals with the appropriate deployment of green technologies. A coordinated energy strategy in the ASEAN will also improve the nuclear prospects, which are complicated by political factors, and public acceptance towards nuclear energy needs to be boosted.

Implementing energy efficiency improvements policy in the ASEAN through policy measures such as attracting Foreign Direct Investment (FDI) and reducing energy consumption in public goods provisions such as streetlights are desirable [60]. Cross-sectoral partnership and international power connectivity in the ASEAN region should be the way forward. The European Union provides a perfect example, whereby their partnership in renewable energy lowered the energy supply from coal by 3% [8]. We propose that short-term and medium-term policies to facilitate decarbonization include boosting public acceptance to nuclear energy, implementing energy efficiency improvement policies and complete elimination of fossil fuel consumption subsidies. The longer-term policies are to deploy CCS technologies as an enabler of hydrogen energy and to increase both the public and private sector energy investments in and development of CCS technologies. These are important policy lessons for the ASEAN governments to accommodate in energy policy crafting and promote sustainable development in the region through green energy development and deployment as a viable strategy to adapt to climate change.

Author Contributions: R.N.—conceptualization, methodology, writing—original draft preparation, writing—review and editing, validation, formal analysis, investigation; H.P.—writing—review and editing, resources, project adminstartion, funding acquisition; A.K.—writing—original draft preparation. All authors have read and agreed to the published version of the manuscript.

Funding: This research paper was carried out under the funding from the Economic Research Institute for ASEAN and East Asia.

Acknowledgments: We also acknowledge the valuable comments received from Fukunari Kimura, Jun Arima and participants at the First Working Group Meeting—Energy Sustainability and Climate Change in ASEAN—held on 14 September 2020. The authors are also grateful to the academic editor and the three anonoymous reviewers for providing helpful comments in improving the paper All remaining errors are the sole responsibility of the authors alone.

Conflicts of Interest: The authors declare no conflict of interest.

References

1. Steffen, W.; Richardson, K.; Rockström, J.; Cornell, S.E.; Fetzer, I.; Bennett, E.M.; Sörlin, S. Planetary boundaries: Guiding human development on a changing planet. *Science* **2015**, *347*. [CrossRef]
2. Raitzer, D.A.; Bosello, F.; Tavoni, M.; Orecchia, C.; Marangoni, G.; Nuella Samson, J.G. *Southeast Asia and The Economics of Global Climate Stabilization*; ADB: Metro Manila, Philipphines, 2015; pp. 1–191.
3. IEA. *Southeast Asia Energy Outlook*; IEA: Paris, France, 2019; pp. 1–18. [CrossRef]

4. IRENA. *REmap-Renewable Energy Outlook for ASEAN*; International Renewable Energy Agency (IRENA): Abu Dhabi, UAE, 2016.
5. Rooney, K.; Writer, S.; Content, F. Can Southeast Asia Keep up with Growing Energy Demand? 2020, pp. 4–7. Available online: https://www.weforum.org/agenda/2019/12/asean-southeast-asia-energy-emissions-renewables/ (accessed on 20 August 2020).
6. International Resources Group. USAID Asia-Pacific Regional Climate Change Adaptation Assessment. Options. 2010. Available online: https://pdf.usaid.gov/pdf_docs/PNADS197.pdf (accessed on 10 July 2020).
7. Phoumin, H.; Kimura, F.; Arima, J. Potential Renewable Hydrogen from Curtailed Electricity to Decarbonize ASEAN's Emissions: Policy Implications. *Sustainability* **2020**, *12*, 560. [CrossRef]
8. IEA. *Nuclear Power in a Clean Energy System*; IEA: Paris, France, 2019. [CrossRef]
9. Louis, J. Rising Temperatures Surge Energy Demand in ASEAN. *The ASEAN Post*. 1–4 March 2020. Available online: https://theaseanpost.com/article/rising-temperatures-surge-energy-demand-asean (accessed on 10 July 2020).
10. Erdiwansyah, R.M.; Sani, M.S.M.; Sudhakar, K. Renewable energy in Southeast Asia: Policies and recommendations. *Sci. Total Environ.* **2019**, *670*, 1095–1102. [CrossRef]
11. Singh, A.; Jamasb, T.; Nepal, R.; Toman, M. Electricity cooperation in South Asia: Barriers to cross-border trade. *Energy Policy* **2018**, *120*, 741–748. [CrossRef]
12. Halawa, E.; Geoffrey, J.; Shi, X.; Novieta, H.S.; Rabindra, N. The Prospect for an Australian–Asian Power Grid: A Critical Appraisal. *Energies* **2018**, *11*, 200. [CrossRef]
13. Nepal, R.; Jamasb, T. Caught between Theory and Practice: Government, Market and Regualtory Failure in Electricity Sector Reforms. *Econ. Anal. Policy* **2015**, *46*, 16–24. [CrossRef]
14. Schmidt, T.S.; Huenteler, J. Anticipating industry localization effects of clean technology deployment policies in developing countries. *Glob. Environ. Chang.* **2016**, *38*, 8–20. [CrossRef]
15. Shi, X.; Malik, C. Assessment of ASEAN Energy Cooperation within the ASEAN Economic Community. In *ERIA Discussion Paper Series*; 2013; Available online: https://www.eria.org/ERIA-DP-2013-37.pdf (accessed on 10 July 2020).
16. Taguibao, J.G. Sustainable Energy Policy in Southeast Asia: Domestic Variation and Congruence in Policy Discourses. *Asian Politics Policy* **2019**, *11*, 327–346. [CrossRef]
17. Li, B.; Nian, V.; Shi, X.; Li, H.; Boey, A. Perspectives of energy transitions in East and Southeast Asia. *Wiley Interdiscip. Rev. Energy Environ.* **2020**, *9*, 1–7. [CrossRef]
18. Lee, H.H.; Iraqui, O.; Wang, C. The Impact of Future Fuel Consumption on Regional Air Quality in Southeast Asia. *Sci. Rep.* **2019**, *9*, 1–20. [CrossRef] [PubMed]
19. Taghizadeh-Hesary, F.; Rasoulinezhad, E. Analyzing Energy Transition Patterns in Asia: Evidence from Countries with Different Income Levels. *Front. Energy Res.* **2020**, *8*, 1–13. [CrossRef]
20. Rehbein, J.A.; Watson, J.E.M.; Lane, J.L.; Sonter, L.J.; Venter, O.; Atkinson, S.C.; Allan, J.R. Renewable energy development threatens many globally important biodiversity areas. *Glob. Chang. Biol.* **2020**, *26*, 3040–3051. [CrossRef] [PubMed]
21. Trueman, C. How Are Countries in Southeast Asia Embrcing Green Technology. 2018. Available online: https://www.cio.com/article/3322530/how-are-countries-in-southeast-asia-embracing-green-technology.html (accessed on 10 July 2020).
22. ASEAN Secretariat. ASEAN Sustainable Urbanisation Strategy. In *Global Environmental Issues*, 2nd ed.; ASEAN: Jakarta, Indonesia, 2012. [CrossRef]
23. IRENA. Renewable Capacity Highlights Renewable Generation Capacity by Energy Source. March 2018. Available online: http://www.irena.org/-/media/Files/IRENA/Agency/Publication/2018/Mar/RE_capacity_highlights_2018.pdf?la=en&hash=21795787DA9BB41A32D2FF3A9C0702C43857B39C (accessed on 10 September 2020).
24. Salleh, S.F.; Roslan, M.E.M.; Shamsuddin, A.H.; Tuan Abdullah, T.A.B.; Sovacool, B.J. Transitioning to a Sustainable Development Framework for Bioenergy in MALAYSIA: Policy Suggestions to Catalyse the Utilisation of Palm Oil Mill Residues. *Energy Sustain. Soc.* **2020**, *10*. [CrossRef]
25. Tun, M.M.; Juchelkova, D.; Win, M.M.; Thu, A.M.; Puchor, T. Biomass Energy: An Overview of Biomass Sources, Energy Potential, and Management in Southeast Asian Countries. *Resources* **2019**, *8*, 81. [CrossRef]
26. Kurniawan, R. Indonesia to Have Its First and Largest Floating Solar Power by 2019–2021. 2020. Available online: https://www.rambuenergy.com/2017/12/indonesia-to-have-its-first-and-largest-floating-solar-power-by-2019/ (accessed on 20 August 2020).
27. Hajramurni, A. Jokowi Inaugurates First Indonesian Wind Farm in Sulawesi. *Jkt. Post*. 2018. Available online: https://www.thejakartapost.com/news/2018/07/02/jokowi-inaugurates-first-indonesian-wind-farm-in-sulawesi.html (accessed on 10 July 2020).
28. Clean Cooking Alliance Philippines. 2020, pp. 1–2. Available online: https://www.cleancookingalliance.org/country-profiles/61-philippines.html (accessed on 10 July 2020).
29. IEA. The Future of Cooling in Southeast Asia–Analysis-IEA. 2018, pp. 1–3. Available online: https://www.iea.org/reports/the-future-of-cooling-in-southeast-asia (accessed on 20 August 2020).
30. IRENA. Global Renewables Outlook: Energy transformation 2050. In *International Renewable Energy Agency*; Available online: https://www.irena.org/publications/2020/Apr/Global-Renewables-Outlook-2020 (accessed on 20 November 2020).
31. Nian, V.; Chou, S.K. The state of nuclear power two years after Fukushima-The ASEAN perspective. *Appl. Energy* **2014**, *136*, 838–848. [CrossRef]
32. Nian, V.; Hari, M.P. Incentivizing the Adoption of Nuclear and Renewable Energy in Southeast Asia. *Energy Procedia* **2017**, *105*, 3683–3689. [CrossRef]

33. Zink, J.C. Trend favors nuclear-hydrogen economy. *Power Eng.* **2003**, *107*, 20.
34. Brown, L.C.; Besenbruch, G.E.; Lentsch, R.D.; Schultz, K.R.; Funk, J.F.; Pickard, P.S.; Showalter, S.K. High efficiency generation of hydrogen fuels using nuclear power. *Gen. At.* **2003**, 29–30. [CrossRef]
35. Thornhill, B.J.; Roston, E. *Persistent Carbon Emissions Signal Global Climate Goal Is Out of Reach. 2019–2021*. 2019. Available online: https://www.bnnbloomberg.ca/persistent-carbon-emissions-signal-global-climate-goal-is-out-of-reach-1.1357053 (accessed on 10 July 2020).
36. Zhang, T. CCS Development in Southeast Asia, Global CCS Isntitute. May 2020. Available online: https://www.globalccsinstitute.com/wp-content/uploads/2020/05/Brief_CCS-Development-in-South-East-Asia_May-2020.pdf (accessed on 20 August 2020).
37. ADB. Asia Wakes Up to CCUS, ADB Knowledge Events. 2019. Available online: https://events.development.asia/learning-events/asia-wakes-ccus (accessed on 10 July 2020).
38. ADB. Prospects for Carbon Capture and Storage in Southeast Asia. 2013. Available online: http://hdl.handle.net/11540/1434 (accessed on 10 July 2020).
39. Evwind. *Hydrogen from Wind Energy and Solar Power, the Future*. 2020, pp. 1–8. Available online: https://www.evwind.es/2020/01/21/hydrogen-from-wind-energy-and-solar-power-the-future/73170 (accessed on 10 July 2020).
40. Muradov, N.Z.; Veziroğlu, T.N. From hydrocarbon to hydrogen-carbon to hydrogen economy. *Int. J. Hydrog. Energy* **2005**, *30*, 225–237. [CrossRef]
41. Barreto, L.; Makihira, A.; Riahi, K. The hydrogen economy in the 21st century: A sustainable development scenario. *Int. J. Hydrog. Energy* **2003**, *28*, 267–284. [CrossRef]
42. Jose, M.B.; Hasegawa, T. The Future of Hydrogen for G20. Seizing today's opportunities. In *Report Prepared by the IEA on Hydrogen*; 2019. Available online: https://www.iea.org/reports/the-future-of-hydrogen (accessed on 10 July 2020).
43. IEA. *Deploying Renewables in Southeast Asia-Trends and Potentials*; IEA: Paris, France, 2010; pp. 1–164. [CrossRef]
44. Lee, D.H.; Park, S.Y.; Hong, J.C.; Choi, S.J.; Kim, J.W. Analysis of the energy and environmental effects of green car deployment by an integrating energy system model with a forecasting model. *Appl. Energy* **2013**, *103*, 306–316. [CrossRef]
45. ASEAN. Asean Community Vision 2025. 2015, p. 8. Available online: http://www.asean.org/wp-content/uploads/images/2015/November/aec-page/ASEAN-Community-Vision-2025.pdf (accessed on 10 July 2020).
46. IRENA. How Falling Costs Make Renewables a Cost-Effective Investment. 2020, pp. 1–8. Available online: https://www.irena.org/newsroom/articles/2020/Jun/How-Falling-Costs-Make-Renewables-a-Cost-effective-Investment (accessed on 30 September 2020).
47. Creamer, T. Replacing World's Costliest 500 GW of Coal with Solar and Wind Would Shave $23bn off System Costs. 2018, pp. 1–10. Available online: https://www.engineeringnews.co.za/article/replacing-worlds-costliest-500-gw-of-coal-with-solar-and-wind-would-shave-23bn-off-system-costs-2020-06-02/rep_id:4136#:~{}:text=Replacingthecostliest500GW,2019%2Canewreportshows (accessed on 10 July 2020).
48. Deloitte China. Fueling the Future of Mobility Hydrogen and fuel cell solutions for transportation. *Financ. Advis.* **2019**, *1*, 1.
49. Bermudez, J.M.; Hasegawa, T. Hydrogen. IEA. 3–4 June 2020. Available online: https://www.iea.org/reports/hydrogen (accessed on 30 September 2020).
50. Sachs, J.D.; Woo, W.T.; Yoshino, N.; Taghizadeh-Hesary, F. Importance of Green Finance for Achieving Sustainable Development Goals and Energy Security. In *Handbook of Green Finance. Sustainable Development*; Sachs, J., Woo, W., Yoshino, N., Taghizadeh-Hesary, F., Eds.; Springer: Singapore, 2019. [CrossRef]
51. Shi, X. The future of ASEAN energy mix: A SWOT analysis. *Renew. Sustain. Energy Rev.* **2016**, *53*. 672–680. [CrossRef]
52. Williams, J.M. Emerging costs of China's belt and road strategy for transboundary water in south and southeast Asia. *Int. J. Energy Water Resour.* **2019**, *3*, 81–92. [CrossRef]
53. IRENA. *Renewable Energy Market Analysis: Southeast Asia*; International Renewable Energy Agency: Abu Dhabi, UAE, 2018; Available online: www.irena.org (accessed on 25 October 2020).
54. Asean, T.; Team, P. *ASEAN's Renewable Energy Challenges*. 2020, pp. 1–5. Available online: https://theaseanpost.com/article/aseans-renewable-energy-challenges (accessed on 25 October 2020).
55. Gnanasagaran, A. *Lao Dam Failure Creates Hydroelectric Woes*. (August 2018). 2020, pp. 1–6. Available online: https://theaseanpost.com/article/lao-dam-failure-creates-hydroelectric-woes (accessed on 10 June 2020).
56. Greening, P.; Joyce, M.; Strachan, E. *Energy in ASEAN: Renewable Energy in Vietnam*. 2020, pp. 1–15. Available online: https://www.lexology.com/library/detail.aspx?g=4e781400-114f-4181-a647-f6c5aa222c7d (accessed on 10 July 2020).
57. Vu, K.; Gloystein, H. In Vietnam's Booming Energy Sector Coal Reigns, but Renewables Play Catch-Up. *Reuters*. 2019. Available online: https://www.reuters.com/article/us-vietnam-energy-analysis/in-vietnams-booming-energy-sector-coal-reigns-but-renewables-play-catch-up-idUSKCN1SU0KJ (accessed on 20 August 2020).
58. IMF. World Economic Outlook Update. *IMF* **2019**, *2*, 6.
59. Birol, F.; IEA. How to Make the Economic Recovery from Coronavirus an Environmentally Sustainable One. IEA Commentary, Prospect Magazine. 2020. Available online: https://www.iea.org/commentaries/how-to-make-the-economic-recovery-from-coronavirus-an-environmentally-sustainable-one (accessed on 10 August 2020).
60. Nepal, R. Overcoming the Challenge of Energy Efficiency Financing in Asia. *Asia Pathways–The Blog of the Asian Development Bank Institute*. 20 November 2020. Available online: https://www.asiapathways-adbi.org/2020/11/overcoming-challenge-energy-efficiency-financing-asia/ (accessed on 20 August 2020).

sustainability

MDPI

Article

An Integrated Approach to the Realization of Saudi Arabia's Energy Sustainability

Mohammed Siddig H. Mohammed [1,*], Abdulsalam Alhawsawi [1,2] and Abdelfattah Y. Soliman [1]

1 Department of Nuclear Engineering, Faculty of Engineering, King Abdulaziz University, P.O. Box 80204, Jeddah 21589, Saudi Arabia; amalhawsawi@kau.edu.sa (A.A.); afattah_y@yahoo.com (A.Y.S.)
2 Center for Training & Radiation Prevention, King Abdulaziz University, P.O. Box 80204, Jeddah 21589, Saudi Arabia
* Correspondence: mshmohamad@kau.edu.sa

Abstract: As system thinking is a recognized approach to the comprehension and realization of energy sustainability, this paper applies a holistic representation to the World Energy Trilemma Index (WETI) key indicators using Bayesian Belief Networks (BBN) to illuminate the probabilistic information of their influences in Saudi Arabia's context. The reached realization is suggested to inform the policies to improve energy sustainability, and thus the country's rank in the WETI. The analysis used two groups of learning cases, one used the energy statistics of the period from 1995 to 2019 to show the outlook of the Business as Usual path, and the other addressed the projected data for the period from 2018 to 2037 to investigate the expected impact of the new policies. For both BAU and new policies, the BBN calculated the improvement, stability, and declining beliefs. The most influential factors on energy sustainability performance were the electricity generation mix, CO_2 emissions, energy intensity, and energy storage. Moreover, the interlinkage between the influential indicators and their causes was estimated in the new policies model. A back-casting analysis was carried out to show the changes required to drive the improvement belief to 100%. The compiled BBN can be used to support structuring policymaking and analyzing the projections' outcomes by investigating different scenarios for improvement probabilities of energy sustainability.

Keywords: Saudi Arabia; energy sustainability; world energy trilemma index; Bayesian Belief Network

Citation: Mohammed, M.S.H.; Alhawsawi, A.; Soliman, A.Y. An Integrated Approach to the Realization of Saudi Arabia's Energy Sustainability. *Sustainability* **2021**, *13*, 205. https://doi.org/10.3390/su13010205

Received: 17 November 2020
Accepted: 17 December 2020
Published: 28 December 2020

Publisher's Note: MDPI stays neutral with regard to jurisdictional claims in published maps and institutional affiliations.

1. Introduction

The United Nations' (UN) 2030 agenda for Sustainable Development was announced in 2015 under the main title and objective of "Transforming Our World," and indeed, the world is witnessing transformations that are implemented and measured by the 17 Sustainable Development Goals (SDGs) and their indicators [1].

The SDG seven aims to ensure access to affordable, reliable, sustainable, and modern energy for all the world's population. The interaction of SDG 7 targets with the targets of SDG 1: No poverty, SDG 2: Zero hunger, SDG 3: Good health and wellbeing, SDG 6: Clean water and sanitation, SDG 8: Decent work and economic growth, and SDG 13: Climate actions have been analyzed and the mutual impacts identified to be reinforcing, enabling, or constraining. In addition, the mutual influence between SDG 7 and the other 16 SDGs has been presented [2], which demonstrated that energy sustainability is crucial for sustainable development.

The discussion and analysis of energy sustainability have been approached in the literature by several methods. A procedure has been proposed to evaluate the Sustainable Useful Index of energy-producing processes. The index assesses the ability to maintain the viability and usefulness of energy sources considering the produced, spent, avoided, and direct energy. The definition does not satisfy the broad concept of energy sustainability [3]. Within the same perspective, energy sustainability analysis has been performed at short-term and long-term levels and applied to a case study on the production of distributed H_2O [4].

Energy sustainability analysis for developing countries in the light of SDGs has been attempted. However, the authors limited the discussion on the role of a hybrid power system in improving energy sustainability in urban areas making a case study based on data from an Iranian city [5].

The authors of [6] have studied non-renewable energy and renewable energy efficiency for simultaneous achievement of economic growth and environmental sustainability in the Middle East and North Africa. The scope of the study has not included some important aspects of energy sustainability, such as the social dimension (energy equity and affordability).

Grigoroudis et al. presented a definition and mathematical model based on various indicators that cover the adequacy, reliability, affordability, social, and environmental requirements. The model rates energy sustainability on a 0 to 1 scale using Fuzzy Logic reasoning. The methodology is more integrated than the previous studies, however, one shortcoming is the potential subjectivity associated with assessment by Fuzzy Evaluation [7].

One of the comprehensive and informative methodologies on the performance of energy sustainability is World's Energy Trilemma Index (WETI) [8], published since 2010 by The World Energy Council, the UN's accredited energy body. Consistent with the World Energy Council definition of energy sustainability, the WETI ranks the countries energy performance on three dimensions: Energy security, energy equity, and environmental sustainability. The index helps to assess the effectiveness of energy policies for enabling balanced transition management, perform a comparative analysis using the experiences of countries with relevant socioeconomics and energy infrastructure, and eventually inform the policies on the required adjustments.

The WETI has been investigated using several methods to assess its reliability. The research in [9] has praised the value of the WETI in guiding countries to address the energy trilemma. Nonetheless, it has argued that the preferences among the trilemma can change from country to country, which requires weighing the trilemma dimensions adaptably. The research suggested the use of interval decision making and stochastic multicriteria acceptability analysis to measure countries' energy performance, and developed an alternative ranking scheme. In another contribution to addressing the preference variation of the trilemma dimensions, the interval decision matrix and the principal component analysis were used to evaluate the top ten performers in the 2015 version of WETI and produce a comparison rank that debated the weights assigned to the WETI indicators [10]. Principal Component Analysis has been applied to assess the methodology of the WETI using Pearson Correlation test, Kaiser–Meyer–Olkin Measure of Sampling Adequacy and Bartlett's Test of Sphericity. The conclusion was made from the results of Cronbach's Alpha test, and the authors deemed the WETI ranking unreliable [11].

Nevertheless, the methodology of WETI has been revised over the years since its first release, the most recent revisions were in the 2019 version that included the data sources, indicators, weighting, and indexation [8] (p. 42) and further refinements in 2020 report [12] (p. 63).

The ranking is based on grading the performance of each of the dimensions from A to D. Although Saudi Arabia is the world's leading oil producer and is among the top 20 countries in terms of Gross Domestic Product (GDP) [13], it was ranked 78 out of 127 in the 2019 WETI, mainly because of the poor performance in the Environmental Sustainability dimension scoring 35/100 (D) and the average performance in the Energy Security 55/100 (C), while scoring 98/100 (A) in the Energy Equity.

Saudi Arabia's ranking in 2019 retreated from positions 47, 53, and 47 in the years 2016, 2017, and 2018, respectively. In the years from 2016 to 2018, Saudi Arabia kept a consistent grade of BAD. Therefore, the comparative ranking fluctuation could be attributed to the performances of other countries as their ranks were rolling up and down. The downgrade in 2019 ranking is mainly due to the decline in the energy security dimension that has consistently been graded with B due to lack of energy diversity in the past three years, to a C grade in 2019, which again can be attributed to the outpace of other countries besides the geopolitical tensions in the region. The strength of Saudi Arabia in the energy trilemma is in the energy equity dimension due to the availability and affordability of fuel and electricity.

The 2020 WETI was released recently [12]. Saudi Arabia is ranked 55 out of 108 countries, however, the progress from 78 in 2019 does not indicate a corresponding improvement in energy sustainability performance since the grade is still BAD. The quasi progress could be because fewer countries were included in 2020 and also because of the performance swings of the other countries.

In 2018, Saudi Arabia's total primary energy supply was 133,291 ktoe of oil and 78,009 ktoe of natural gas, 221,836 GWh of the electricity was generated from natural gas, 125,860 GWh from oil, and 155 GWh from solar Photovoltaic (Solar PV) [14]. The access to electricity covers 100% of the population [15], and the fuel and electricity prices, although they have recently been witnessing subsidies reforms, are still affordable. The presented figures explain Saudi's Energy Trilemma Index high score in energy equity and the low score of environmental sustainability. Moreover, the statistics show the reliance of the Saudi economy and the energy sector on oil and gas consumption and exports, which impacts the energy security score.

However, in recent years, and simultaneously with the efforts of the world to undergo an unprecedented transition to sustainable development, Saudi Arabia has announced an ambitious transformation plan known as Vision 2030 [16] that was built around three pillars: A vibrant society, a thriving economy, and an ambitious nation. One of the objectives of the vibrant society pillar is the maintenance of environmental sustainability, with one of its measures being the reduction of air pollution. The thriving economy pillar mainly aims at economic diversification, and it includes the objective of introducing renewables to the country's energy mix, increasing the production of natural gas, and controlling energy consumption by introducing plans for fuel-targeted subsidies. The objectives mentioned above support the implementation of programs that can lead to enhancing energy security and environmental sustainability.

There are two main trends in the literature about the energy sector in Saudi Arabia, one is about alternative energy sources, mainly renewables, and the other is on energy economics. The status and potential of renewable energy resources in Saudi Arabia have been reviewed, and the possible roles of renewable energy in developing policies for secured and cost-effective energy have been examined [17]. Renewable energy solutions for the challenge of increasing oil consumption in Saudi Arabia have been discussed as well as the outlook of energy cost and clean environment [18]. The human resources requirements to meet the future of renewables in Saudi Arabia have been presented [19].

Regarding energy economics, different policy scenarios to decouple the reduction in fuel consumption and energy cost increase or optimizing the prices of industrial fuels and household electricity [20–22] to seek a more efficient energy system have been discussed.

The presented literature gives useful insights and solutions, however, they considered siloed elements of the energy system, which have not addressed the holism of energy sustainability.

This work proposes to apply the holistic approach of the system thinking [23] utilizing Bayesian Belief Network (BBN) to examine the influences of the indicators underpinning the implementation of energy security, energy equity, and environmental sustainability in Saudi Arabia's context. The reached comprehension uncovers the probabilities of the impact and mutual interactions between the indicators and the likelihood of changes. The proposed method can support decision-making in energy policy prioritization, schedule, or amendments that can result in the improvement of Saudi's energy sustainability and WETI rank.

2. Materials and Methods

BBN represents the probabilistic relationships between a set of variables in a Directed Acyclic Graph (DAG). The DAG is composed of nodes to denote the variables and links (arrows) to represent the causal connection between the variables. The relationship between the causes and effects is described by Conditional Probability Tables (CPT) to identify the belief that the effect variable will be in a specific state given the state of the cause variable.

If a state of a variable is changed, the change is transmitted through the links, and the network is solved using Bayes' theorem.

$$P(A|B) = \frac{P(B|A)\,P(A)}{P(B)} \tag{1}$$

where P(A) is the prior distribution of variable A, P(A | B) is the posterior distribution (the probability of A given new data B), and P(B | A) the likelihood function [24] (p. 6).

Introductions and a detailed formal definition of BBN are given in [25–27].

In energy systems and energy policy, the BBN has been for providing a tool for policymaking in the renewable energy sector [28], decision-making in clean energy investment [29], assessment of power systems [30], and the integration of renewables into the grids [31].

BBN is used in this paper to examine the influences of some WETI indicators on energy sustainability in Saudi Arabia.

The calculation of the WETI is based on 32 indicators, however, twelve key metrics are used in the countries' profiles to exhibit the performance. This paper considers nine key metrics shown in Figure 1, generated using Vensim system dynamics simulation software. The remaining three key metrics of the fourth dimension, the country context, are beyond the current scope.

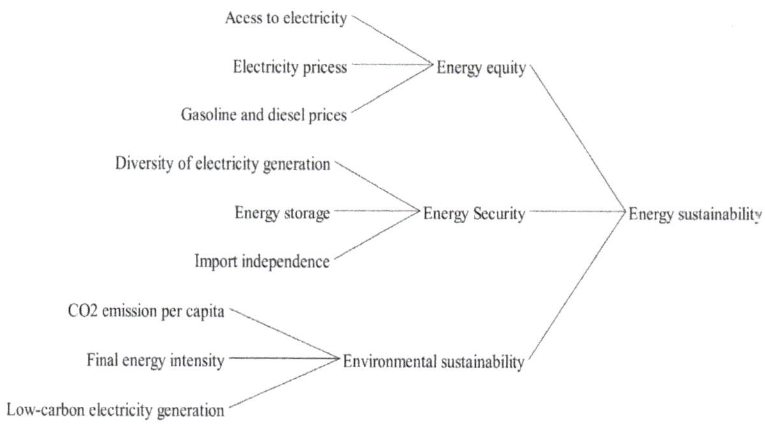

Figure 1. Causes tree energy sustainability using the selected key metrics.

Cases of examples or experiences data are provided to train the BBN to capture the believed states in different scenarios. The datasets measuring the indicators between the years 1990 and 2019 were obtained from various sources cited in Table 1, used to train the constructed BBN. Table 1 contains 25 cases, with each row being a case.

Table 1. Energy statistics for the period from 1990 to 2019. The asterisks mean missing data.

	Energy Imports % of Energy Use [32]	Oil Refinery Capacities (Million Barrels/Day) [33]	Share of Renewables in the Electricity Mix % [34]	Energy Intensity (kW-h/2011$) [35]	CO_2 (Tons/Capita [36]	Elect. Prices SR/kWh [37]	Gasoline Prices SR/Liter [38–40]	Access to Elect. (% of Population) [15]
1990	−535.188	1.86	*	3.212282	11.42588	0.07	*	100
1991	−586.682	1.645	*	3.549141	15.94175	0.05	*	100
1992	−520.327	1.66	*	2.773358	16.49478	0.05	*	100
1993	−490.736	1.67	*	3.076266	17.63903	0.05	*	100
1994	−449.632	1.683	*	3.801903	16.88071	0.05	*	100
1995	−449.353	1.692	*	3.717125	12.59267	0.05	0.16	100
1996	−418.536	1.699	*	3.617888	13.56865	0.05	*	100
1997	−438.437	1.704	*	3.640062	11.11945	0.05	*	100
1998	−420.669	1.762	*	4.066808	10.47502	0.05	0.16	100
1999	−372.668	1.808	*	3.592834	11.18999	0.05	*	100
2000	−386.248	1.798	*	3.155803	14.34164	0.05	0.24	100
2001	−361.93	1.805	*	3.511994	13.98764	0.05	0.24	100
2002	−290.774	1.809	*	3.292017	14.93586	0.05	0.24	100
2003	−349.868	2.049	*	3.255635	14.53989	0.05	0.24	100
2004	−352.903	2.074	*	2.945454	17.05746	0.05	0.24	100
2005	−365.873	2.102	*	2.584772	16.62125	0.05	0.24	100
2006	−316.28	2.102	*	2.372616	17.60497	0.05	0.16	100
2007	−289.452	2.102	*	2.291149	15.34697	0.05	0.16	100
2008	−266.008	2.102	0.000277	2.04159	16.69991	0.05	0.16	100
2009	−211.609	2.109	0.000262	2.424205	17.49302	0.05	0.16	100
2010	−186.51	2.109	0.001631	2.272282	18.87995	0.05	0.16	100
2011	−232.737	2.107	0.001907	1.866232	17.60523	0.05	0.16	100
2012	−211.956	2.107	0.008289	1.869318	19.31661	0.05	0.16	100
2013	−219.74	2.507	0.013063	1.903283	17.99534	0.05	0.16	100
2014	−191.524	2.899	0.012645	2.054486	19.46813	0.05	0.16	100
2015	*	2.899	0.035919	2.034974	19.5753	0.05	*	100
2016	*	2.901	0.034878	2.032694	19.46668	0.05	0.2	99.9
2017	*	2.826	0.037127	*	19.12645	0.05	*	99.93
2018	*	2.835	0.040355	*	18.43629	0.18	0.36	100
2019	*	2.835	0.198166	*		0.18	0.36	100

The oil refinery capacities were used to indicate the energy storage capacity, and the share of renewables in the electricity mix was used as an input for two indicators: Diversity of electricity generation and low-carbon electricity generation.

The selected metrics were represented using BBN. The states of the variables in the BBN were drawn from the data in Table 1. The states were described as declining, stable, or improving according to the comparison of the measurement of the specified year to the average of the preceding five years. If the change percent is zero, the state is named stable. Positive and negative percentages are named improving or declining depending on the specific indicator, for example, a negative change in CO_2 emission is an improvement. The states of the indicators are given in Table 2.

However, the changes in the electricity generation sources were treated differently because the growth in the renewables shares in the electricity mix was negligible from 2008 to 2014, so they were not considered improvements. Then, from 2015 to 2019, the states were based on calculating the annual growth.

The states of the nodes in Figure 2 are shown with equal probability distributions indicating that the BBN needs CPTs and training data to be fully functional. The belief networks are implemented using the Netica toolkit [41].

Table 2. States of the key indicators for the period from 1995 to 2019. The asterisks mean missing data

	Imports	Storage	Elec. Gen. Diversity	Energy Intensity	CO_2/Capita	Access to Elec.	Elec. Prices	Gas Prices
1	Stable	Declining	Stable	Declining	Declining	Stable	Declining	*
2	Stable	Improving	Stable	Declining	Declining	Stable	Stable	*
3	Stable	Improving	Stable	Declining	Declining	Stable	Stable	*
4	Stable	Improving	Stable	Declining	Declining	Stable	Stable	*
5	Stable	Improving	Stable	Improving	Declining	Stable	Stable	*
6	Stable	Improving	Stable	Improving	Declining	Stable	Stable	Declining
7	Stable	Improving	Stable	Improving	Declining	Stable	Stable	Stable
8	Stable	Improving	Stable	Improving	Declining	Stable	Stable	Stable
9	Stable	Improving	Stable	Improving	Declining	Stable	Stable	Stable
10	Stable	Improving	Stable	Improving	Declining	Stable	Stable	Stable
11	Stable	Improving	Stable	Improving	Declining	Stable	Stable	Stable
12	Stable	Improving	Stable	Improving	Declining	Stable	Stable	Improving
13	Stable	Improving	Stable	Improving	Declining	Stable	Stable	Stable
14	Stable	Improving	Stable	Improving	Declining	Stable	Stable	Stable
15	Stable	Improving	Stable	Improving	Declining	Stable	Stable	Stable
16	Stable	Stable	Stable	Improving	Declining	Stable	Stable	Stable
17	Stable	Stable	Stable	Improving	Declining	Stable	Stable	Stable
18	Stable	Stable	Stable	Improving	Declining	Stable	Stable	Stable
19	Stable	Improving	Stable	Improving	Declining	Stable	Stable	Stable
20	Stable	Improving	Stable	Improving	Declining	Stable	Stable	Stable
21	Stable	Improving	Stable	Declining	Declining	Stable	Stable	Stable
22	Stable	Improving	Stable	Declining	Declining	Stable	Stable	Declining
23	Stable	Improving	Stable	*	Declining	Stable	Stable	Declining
24	Stable	Improving	Stable	*	Improving	Stable	Declining	Declining
25	Stable	Stable	Improving	*	*	Stable	Declining	Declining

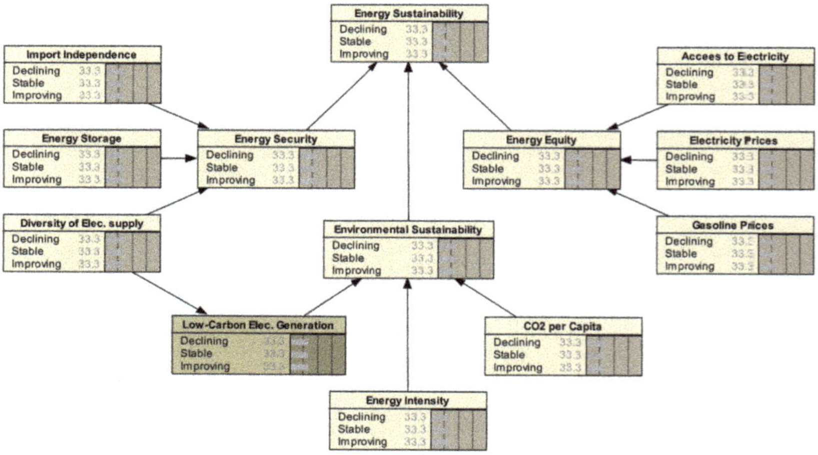

Figure 2. Uncompiled BBN of energy sustainability.

The CPTs for energy security, energy equity, environmental sustainability, and overall energy sustainability were created based on the weights of the variables given in the description of the WETI in Annex A of [8]. For example, the WETI gives the energy security dimension a weight of 30% distributed equally between five indicators, 6% each. Therefore, in CPT of energy security's three indicators used in this paper, equal probabilities of 0.33 were given to the declining, stable and improving statuses, which means, for instance, if the import independence and energy storage are declining but electricity generation diversity is improving, there will be 0.66 declining, 0.33 improving, and 0% stable probabilities.

For the effect variables, the size of the CPTs is the multiplication product of the numbers of the states of the effect and all its cause nodes. For the causes, it can be described by a marginal probability distribution. The probability tables are given in Appendix A.

The available data for the new policies for the period from 2018 to 2037 was mainly obtained from the energy policy simulator [42] jointly developed by Energy Innovation Policy and Technology LLC and King Abdullah Petroleum Studies and Research Center. The energy policy simulator presents data till 2050, nevertheless, the data of 20 cases (2018–2037) were used to provide a statistically acceptable representation for the near future period. The evaluation of the data to draw the states of the variables was done case by case. For example, the obtained electricity prices were for 2020, 2030, and 2035, so the periods between each interval were split between two states to describe the gradual increase or decrease.

The interconnections between the indicators were considered to recoup the missing data. The energy intensity and the CO_2 emission per capita were calculated based on energy efficiency, energy equity, and low-carbon electricity generation. The information of the interconnections is based on an analysis of Saudi Arabia's CO_2 emissions drop in 2018 [43]. The energy efficiency was given improving states until 2030 and then stable states to 2037 based on the information in the Saudi Energy Efficiency Program that efficiency will be improved to reach a 20% consumption reduction by 2030 [44].

The states of the indicators are given in Table 3. The CPTs for energy intensity and CO_2 per capita are given in Appendix B.

Table 3. States of the key indicators for the period from 2018 to 2037.

	Imports	Storage	Elec. Gen. Diversity	Energy Efficiency	Access to Elec.	Elec. Prices	Gas Prices
1	Stable	Improving	Improving	Improving	Stable	Declining	Declining
2	Stable	Declining	Improving	Improving	Stable	Declining	Declining
3	Improving	Improving	Stable	Improving	Stable	Declining	Stable
4	Improving	Improving	Improving	Improving	Stable	Declining	Stable
5	Improving	Improving	Improving	Improving	Stable	Declining	Stable
6	Improving	Improving	Improving	Improving	Stable	Declining	Stable
7	Improving	Improving	Improving	Improving	Stable	Declining	Stable
8	Improving	Improving	Improving	Improving	Stable	Declining	Stable
9	Improving	Improving	Improving	Improving	Stable	Stable	Stable
10	Improving	Improving	Improving	Improving	Stable	Stable	Stable
11	Declining	Improving	Improving	Improving	Stable	Stable	Stable
12	Declining	Improving	Improving	Improving	Stable	Stable	Stable
13	Declining	Stable	Stable	Improving	Stable	Stable	Stable
14	Declining	Stable	Stable	Stable	Stable	Stable	Stable
15	Declining	Stable	Stable	Stable	Stable	Stable	Stable
16	Declining	Stable	Stable	Stable	Stable	Improving	Stable
17	Stable	Stable	Stable	Stable	Stable	Improving	Stable
18	Stable	Stable	Stable	Stable	Stable	Improving	Stable
19	Stable	Stable	Stable	Stable	Stable	Improving	Stable
20	Stable	Stable	Stable	Stable	Stable	Improving	Stable

3. Results

Figure 3 shows the results of compiling the BBN using the 25 cases from Table 2, which reveal that the likelihood of improvement in energy sustainability was 25.5%, which is comparable to the declining likelihood of 23.8%, while the most likely prospect was the stability of the existing situation with a 50.6% chance.

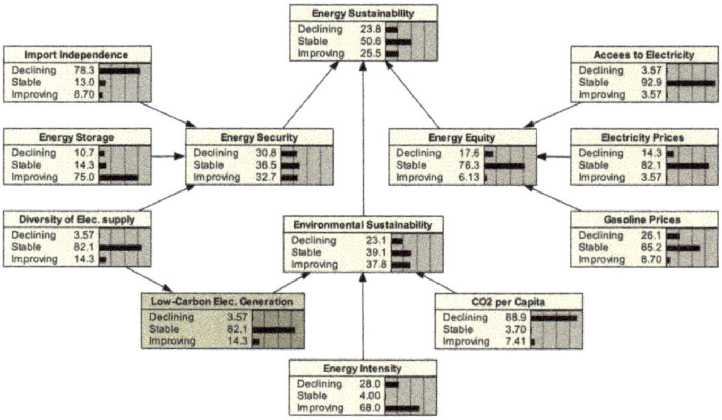

Figure 3. Compiled energy sustainability network for the period from 1995 to 2019.

Netica was used to carry out a sensitivity analysis, which revealed that the dependence of energy sustainability on energy security is comparable to that on environmental sustainability, and the strength of the energy equity effect is half of that of the other two dimensions. By taking the analysis to the level of the indicators, the most influential indicators were the diversity of the electricity supply followed by the energy intensity and then energy storage.

The used toolkit allows examining different scenarios by altering the states of the different variables. For example, a back-casting scenario was created by setting the improvement probability for energy sustainability to 100% and looking at changes imposed in the probabilities of the states of the cause variables (Figure 4). The back-casting results reaffirmed the previous sensitivity analysis. They showed that the most required improvement should be by further 12.5% in the diversity of electricity generation, which tacitly drives another 12.5% improvement in the share of low-carbon electricity generation, then 12.8% in energy intensity, and 8.2% in energy storage.

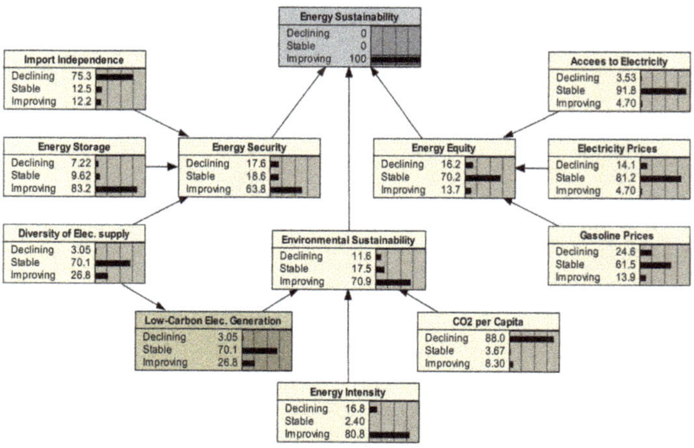

Figure 4. Back-casting 100% improvement probability of energy sustainability network for the period from 1995 to 2019.

Similarly, many small incremental changes in the states of any of the variables can be attempted to examine the most probable routes for the improvement of each dimension or that of the overall energy sustainability.

The compiled BBN for the BAU path did not account for the interconnections between the variables. For example, the impact of low-carbon electricity generation on the CO_2 emissions, fuel prices on energy intensity, and electricity generation mix on the affordability. The reason is that the paper studies the specific case of Saudi Arabia's performance, where actual data that already represent the sum measurements are available and do not need further calculation or elicitation.

However, some of the interconnections were estimated to assess the energy sustainability landscape in the light of policy changes in the Saudi 2030 Vision (Figure 5). The improvement likelihood of energy sustainability was 33.6%, with a 53% probability that the performance will be steady during the specified period. The back-casting (Figure 6) and sensitivity analysis showed that the most influential group of indicators is that composed of the diversity of electricity generation, CO_2 emission per capita, and energy intensity, respectively, in terms of the magnitudes of their strength. The groups in the second order of influence were energy storage and import independence with comparable strengths.

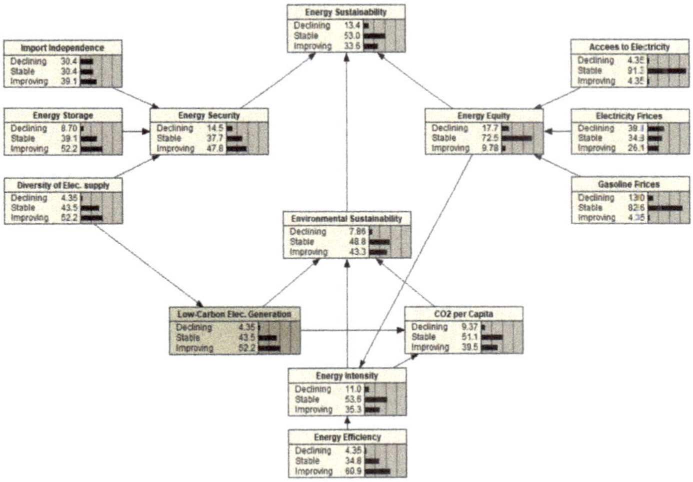

Figure 5. Compiled energy sustainability network for the period from 2018 to 2037.

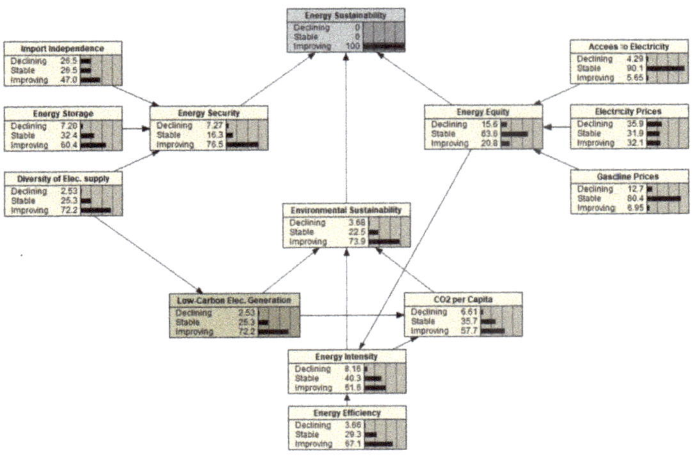

Figure 6. Back-casting 100% improvement probability of energy sustainability network for the period from 2019 to 2030.

4. Discussion

System thinking is an appropriate approach to study energy sustainability policies. Probabilistic and mathematical modeling enables a formal realization of energy sustainability dynamics.

Applying Bayesian Belief Networks to Saudi Arabia's context quantified the probability of improvement, decline, and steadiness of the Business as Usual scenario pertaining to the identified energy sustainability dimensions. A similar method was used by Daim et al. to support policy design in the state of Oregon, USA. It differs from the method in this paper in that the BBNs have been constructed by conversion from Causal Maps, which were developed by consulting experts and energy authorities in Oregon. The step of generating the Causal Maps is analogous to the adoption of the WETI indicators (causes) and dimensions (effects) in this work. The results offered networks with probabilities like those shown in Section 3, Figures 3–6, describing the different states of the correlated factors [29].

The BBN application presented in [28] has sought a higher accuracy in the construction of the BBN by applying an augmented naive model to the quantitative data and k-folds analysis of the Bayesian models. The researchers examined the best scenarios to inform policymaking in Italy and Germany concerning geothermal energy and hydro energy. Another study aimed at computing probabilities of power system states to enable renewable energy integration into smart grids and improve power flow control. The approach addressed a more sophisticated issue of real-time modeling of power systems [31].

The scope of the previous studies was limited to exploring the scenarios of nuclear energy, renewable energy, and investments in different contexts. The contribution of this work is the use of BBN to model the entire energy sustainability system covering energy security, energy equity, and environmental sustainability.

This research was performed before the release of 2020 WETI. Saudi Arabia's 2020 profile shows slight changes compared to 2019 scores: From 55 to 59.9 in energy security, from 98 to 99 in energy equity, and from 35 to 44.3 in environmental sustainability [45]. The change extent supports the findings of the research depicted in Figure 3 and described in Section 3 concerning the energy policy Business as Usual scenario.

For the 2030 policies, the study provided a tool to devise numerous states of each policy and evaluate the propagation of their impact to assist in identifying the critical engagements to achieve energy sustainability. For instance, CO_2 per capita and energy intensity are interlinked with the parent variables being energy efficiency and energy equity, and the impact of the latter was mainly due to the fluctuations in the electricity prices. Therefore, the joint impact of energy efficiency and energy prices reform needs equal consideration to that given to energy resources diversification.

BBN is a useful decision support tool that can give insights and an improved understanding of the policy context and allows the examination of several alternatives. The sensitivity analysis that measures the interdependencies between the BBN nodes gives specific information that the policymakers can use to plan the desired adjustments for improved sustainability.

For enhanced accuracy in analyzing the new energy policies, a more complex BBN comprising the 32 indicators of the WETI and their causes is suggested. The most influential indicators identified in the suggested complex BBN can be further investigated to disclose the required finer interventions. Moreover, other system dynamic methods can be applied to attain different perceptions of interdependence.

The projected data can also be reinforced by experts' judgment and stakeholder consultations to assist the policymakers in optimizing the options. This type of qualitative data can be used in BBN following methods like those described in [16,46] (pp. 11–12).

Author Contributions: Conceptualization, M.S.H.M., A.A. and A.Y.S.; sata curation, M.S.H.M.; investigation, M.S.H.M.; methodology, M.S.H.M.; project administration, A.A.; supervision, M.S.H.M.; validation, M.S.H.M.; writing—original draft, M.S.H.M.; writing—review & editing, M.S.H.M. and A.A. All authors have read and agreed to the published version of the manuscript.

Funding: This project was funded by the Deanship of Scientific Research (DSR) at King Abdulaziz University, Jeddah, under grant No. (J: 65-135-1441). The authors, therefore, acknowledge with thanks DSR for technical and financial support.

Institutional Review Board Statement: Not applicable.

Informed Consent Statement: Not applicable.

Data Availability Statement: Data set available on request to corresponding authors.

Conflicts of Interest: The authors declare no conflict of interest.

Appendix A. Conditional Probability Tables (CPTs) for the Energy Sustainability Network for the Period from 1995 to 2019

Table A1. Energy security CPT.

Import Independence	Energy Storage	Diversity of Elec. Supply	Declining	Stable	Improving
Declining	Declining	Declining	100	0	0
Declining	Declining	Stable	66.666	33.334	0
Declining	Declining	Improving	66.666	0	33.334
Declining	Stable	Declining	66.666	33.334	0
Declining	Stable	Stable	33.334	66.666	0
Declining	Stable	Improving	33.333	33.333	33.333
Declining	Improving	Declining	66.666	0	33.334
Declining	Improving	Stable	33.333	33.333	33.333
Declining	Improving	Improving	33.333	0	66.667
Stable	Declining	Declining	66.666	33.334	0
Stable	Declining	Stable	33.334	66.666	0
Stable	Declining	Improving	33.333	33.333	33.333
Stable	Stable	Declining	33.334	66.666	0
Stable	Stable	Stable	0	100	0
Stable	Stable	Improving	0	66.666	33.334
Stable	Improving	Declining	33.333	33.333	33.333
Stable	Improving	Stable	0	66.666	33.334
Stable	Improving	Improving	0	33.333	66.667
Improving	Declining	Declining	66.666	0	33.334
Improving	Declining	Stable	33.333	33.333	33.333
Improving	Declining	Improving	33.333	0	66.667
Improving	Stable	Declining	33.333	33.333	33.333
Improving	Stable	Stable	0	66.666	33.334
Improving	Stable	Improving	0	33.333	66.667
Improving	Improving	Declining	33.333	0	66.667
Improving	Improving	Stable	0	33.333	66.667
Improving	Improving	Improving	0	0	100

Table A2. Energy equity CPT.

Gasoline Prices	Electricity Prices	Access to Electricity	Declining	Stable	Improving
Declining	Declining	Declining	100	0	0
Declining	Declining	Stable	75	25	0
Declining	Declining	Improving	75	0	25
Declining	Stable	Declining	75	25	0
Declining	Stable	Stable	50	50	0
Declining	Stable	Improving	50	25	25
Declining	Improving	Declining	75	0	25
Declining	Improving	Stable	50	25	25
Declining	Improving	Improving	50	0	50
Stable	Declining	Declining	50	50	0
Stable	Declining	Stable	25	75	0
Stable	Declining	Improving	50	25	25
Stable	Stable	Declining	25	75	0
Stable	Stable	Stable	0	100	0
Stable	Stable	Improving	0	75	25
Stable	Improving	Declining	25	50	25
Stable	Improving	Stable	0	75	25
Stable	Improving	Improving	0	50	50
Improving	Declining	Declining	50	0	50
Improving	Declining	Stable	25	25	50
Improving	Declining	Improving	25	0	75
Improving	Stable	Declining	25	25	50
Improving	Stable	Stable	0	50	50
Improving	Stable	Improving	0	25	75
Improving	Improving	Declining	25	0	75
Improving	Improving	Stable	0	25	75
Improving	Improving	Improving	0	0	100

Table A3. Environmental sustainability CPT.

Low-Carbon Elec. Generation	Energy Intensity	CO_2 per Capita	Declining	Stable	Improving
Declining	Declining	Declining	100	0	0
Declining	Declining	Stable	90	10	0
Declining	Declining	Improving	90	0	10
Declining	Stable	Declining	55	45	0
Declining	Stable	Stable	45	55	0
Declining	Stable	Improving	45	45	10
Declining	Improving	Declining	55	0	45
Declining	Improving	Stable	45	10	45
Declining	Improving	Improving	45	0	55
Stable	Declining	Declining	55	45	0
Stable	Declining	Stable	45	55	0
Stable	Declining	Improving	45	45	10
Stable	Stable	Declining	10	90	0
Stable	Stable	Stable	0	100	0
Stable	Stable	Improving	0	90	10
Stable	Improving	Declining	10	45	45
Stable	Improving	Stable	0	55	45
Stable	Improving	Improving	0	45	55
Improving	Declining	Declining	55	0	45
Improving	Declining	Stable	45	10	45
Improving	Declining	Improving	45	0	55
Improving	Stable	Declining	10	45	45
Improving	Stable	Stable	0	55	45
Improving	Stable	Improving	0	45	55
Improving	Improving	Declining	10	0	90
Improving	Improving	Stable	0	10	90
Improving	Improving	Improving	0	0	100

Table A4. Overall energy sustainability CPT.

Energy Equity	Energy Security	Environmental Sustainability	Declining	Stable	Improving
Declining	Declining	Declining	100	0	0
Declining	Declining	Stable	66.667	33.333	0
Declining	Declining	Improving	66.667	0	33.333
Declining	Stable	Declining	66.667	33.333	0
Declining	Stable	Stable	33.333	66.667	0
Declining	Stable	Improving	33.333	33.333	33.333
Declining	Improving	Declining	66.667	0	33.333
Declining	Improving	Stable	33.333	33.333	33.333
Declining	Improving	Improving	33.333	0	66.667
Stable	Declining	Declining	66.667	33.333	0
Stable	Declining	Stable	33.333	66.667	0
Stable	Declining	Improving	33.333	33.333	33.333
Stable	Stable	Declining	33.333	66.337	0
Stable	Stable	Stable	0	100	0
Stable	Stable	Improving	0	66.667	33.333
Stable	Improving	Declining	33.333	33.333	33.333
Stable	Improving	Stable	0	66.667	33.333
Stable	Improving	Improving	0	33.333	66.667
Improving	Declining	Declining	66.667	0	33.333
Improving	Declining	Stable	33.333	33.333	33.333
Improving	Declining	Improving	33.333	0	66.667
Improving	Stable	Declining	33.333	33.333	33.333
Improving	Stable	Stable	0	66.667	33.333
Improving	Stable	Improving	0	33.333	66.667
Improving	Improving	Declining	33.333	0	66.667
Improving	Improving	Stable	0	33.333	66.667
Improving	Improving	Improving	0	0	100

Appendix B. CPTs for the Energy Intensity and CO_2 per Capita for the Period from 2018 to 2037

Table A5. Energy intensity CPT.

Energy Equity	Energy Efficiency	Declining	Stable	Improving
Declining	Declining	100	0	0
Declining	Stable	50	50	0
Declining	Improving	50	0	50
Stable	Declining	50	50	0
Stable	Stable	0	100	0
Stable	Improving	0	50	50
Improving	Declining	50	0	50
Improving	Stable	0	50	50
Improving	Improving	0	0	100

Table A6. CO_2 emissions per capita CPT.

Energy Equity	Energy Efficiency	Declining	Stable	Improving
Declining	Declining	100	0	0
Declining	Stable	75	25	0
Declining	Improving	75	0	25
Stable	Declining	25	75	0
Stable	Stable	0	100	0
Stable	Improving	0	75	25
Improving	Declining	25	0	75
Improving	Stable	0	25	75
Improving	Improving	0	0	100

References

1. UN General Assembly. Transforming our World: The 2030 Agenda for Sustainable Development. In Proceedings of the Transforming Our World: The 2030 Agenda for Sustainable Development, New York, NY, USA, 21 October 2015; A/RES/70/1.
2. Griggs, D.; Nilsson, D.J.; Stevance, M.; McCollum, A. *A Guide to SDG Interactions: From Science to Implementation*; International Council for Science: Paris, France, 2017.
3. Di Addario, M.; Malavè, A.C.L.; Sanfilippo, S.; Fino, D.; Ruggeri, B. Evaluation of sustainable useful index (SUI) by fuzzy approach for energy producing processes. *Chem. Eng. Res. Des.* **2016**, *107*, 153–166. [CrossRef]
4. Gómez-Camacho, C.E.; Ruggeri, B. Energy Sustainability Analysis (ESA) of Energy-Producing Processes: A Case Study on Distributed H2 Production. *Sustainability* **2019**, *11*, 4911. [CrossRef]
5. Armin Razmjoo, A.; Sumper, A.; Davarpanah, A. Energy sustainability analysis based on SDGs for developing countries. *Energy Sour. Part A Recover. Util. Environ. Eff.* **2020**, *42*, 1041–1056. [CrossRef]
6. Ibrahim, M.D.; Alola, A.A. Integrated analysis of energy-economic development-environmental sustainability nexus: Case study of MENA countries. *Sci. Total Environ.* **2020**, *737*, 139768. [CrossRef] [PubMed]
7. Grigoroudis, E.; Kouikoglou, V.S.; Phillis, Y.A.; Kanellos, F.D. Energy sustainability: A definition and assessment model. *Oper. Res.* **2019**, 1–41. [CrossRef]
8. Council, W.E.; Wyman, O. *World Energy Trilemma Index 2019*; World Energy Council: London, UK, 2019.
9. Song, L.; Fu, Y.; Zhou, P.; Lai, K.K. Measuring national energy performance via Energy Trilemma Index: A Stochastic Multicriteria Acceptability Analysis. *Energy Econ.* **2019**, *66*, 313–319. [CrossRef]
10. Al Asbahi, A.A.M.H.; Gang, F.Z.; Iqbal, W.; Abass, Q.; Mohsin, M.; Iram, R. Novel approach of Principal Component Analysis method to assess the national energy performance via Energy Trilemma Index. *Energy Rep.* **2019**, *5*, 704–713. [CrossRef]
11. Šprajc, P.; Bjegović, M.; Vasić, B. Energy security in decision making and governance—Methodological analysis of energy trilemma index. *Renew. Sustain. Energy Rev.* **2019**, *114*, 109341. [CrossRef]
12. Council, W.E.; Wyman, O. *World Energy Trilemma Index 2020*; World Energy Council: London, UK, 2020.
13. The World Bank. *World Bank Data, Gross Domestic Product 2019*. 2019. Available online: https://data.worldbank.org/indicator/NY.GDP.MKTP.CD (accessed on 3 June 2020).
14. IEA. *World Energy Balances*; International Energy Agency: Paris, France, 2019.
15. The World Bank. Access to Electricity (% of Population), World Bank Data. Available online: https://data.worldbank.org/indicator/EG.ELC.ACCS.ZS (accessed on 3 June 2020).
16. Kingdom of Saudi Arabia Vision 2030. 2017. Available online: Vision2030.gov.sa (accessed on 1 March 2020).
17. Amran, Y.H.A.; Amran, Y.H.M.; Alyousef, R.; Alabduljabbar, H. Renewable and sustainable energy production in Saudi Arabia according to Saudi Vision 2030; Current status and future prospects. *J. Clean. Prod.* **2020**, *247*, 119602. [CrossRef]
18. Al-Douri, Y.; Waheeb, S.A.; Voon, C.H. Review of the renewable energy outlook in Saudi Arabia. *J. Renew. Sustain. Energy* **2019**, *11*, 015906. [CrossRef]
19. Barhoumi, E.M.; Okonkwo, P.C.; Zghaibeh, M.; Belgacem, I.B.; Alkanhal, T.A.; Abo-Khalil, A.G.; Tlili, I. Renewable energy resources and workforce case study Saudi Arabia: Review and recommendations. *J. Therm. Anal. Calorim.* **2020**, *141*, 221–230. [CrossRef]
20. Matar, W.; Murphy, F.; Pierru, A.; Rioux, B. Lowering Saudi Arabia's fuel consumption and energy system costs without increasing end consumer prices. *Energy Econ.* **2015**, *49*, 558–569. [CrossRef]
21. Matar, W.; Anwer, M. Jointly reforming the prices of industrial fuels and residential electricity in Saudi Arabia. *Energy Policy* **2017**, *109*, 747–756. [CrossRef]
22. Matar, W.; Murphy, F.; Pierru, A.; Rioux, B.; Wogan, D. Efficient industrial energy use: The first step in transitioning Saudi Arabia's energy mix. *Energy Policy* **2017**, *105*, 80–92. [CrossRef]
23. Donella, H.M. *Thinking in Systems—A Primer*; Earthscan: London, UK, 2009.
24. Pollino, C.A.; Henderson, C. *Bayesian Networks: A Guide for Their Application in Natural Resource Management and Policy*; Landscape Logic: San Diego, CA, USA, 2010.
25. Jensen, F.V. *An Introduction to Bayesian Networks*; UCL Press: London, UK, 1996.
26. Holmes, D.E.; Jain, L.C. Introduction to Bayesian Networks. In *Innovations in Bayesian Networks*; Springer: Berlin/Heidelberg, Germany, 2008; pp. 1–5.
27. Niedermayer, D. An Introduction to Bayesian Networks and Their Contemporary Applications. In *Innovations in Bayesian Networks*; Springer: Berlin/Heidelberg, Germany, 2008; pp. 117–130.
28. Moldir Zholdasbayeva, S.P.; Vasilios, Z. Bayesian Networks based Policy Making in the Renewable Energy Sector. In Proceedings of the ICAART 2020, Valleta, Malta, 22–24 February 2020.
29. Daim, T.; Kayakutlu, G.; Suharto, Y.; Bayram, Y. Clean energy investment scenarios using the Bayesian network. *Int. J. Sustain. Energy* **2014**, *33*, 400–415. [CrossRef]
30. Yu, D.C.; Nguyen, T.C.; Haddawy, P. Bayesian network model for reliability assessment of power systems. *IEEE Trans. Power Syst.* **1999**, *14*, 426–432. [CrossRef]
31. Mengshoel, O.J.; Sundararajan, P.K.; Reed, E.; Piao, D.; Johnson, B. Renewable Energy Integration: Bayesian Networks for Probabilistic State Estimation. In *International Workshop on Data Analytics for Renewable Energy Integration*; Springer: Cham, Switzerland, 2018; pp. 63–82.

32. The World Bank. Energy Imports, Net (% of Energy Use)—Saudi Arabia, World Bank Data. 2014. Available online: https://data.worldbank.org/indicator/EG.IMP.CONS.ZS?locations=SA (accessed on 1 July 2020).
33. Saudi Arabia Oil Refinery Capacities. 2019. Available online: https://ycharts.com/indicators/saudi_arabia_oil_refinery_capacities (accessed on 19 August 2020).
34. Our World in Data, Electricity Generation from Low-Carbon Sources. 2019. Available online: https://ourworldindata.org/grapher/low-carbon-electricity?tab=chart&year=latest&country=~{}SAU (accessed on 30 June 2020).
35. Our World in Data, Energy Intensity. 2016. Available online: https://ourworldindata.org/grapher/energy-intensity?tab=chart&year=latest&country=~{}SAU (accessed on 30 June 2020).
36. Saudi Arabia—CO_2 Emissions Per Capita, World Data Atlas. 2018. Available online: https://knoema.com/atlas/Saudi-Arabia/CO2-emissions-per-capita#:~{}:text=In2018%2CCO2emissionsper,averageannualrateof2.20%25knoema.com (accessed on 20 August 2020).
37. KAPSARC. Electricity Tariffs in Saudi Arabia, Data Portal. 2018. Available online: https://datasource.kapsarc.org/explore/dataset/electricity-prices-in-saudi-arabia0/table/?disjunctive.consumption_slab&disjunctive.sector&sort=consumption_slab (accessed on 20 August 2020).
38. KAPSARC. Gasoline Prices, Data Portal. 2014. Available online: https://datasource.kapsarc.org/explore/dataset/gasoline-prices/table/?disjunctive.country&disjunctive.indicator&sort=time_period&refine.country=Saudi+Arabia&refine.indicator=Gasoline+Price+US$%2F+Liter (accessed on 20 August 2020).
39. Fuel Price Data Saudi Arabia. 2014. Available online: https://energypedia.info/index.php?title=Fuel_Price_Data_Saudi_Arabia&action=formeditenergypedia.com (accessed on 20 August 2020).
40. Fuel Prices in Saudi Arabia. 2020. Available online: https://www.thefuelprice.com/Fsa/en (accessed on 20 August 2020).
41. Norsys Software Corp. Netica Application. Available online: https://www.norsys.com/netica.html (accessed on 1 August 2020).
42. Energy Innovation Policy and Technology LLC and KAPSRC, Policy Solutions, The Energy Policy Simulator. 2020. Available online: https://saudiarabia.energypolicy.solutions/ (accessed on 24 June 2020).
43. Howarth, N.; Lanza, A.; Al Shehri, T. *What Is Behind the Recent Fall in Saudi Arabia's CO_2 Emissions?* KAPSARC: Riyadh, Saudi Arabia, 2019.
44. Saudi Energy Efficiency Center. Saudi Energy Efficiency Program. Available online: https://seec.gov.sa/en/about/saudi-energy-efficiency-program/ (accessed on 15 August 2020).
45. World Energy Council. Saudi Arabia Country Profile. 2020. Available online: https://trilemma.worldenergy.org/#!/country-profile?country=SaudiArabia&year=2020 (accessed on 16 November 2020).
46. Cain, J. *Planning Improvements in Natural Resources Management Guidelines for Using Bayesian Networks to Support the Planning and Management of Development Programmes in the Water Sector and Beyond*; Centre for Ecology and Hydrology: Wallingford, UK, 2001.

sustainability

MDPI

Article

Blockchain-Based Securing of Data Exchange in a Power Transmission System Considering Congestion Management and Social Welfare

Moslem Dehghani [1], Mohammad Ghiasi [1], Taher Niknam [1,*], Abdollah Kavousi-Fard [1], Mokhtar Shasadeghi [1], Noradin Ghadimi [2,3,*] and Farhad Taghizadeh-Hesary [4]

1 Department of Electrical and Electronics Engineering, Shiraz University of Technology, Shiraz 71555/313, Iran; dehghani.kau@gmail.com (M.D.); m.ghiasi1@gmail.com (M.G.); abdollah.kavousifard@gmail.com (A.K.-F.); shasadeghi@sutech.ac.ir (M.S.)

2 Young Researchers and Elite Club, Ardabil Branch, Islamic Azad University, Ardabil 19585/466, Iran

3 Department of Industrial Engineering, Ankara Yıldırım Beyazıt University (AYBU), 06760 Ankara, Turkey

4 Social Science Research Institute, Tokai University, Kanagawa, Hiratsuka 259-1292, Japan; farhad@tsc.u-tokai.ac.jp

* Correspondence: niknam@sutech.ac.ir (T.N.); nghadimi@ybu.edu.tr (N.G.)

Abstract: Using blockchain technology as one of the new methods to enhance the cyber and physical security of power systems has grown in importance over the past few years. Blockchain can also be used to improve social welfare and provide sustainable energy for consumers. In this article, the effect of distributed generation (DG) resources on the transmission power lines and consequently fixing its conjunction and reaching the optimal goals and policies of this issue to exploit these resources is investigated. In order to evaluate the system security level, a false data injection attack (FDIA) is launched on the information exchanged between independent system operation (ISO) and under-operating agents. The results are analyzed based on the cyber-attack, wherein the loss of network stability as well as economic losses to the operator would be the outcomes. It is demonstrated that cyber-attacks can cause the operation of distributed production resources to not be carried out correctly and the network conjunction will fall to a large extent; with the elimination of social welfare, the main goals and policies of an independent system operator as an upstream entity are not fulfilled. Besides, the contracts between independent system operators with distributed production resources are not properly closed. In order to stop malicious attacks, a secured policy architecture based on blockchain is developed to keep the security of the data exchanged between ISO and under-operating agents. The obtained results of the simulation confirm the effectiveness of using blockchain to enhance the social welfare for power system users. Besides, it is demonstrated that ISO can modify its polices and use the potential and benefits of distributed generation units to increase social welfare and reduce line density by concluding contracts in accordance with the production values given.

Keywords: FDIA; blockchain; data exchanging; under-operating agents; ISO; electricity market

Citation: Dehghani, M.; Ghiasi, M.; Niknam, T.; Kavousi-Fard, A.; Shasadeghi, M.; Ghadimi, N.; Taghizadeh-Hesary, F. Blockchain-Based Securing of Data Exchange in a Power Transmission System Considering Congestion Management and Social Welfare. *Sustainability* **2021**, *13*, 90. https://dx.doi.org/10.3390/su13010090

Received: 28 November 2020
Accepted: 21 December 2020
Published: 23 December 2020

Publisher's Note: MDPI stays neutral with regard to jurisdictional claims in published maps and institutional affiliations.

1. Introduction

As a result of the advancement of technology in power systems, the importance of using new methods to protect smart grids (SGs) becomes more apparent, and one of these methods is blockchain. In October 2008, in the reference [1], blockchain technology distribution was seen under the alias "Satoshi Nakamoto", with the aim of supporting the first Bitcoin cryptocurrency, and it resulted in the start of the Bitcoin network in January 2009. After that, Bitcoin has arrived inchmeal at the financial industry systems, and is recognized as the most influential and important cryptocurrency. Besides, blockchain technology after Bitcoin has become a game-variable innovation around the world, and lots of industries exist which will be interrupted via blockchain, including the life sciences, legal industry, health care, financial services, cyber security, supply series management, cloud storage, charity, electing, government, social interests, energy management, private transport and ride sharing, retail, and actual estate between others [2–4]. Blockchain usages in carbon credits, distributed energy resources, and renewable and power system data security versus cyber-attacks are really encouraging [5,6].

So, the available electricity market must be modified to benefit from this novel technology. In this study, there are two purposes, including examining how a great industrial user is able to manage best under such a novel blockchain on the basis of an electricity market as well as how to be safe from cyber-attacks like false data injection attack (FDIA).

1.1. Background

Networking, informatization, and intelligentization have been gradually realized by power systems [7] with the help of the usage of advanced information technology like calculation, communication, control, and perception, as well as using the expansion of the cyber-physical power system (CPPS). Meanwhile, real-time analytical promotion, efficient allocation of power systems and scientific decisions, interface terminals, and the open communication networks also carry potential security risks [8,9]. In comparison with the relatively strong initial power system, investigations on the security protection of electrical power information networks are in the initial stages, with a great deal of security vulnerabilities unknown [10]. Due to the remarkable interest relevance and high transmissibility of an electrical power grid, after the attack, it will have a rigid effect on industrial production, power security, and the livelihood of people, which has attracted considerable regard [11]. As a new attack method for major industrial facilities, safe and stable power system operation, whose defense and attack structure needs further research and investigation [12], have been threatened by cyber-attack.

Pursuant to the purposes of the attack, cyber-attacks versus power systems are able to be categorized as integrity, confidentiality of information, and destruction of availability. The availability destruction is obtained in attacks modifying network topology, black hole attacks, and unavailable information due to communication interruption, whose common layouts are denial of service (DOS) attacks. The integrity destruction is obtained in replay attack, man-in-the-middle attack, and wrong information due to false data injection (FDI), whose common layouts are FDI attacks. The confidentiality destruction is obtained in internal employee attack, utilization of malware, and illegal usage and data leakage, whose common layout is brute force password cracking [13,14]. As a common way to destroy the integrity of information, the FDIA is able to interrupt the outcomes of state estimation analysis, therefore misleading the control center's decision. The FDIA was first introduced by Liu in 2009, and the FDIA's point is to conduct coordinated attacks on sensors, falsify specific measurements, and manipulate particular state estimation information [15,16].

Over the past decade, the concept of the micro-grid (MG) is gaining more popularity resulting from its economic and technical benefits like higher reliability and resilience, closeness to the users, higher power quality, lower power losses, less operation cost, and self-healing capability [17,18]. Nevertheless, along with these benefits, several significant challenges exist in the management and operation of the micro-grids (MGs), which have attracted the attention of plenty of researchers [19]. In the ordinary power network, the distributed system operator (DSO) is responsible for applying the optimal energy management throughout the grid.

Nevertheless, in recent power networks, DSOs and MGs might have various policies and owners. As the network is able to be formed of some intercoupled MGs, the total network operation can be significantly affected by any change in any of the subsystems. So, a coordinator of a strong system level is required for the total network energy management.

Developments in management and economic laws in the electricity industry have long been taking place in this large industry under the congestion of the power system structure. The problem of congestion of the transmission network is considered as one of the most important issues in the discussion of the full implementation of the restructuring of power systems. Congestion means using a transmission network outside the operating range.

Restrictions such as transmission line capacity and transformers, maximum and minimum voltage values, and maximum voltage angles of the busbars (which are determined by various studies on the network) are among the limitations of operation. The consequences of congestion in power systems include sudden price jumps in some areas, increased market power, increased electricity prices, reduced efficiency of the electricity network, reduced competition, etc. [20,21]. When the power grid is congested, the capacity of some transmission lines no longer meets the needs of all customers. In this case, the independent system operator, as the main institution for maintaining the security of the power system, acts in different ways to manage the network's congestion. Independent system operation (ISO) is an institution that is responsible for coordinating and maintaining the security and reliability of power systems; in this regard, density management is the main task of this institution, which always tries to encourage investors and private owners of electric generators to participate in density management [22].

Congestion may be demonstrated during operation or in network operation planning. When using the power system, factors such as the sudden departure of one or more transmission lines or transformers in the network, the unexpected cessation of production in one or more generators in the system, and unforeseen changes in energy consumption, as well as uncoordinated transactions in electricity sales, can lead to network congestion. In planning the operation of a power system, the provision of an inappropriate program for the generation and consumption of electricity, the implementation of which violates the restrictions on the operation of the transmission network, is the main cause of network congestion [23].

Lately, blockchain has been promoted as a reliable and effective technology for online financial operations via communication only between peer-to-peer transaction networks and without the intervention of third parties [24]. The data are able to be reserved in the distributed databases, which are generally small, instead of reserving all data in a central data center by using blockchain. This might result in increasing the total cloud system security, since more of the losses from attacks on such databases are simply able to be locally prevented. So, blockchain is able to be successfully used in different areas, like the Internet of Things (IoT) and the financial sector.

Peer-to-peer (P2P) energy trading research based on blockchain has been studied. In [25], a blockchain with seven components based on MG energy market was offered and a smart contract was used to make a high-performance information system on the basis of the blockchain strategy. In [26], blockchain has been used to constitute a machine to machine electricity market in a chemical industry context, and a private blockchain on the basis of software system multichain was applied to validate energy transaction.

With the aim of facilitating a P2P market, a two-layer energy market system on the basis of multi-agent and blockchain technology has been suggested in [27]. For electric vehicles, also, a blockchain-based consortium for local aggregators has been proposed in [28] with the aim of auditing and validating electricity market plug-in hybrid electric vehicles (PHEVs). A novel blockchain-based energy system has been suggested in [29] with the aim of enabling electric vehicles to share transactions and publicly audit without the support of any reliable interposition. Plus, the consortium blockchain method was expanded to generic energy blockchain transactions with the aim of transaction security and credit-based payment. In [30], a decentralized energy dealing system using blockchain on the basis of tokens, streams of unknown encrypted messaging, and multi-signatures is proposed with the aim of solving issues of the privacy and security of information of demand and dealing.

In [31], forward and real-time markets have been proposed for bilateral agreements in P2P energy dealing. In [32], the energy broker's role has been proposed via amplification learning for indirect energy trading between customers. For electric vehicles, [33], a special P2P trading system was studied with the aim of reducing the effect of the charging procedure during the working hours of power systems. Additionally, in [34], in P2P energy trading, different game– and auction–theoretic methods were studied, and in [35], especially, a Nash bargaining view was applied with the aim of developing a frame of bilateral trans-active energy dealing for numerous contributors. For P2P trading, in [36], restrictions of physical low-voltage networks have been investigated by applying sensitivity analysis. Furthermore, in [37], power damages were assigned for MG peer-to-peer blockchain market contributors. At the power distribution system level, [38] suggested a day-ahead forecasting energy market strategy to help distribution system operators in order to optimize distributed energy resource applications; however, in [39], a novel P2P energy market on the basis of the content of multi-class energy management with the aim of coordinating dealing amongst prosumers with heterogeneous preferences was introduced.

Blockchain can be considered as the main technology of Bitcoin and some other types of cryptocurrencies, making this one of the world's most breathtaking technologies in 2010. The problem of transmission network conjunction has always been one of the serious obstacles against the full implementation of the restructuring of power systems, the correct and free communication of the producers and consumers, and the main challenge of independent system operation. Therefore, the policy of using distributed generation resources to manage the conjunction of transmission lines has been of particular importance. The policy of applying the transparent blockchain technology will aid in reducing risks and provide superior security to the grid, hence financial fraud is eliminated and the entire operational charge is attenuated. In order to illustrate the matters occurring in the usual blockchain layouts, mainly because of the storage and high level of elaborations of hash address computations, a blockchain technology is proposed in this article. Plus, a new data restoration technique is expanded with the aim of providing an approach to restore the appropriate and accurate data.

Nowadays, with the use of communication networks to exchange information between ISO and under-operating agents, sabotage cyber-attacks, including the false data injection attack (FDIA), are on the rise in order to destroy network stability and inflict financial losses. Therefore, providing a solution and policy to secure the information exchanged between systems is of great importance, so the use of blockchain technology as a solution to secure data is essential.

In this paper, the system under study is first examined under various conditions, including normal mode and false data injection into exchanged data such as loads, prices, and productions, and it is shown that this cyber-attack disrupts network congestion, reduces welfare, increases costs, and upsets the balance of production and demand.

1.2. Paper Structure

The remainder of the article is presented as follows: Section 2 defines the principles of blockchain and false data injection attacks. Problem theory, along with the formulation and the network under study and the system parameters, are presented in Section 3. The simulation results are analyzed under different normal scenarios and false data attacks and blockchain techniques to secure the messages and data in Section 4. Finally, the final conclusion is given in Section 5.

2. Basic Concept

2.1. The Data-Sharing Structure According to Blockchain Technology

2.1.1. Framework Overview

In order to present services based on a fine-grained and data-sharing structure, the transactions saved in the blockchain database based on the privacy level have been classified. The level of privacy contains community data, encrypted data, and public data. Public data, here, refer to a datum which is able to be observed through entire nodes. In addition, community public data give information which is able to be recognized through entire nodes pertaining to the similar community, and encrypted information basically refers to the private data and those which users are willing to purchase/sell.

In general, when users share professional information, it is suggested that they adjust the amount of information privacy on public data in the community so that this information is shared by more users who actually access it, and that need it to be visible. As a result, the main purpose can be to apportion the community reasonably through gaining a community diagnosing technique so that the public information of the community is able to be divided between more users who actually require it. Figure 1 depicts the data-sharing structure according to the blockchain method. We have three different layers in our presented data-sharing strategy, such as Data Layer, Blockchain Layer, and Detection Layer. Information is gathered by the Data Layer and sent to the Detection Layer. The Detection Layer implements a community detection technique that divides customers into various communities and limits the domain of data sharing. The Blockchain Layer also has the responsibility of keeping the result of community detection and transaction records secure.

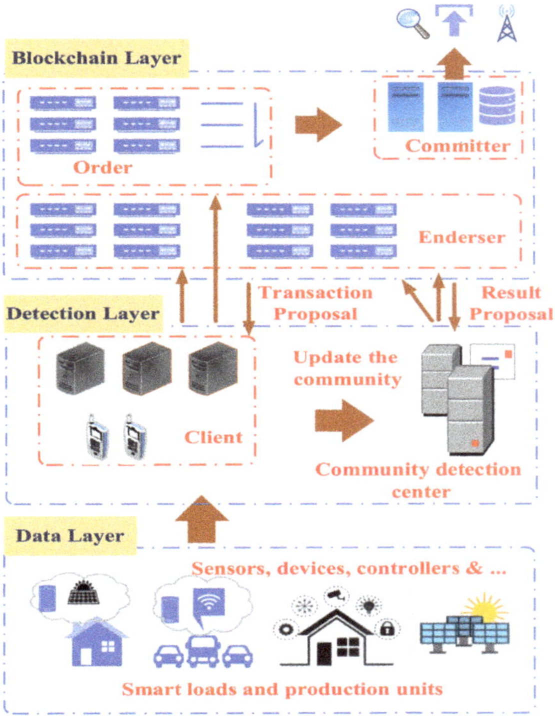

Figure 1. The data-sharing structure according to the blockchain technology.

2.1.2. Data Layer

The Data Layer contains data perceived through large-scale sensors [40] such as product type and product quantity, device performance status, and other data of several parameters.

Perception data are obtained by sensors and uploaded to the client server for comprehensive analysis and sharing. Perception information, after being obtained from the sensors, is sent and shared on the client's server for analysis.

2.1.3. Detection Layer

The Detection Layer includes the clients, the client server, and also the community recognition server. The client server basically has the responsibility of getting the perception data and then sending them to the blockchain network to be uploaded, and also to the community detection server. Additionally, label data are also produced when the user links the dependency chain. The community detection server has the responsibility of gathering entire label data and carrying out the community detection technique, as well as producing the community detection outcomes, and also sending such data to the blockchain network. When the community identification outcomes have been successfully saved to the blockchain database, the customer executes the smart contract in order to search for divided information from the community.

2.1.4. Blockchain Layer

Since the blockchain layer is made based on the hyperledger fabric structure, it consists of confirmation, order, and also committer nodes. The confirmation node has the responsibility of ensuring the transaction is offered through the customer server. The order node has the responsibility of classifying and packing the transactions into the blocks. The committer node also has the responsibility of validating and adding blocks into the database's blockchain. The common transaction procedure in the fabric is designed as the following stages:

- The customer side makes the transaction offer.
- The transaction execution is simulated by the node.
- The customer transmits the transaction into the consensus service.
- The customer orders the transactions with consensus, create new blocks, and renders the transactions.

In this article, the data-sharing structure applies to this transaction procedure in the phase of data sharing.

2.2. FDIA

The invaders attack the communication network, where invaders perfuse false data to messages or mensuration, which is shown in Figure 2. The FDI invader is able to perfuse the predetermined attack vector a by manipulating specific mensuration or messages. An invader is able to perfuse an attack vector to compromise the main mensuration which is shown in Equation (1); e represents the error vector; z defines the vector of measurements containing measurement readings from the sensor and ISO; H defines the Jacobian matrix with respect to x.

$$\hat{z}_{\mathrm{a}} = z + \mathrm{a} = Hx + \mathrm{a} + e \quad (1)$$

where: $\mathrm{a} = (\mathrm{a}_1, \mathrm{a}_2, \ldots, \mathrm{a}_\mathrm{m})^\mathrm{T} \cap \mathrm{a} \neq 0$. According to [14], the attack vector a is able to be adjusted by Equation (2).

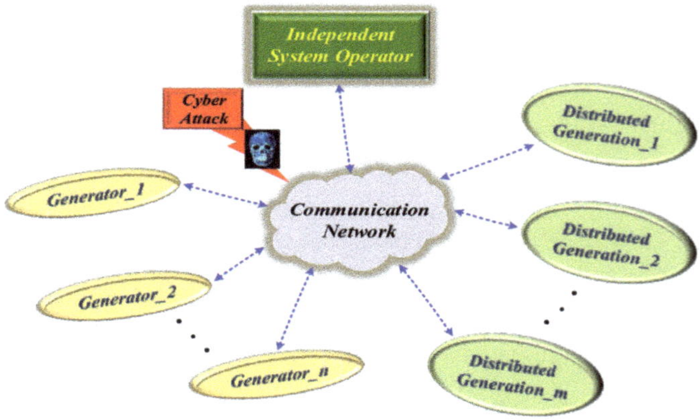

Figure 2. False data injection attack.

$$\mathrm{a} = \mathrm{Hc}, \quad (2)$$

where: $c = (c_1, c_2, \ldots, c_n)^T \cap c \neq 0$ is an arbitrary vector. The estimated state vector $\hat{x}$ is altered in Equation (3) with the false data injection.

$$\hat{x} = \left(H^T \Lambda H\right)^{-1} H^T \Lambda z_a = \hat{x} + c, \tag{3}$$

However, the residue r stays unaltered after attack and it still less than threshold τ;

$$r_a = ||\hat{z}_a - H\hat{x}_a||_2 = ||z + a - H(\hat{x} + c)||_2 = ||z - H\hat{x} + a - Hc)||_2 = ||z - H\hat{x}||_2 = r, \tag{4}$$

So the FDI invaders are able to circumvent and be undiscoverable to the traditional dab data detection.

3. Problem Theory

Limit pricing is one of the most well-known procedures of pricing congestion transmission networks, which is derived from the cost limit. The limit pricing is equal to the ratio of the increase in production costs to the increase in one megawatt of system load. If the limit pricing is calculated for increasing the load in a particular bus, it is called the node or local price, and if it is calculated for increasing the load in a particular area, it is called the regional price [41]. Node values for all busbars and regional prices for all zones will be equal when there is no network congestion.

Under normal power system conditions (lack of congestion and loss), logical marginal pricing (LMP) is the same in all busbars and there is a financial balance in the network busbars, in which case, this uniform price (called market clearing price (MCP)), as the purchase price and electricity sales, is used throughout the network, but when network congestion occurs, the LMP will be different on all busbars, which must be fixed using the methods mentioned in the corrective clogging management to fix the LMP busbars (fix the congestion of transmission lines).

A good way to manage network congestion is to put the market on a pricing basis. In this type of pricing, the ISO receives voluntary offers from market participants, selects the best solution with the lowest price, and finally performs the best load distribution; ultimately, all transmission line limitations are met and the system is balanced at the lowest price. Local price limits are obtained. In this paper, DC optimal load distribution (network losses are not considered) is used to manage clogging combined with social welfare (SW) maximization. To solve the DC optimal power flow (DCOPF) problem, equations have been used in the generalized algebra modeling system (GAMS) program. The discontinuous nonlinear program (DNLP) solver is used to solve the nonlinear program. The constraints of optimization include real equations of real power in all network buses, power transmission limitations, bus voltage limits, and production limits. The optimization of this issue is the amount of production of each generator (G) and the local limit price in all network buses. In this paper, in order to eliminate line congestion, with optimal use of distributed generation (DG), LMP stabilization of network buses in a certain value has been established. By stabilizing the local price limit of network buses in a fixed amount, the creation of market power by expensive generators is prevented. Under these conditions, the market is approaching full competition, and in general, the disadvantages caused by network congestion will be eliminated.

Formulation

The consumer interest function and the producer cost function are in accordance with relationships (5) to (8), respectively, in dollars per hour [41].

$$B_{dj}\left(P_{dj}\right) = -\frac{1}{2}c_{dj}{P_{dj}}^2 + d_{dj}P_{dj} + m_i \tag{5}$$

$$C_{gi}(p_{gi}) = \frac{1}{2}a_{gi}{P_{gi}}^2 + b_{gi}P_{gi} + m_i \tag{6}$$

$$C_k\left(P_{DG_k}\right) = \frac{1}{2}a_{DGk}P_{DGk}^2 + b_{DGk}P_{DGk}^2 + c_{DGk} \tag{7}$$

$$C_w = \gamma_w \times d_{wind} \times P_w \tag{8}$$

In these relations, c_{dj} and d_{dj} are slope and width from the origin of the uniform curve of $j - th$ consumer demand, a_{gi} and b_{gi} are slope and width from the origin of the uniform curve suggested by the generator, m_j and m_i are constant coefficients of profit and consumption functions of the $j - th$ generator and $i - th$ generator, P_{dj} and P_{gi} are the real power of the $j - th$ consumption and $i - th$ generator, $C_k(P_{DG_k})$ provides the cost function of the $k - th$ DG number, N_m is the number of DGs connected to the network and P_{DGk} represents the active power generation of the $k - th$ DG number, C_w represents the cost of wind turbine production in each scenario, d_{wind} is the recommended price of wind turbine per MW/h of power generation, P_w gives the production power of the wind turbine unit in each scenario (MW), w is considered as low and high scenarios for the power production of wind turbines in the set of Ω, and γ_W represents the probabilities for the two scenarios of wind power production, 0.4 and 0.6, that are chosen for the low and high scenarios, respectively.

The formulation of the problem of density management in reconstructed power systems is the maximization of social welfare by considering the power balance constraints and the density of transmission lines. Mathematically, the objective function of the problem (maximizing social welfare) is a nonlinear relation (9):

$$\text{Max } SW = \sum_{i=1}^{N_d} B_{di}(P_{di}) - \sum_{i=1}^{N_g} C_{gi}(P_{gi}) - \sum_{k=1}^{N_m} C_k(P_{DGk}) - \sum_{w \in \Omega} \gamma_w \times d_{wind} \times P_w \tag{9}$$

According to the formulation in relation to (9), the objective function, which is social welfare, is equal to the total profit of consumers minus the total cost of producers, which should be maximized according to the ISO view of this objective function. The equal and unequal constraints are given below. The profit of each manufacturer is obtained according to the following relationship:

$$\text{profit}_q = (LMP_{q\_n} \times P_{q\_n}) - C_q \tag{10}$$

where: profit_q is the profit of the producer q, LMP_{q_n} is the local limit pricing in the $n -$ th bus where the $q -$ th producer is located, P_{q_n} is the production capacity of the q $-$ th producer in the $n -$ th bus, and C_q is the cost of the q $-$ th producer. Hence, we have:

Power balance constraints as:

$$P_{gi} + P_{DGi} + P_W - P_{di} = \sum_{j=1}^{N} \frac{1}{x_{i-j}}(\delta_i - \delta_j), \text{ for } i \in u_{DG} \tag{11}$$

Maximum power limitation as:

$$|Pl_{i-j}| \leq Pl_{i-j}{}^{max,\ l} \tag{12}$$

Range of variables as:

$$0 \leq P_{gi} \leq P_{gi}{}^{max,\text{ for } g} \tag{13}$$

$$0 \leq P_{DGk} \leq P_{DG}^{max} \tag{14}$$

$$0 \leq P_{dj} \leq P_{dj}{}^{max,\text{ for } d} \tag{15}$$

$$\delta_i^{min_i{}_i^{max \text{ for}}} \tag{16}$$

In these relations, N and N_L are the number of system busbars and the number of lines, respectively. δ_i is voltage angle in the $i - th$ busbar; x_{i-j} gives the inductive reactance of the connecting line series amongst i and j buses; u_{DG} provides the network DG set and P_{DGk}^{max} is the operating rate of the $k -$ th DG. Pl_{i-j} and $Pl_{i-j}{}^{max}$ are the active power and maximum active power at the connection line between the buses i and j, respectively.

$P_{gi}{}^{max}$ and $P_{dj}{}^{max}$ represent maximum values of P_{gi} and P_{dj}. δ_i^{min} and δ_i^{max} are the minimum and maximum values of δ_i. The local limit value is also obtained from the power balance equilibrium in each bus.

Information of line parameters, generator cost coefficients, and consumption rates in different busbars is given in Tables 1–3, respectively.

Table 1. Grid line details.

Line	From	To	X (p.u)	Pl^{max} (MW)
1	1	2	0/0592	60
2	1	5	0/2230	100
3	2	3	0/1980	100
4	2	4	0/1763	100
5	2	5	0/1739	50
6	3	4	0/1710	50
7	4	5	0/0421	100
8	4	7	0/2091	50
9	4	9	0/5562	50
10	5	6	0/2520	50
11	6	11	0/1989	50
12	6	12	0/2558	50
13	6	13	0/1303	50
14	7	8	0/1762	100
15	7	9	0/1100	100
16	9	10	0/0845	50
17	9	14	0/2704	50
18	10	11	0/1921	50
19	12	13	0/1999	50
20	13	14	0/3480	50

Table 2. Grid generator details.

Gen No.	Bus No.	P_g^{max}(MW)	a (USD/(MWh)^2)	b (USD/MWh)	m (USD/h)
1	1	250	0/43	20	0
2	2	200	0/25	20	0
3	3	60	0/01	40	0
4	6	50	0/01	40	0
5	8	60	0/01	40	0

Table 3. Consumption details in busbars.

Load No.	Bus No.	P_d (MW)
1	2	8/17
2	3	42/89
3	4	56/05
4	5	26/52
5	6	35/6
6	9	26/26
7	10	12/65
8	11	39/68
9	12	11/8
10	13	10/7
11	14	33/91

In this work, it is assumed that four distributed generation units without uncertainty and one wind turbine with two high and low production scenarios, according to Table 4, are connected to the 14-bus IEEE test network. The specifications of the cost function of these units and their capacity and installation location are given in Table 4. Additionally, the network of 14 buses that the IEEE studied is shown in Figure 3.

Table 4. Distributed generation (DG) unit details.

DG	Bus No.	P_{DG}^{max} (MW)	b_DG (USD/MWh)
DG1	14	20	30
DG2	12	30	30
DG3	9	20	30
DG4	4	25	30
wind	10	5,20	15

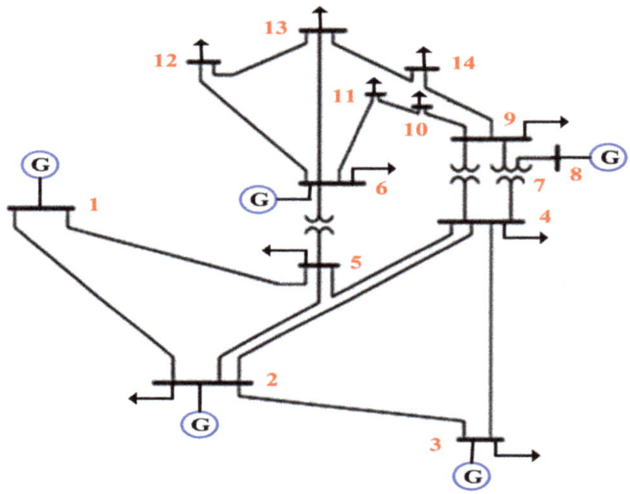

Figure 3. The studied network of 14 buses by the IEEE.

4. Simulation Results and Analysis

The studied network is the 14-bus IEEE network; the network consists of 12 loads, five generators, one uncertainty (wind) source, and three distributed generation (no uncertainty) sources for an hour of the day ahead of the electricity market. It is assumed that the wind would work in two scenarios: high and low. Wind generation in the low scenario is 5 MW and in the high scenario, it is 20 MW. The independent system operator (ISO), as an upstream entity, wants distributed generation sources (DGSs) available in the network to manage network congestion and maximize the social welfare and, ultimately, the ISO contract with the DGSs corresponds to their optimal generation for an hour of the day ahead. In the meantime, the wind turbine is presented as a balancing generator for the system. Without DGSs on the network, locational marginal prices (LMPs) are different in buses; in this case, there is congestion in the network. As shown in Figure 4, with the presence of DGSs, these prices are stabilized and network congestion is managed. As a result, social welfare is increased and operating costs are reduced.

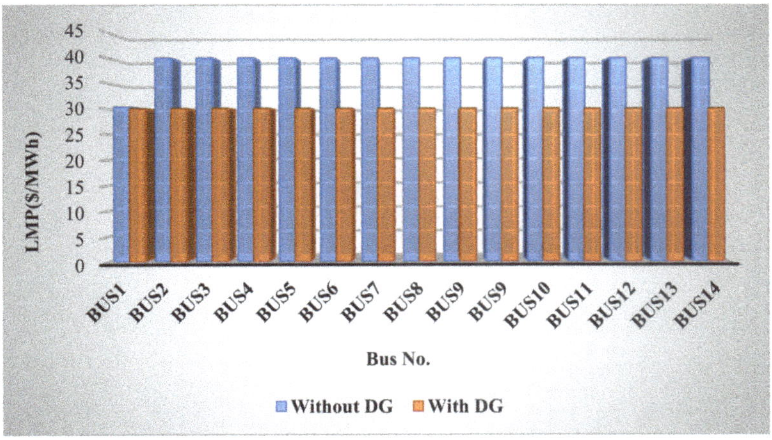

Figure 4. Locational marginal prices (LMPs) in different buses, with or without distributed generation sources in the network.

Next, it is been shown what will happen to the power system and electricity market situation with DGSs on the network by injecting different cyber-attacks (false data injection attacks), including load increase (LI), load decrease (LD), DGS price changes (DGPCH), and generator price changes (GPCH). In the FDI attacks, the attacker is able to access the data of the communication links, sensors, local controllers, and central control units so, to simulate the FDI attack, it has been assumed that the attacker can manipulate the data, therefore at the time of the attack, the data has been manipulated to show the attack outcomes.

4.1. Scenario 1: Normal

In this case, it is assumed that the network is in operation with the presence of DGSs, and no false data injection attack has been performed on the network load and unit prices. According to Table 5, the optimal production and profit of each of the operating units in the 14-bus network is given. In this case, cheap generators 1 and 2 in both cases are wind production in the production line, while the more expensive generators 3, 4 and 5 are not used. Figure 5 shows LMPs on different buses without network attacks.

Table 5. Status of units in normal mode.

Unit	Bus No.	Generation (MW)		Profit (USD)	
		Low	High	Low	High
G1	1	115.314	100	2378.451	2439.277
G2	2	20	20	500	500
G3	3	0	0	0	0
G4	6	0	0	0	0
G5	8	0	0	0	0
		Total		3378.451	2939.277
Unit	**Bus No.**	**Generation (MW)**		**Profit (USD)**	
		Low	**High**	**Low**	**High**
DG1	14	15.604	6.054	468.123	468.123
DG2	12	0	4.613	0	0
DG3	9	6.470	2.535	194.103	194.103
DG4	4	4.108	4.108	123.231	123.231
		Total		785.457	519.309

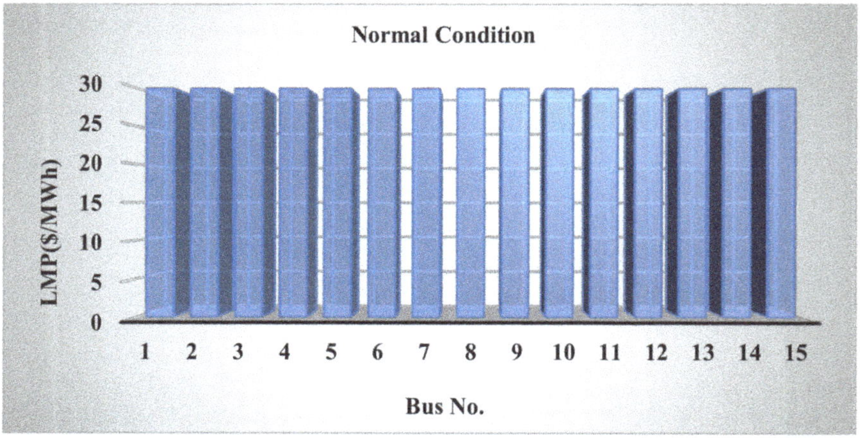

Figure 5. LMPs on different buses without network attacks.

DGSs are also in operation, as shown in Table 5. It should be noted that these optimal results will lead to optimal network status. Bus node prices are reduced and stabilized, network congestion is managed, and network social welfare is maximized. As a result, ISO can contract with distributed generation units in accordance with these optimal production values. The results are obtained by the powerful GAMS optimization software. In this case, social welfare is 3554.

4.2. Scenario 2: Incremental Attack on Load

Sometimes the attacker tries to break the market and disrupt the power supply for various reasons. One of these types of attacks is the false data injection attack on the smart power network components, such as loads, measurements, detectors, and sensors.

In Scenario 2, it is assumed that the attacker has access to the loads of one hour from the day ahead of the market and by virtually changing the loads by 1.8 times of the main load, the attacker can change the production conditions and profit of the units in the market and create congestion in the network. In this case, LMPs increase in all buses, so this price in bus 1 is different from other buses, in which case it can be said that there is congestion in the network. According to Figure 6, as the load increases, according to Table 5, all units increase their production. DGSs generate electricity at their maximum capacity, and cheap and even expensive generators are on the production line. Given that in reality there has been no increase in load and the attacker has changed the network load information, additional production takes place in the network, which will lead to serious damage to the network and also increase production costs and reduce social welfare. In this case, social welfare is 3148.

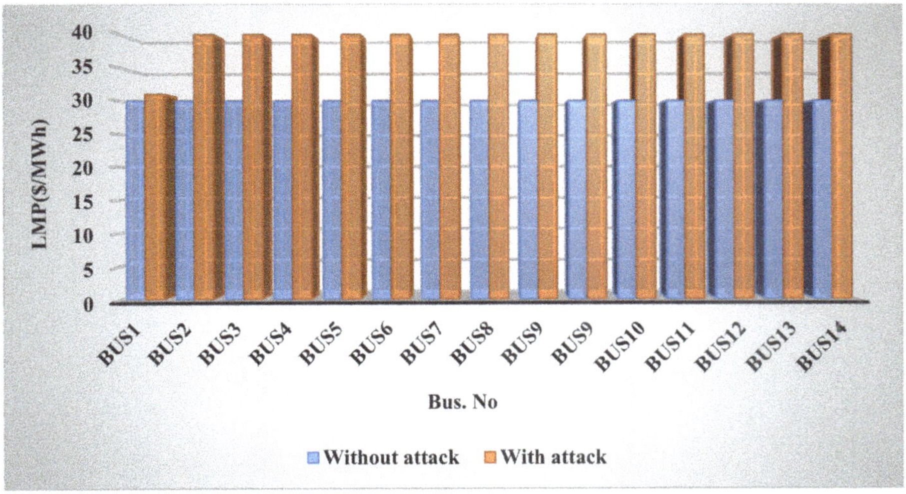

Figure 6. LMPs at different buses in the case of an incremental load attack.

4.3. *Scenario 3: Decreasing Attack on Load*

In scenario 3, it is assumed that the attacker has access to the loads of one hour from the day ahead of the market and by virtually reducing the loads by 0.5 times of the main load. the attacker can change the production conditions and profit of the units in the market.

The LMPs have dropped, as shown in Figure 7, but because this is not the case in reality and the load has not been reduced, generators have minimized their production and DGSs have very little production (according to Table 6), which causes failure to provide real loads and a large reduction in unit profits, especially production units, will be dispersed. Additionally, due to the imbalance of load and production in the network, blackouts occur in various sections of the grid. Social welfare in this case is 2536, which is a decrease compared to the normal state.

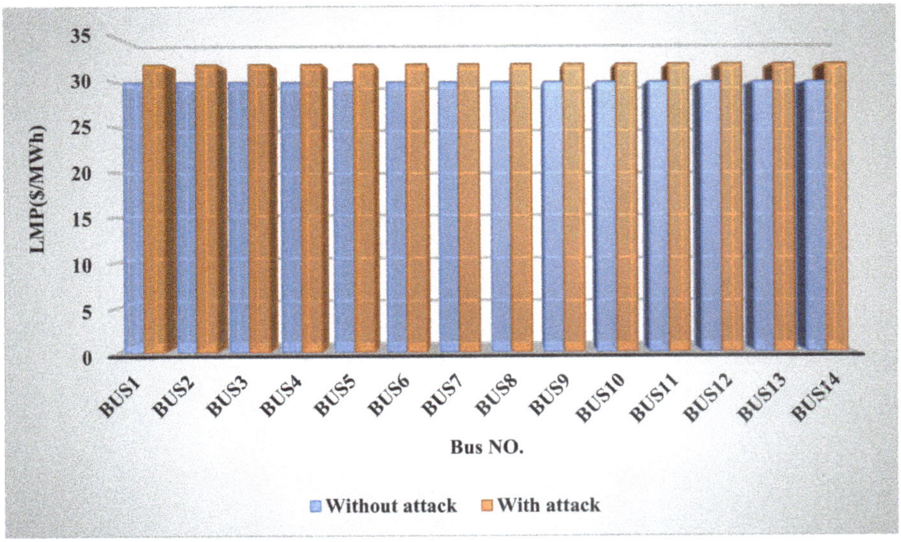

Figure 7. LMPs at different buses in the case of a reduced load attack.

Table 6. Status of units in incremental load attack mode.

Unit	Bus No.	Generation (MW)		Profit (USD)	
		Low	High	Low	High
G1	1	115.314	100.314	2878.451	2439.277
G2	2	40.479	40.479	1219.212	1219.212
G3	3	11.972	11.972	480.303	480.303
G4	6	11.972	11.972	387.669	387.669
G5	8	11.972	11.972	387.669	387.669
	Total			5538.572	5099.398
Unit	**Bus No.**	**Generation (MW)**		**Profit (USD)**	
		Low	**High**	**Low**	**High**
DG1	14	20	20	600	600
DG2	12	30	30	900	900
DG3	9	20	20	600	600
DG4	4	25	25	750	750
	Total			2850	2850

4.4. *Scenario 4: Attacking Generator Bid Prices*

It is supposed that the attacker has access to the generator's suggested prices. The attacker can change the suggested prices of the generators (in this scenario, the prices of cheap and expensive production units were shifted together). In this case, according to Table 7, cheap generators and even DGSs have no production, and this will seriously damage the interests of these units. In this case, the power supply of the network will only be the responsibility of expensive generators and will disrupt the financial balance of the market. Table 8 provides the status of units in attack mode at generator prices.

Table 7. Status of units in the case load reduction attack mode.

Unit	Bus No.	Generation (MW)		Profit (USD)	
		Low	High	Low	High
G1	1	58.467	58.467	1316.427	1434.832
G2	2	10.063	10.063	226.580	246.959
G3	3	0	0	0	0
G4	6	0	0	0	0
G5	8	0	0	0	0
	Total			1543.007	1681.791
Unit	**Bus No.**	**Generation (MW)**		**Profit (USD)**	
		Low	**High**	**Low**	**High**
DG1	14	0.002	0	0.006	0
DG2	12	0.002	0	0.006	0
DG3	9	0.002	0	0.006	0
DG4	4	0.002	0.002	0.006	0.006
	Total			0.024	0.006

Table 8. Status of units in attack mode at generator prices.

Unit	Bus No.	Generation (MW)	
		Low	High
G1	1	0	0
G2	2	0	0
G3	3	49.853	49.853
G4	6	49.853	49.853
G5	8	49.853	49.853
Unit	**Bus No.**	**Generation (MW)**	
		Low	**High**
DG1	14	0.07	0
DG2	12	0	0
DG3	9	0	0
DG4	4	0.003	0

4.5. Scenario 5: Attack on Distributed Generation Prices

In this case, also, the attacker manipulates the proposed prices of DGSs (price increases by 1.4 times the original prices). As a result, according to Table 9, the generation of DGSs reaches zero and does not benefit them. Thus, operating costs increase and social welfare decreases (3320).

Table 9. Status of units in attack mode at DG prices.

Unit	Bus No.	Generation (MW)	
		Low	High
G1	1	115.314	100.314
G2	2	40.023	40.023
G3	3	0.574	0.574
G4	6	0.574	0.574
G5	8	0.574	0.574
Unit	**Bus No.**	**Generation (MW)**	
		Low	**High**
DG1	14	0	0
DG2	12	0	0
DG3	9	0	0
DG4	4	0	0

4.6. Generation of Units in Different Scenarios

Figure 8 shows the production of generators in different scenarios. This figure shows that all generators in scenario 2 due to increased load are on the network production line. The lowest generator output was related to the load reduction data attack scenario. Figure 9 also shows the production of DGSs in different scenarios. In this case, the highest production is related to scenario 1.

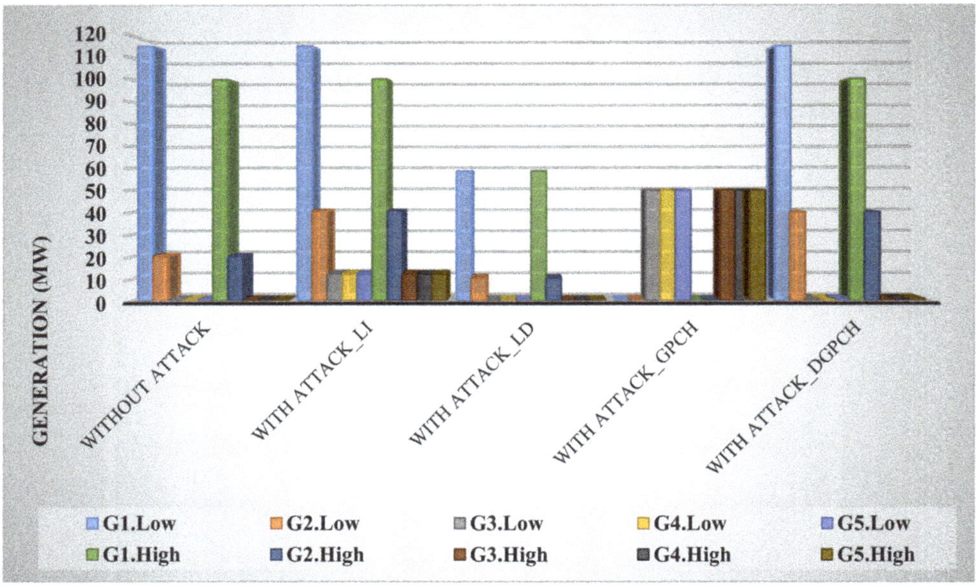

Figure 8. Production of generators in different scenarios.

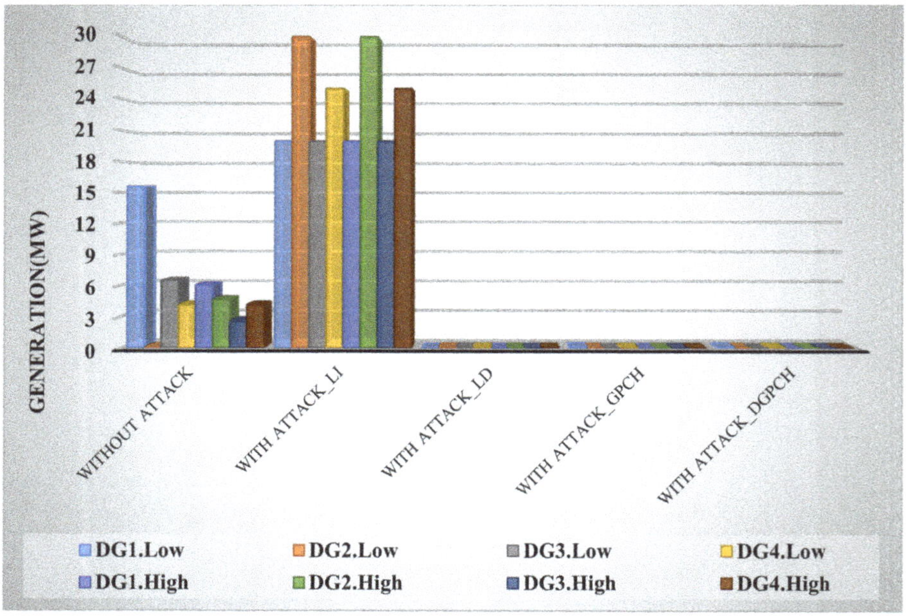

Figure 9. Production of distributed generation sources (DGSs) in different scenarios.

4.7. Implementation of Blockchain for Secure Exchange of Messages amongst ISO and Under-Operating Agents

In this scenario, in order to secure the data exchanged between ISO and under-operating agents, the blockchain technique has been used to exchange information, and for this purpose, unit prices are sent to ISO along with the production rate based on the system conditions. Confirmation of the validity of the sent data in the blockchain platform and non-destructive manipulation of information, such as FDI attack, and the amount of production of units in the desired time frame are sent to the units in the blockchain platform and then the units, after validating the sent data by ISO, produce the desired amount of power, which makes the data safe to send and any attack based on the messages is detected. It should be noted that the main objectives of the independent system operator, as in the above institution, should be met, such as reducing the LMP of the buses (reducing the congestion of transmission lines) and maximizing the social welfare of the network. DGSs should be properly closed. Table 10 shows examples of exchanged messages between ISO and under-operating agents based on blockchain technology.

Table 10. Transaction blockchain description.

Block Index		Block Information		
Index		1		Description
Time		1		
Transaction Message	Sender	Receiver	Power (MW)	
(Data)	ISO	DG3	6.470	
Previous HA		6ace5c77c9f1abba4f25f81f5557ca3d		
Self HA		7ab52efec4674cbf34f97e2a72d5e753		
Index		2		Description
Time		2		
Transaction Message	Sender	Receiver	Power (MW)	
(Data)	ISO	DG4	4.108	
Previous HA		7ab52efec4674cbf34f97e2a72d5e753		
Self HA		266477dd8561492cfa2bef961485b8a4		
Index		4		Description
Time		4		
Transaction Message	Sender	Receiver	Power (MW)	
(Data)	ISO	G1	115.314	
Previous HA		098e81b5609be3f39bdb61c2e6d2c67d		
Self HA		de3a8a139ecee8aaf023fc914417d748		
Index		5		Description
Time		4		
Transaction Message	Sender	Receiver	Power (MW)	
(Data)	ISO	G1	80	Cyber-attack has occurred
Previous HA		098e81b5609be3f39bdb61c2e6d2c67d		
Self HA		029e6f1af9dacfe2f3e01fda58634a00		
Index		6		Description
Time		5		
Transaction Message	Sender	Receiver	b (USD/MWh)	
(Data)	G4	ISO	40	
Previous HA		de3a8a139ecee8aaf023fc914417d748		
Self HA		3490acda48423d504c295da91fa2aa92		
Index		7		Description
Time		5		
Transaction Message	Sender	Receiver	b (USD/MWh)	
(Data)	G4	ISO	20	Cyber-attack has occurred
Previous HA		de3a8a139ecee8aaf023fc914417d748		
Self HA		706ef584b4a6529f9af410c950fcecc9		

The generic blockchain relevant to ISO is demonstrated in Table 10 and shows that the information is in accordance with the specific blockchain of ISO. This procedure is able to aid the recovery of the information contained in a cyber-attack or package in private and public blockchains.

Nevertheless, it is noteworthy that the output powers of agent generation and other data, as well as the costs of the generation of every DG and G, do not exist in the blockchain, which is able to raise the privacy and the security of the messages and network.

The transaction blocks are presented by Table 10. Pursuant to this table, for example, at t = 1, DG3 gets a message from ISO. In addition, the value of the provided power in megawatts, as well as the generating DG, are demonstrated in this table. Like the specific blockchain, any block includes the blockchain hash algorithm (HA), recognized as the self HA, that is able to chain to the prior block through utilizing the prior HA. Plus, if a cyber-attack has happened in the message (see indexes 7 and 5), the HA is altered and the HA is not the similar, so the multiple message is defined.

5. Conclusions

DG resources have been applied to reduce LMPs and maximize the social welfare of the network to reach ISO's main objectives. It has been shown that by applying DGs in power grids and stabilizing the LMP of the busbars, the congestion was managed and the production of more expensive units was minimized. Additionally, the social welfare of the network was maximized for the seasons under contract with DG. The blockchain technology has been used to secure messages and exchanged data between ISO and under-operating agents.

The results demonstrated that ISO can modify its polices and use the potential and benefits of DGSs to increase social welfare and reduce line density by concluding contracts in accordance with the production values given. In addition to the cyber security reinforcement, the considered policy sample is decentralized, transparent, and secured, and is able to decrease the risks to the network, remove the financial spoof, and reduce the whole cost of operation.

The outcomes of simulation on a trial system confirmed the great effectiveness and performance of the considered policy frame, particularly in the presence of a cyber-attack where the information is not available for outside unwarranted parts of the system. This is chiefly because the HAs are altered in any repeat.

Author Contributions: M.D. and M.G. proposed the idea, developed the model, and performed the simulation works and also wrote the paper. T.N. and A.K.-F. led the project. F.T.-H., M.S. and N.G. were in charge of reviewing and editing the paper. This work was conducted under the supervision of T.N., F.T.-H. All authors have read and agreed to the published version of the manuscript.

Funding: This research received no external funding.

Data Availability Statement: Data is contained within the article.

Conflicts of Interest: The authors declare no conflict of interest.

Nomenclature

$\mathbf{z}$	Vector of measurements
$\mathbf{H}$	Jacobian matrix
$\mathbf{a}$	Attack vector
$\mathbf{c}$	Arbitrary vector
$\hat{\mathbf{x}}$	Estimated state vector
$\mathbf{r}$	Residue
τ	Threshold
c_{dj}	Slope from the origin of the uniform curve of $j-$th consumer demand
d_{dj}	Width from the origin of the uniform curve of $j-$th consumer demand
a_{gi}	Slope from the origin of the uniform curve suggested by the generator
b_{gi}	Width from the origin of the uniform curve suggested by the generator

$\mathbf{m_j}$	Constant coefficients of profit and consumption functions of $j-$ th G
$\mathbf{m_i}$	Constant coefficients of profit and consumption functions of $i-$ th G
$\mathbf{P_{dj}}$	Real power of $j-$ th consumption
$\mathbf{P_{gi}}$	Real power of $i-$ th generator
$\mathbf{C_k(P_{DG_k})}$	Cost function of $k-$ th DG number
$\mathbf{N_m}$	Number of DGs connected to the network
$\mathbf{P_{DGk}}$	Active power generation of $k-$ th DG number
$\mathbf{C_w}$	Cost of wind turbine production
$\mathbf{d_{wind}}$	Recommended price of wind turbine
$\mathbf{P_w}$	Production power of wind turbine unit
$\mathbf{w}$	Considered as low and high scenarios for power production of wind turbine
γ_w	Probabilities for the two scenarios of wind power production
$\mathbf{Profit_q}$	Profit of the producer q
$\mathbf{LMP_{q\_n}}$	Local limit pricing in $n-$ th bus where the $q-$ th producer is located
$\mathbf{P_{q\_n}}$	Production capacity of the $q-$ th producer in $n-$ th bus
$\mathbf{C_q}$	Cost of $q-$ th producer
N	Number of system busbars
$\mathbf{N_L}$	Number of lines
δ_i	Voltage angle in $i-$ th busbar
x_{i-j}	Inductive reactance of the connecting line series amongst i and j buses
u_{DG}	Network DG set
$\mathbf{P_{DGk}^{max}}$	Operating rate of $k-$ th DG
$\mathbf{Pl_{i-j}}$	Active power at the connection line between the buses i and j
$\mathbf{Pl_{i-j}}^{max}$	Maximum active power at the connection line between buses i and j
$\mathbf{P_{gi}}^{max}$	Maximum values of P_{gi}
$\mathbf{P_{dj}}^{max}$	Maximum values of P_{dj}
δ_i^{min}	Minimum values of δ_i
δ_i^{max}	Maximum values of δ_i
List of abbreviations	
DG	Distributed generation
FDIA	False data injection attack
ISO	Independent system operation
SGs	Smart grids
CPPS	Cyber-physical power system
DOS	Denial of service
FDI	False data injection
MG	Micro-grid
MGs	Micro-grids
DSO	Distributed system operator
IoT	Internet of Things
P2P	Peer-to-peer
PHEVs	Plug-in hybrid electric vehicles
GPS	Global positioning system
LMP	Logical marginal pricing
MCP	Market clearing price
SW	Social welfare
DCOPF	DC optimal power flow
GAMS	Generalized algebra modeling system
G	Generator
DNLP	Discontinuous nonlinear program
DGSs	Distributed generation sources
LI	Load increase
LD	Load decrease
DGPCH	Distributed generation source price changes
GPCH	Generator price changes
LMPs	Locational marginal prices
DGs	Distributed generations
HA	Hash algorithm

References

1. Nakamoto, S. Bitcoin: A Peer-to-Peer Electronic Cash System. 2008. Available online: http://bitcoin.org/bitcoin.pdf (accessed on 17 November 2020).
2. Chapron, G. The environment needs cryptogovernance. *Nat. Cell Biol.* **2017**, *545*, 403–405. [CrossRef]
3. Yu, D.; Zhang, T.; He, G.; Nojavan, S.; Jermsittiparsert, K.; Ghadimi, N. Energy management of wind-PV-storage-grid based large electricity consumer using robust optimization technique. *J. Energy Storage* **2020**, *27*, 101054. [CrossRef]
4. Abedinia, O.; Zareinejad, M.; Doranehgard, M.H.; Fathi, G.; Ghadimi, N. Optimal offering and bidding strategies of renewable energy based large consumer using a novel hybrid robust-stochastic approach. *J. Clean. Prod.* **2019**, *215*, 878–889. [CrossRef]
5. Ashley, M.J.; Johnson, M.S. Establishing a Secure, Transparent, and Autonomous Blockchain of Custody for Renewable Energy Credits and Carbon Credits. *IEEE Eng. Manag. Rev.* **2018**, *46*, 100–102. [CrossRef]
6. Kavousi-Fard, A.; Su, W.; Jin, T. A Machine-Learning-Based Cyber Attack Detection Model for Wireless Sensor Networks in Microgrids. *IEEE Trans. Ind. Inform.* **2020**, *17*, 650–658. [CrossRef]
7. Ghiasi, M.; Dehghani, M.; Niknam, T.; Kavousi-Fard, A. Investigating Overall Structure of Cyber-Attacks on Smart-Grid Control Systems to Improve Cyber Resilience in Power System. *IEEE Smart Grid Newsl.* **2020**, *1*, 1. Available online: https://smartgrid.ieee.org/newsletters/march-2020/investigating-overall-structure-of-cyber-attacks-on-smart-grid-control-systems-to-improve-cyber-resilience-in-power-system (accessed on 17 November 2020).
8. Rahmani, M.; Hosseinian, S.; Abedi, M. Stochastic two-stage reliability-based Security Constrained Unit Commitment in smart grid environment. *Sustain. Energy Grids Netw.* **2020**, *22*, 100348. [CrossRef]
9. Rezaee, M.; Moghaddam, M.S.; Ranjbar, S.; Shokouhinia, M. Online estimation of power system separation as controlled islanding scheme in the presence of inter-area oscillations. *Sustain. Energy Grids Netw.* **2020**, *21*, 100306. [CrossRef]
10. Kamil, I.A.; Ogundoyin, S.O. A big data anonymous batch verification scheme with conditional privacy preservation for power injection over vehicular network and 5G smart grid slice. *Sustain. Energy Grids Netw.* **2019**, *20*, 100260. [CrossRef]
11. He, H.; Yan, J. Cyber-physical attacks and defences in the smart grid: A survey. *IET Cyber Phys. Syst. Theory Appl.* **2016**, *1*, 13–27. [CrossRef]
12. Hahn, A.; Govindarasu, M. Cyber Attack Exposure Evaluation Framework for the Smart Grid. *IEEE Trans. Smart Grid* **2011**, *2*, 835–843. [CrossRef]
13. Chen, L.; Yue, D.; Dou, C.; Chen, J.; Cheng, Z. Study on attack paths of cyber attack in cyber-physical power systems. *IET Gener. Transm. Distrib.* **2020**, *14*, 2352–2360. [CrossRef]
14. Foroutan, S.A.; Salmasi, F.R. Detection of false data injection attacks against state estimation in smart grids based on a mixture Gaussian distribution learning method. *IET Cyber Phys. Syst. Theory Appl.* **2017**, *2*, 161–171. [CrossRef]
15. Dehghani, M.; Ghiasi, M.; Niknam, T.; Kavousi-Fard, A.; Padmanaban, S. False Data Injection Attack Detection based on Hilbert-Huang Transform in AC Smart Islands. *IEEE Access* **2020**, *8*, 179002–179017. [CrossRef]
16. Dabbaghjamanesh, M.; Mehraeen, S.; Kavousi-Fard, A.; Ferdowsi, F. A new efficient stochastic energy management technique for interconnected AC microgrids. In Proceedings of the 2018 IEEE Power Energy Society General Meeting (PESGM), Portland, OR, USA, 5–10 August 2018; pp. 1–5. [CrossRef]
17. Ghiasi, M. Detailed study, multi-objective optimization, and design of an AC-DC smart microgrid with hybrid renewable energy resources. *Energy* **2019**, *169*, 496–507. [CrossRef]
18. Aghajani, G.; Ghadimi, N. Multi-objective energy management in a micro-grid. *Energy Rep.* **2018**, *4*, 218–225. [CrossRef]
19. Fathi, M.; Ghiasi, M. Optimal DG Placement to Find Optimal Voltage Profile Considering Minimum DG Investment Cost in Smart Neighborhood. *Smart Cities* **2019**, *2*, 20. [CrossRef]
20. Kumar, P.; Nikolovski, S.; Ali, I. Introduction to the Special Section on Innovative Technologies for Microgrid and Smart-grid Systems. *Comput. Electr. Eng.* **2019**, *77*, 414–415. [CrossRef]
21. Jaghargh, M.J.P.; Mashhadi, H.R. Structural and behavioural evaluation of renewable energy power plants' impacts on transmission network congestion using an analytical approach. *IET Renew. Power Gener.* **2020**, *14*, 1164–1173. [CrossRef]
22. Biskas, P.; Bakirtzis, A.G. Decentralised congestion management of interconnected power systems. *IEE Proc. Gener. Transm. Distrib.* **2002**, *149*, 432. [CrossRef]
23. Zheng, Z.; Xie, S.; Dai, H.; Chen, X.; Wang, H. An overview of blockchain technology: Architecture, consensus, and future trends. In Proceedings of the 2017 IEEE International Congress on Big Data (BigData Congress), Honolulu, HI, USA, 25–30 June 2017; pp. 557–564.
24. Mengelkamp, E.; Gärttner, J.; Rock, K.; Kessler, S.; Orsini, L.; Weinhardt, C. Designing microgrid energy markets: A case study: The Brooklyn Microgrid. *Appl. Energy* **2018**, *210*, 870–880. [CrossRef]
25. Sikorski, J.J.; Haughton, J.; Kraft, M. Blockchain technology in the chemical industry: Machine-to-machine electricity market. *Appl. Energy* **2017**, *195*, 234–246. [CrossRef]
26. Luo, F.; Dong, Z.Y.; Liang, G.; Murata, J.; Xu, Z. A Distributed Electricity Trading System in Active Distribution Networks Based on Multi-Agent Coalition and Blockchain. *IEEE Trans. Power Syst.* **2019**, *34*, 4097–4108. [CrossRef]
27. Kang, J.; Yu, R.; Huang, X.; Maharjan, S.; Zhang, Y.; Hossain, E. Enabling Localized Peer-to-Peer Electricity Trading Among Plug-in Hybrid Electric Vehicles Using Consortium Blockchains. *IEEE Trans. Ind. Informatics* **2017**, *13*, 3154–3164. [CrossRef]
28. Su, Z.; Wang, Y.; Xu, Q.; Fei, M.; Tian, Y.-C.; Zhang, N. A Secure Charging Scheme for Electric Vehicles with Smart Communities in Energy Blockchain. *IEEE Internet Things J.* **2019**, *6*, 4601–4613. [CrossRef]

29. Aitzhan, N.Z.; Svetinovic, D. Security and Privacy in Decentralized Energy Trading Through Multi-Signatures, Blockchain and Anonymous Messaging Streams. *IEEE Trans. Dependable Secur. Comput.* **2016**, *15*, 840–852. [CrossRef]
30. Morstyn, T.; Teytelboym, A.; McCulloch, M. Bilateral Contract Networks for Peer-to-Peer Energy Trading. *IEEE Trans. Smart Grid* **2019**, *10*, 2026–2035. [CrossRef]
31. Chen, T.; Su, W. Indirect Customer-to-Customer Energy Trading With Reinforcement Learning. *IEEE Trans. Smart Grid* **2019**, *10*, 4338–4348. [CrossRef]
32. Álvaro-Hermana, R.; Ardanuy, J.F.; Zufiria, P.J.; Knapen, L.; Janssens, D. Peer to Peer Energy Trading with Electric Vehicles. *IEEE Intell. Transp. Syst. Mag.* **2016**, *8*, 33–44. [CrossRef]
33. Rana, R.; Bhattacharjee, S.; Mishra, S. Energy trading framework for electric vehicles: An assignment matching-theoretic game. *IET Smart Grid* **2019**, *2*, 371–380. [CrossRef]
34. Li, J.; Zhang, C.; Xu, Z.; Wang, J.; Zhao, J.; Zhang, Y.J.A. Distributed transactive energy trading framework in distribution networks. *IEEE Trans. Power Syst.* **2018**, *33*, 7215–7227. [CrossRef]
35. Guerrero, J.; Chapman, A.C.; Verbič, G. Decentralized P2P Energy Trading Under Network Constraints in a Low-Voltage Network. *IEEE Trans. Smart Grid* **2019**, *10*, 5163–5173. [CrossRef]
36. Di Silvestre, M.L.; Gallo, P.; Ippolito, M.G.; Sanseverino, E.R.; Zizzo, G. A Technical Approach to the Energy Blockchain in Microgrids. *IEEE Trans. Ind. Inform.* **2018**, *14*, 4792–4803. [CrossRef]
37. Ghorani, R.; Farzin, H.; Fotuhi-Firuzabad, M.; Wang, F. Market design for integration of renewables into transactive energy systems. *IET Renew. Power Gener.* **2019**, *13*, 2502–2511. [CrossRef]
38. Morstyn, T.; McCulloch, M. Multiclass Energy Management for Peer-to-Peer Energy Trading Driven by Prosumer Preferences. *IEEE Trans. Power Syst.* **2019**, *34*, 4005–4014. [CrossRef]
39. Gai, K.; Wu, Y.; Zhu, L.; Qiu, M.; Shen, M. Privacy-Preserving Energy Trading Using Consortium Blockchain in Smart Grid. *IEEE Trans. Ind. Inform.* **2019**, *15*, 3548–3558. [CrossRef]
40. Qiu, T.; Liu, J.; Si, W.; Wu, D.O. Robustness Optimization Scheme with Multi-Population Co-Evolution for Scale-Free Wireless Sensor Networks. *IEEE ACM Trans. Netw.* **2019**, *27*, 1028–1042. [CrossRef]
41. Singh, K.; Yadav, V.K.; Padhy, N.P.; Sharma, J. Congestion Management Considering Optimal Placement of Distributed Generator in Deregulated Power System Networks. *Electr. Power Compon. Syst.* **2014**, *42*, 13–22. [CrossRef]

sustainability

MDPI

Article

Potential Renewable Hydrogen from Curtailed Electricity to Decarbonize ASEAN's Emissions: Policy Implications

Han Phoumin [1,*], Fukunari Kimura [1,2] and Jun Arima [1,3]

[1] Economic Research Institute for ASEAN and East Asia (ERIA), Think Tank, Jakarta 10270, Indonesia; vzf02302@nifty.ne.jp (F.K.); junarima@g.ecc.u-tokyo.ac.jp (J.A.)
[2] Faculty of Economics, Keio University, Tokyo 108-8345, Japan
[3] Graduate School of Public Policy, Tokyo University, Tokyo 113-0033, Japan
* Correspondence: han.phoumin@eria.org

Received: 19 November 2020; Accepted: 5 December 2020; Published: 17 December 2020

Abstract: The power generation mix of the Association of Southeast Asian Nations (ASEAN) is dominated by fossil fuels, which accounted for almost 80% in 2017 and are expected to account for 82% in 2050 if the region does not transition to cleaner energy systems. Solar and wind power are the most abundant energy resources but contribute negligibly to the power mix. Investors in solar or wind farms face high risks from electricity curtailment if surplus electricity is not used. Employing the policy scenario analysis of the energy outlook modelling results, this paper examines the potential scalability of renewable hydrogen production from curtailed electricity in scenarios of high share of variable renewable energy in the power generation mix. The study found that ASEAN has high potential in developing renewable hydrogen production from curtailed electricity. The study further found that the falling cost of renewable hydrogen production could be a game changer to upscaling the large-scale hydrogen production in ASEAN through policy support. The results implied a future role of renewable hydrogen in energy transition to decarbonize ASEAN's emissions.

Keywords: energy transition; renewables; hydrogen; fossil fuels; emissions

1. Introduction

The economic, social, and political dynamics of the Association of Southeast Asian Nations (ASEAN) have made it one of the fastest-growing regions. However, Southeast Asia faces great challenges in matching its energy demand with sustainable energy supply as the region transitions to a lower-carbon economy. The transition requires development and deployment of green energy sources. Growing energy demand can be met by energy supply produced by renewables and other clean energy alternatives such as hydrogen and by clean technologies [1]. Whilst Organization for Economic Co-operation and Development (OECD) countries have quickly reduced greenhouse gas emissions in response to the commitments of the Paris Climate Conference or the 21st Conference of the Parties (COP 21), developing Asia has some way to go to balance economic growth and affordable and available energy. Much of the future energy mix of emerging ASEAN countries will rely on fossil fuel to power economic development. However, they can follow a renewable energy path to economic growth, social well-being, and environmental sustainability.

The power generation mix of the Association of Southeast Asian Nations (ASEAN) is dominated by fossil fuels, which accounted for almost 80% in 2017 and are expected to account for 82% in 2050 if the region does not transition to cleaner energy systems [2].

Reducing greenhouse gas emissions is high on the global agenda under COP 21 and the upcoming COP 26, which will require leaders to pursue alternative fuel pathways, shifting from fossil fuel–based to clean energy systems. In this regard, hydrogen fuel represents growth potential as world leaders start to see the great benefit and promise of its use to abate climate change. In many ASEAN countries, hydrogen as an alternative fuel is not yet on the policy agenda. The ASEAN Plan of Action for Energy Cooperation (APAEC) Phase 2, however, will include policy measures to encourage emerging and alternative technologies such as hydrogen and energy storage.

The potential use of hydrogen in transport, power generation, and industry has been proven by projects around the world. Renewable hydrogen has attracted leaders' attention as an option to increase the share of renewables in electrical grids amidst the falling cost of renewable electricity from wind and solar energy. The International Renewable Energy Agency (IRENA) [3] predicted that the cost of electrolyzers, the devices used to produce hydrogen from water, will halve from US$840 now to US$420 per kW by 2040. Renewable hydrogen production could be the cheapest energy option in the foreseeable future. The cost-competitiveness of producing renewable H_2 is key for the wide adoption of hydrogen. Renewable H_2 production costs dropped drastically from US$10–US$15/kg in 2010 to US$4–US$6/kg in 2020 [4]. Costs are expected to decrease to US$2.00–US$2.50/kg of H_2 in 2030, which is competitive with hydrogen production using natural gas through steam methane reforming with carbon capture, sequestration, and storage (CCS) [5].

Hydrogen is a clean energy carrier and can be stored and transported for use in hydrogen-run vehicles, synthetic fuels, upgrading of oil and/or biomass, ammonia and/or fertilizer production, metal refining, heating, and other end uses. Developing hydrogen, therefore, is an ideal pathway to sustainable clean energy systems and can help scale renewables such as solar and wind energy. Adopting renewable hydrogen would bring more renewables into the energy mix and could be a game changer in the transition from fossil dependence to a cleaner energy system in ASEAN. Hydrogen could help integrate the current electricity system with wind and solar energy. Solar and wind penetration of the electrical grid is hindered by the high intermittency of electricity from wind and solar energy, and many grid operators in ASEAN are, therefore, hesitant to include a large share of it.

The Economic Research Institute for ASEAN and East Asia's research on hydrogen energy since 2017 has identified the significant potential of hydrogen energy supply and demand in East Asia. By 2040, the cost of hydrogen will decrease by more than 50% if it is adopted in all sectors. The target price of US$2.00–US$2.50/kg of H_2 in 2040 is competitive with the price of gasoline. The cost of supplying hydrogen is about 3–5 times higher than that of gas, mainly due to limited investment in hydrogen supply chains and the lack of a strategy to widely adopt hydrogen usage. The wide adoption and usage of hydrogen will need time to ensure cost-competitiveness and safety, especially for automobiles. The large-scale hydrogen-based energy transition from 'grey' and 'blue' to 'green' hydrogen will happen concurrently with a global shift to renewables. 'Green' hydrogen can face current system integration challenges that have blocked increasing the share of wind and solar energy.

In ASEAN, Brunei Darussalam leads in the hydrogen supply chain and has supplied liquefied hydrogen from Muara port to Japan since late 2019 [6]. However, the liquefied hydrogen process consumes a great deal of energy to cool gaseous hydrogen into liquid hydrogen at temperatures of −253 degrees Celsius and lower. The hydrogen supply chain demonstration project, in cooperation with Japan's government, explored an alternative way of shipping hydrogen using a new technology called liquid organic hydrogen carrier. If the technology is economically viable, it will pave the way for market access worldwide and overcome hydrogen supply chain barriers.

In many ASEAN countries, hydrogen is not yet on the policy agenda as an alternative fuel. However, APAEC, which is under preparation for endorsement at the ASEAN Ministers on Energy Meeting in November 2020, will include policy measures to promote emerging and alternative technologies such as hydrogen and energy storage [7]. APAEC will help AMS increase their adoption of hydrogen to enlarge the share of hydrogen in the energy mix.

The study investigates the potential of renewable hydrogen as a clean energy source for ASEAN's energy mix, which will need huge investment in hydrogen energy–related industries. The paper aims to do the following:

1. Use energy modelling scenarios to explore policy options of increasing the share of renewables, particularly wind and solar energy, in the power mix, and explore the possibility of electricity curtailment resulting from the high share of renewables that can be converted to hydrogen production.
2. Estimate the potential emission abatement resulting from the introduction of hydrogen produced using curtailed renewable electricity.
3. Review scalable renewable electricity from wind and solar energy from a cost reduction perspective, considering global experience.
4. Review technologies and cost perspectives of hydrogen produced using curtailed electricity.
5. Review a hydrogen policy and road map that can be applied to ASEAN.

Hydrogen adoption and development could be highly beneficial for ASEAN. Renewable hydrogen will enable the deployment of variable renewable energy (VRE) such as wind and solar and will be a game changer by breaking the barrier of integrated traditional power systems, which cannot absorb a high share of wind and solar energy. The paper is organized as follows: Section 2 reviews the pathways of hydrogen production processes; Section 3 explains the methodological approaches; Section 4 discusses the study's results; and Section 5 draws conclusions and policy implications.

2. Selected Pathways of Hydrogen Production Processes

Hydrogen emits zero emissions when used in combustion for heat and energy. If pure hydrogen (H_2) combusts by reacting with oxygen (O_2), it will form water (H_2O) and release energy that can be used as heat, in thermodynamics, and for thermal efficiency. Hydrogen is the most abundant chemical substance in the universe, but it is rarely found in pure form (H_2) because it is lighter than air and rises into the atmosphere. Hydrogen is found as part of compounds such as water and biomass and in fossil fuels such as coal, gas, and oil [8]. Several ongoing researches use two processes to extract hydrogen fuel: steam methane reforming, mainly applied to extract hydrogen from fossil fuels, and electrolysis of water, applied to extract hydrogen from water using electricity.

Steam methane reforming extracts hydrogen from methane using high-temperature steam (700–1000 °C). The product of steam methane reforming is hydrogen, carbon monoxide, and a small amount of carbon dioxide [9]. Most hydrogen is produced through this process, which is the most mature technology. Given how cheap natural gas is in the US and other parts of the world, hydrogen is one pathway to transition to a cleaner economy if steam methane reforming can be augmented with CCS. Technically, the chemical reaction process can be written as follows.

Steam methane reforming reaction (heat must be supplied through an endothermic process):

$$CH_4 + H_2O\ (+heat) \rightarrow CO + 3H_2, \tag{1}$$

Applying water-gas shift reaction (1) produces more hydrogen:

$$CO + H_2O \rightarrow CO_2 + H_2(+small\ amount\ of\ heat), \tag{2}$$

At this stage, carbon dioxide and other impurities are removed from the gas stream, so the final product is pure hydrogen.

Instead of steam methane reforming, partial oxidation can be applied to methane gas to produce hydrogen. However, the partial oxidation reaction produces less hydrogen fuel than does steam methane reforming. Technically, partial oxidation is an exothermic process, producing carbon monoxide and hydrogen and giving off heat:

$$CH_4 + \frac{1}{2}O_2 \rightarrow CO + 2H_2\ (+heat), \tag{3}$$

Applying a water-gas shift reaction in (3) produces more hydrogen:

$$CO + H_2O \rightarrow CO_2 + H_2(+small\ amount\ of\ heat), \tag{4}$$

Electrolysis can produce hydrogen by splitting water into hydrogen and oxygen in an electrolyzer, which consists of an anode and a cathode. Electrolyzers may have slightly different functions depending on the electrolyte material used for electrolysis.

The polymer electrolyte membrane (PEM) electrolyzer is an electrochemical device to convert electricity and water into hydrogen and oxygen. The PEM electrolyte is solid plastic. The half reaction that takes place on the anode side forms oxygen, protons, and electrons:

$$2H_2O \rightarrow O_2 + 4H^+ + 4e^-, \tag{5}$$

The electrons flow through the external circuit and the hydrogen ions move across the PEM to the cathode, in which hydrogen ions combine with electrons from the external circuit to form hydrogen gases:

$$4H^+ + 4e^- \rightarrow 2H_2, \tag{6}$$

PEM electrical efficiency is about 80% in terms of hydrogen produced per unit of electricity used to drive the reaction. PEM efficiency is expected to reach 86% before 2030.

Another method is alkaline water electrolysis, which takes place in an alkaline electrolyzer with alkaline water (pH > 7) with an electrolyte solution of potassium hydroxide (KOH) or sodium hydroxide (NaOH). In the alkaline electrolyzer, the two electrodes are separated. Hydroxide ions (OH−) are transported through the electrolyte from cathode to anode, with hydrogen generated on the cathode side. This method has been commercially available for many years, and the new method of using solid alkaline exchange membrane is promising as it is working in a laboratory environment.

3. Methodology and Scenario Assumptions

Hydrogen is used mainly to produce petrochemicals and ammonia. The potential of hydrogen, however, clearly remains untapped in ASEAN countries because it is a clean energy carrier that can be produced from various sources using fossil fuel and renewable energy. To build a hydrogen society, the cost of producing hydrogen must be competitive with that of conventional fuels, such as gas, for transport and power generation.

Renewable or 'green' hydrogen must be produced using renewable electricity from wind, solar, hydropower, and geothermal energy. Excess electricity from nuclear power, however, could be used to produce hydrogen as nuclear power plants provide base-load power and cannot be easily ramped up and down. During low demand, electricity from nuclear energy and VRE could be used to produce hydrogen. To produce renewable hydrogen using VRE, it is important to know the predicted available curtailed electricity resulting from power system integration challenges due to higher share of renewables.

Two components determine the cost to produce 'green' hydrogen: electricity cost from renewables and the cost of electrolysis. If these costs could be reduced significantly to allow the cost of hydrogen production to be competitive with that of natural gas, then hydrogen adoption and usage could be accelerated. This study reviews the falling cost of VRE and electrolysis to see how their current and future cost could allow a competitive hydrogen production cost. High VRE penetration of the electrical grid is the biggest challenge for the grid operator as electricity from VRE is variable and intermittent. Upgrading the grid system with the Internet of Things to create a smart grid could allow more penetration by VRE; otherwise, VRE electricity would be greatly curtailed due to a weak power grid system. This study calculates potential renewable hydrogen production and potential emission abatement under various scenarios assuming the following:

- Under current grid system integration, curtailment is likely to be 20–30% if the VRE share in the power mix exceeds more than 10%. Given the large potential of hydropower, geothermal, wind, and solar energy, increasing the share of renewables is technically possible using hydrogen storage. The study assumes the following scenarios: replacement by renewables of total combined fossil fuel generation (coal, oil, and gas) by 10%, 20%, and 30% by 2050, or, Scenario1 = 10%, Scenario2 = 20%, and Scenario3 = 30%.
- In Scenario1, Scenario2, and Scenario3, renewable hydrogen production using curtailed electricity is calculated based on assumptions of curtailed electricity generated from renewables at the rate of 20–30% of total generation from renewables. Potential renewable hydrogen produced using curtailed electricity in Scenario1, Scenario2, and Scenario3 is expressed as Scenario1H_2, Scenario2H_2, and Scenario3H_2.
- The formulas to calculate potential renewable hydrogen production in the renewable scenarios are as follows:
    - Scenario1H_2 (Mt-H_2) = [Scenario1 (TWh) × (Percentage of curtailed electricity)/48 (TWh)].
    - Scenario2H_2 (Mt-H_2) = [Scenario2 (TWh) × (Percentage of curtailed electricity)/48 (TWh)].
    - Scenario3H_2 (Mt-H_2) = [Scenario3 (TWh) × (Percentage of curtailed electricity)/48 (TWh)].

Mt-H_2 stands for million tonnes of hydrogen; TWh is terawatt-hour; and percentage of curtailed electricity is 20–30% of total generation from renewables. The study also applies the conversion factor of 48 kilowatt-hours (kWh) of electricity needed to produce 1 kg H_2 [10].

The potential emission abatement is the difference between (a) the business as usual (BAU) scenario and (b) the alternative policy scenario (APS) and other high-renewable-share scenarios such as Senario1, Scenario2, and Scenario3.

To estimate potential hydrogen produced using curtailed electricity, the power generation mix for the BAU and APS is estimated using ASEAN countries' energy models by applying the Long-range Energy Alternative Planning System (LEAP) software, an accounting system to project energy balance tables based on final energy consumption and energy input and/or output in the transformation sector. The LEAP software has been chosen in this study to estimate the future demand in power generation mix because the input of energy data provided by experts from ASEAN member states adopted their energy demand and supply modelling based on the LEAP modelling structure. Thus, the forecast of power generation demand is based on energy demand equations by energy and sector and future macroeconomic assumptions.

In the modelling work applying LEAP, the baseline of 10 AMS was 2017, the real energy data available in 2017, which are the latest that the study employed. Projected demand growth is based on government policies, population, economic growth, and other key variables, such as energy prices used by the International Energy Agency energy demand model [11]. BAU is in line with current energy policy in the baseline information, which is used to predict future energy demand growth. However, APS differs from BAU in policy changes and targets, with a greater share of renewables, including possible nuclear uptake based on an alternative policy for energy sources and more efficient power generation and energy in final energy consumption.

For electricity generation, experts from 10 AMS specified assumptions based on their national power development plans and used the assumptions to predict ASEAN's power generation mix. For renewable hydrogen production, the study applies a conversion factor of 48 kWh of electricity needed to produce 1 kg of hydrogen [10].

4. Results and Discussion

The potential of renewable hydrogen produced using curtailed electricity in Scenario1, Scenario2, and Scenario3 is quantified according to a renewable curtailment rate of 20–30% for the high share of renewables in 2050. Emission abatement—the difference between (i) BAU and (ii) APS, Scenario1,

Scenario2, and Scenario3—is calculated. The higher share of renewables under Scenario1, Scenario2, and Scenario3 could only happen if hydrogen is developed as an energy storage by utilizing curtailed renewable electricity. The study discusses hydrogen as an enabler of higher shares of renewables, the need to reduce the cost of renewable hydrogen production by reducing the cost of electrolysis and renewables, and the need to develop a hydrogen road map for ASEAN to guide industry and key investors in renewable hydrogen development. The road map will help create a large-scale ASEAN hydrogen society.

4.1. Potential Renewable Hydrogen from Curtailed Electricity

ASEAN's power generation is dominated by fossil fuel (coal, oil, and gas), the share of which in the power mix was 79% (equivalent to 1041 TWh) in 2017 and is predicted to be 82% (2826 TWh) and 72% (2087 TWh) in 2050 for BAU and APS, respectively (Figure 1). The share of combined fossil fuel (coal, oil, and gas) in the power generation mix is expected to reduce drastically from 82% in BAU to 65%, 58%, and 51% in Scenario1, Scenario2, and Scenario3, respectively, in 2050 (Figure 2). The share of combined renewables is expected to increase from 18% in BAU to 35%, 42%, and 49% in Scenario1, Scenario2, and Scenario3, respectively, in 2050. The higher share of renewables in the power generation mix is desirable to decarbonize emissions in ASEAN's future energy system. However, the high share of renewables can only happen with bold policy actions to develop and deploy renewable hydrogen to support the power integration system, which has a higher penetration of renewables. Utilizing unused electricity and/or curtailed renewable electricity to produce hydrogen could be ideal to tap the maximum potential of renewables.

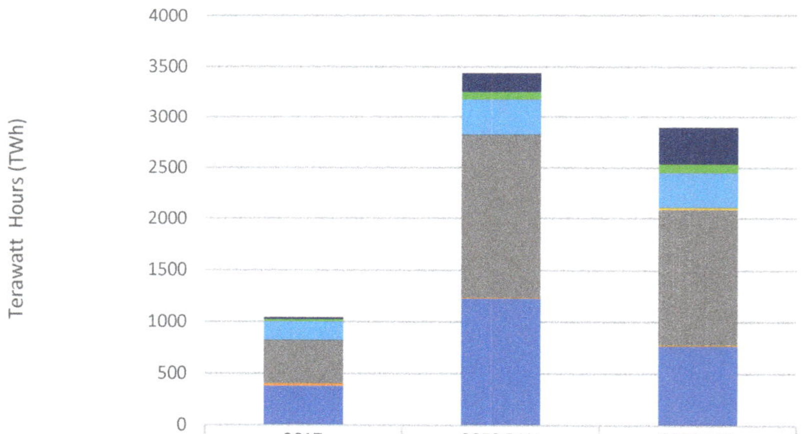

	2017	2050 BAU	2050 APS
Solar, Wind, Biomass	14	185	356
Geothermal	23	73	86
Hydro	183	356	344
Nuclear	0	0	21
Natural gas	414	1582	1303
Oil	26	12	12
Coal	381	1232	772

Figure 1. ASEAN's Power Generation Mix in Business as Usual and Alternative Policy Scenario by Source. BAU = business as usual, APS = alternative policy scenario. Source: Authors.

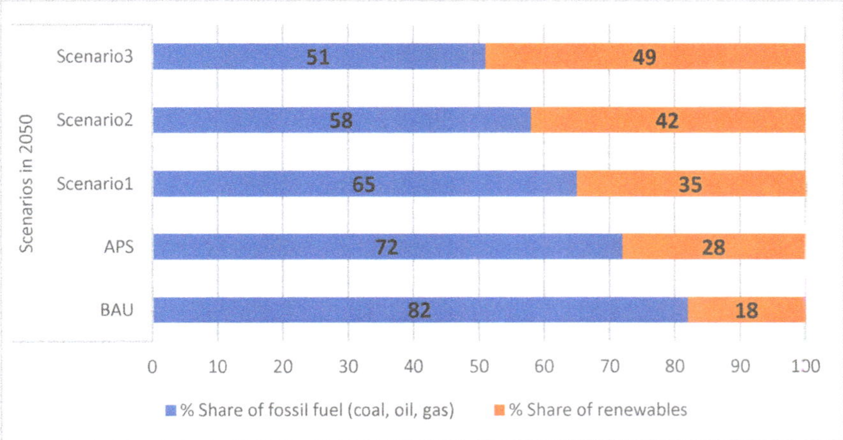

Figure 2. Share of Combined Fossil Fuels (coal, oil and gas) vs Renewables under Various Scenarios. APS= alternative policy scenario, BAU = business as usual. Note: Scenario1, Scenario2, and Scenario3 envision replacing combined fossil fuel (coal, oil, and natural gas) power generation with renewables (mainly variable renewable energy) at 10%, 20%, and 30%, respectively, in 2050. Source: Authors.

Scenario1, Scenario2, and Scenario3 assume the replacement of combined fossil fuel (coal, oil, and gas) power generation in 2050 with 10%, 20%, and 30% of power generation from renewables. Renewable power generation amounts in 2050 are 1016 TWh, 1224 TWh, and 1433 TWh for Scenario1, Scenario2, and Scenario3, respectively (Table 1).

Table 1. ASEAN's Power Generation Mix under Various Scenarios of Share of Renewables (TWh).

	2050 APS	Replacement of Coal, Oil, and Natural Gas by Renewables		
		Scenario1 = 10%	Scenario2 = 20%	Scenario3 = 30%
Coal	772	698.8	618	540
Oil	12	11	10	8
Natural gas	1303	1173	1042	912
Renewables (wind, solar, hydro, geothermal, and/or biomass)	807	1016	1224	1433

APS = alternative policy scenario. Note: Scenario1, Scenario2, and Scenario3 envision replacing combined fossil fuel (coal, oil, and natural gas) power generation with renewables (mainly variable renewable energy) at 10%, 20%, and 30%, respectively, in 2050. Source: Authors.

In Scenario1, Scenario2, and Scenario3, the shares of renewables in the power mix will be 35%, 42%, and 49%, respectively, in 2050. Because of higher shares of renewables in the power mix, renewable energy generation will be highly curtailed. The curtailed electricity rate could vary from 20% to 30%, depending on the power grid infrastructure in AMS. Based on this curtailed electricity, with varying shares of renewables in Scenario1, Scenario2, and Scneario3, hydrogen production scenarios are created— Scenario1H_2, Scenario2H_2, and Scenario3H_2. Potential renewable hydrogen from curtailed electricity in scenarios in AMS range from 4.23 to 8.96 million tonnes hydrogen (Table 2).

Table 2. ASEAN's Potential Renewable Hydrogen (H_2) from Curtailed Electricity.

Hydrogen Production	Potential Renewable Hydrogen Production		
	Scenario1H_2 (Million Tonnes H_2)	Scenario2H_2 (Million Tonnes H_2)	Scenario3H_2 (Million Tonnes H_2)
Of 20% curtailed renewables	4.23	5.10	5.97
Of 30% curtailed renewables	6.35	7.65	8.96

H_2 = hydrogen, Scenario1H_2 = hydrogen production in Scenario1, Scenario2H_2 = hydrogen production in Scenario2, Scenario3H_2 = hydrogen production in Scenario3. Note: 20–30% curtailed electricity applied for combined renewable power generation in 2050. The study applied a conversion factor of 48 kilowatt-hours (kWh) of electricity needed to produce 1 kilogram (kg) H_2 [10]; 1 kg of H_2 could generate 33.3 kWh [10]. Source: Authors.

The higher share of renewables under various scenarios such as APS, Scenario1, Scenario2, and Scenario3 will see a large reduction in carbon dioxide emissions (CO_2), which could result in decarbonizing emissions and contribute to COP commitments. Potential emission abatement ranges from −340 million tonnes carbon (Mt-C) in APS to −648 Mt-C, −710 Mt-C, and −774 Mt-C in Scenario1, Scenario2, and Scenario3, respectively (Table 3). Emissions were cut by 28% from BAU to APS, 53% from BAU to Scenario1, 58% from BAU to Scenario2, and 64% from BAU to Scenario3.

Table 3. Potential Emission Reduction under Various Scenarios (Mt-C).

	2017	2050	2050	2050
	Baseline	Emissions under Various Scenarios	Emission Abatement Potential	% Emission Reduction from BAU
BAU	376	1216		
APS	376	876	−340	28%
Scenario1	376	568	−648	53%
Scenario2	376	506	−710	58%
Scenario3	376	442	−774	64%

APS = alternative policy scenario, BAU = business as usual, Mt-C = million tonnes carbon. Note: Emission abatement potential is change of emissions from BAU to APS and other scenarios in 2050 under high renewables in Scenario1, Scenario2, and Scenario3. Source: Authors.

4.2. Hydrogen, an Enabler to Scale up Variable Renewable Energy

In ASEAN, power generation is dominated by coal, gas, and hydropower. Intermittent renewables from solar and wind energy contributed a negligible amount (14.47 TWh) or about 1.4% in 2017. However, the most optimistic prediction is that ASEAN will increase the share of wind and solar energy in the power generation mix to about 12.3% by 2050 (calculated from Figure 1). The inclusion of the share of hydro (17.6%) and geothermal (2.2%) energy in the power generation mix contributed to the overall renewable share of 21.2% in 2017. However, future abundant resources are wind and solar energy, the current share of which is negligible. Grid operators had many misperceptions of VRE such as wind and solar energy, although its production cost has drastically dropped in recent years; solar photovoltaic farms' levelized cost of electricity (LCOE) dropped from US$0.378/kWh in 2010 to US$0.043/kWh in 2020 in some places [12]. Similarly, all LCOE cost trends for wind energy and concentrated solar power dropped drastically in 2010–2020 and will continue to drop in 2021 (Figure 3), but their share in the power generation mix remains small. Misperceptions stemmed from the concern that VRE production is variable and intermittent, and that its higher share in the grid will add costs as it will require backup capacity from conventional gas power plants [12].

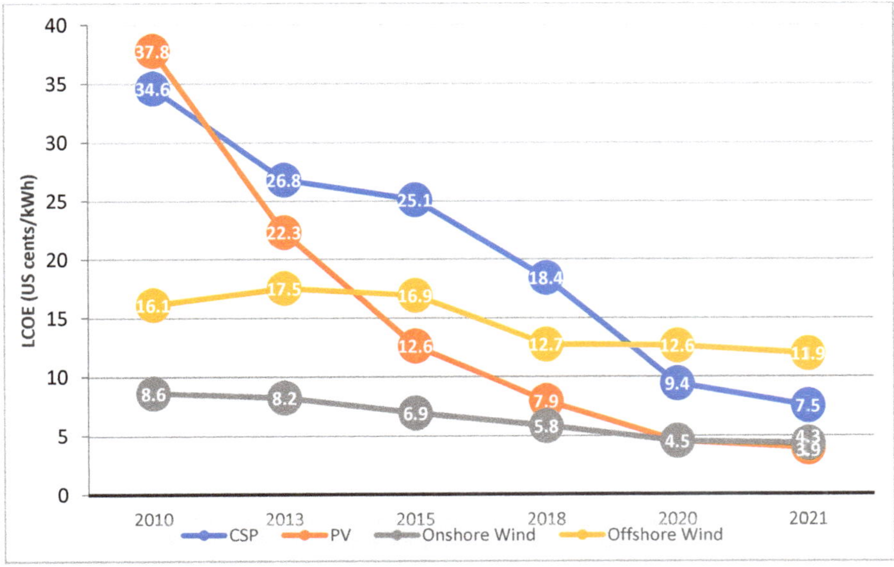

Figure 3. Falling Costs of Renewables in terms of Levelized Cost of Electricity (LCOE). CSP = concentrated solar power, kWh = kilowatt-hour, LCOE = levelized cost of electricity, PV = photovoltaic. Source: IRENA [13].

Technically, VRE power production output varies within a few seconds depending on wind or sunshine. However, the risk of variable energy output can be minimized if the power system is largely integrated within the country and within the region. The aggregation of output from solar and wind energy from different locations has a smoothing effect on net variability [12]. However, the ASEAN power grid is progressing slowly, and the integrated ASEAN power market might be far off because of several reasons, such as regulatory and technical harmonization issues within ASEAN power grids and utilities.

Scalable electricity production from wind and solar energy faces tremendous challenges from the current practice of system integration in ASEAN. Investors in solar or wind farms will confront high risks from electricity curtailment if surplus electricity is not used. Many countries have advanced research and technologies for battery storage (lithium-ion batteries) for surplus electricity produced from wind and solar energy, but advanced battery storage remains costly. Produced from electrolysis using surplus electricity, hydrogen has many advantages as it can be stored as liquid gas, which is suitable for numerous uses and easy to transport. Many ASEAN countries could produce wind, solar, hydropower, or geothermal electricity. Their resources, however, are far from demand centers and developing the resources would require large investments in undersea transmission cables. A solution would be to turn renewables into easily shipped hydrogen.

Hydrogen is a potential game changer for decarbonizing emissions, especially in sectors where they are hard to abate, such as cement and steel. Scalable resources from wind and solar energy and other renewables can be fully developed by widely adopting the hydrogen solution. The more electricity produced from wind and solar energy, the higher the penetration by renewables of the grid; at the same time, surplus electricity during low demand hours can be used to produce hydrogen. The more power generated from wind and solar energy and other renewables, the greater the possibility to increase the efficiency of electrolysis to produce hydrogen. On-site hydrogen production from wind and solar farms will solve the issue of curtailed wind and solar electricity. To increase the efficiency of electrolysis and allow further penetration by renewables of grids, a hybrid energy system including hydropower, geothermal, or nuclear plants, for example, would be the perfect energy choice. Since hydrogen is a clean energy carrier and can be stored and transported for use in, amongst others, hydrogen vehicles, synthetic fuels, upgrading of oil and/or biomass, ammonia and/or fertilizer production, metal refining, heating, and other end uses, hydrogen development is an ideal pathway to a sustainable clean energy system and enables scalable VRE such as solar and wind energy.

4.3. Need to Reduce Renewable Hydrogen Production Cost

Cost-competitiveness of producing renewable hydrogen is key for the wide adoption of hydrogen uses. The upfront costs of renewable hydrogen such as electrolyzers, transport infrastructure, and storage, and the varying costs of electricity tariffs are key factors contributing to the high production cost of renewable hydrogen (Figure 4). 'Green' hydrogen production costs dropped drastically from US$10–US$15/kg of H_2 in 2010 to US$4–US$6/kg of H_2 in 2020, with varying assumptions of lower and higher upfront costs of electrolyzers with 20 MW and producing capacity of 4000 normal cubic meters per hour [3,13]. The costs are expected to reduce to US$2.00–US$2.60/kg of H_2 in 2030, which is competitive with steam methane reforming with CCS.

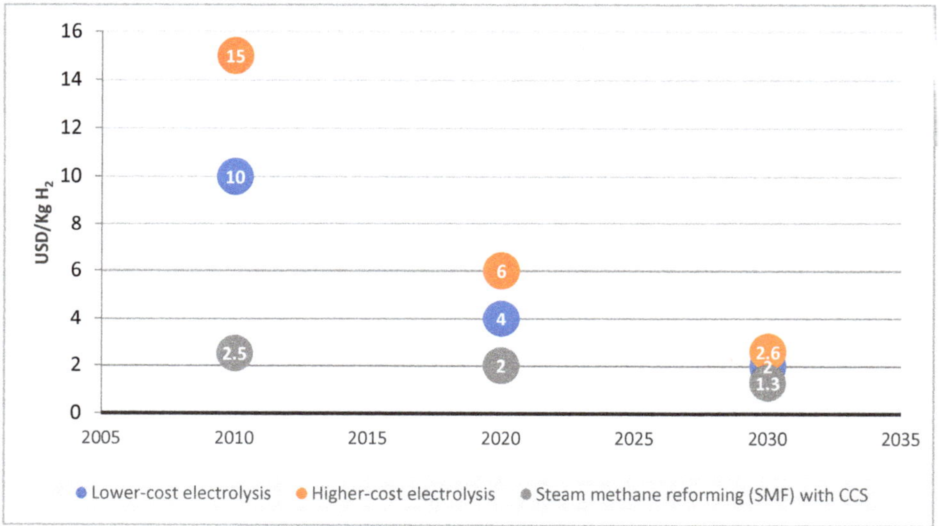

Figure 4. Hydrogen Production Cost Trends with Upfront Cost of Electrolyzers. CCS = carbon capture, sequestration, and storage, SMF = steam methane reforming. H_2 = hydrogen. Note: Assumption: 4000–normal cubic meter per hour (20 MW) polymer electrolyte membrane electrolyzers connected to offshore wind. The lower-cost electrolysis case is US$200/kilowatt (kW). The middle-cost electrolysis case is US$400/kW. The higher-cost electrolysis case is US$600/kW. Source: Authors, based on Hydrogen Council [4], DOE [14], and IRENA [15].

Considering the electricity tariffs of up to US$0.10/kWh with varying load factors of 10–50%, the cost of producing hydrogen ranged from US$0.90–US$5.50/kg of H_2 to US$4.20–US$8.90/kg of H_2 (Figure 5), meaning that electricity tariff is the major cost of producing hydrogen using electrolysis. At zero electricity tariff or when VRE is expected to be curtailed, the cost of producing hydrogen can be as low as US$0.90/kg of H_2 at an electrolyzer's load factor of 50%, and US$5.50/kg of H_2 at an electrolyzer's load factor of 10%. The International Renewable Energy Agency's target of cost-competitiveness of producing renewable hydrogen is US$2.00–US$2.50/kg of H_2 [16]. In this case, an electricity tariff of US$0.03/kWh with an electrolyzer's load factor of 30% is the most practical given all the constraints.

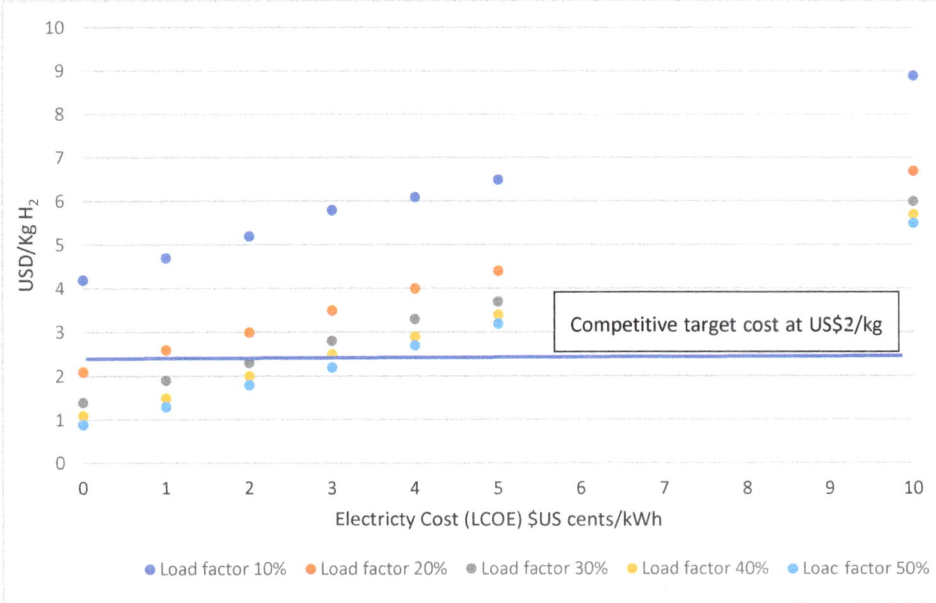

Figure 5. Hydrogen Production Cost with Varying Electricity Cost and Electrolysis Load Factors. H_2 = hydrogen, kWh = kilowatt-hour, LCOE = levelized cost of electricity. Note: Assumption: The polymer electrolyte membrane electrolyzer is connected with the grid. Source: Authors, based on Hydrogen Council [4], DOE [14], and IRENA [15].

The solar photovoltaic farm and onshore wind already cost US$0.02–US$0.03/kWh in some locations [16]. Even the target cost of US$2.00–2.50/kg of H_2 to produce 'green' hydrogen, however, would not be competitive with low-cost natural gas at US$5 per gigajoule (GJ) (Conversion factor: US$0.01/kWh = US$2.80/GJ.) (US$0.018/kWh), but would be with natural gas, which costs US$10–US$16/GJ (US$0.036–US$0.057/kWh).

Technically, if renewable hydrogen production uses only curtailed electricity from renewables, the operating load factor of electrolysis, which contributes the most to the cost of producing hydrogen, will likely be low at 10% or less. According to the Hydrogen Council [4], the electrolyzer will need to run at a load factor of at least 30% or more to lower the cost of producing hydrogen to US$2.00–2.50/kg of H_2, which is competitive with the natural gas grid price.

Electrolysis facilities must have a load factor above 30% to ensure the cost-competitiveness of producing renewable hydrogen, and other capital expenditures such as the electrolyzer's upfront cost must be reduced by 50% from US$840 today to US$420 per kilowatt by 2040. As wind and solar energy is expected to increase its share in the power generation mix, expected curtailed electricity from renewables will be higher by 10–30%. By 2030, the share of VRE curtailment will be 10–30% in Sweden, which provides the most incentives for renewable hydrogen [16]. In 2020, Chile, Australia, and Saudi Arabia have achieved the target cost of US$2.50/kg to produce 'green' hydrogen because of cheap access to electricity from wind and solar energy. The cost is expected to drop further to US$1.90/kg in 2025 and to US$1.20/kg in 2030, which is highly competitive with the cost of 'grey' hydrogen production.

Effective policies and incentives to develop and adopt hydrogen can promote economies of scale and cost-competitiveness in producing hydrogen, encouraging investors to manufacture electrolyzers; improve their efficiency, operation, and maintenance; and use low-cost renewable power such as hydrogen to enable scaling VRE penetration of the power grid. 'Green' hydrogen production cost could decline even faster and go even lower than US$2/kg of H_2 if governments, business, and stakeholders join hands to adopt the wider use of 'green' hydrogen and increase investment and R&D in hydrogen fuels. Australia, Chile, and Saudi Arabia have achieved cost-competitiveness in wind and solar energy generation.

The energy transition will largely depend on the clean use of fossil fuel leading to a clean energy future. Although hydrogen is a clean fuel, the way it is produced matters. Almost 95% of hydrogen production is from natural gas with or without CCS. The gasification of coal can be used as feedstock for producing hydrogen, but it emits roughly four times more CO_2/kg of H_2 produced than natural gas feedstock does. The production cost of low-carbon 'blue' hydrogen depends on feedstock cost and suitable geographical CCS storage. IRENA (2019a) estimated that 'blue' hydrogen production in China and Austria with current CCS infrastructure could realize a production cost of about US$2.10/kg of H_2 for a cost of coal of about US$60 per ton. In the US, where natural gas is below US$3 per million British thermal units and has large-scale CO_2 storage such as depleted gas fields and suitable rock formations, 'blue' hydrogen cost could drop below US$1.50/kg in some locations. If the carbon cost of about US$50 per ton of CO_2 is considered, low-carbon hydrogen could reach parity with 'grey' hydrogen. 'Blue' hydrogen cost in the US and the Middle East could drop further to about US$1.20/kg in 2025 if economies of scale prevail.

World leaders need to provide a clear policy to develop and adopt hydrogen. The right policy will enable economies of scale for producing hydrogen cost-competitively, inducing investors to explore electrolyzer manufacturing; improve electrolyzer efficiency, operation, and maintenance; and use low-cost renewable power. With the full participation of governments, business, and stakeholders, hydrogen can become the fuel that enables scaling up renewable energy penetration in all sectors, decarbonizing global emissions.

4.4. Need for Renewable Hydrogen Development Policies in ASEAN

Until 2020, ASEAN did not have a hydrogen road map. APAEC, however, mentions alternative technologies and clean fuels such as hydrogen and energy storage. APAEC will help AMS increase the share of hydrogen in the energy mix. An ASEAN hydrogen road map is needed to guide national road maps. Based on the analysis of the drastic drop in the cost of VRE and electrolyzers, opportunities to introduce 'green' hydrogen produced using curtailed electricity will be plentiful. The hydrogen road map should include hydrogen development and penetration in transport, power generation, and industry. To guide investment, hydrogen penetration policies and targets must be set up. This study, however, can only suggest policies to develop, adopt, and use hydrogen. The study adopts Australia's hydrogen road map, especially its key polices [16,17], and tailors them to ASEAN's energy landscape. In developing the ASEAN's hydrogen roadmap, four key policies are highlighted below.

The first policy area is on financing which aims to provide access to lower-cost financing for hydrogen development and low-emission projects. In this regard, the government in ASEAN may need to consider providing fiscal policy incentives for local manufacturing for hydrogen development and financing incentives for low-emission electricity.

Another policy is on regulations which aims to set up targeted policies to stimulate hydrogen demand. In this regard, the government in ASEAN may need to consider developing hydrogen-specific regulations across AMS to support hydrogen development in power generation, transport, and industry. The regulation should allow grid-firming services from electrolyzers to be compensated, and allows for on-site hydrogen production and, where possible, position plants close to where the hydrogen will be used. Furthermore, the gas pipeline regulations should be reviewed to consider including gaseous hydrogen.

Thirdly, the policy is on research and development (R&D) which aims to establish demonstration projects for mature hydrogen technologies. The government in ASEAN should also consider setting up a hydrogen center of excellence as a research body to bring in all parties to work on technologies and policy coordination. The center should also conduct research and development in plant efficiency and safety, and in hydrogen shipment, pipeline, and storage.

The fourth policy target is on social acceptance which aims to develop a public engagement plan and strategy to support clean fuels such as hydrogen and ensures that communities understand all aspects of its use. The social acceptance is key for promoting willingness to pay for clean fuels.

To promote the future hydrogen society, ASEAN will have to develop a comprehensive hydrogen road map that includes a policy framework supporting hydrogen production, storage, and transport. The policy framework also needs to support hydrogen utilization in power generation, transport, heat production, industrial feedstock, and import and export. In developing the hydrogen road map, the governments in the countries of ASEAN should consult industrial, financial, and banking stakeholders. The road map will need to cultivate people's willingness to support a hydrogen society.

5. Conclusions and Policy Implications

ASEAN's energy transition will largely depend on increasing the share of renewables and clean fuels such as hydrogen and the clean use of fossil fuel to create a clean energy future. Fossil fuel (coal, oil, and gas) accounted for almost 80% of ASEAN's energy mix in 2017, a share that is expected to rise to 82% in BAU. Transitioning from a fossil fuel–based energy system to a clean energy system requires drastic policy changes to encourage embracing renewables and clean fuels whilst accelerating the use of clean technologies in employing fossil fuel (coal, oil, and natural gas). The study used energy modelling scenarios to explore policy options to abate emissions in ASEAN by giving wind and solar energy a high share of the energy mix and using electricity curtailment to promote renewable hydrogen production. The findings found that ASEAN has high potential to produce renewable hydrogen using curtailed electricity. The higher share of renewables under various policy scenarios will see a large reduction in CO_2 emissions, which could lead to decarbonizing emissions and contribute to abating global climate change. The potential emission abatement ranges from −340 Mt-C in APS to −648 Mt-C, −710 Mt-C, and −774 Mt-C in Scenario1, Scenario2, and Scenario3, respectively. Emissions will be cut by 28% from BAU to APS, 53% from BAU to Scenario1, 58% from BAU to Scenario2, and 64% from BAU to Scenario3.

The results of the study imply policy implications for ASEAN's energy policy reforms to ensure that clean fuels such as hydrogen and renewables and clean technologies will have a big role to play to decarbonize ASEAN's emissions. The below policy implications are derived from this study for the hydrogen adoption in ASEAN:

- ASEAN leaders must strongly commit to promoting a hydrogen society. ASEAN Ministers on Energy Meetings, facilitated by the ASEAN Secretariat, are an excellent platform for drafting a clear and actionable hydrogen development road map.
- ASEAN energy leaders must develop a clear strategy to promote hydrogen use in transport; power generation; and other sectors where emissions are hard to abate, such as the iron and steel industries. Singapore, Malaysia, Thailand, Indonesia, and the Philippines could take the lead by investing in R&D on hydrogen produced from renewables and non-renewables and by setting investment targets adapted from OECD countries. Investment in industries that can adopt hydrogen energy has strong potential, but to realize it ASEAN must accelerate its plans and strategies to embrace hydrogen use.
- Leaders in ASEAN and around the world must provide a clear investment policy to develop and adopt hydrogen as a fuel. The policy must enable economies of scale in cost-competitive production of hydrogen to induce investors to consider electrolyzer manufacturing; improvements in electrolyzer efficiency, operation, and maintenance; and the use of low-cost renewable power. With the full participation of governments, business, and stakeholders, hydrogen can become the fuel that enables scaling up renewable energy penetration in all sectors, decarbonizing global emissions.
- Governments must engage the public, build its awareness of the many benefits of a hydrogen society, and ensure that the public is willing to pay for them. The success of introducing hydrogen on a large scale needs the participation of all stakeholders, including governments and public

and private companies. Financing mechanisms such as banks must create favorable conditions to finance facilities such as electrolyzers. Governments must provide financial incentives to invest in developing hydrogen.

- Improving the electricity governance system in ASEAN developing countries will help reduce the cost of managing energy systems, allow the uptake of clean energy technology investment, and upgrade the grid system to bring in more renewables. The energy sector must be reformed; rules and procedures must allow more advanced and competitive technologies to enter the market. Electricity reform will attract foreign investment to modernize electricity infrastructure, including by making power systems more efficient and phasing out inefficient power generation and technologies.
- Unbundling of ownership in the electricity market, non-discriminatory third-party access to transmission and distribution networks, and the gradual removal of subsidies for fossil fuel–based power generation are key to ensure market competition. Other policies to attract foreign investment include tax holidays; reduction of market barriers and regulatory burdens; and plans to reduce the upfront cost investment, such as a rebate payment system through government subsidies and government guarantees that investment will be feasible and low risk.

Author Contributions: Conceptualization, F.K. and J.A.; methodology, H.P.; software, H.P.; validation, H.P., F.K. and J.A.; formal analysis, H.P.; investigation, H.P.; resources, H.P.; data curation, H.P.; writing—original draft preparation, H.P.; writing—review and editing, H.P.; visualization, H.P.; supervision, H.P.; project administration, H.P.; funding acquisition, H.P. All authors have read and agreed to the published version of the manuscript.

Funding: This research received no external funding.

Conflicts of Interest: The authors declare no conflict of interest.

References

1. Hydrogen the Missing Piece of ASEAN Green Energy. Available online: https://www.thejakartapost.com/academia/2020/07/30/hydrogen-the-missing-piece-of-asean-green-energy.html (accessed on 15 October 2020).
2. *Energy Outlook and Saving Potential in the East Asia*; Economic Research Institute for ASEAN and East Asia: Jakarta, Indonesia, 2019. Available online: https://www.eria.org/publications/energy-outlook-and-energy-saving-potential-in-east-asia-2019/ (accessed on 5 May 2020).
3. *Hydrogen from Renewable Power: Technology Outlook for the Energy Transition*; International Renewable Energy Agency (IRENA): Abu Dhabi, UAE, 2018. Available online: https://irena.org/-/media/Files/IRENA/Agency/Publication/2018/Sep/IRENA_Hydrogen_from_renewable_power_2018.pdf (accessed on 10 June 2020).
4. Path to Hydrogen Competitiveness: A Cost Perspective. Hydrogen Council. 2020. Available online: https://hydrogencouncil.com/wp-content/uploads/2020/01/Path-to-Hydrogen-Competitiveness_Full-Study-1.pdf (accessed on 10 June 2020).
5. ERIA. *Demand and Supply Potential of Hydrogen Energy in East Asia*; Kimura, S., Li, Y., Eds.; Economic Research Institute for ASEAN and East Asia: Jakarta, Indonesia, 2019; Available online: https://www.eria.org/publications/demand-and-supply-potential-of-hydrogen-energy-in-east-asia/ (accessed on 10 June 2020).
6. Hydrogen: A Game-Changer for ASEAN. Available online: https://borneobulletin.com.bn/2020/07/hydrogen-a-game-changer-for-asean/ (accessed on 16 October 2020).
7. Zamora, C.G.; Asean Centre for Energy, Jakarta, Indonesia; Han, P.; Economic Research Institute for ASEAN and East Asia, Jakarta, Indonesia. Personal communication, 2019.
8. *Hydrogen Production and Distribution: Alternative Fuel Data*; US Department of Energy [DOE]: Washington, DC, USA, 2020. Available online: https://afdc.energy.gov/fuels/hydrogen_production.html (accessed on 26 June 2020).
9. *Hydrogen Production: Natural Gas Reforming*; Energy [DOE]: Washington, DC, USA, 2020. Available online: https://www.energy.gov/eere/fuelcells/hydrogen-production-natural-gas-reforming (accessed on 26 June 2020).
10. Renewable Transformation Challenge: Hydrogen Production from Renewables. International Solar Energy Society (ISES). 2020. Available online: http://www.renewableenergyfocus.com/view/3157/hydrogen-production-from-renewables/ (accessed on 27 September 2020).

11. *World Energy Model*; International Energy Agency [IEA]: Paris, France, 2020. Available online: https://www.iea.org/reports/world-energy-model/documentation (accessed on 5 September 2020).
12. Wind and Solar on the Power Grid: Myths and Misperceptions. National Renewable Energy Laboratory (NREL). 2020. Available online: https://www.nrel.gov/docs/fy15osti/63045.pdf (accessed on 1 July 2020).
13. *How Falling Costs Make Renewables a Cost-Effective Investment*; The International Renewable Energy Agency (IRENA): Abu Dhabi, UAE, 2020. Available online: https://www.irena.org/newsroom/articles/2020/Jun/How-Falling-Costs-Make-Renewables-a-Cost-effective-Investment (accessed on 20 June 2020).
14. DOE Hydrogen and Fuel Cells Program Record. United States (US) Department of Energy (DOE). 2014. Available online: https://www.hydrogen.energy.gov/pdfs/18003_current_status_hydrogen_delivery_dispensing_costs.pdf (accessed on 14 July 2020).
15. Renewable Power-to-Hydrogen: Innovation Landscape Brief. Abu Dhabi. 2019. Available online: https://www.irena.org/-/media/Files/IRENA/Agency/Publication/2019/Sep/IRENA_Power-to-Hydrogen_Innovation_2019.pdf?la=en&hash=C166B06F4B4D95AA05C67DAB4DE8E2934C79858D (accessed on 20 June 2020).
16. IRENA. *Hydrogen: A Renewable Energy Perspective*; IRENA: Abu Dhabi, UAE, 2019; Available online: https://www.irena.org/-/media/Files/IRENA/Agency/Publication/2019/Sep/IRENA_Hydrogen_2019.pdf (accessed on 20 June 2020).
17. Bruce, S.; Temminghoff, M.; Hayward, J.; Schmidt, E.; Munnings, C.; Palfreyman, D.; Hartley, P. *National Hydrogen Roadmap*; Commonwealth Scientific and Industrial Research Organisation: Canberra, Australia, 2018; Available online: file:///C:/Users/h.phoumin/Downloads/18-00314_EN_NationalHydrogenRoadmap_WEB_180823.pdf (accessed on 27 April 2020).

Publisher's Note: MDPI stays neutral with regard to jurisdictional claims in published maps and institutional affiliations.

sustainability

MDPI

Article

Reassessing the Environmental Kuznets Curve in Relation to Energy Efficiency and Economic Growth

Jie Zhang [1], Majed Alharthi [2], Qaiser Abbas [3,*], Weiqing Li [4,*], Muhammad Mohsin [5], Khan Jamal [6] and Farhad Taghizadeh-Hesary [7]

1 School of Economics and Resource Management, Beijing Normal University, Beijing 100875, China; zhangjie0190@hotmail.com
2 Finance Department, College of Business, King Abdulaziz University, P.O. BOX 344, Rabigh 21911, Saudi Arabia; mdalharthi@kau.edu.sa
3 Department of Economics, Ghazi University, Dera Ghazi Khan 32200, Pakistan
4 School of International Economics and Tourism Management, Zhejiang International Studies University, Hangzhou 310023, China
5 School of Finance and Economics, Jiangsu University, Zhenjiang 212013, China; m.mohsin3801@yahoo.com
6 Institute for Region and Urban-Rural Development, Wuhan University, Wuhan 430072, China; jamalkhan_87@yahoo.com
7 Social Science Research Institute, Tokai University, Hiratsuka 259–1292, Kanagawa, Japan; farhad@tsc.u-tokai.ac.jp
* Correspondence: Qabbas@gudgk.edu.pk (Q.A.); 88924381@163.com (W.L.)

Received: 22 August 2020; Accepted: 29 September 2020; Published: 10 October 2020

Abstract: Energy consumption and its efficiency are significant factors for economic growth and environmental stress. This study postulates the occurrence of the Environmental Kuznets Curve hypothesis (EKC) by using the Autoregressive-Distributed Lag (ARDL) model. Furthermore, a data envelopment analysis (DEA) model is used to measure energy efficiency, energy intensity, and environment to view the trajectory of EKC for the underline economies. For this purpose, a panel dataset from 1990–2013 of 15 developing countries is analyzed to verify the objectives mentioned above. The results of the panel ARDL support EKC's theory for underline economies, as GDP positively impacts carbon emissions, while the square of GDP is negatively related. The DEA-based results found relatively low environmental conditions in these emerging economies due to high energy intensity and low energy efficiency. This outcome suggests that renewable energy sources must be treated as an essential factor for achieving sustainable economic goals without environmental degradation.

Keywords: environmental Kuznets curve; CO_2 emission; energy efficiency; economic growth; panel ARDL; DEA

---

1. Introduction

Climate change is a devastating phenomenon that people have experienced for the last few decades. Excessive greenhouse gases (GHGs) in the atmosphere, specifically nitrous oxide (N_2O), carbon dioxide (CO_2), methane (CH_4), and chlorofluorocarbons (CFC), are the major causes of global climate change [1,2]. This phenomenon will cause a dramatic change in our world in the coming years, as greenhouse gases absorb heat from the sun and capture it in the atmosphere, causing the Earth's surface temperature to rise. The fifth assessment report of the International Panel on Climate Change (IPCC) concluded that the climate system's human impact is visible [3]. After the Industrial Revolution, population growth and economic development have led to an increase in greenhouse gas emissions and climate because of human interference on the Earth. Among the several greenhouse gases that cause global climate change, carbon dioxide is the most abundant [4]. Its excess may be directly associated with human activities, such as burning fossil fuels, transportation, deforestation, land reclamation, and cement production for agricultural purposes, which increased after the industrial revolution.

In this modern age, enhanced development is the key to the progress of each nation. Developing nations focus on their growth process to boost productivity and grow early. These nations are also attempting to improve their living standards by raising their per capita incomes, and this is being made possible with the help of enhanced growth and development. Emerging economies such as developing Asian nations are now relying on industrialization for their rapid development. In this regard, these developing Asian economies rely on the rapid productivity energy cycle to attain the desired economic upswing. The environment is being affected by CO_2 emissions due to the energy process, as industrialization had led to environmental degradation. Thus, the swift economic upswing gives rise to environmental depletion in developing economies [5]. These Asian economies have been trying to achieve the desired economic upswing for the last two decades but have failed to clean their natural environment. Inadequate policies and limited resources fail to coordinate with the harmful environment, which is causing ecological disorder. Therefore, the developing Asian nations have compromised their environmental conditions and focus on their per capita incomes to increase the so-called standard of living. Therefore, it can be rightly claimed that ecological disorder is rising with the growing rate of economic upswing, which is reflected in the Environmental Kuznets Curve (EKC) [6].

The world energy consumption crossed BTU 583.57 quadrillions in 2017, at a 2.27 percent annual growth rate, against BTU 381.49 quadrillions in 1998. The share of underline developing economies crossed BTU 213.96 quadrillions in 2017, with thirty-seven percent of the world's energy consumption [7]. Thus, energy consumption has increased considerably in these developing economies, with swift economic growth as the desired output. However, environmental degradation has a peak off as the undesired output due to dirty energy sources such as fossil fuel. Most of the energy sources of the underline countries are import-based, which is hurting their economic progress and exchange rate. The world is turning the sources of energy into renewable ones, with 570.96 million tons of renewable energy consumed. However, the share of developing economies is less than twenty-five percent [8].

The current COVID-19 condition has changed the economic and energy scenario. Renewable energy projects have been delayed in developing economies to meet the current financial requirements. The oil price fluctuations during COVID-19 and its impact on the exchange rate have opened a new debate on energy efficiency and economic growth. In this case, the occurrence of EKC theory and its smooth trajectory is debatable, as economic growth is not the only independent factor responsible for it.

There were nearly 20.06 billion units of gross fixed capital formation worldwide in 2018. The underline economies hold 7.77 billion, which is nearly thirty-nine percent of the total [9]. Unfortunately, these developing economies accumulated capital, which is not a technological advance and is thus less efficient in production. It is not only a source of high per-unit cost but can also be a source of environmental degradation due to the high usage of fossil fuel energy, which is imported and dirty.

Developing Asian nations have specific problems, like poverty, unemployment, and a high population growth rate with a low per capita income growth rate. That is why these nations try to boost their development to resolve these issues. However, due to the scarcity of resources and outdated technology, it is difficult to control environmental degradation. In per capita terms, China is responsible for 7.95 metric tons of emissions in 2018 compared to 2.69 metric tons of emissions in 1999, at an annual growth rate of 5.97 percent [10]. Likewise, Malaysia faced 8.02 metric tons of emissions in 2013 compared to 7.76 metric tons in 2010. Finally, Mongolia recorded 14.54 metric tons of emissions in 2013 compared to 9.09 metric tons in 2010. Mongolia is considered the largest carbon dioxide emitter among developing countries in the Asian region [11]. Energy is regarded as the primary source of development, but it is essential not to consider improper planning, scarce resources, and outdated technology. Therefore, to control CO_2 emissions, it is essential to apply the policies about utilizing energy sources. Governments and policymakers need better strategies to use energy efficiently to boost economic activity and control carbon dioxide emissions, especially in developing economies [12].

The current work includes the data of those developing Asian nations that share common social issues. They are also on the same page regarding geographical, financial, political, and ecological circumstances with a higher population growth rate. They are considered highly ranked as carbon dioxide emitters since the 1990s due to their will to become industrialized as quickly as possible. The panel Autoregressive-Distributed Lag (ARDL) and Data Envelopment Analysis (DEA) techniques were utilized to analyze the under-considered data for the following objectives:

1. Reassessing the occurrence of EKC by observing the influence of economic upswing on carbon emission evolving Asian economies.
2. Calculating the association between energy (renewable-renewable) and carbon emission.
3. Examining the role of capital formation in CO_2 emissions and finally providing a policy to overcome the environmental challenges emerging due to high carbon emission.
4. Measuring energy efficiency, energy intensity, and environmental conditions of the underlined economies by using DEA.

The analysis is organized as follows: the remainder of the introduction depicts the literature review, section two describes the materials and methods, results are provided in section three, discussions are in section four, and the conclusion is given in section five.

The Review of Literature

Research is conducted to evaluate the correlation between environmental issues and macroeconomic variables in recent decades. Numerous studies regarding this issue reviewed to verify the influence of energy usage and economic upswing on the developing economies' ecological disorder. Most of the studies focused on a panel of mixed-income nations such as upper income, middle-income, and lower-middle-income; few studies focused on regions such as the Association of Southeast Asian Nations (ASEAN) [13], Organization for Economic Co-operation and Development (OECD), and Organization of the Petroleum Exporting Countries (OPEC) [14]. In contrast, the panel of developing nations from the overall world was also a part of this literature [15], while Ref. [16] verified EKC was part of the developing one belt one road initiative. This literature review showed that energy usage and economic upswing positively correlate with the ecological disorder [17].

The Environmental Kuznets Curve hypothesis asserts a definite link among population growth, GDP, energy usage, and carbon emission. The positive association among these variables verified a specific rise in carbon dioxide emission when economies are developing. As the developing economies are in the development phase, EKC's presence witnessed and confirmed the constructive outcome of an economic upswing on the ecological disorder [18]. The authors [19] also proved the existence of the EKC hypothesis in thirteen nations by analyzing the association of eighteen economic indicators and ecological disorder. Moreover, population growth and economic uncertainty provide an essential answer to all the aforementioned variables for environmental depletion [20].

The authors of [21] verified the EKC theory's occurrence and demonstrated a negative effect of renewable energy on the environment. For the possible occurrence of EKC, environmental efficiency is very vital. It could be attained through energy efficiency, energy pricing, energy intensity, technological innovation, or building high-tech industries. The authors of [22] investigated the role of efficiency growth and convergence to enhance economic productivity using inputs and embrace technologies in 104 countries for a thirty-six-year dataset. This study found that environmental efficiency improved approximately 1.3 percent globally due to energy pricing, restructuring industrial setup, or globalization. The authors of [23] suggested that energy transactions can help to enhance economic and environmental efficiency. The transaction of energy in the different income level nations is different, which may change EKC's pattern or speed. The authors of [24] found that even global crises such as COVID-19 have changed the entire energy pricing mechanism and lead to the collapse of the energy market and the competitiveness of renewable energy projects. Likewise, to view the EKC occurrence, the technological innovations in emission reduction and carbon transfer strategies based on the low carbon preference are deemed necessary. Ref. [25] suggested that low carbon preference can be an excellent source to improve environmental conditions without compromising economic growth. Some of the researchers attempted to verify the notion of EKC through capital formation. For example, Ref. [26] found that capital formation is a source of environmental degradation in G-7 countries. Therefore, the role of energy consumption, fossil fuel or renewable energy preferences, energy innovations to enhance the energy efficiency, or reduce its intensity, innovations for high-tech industrialization are a few core indicators to decide EKC's time and speed.

The existing literature is divided into two different aspects. Many attempted to view the EKC occurrence in developing or developed. For example, Ref. [27–29] analyzed the EKC theory without considering EKC's trajectory in routine or emergency conditions. Other studies like [30–32] attempted to measure the economic, energy, and environmental efficiency via indexing the variables of said field. These studies did nothing for EKC theory and its speed of occurrence.

In conclusion, some studies presented the assenting linkages of an economic upswing with carbon emission, while others delivered a negative association between these variables. The same is the case for energy, carbon emission, and economic upswing. Much of the research work evidenced the EKC hypothesis and established a panel of developed and developing economies. However, only a few have tried to fix the three-dimensional energy effect on economic growth and environmental stress. Moreover, the EKC literature focused on GDP and the conversion of GDP square term but not considering the other variables, such as capital formation, growth rate, and renewable energy consumption. Although these two also have an independent effect on economic growth and environmental condition, they can play an essential role in EKC trajectory and speed. Some other studies attempted to measure energy efficiency in economic cost and environment [33,34]. However, they did not explain EKC to view the real impact of energy efficiency on economic growth and the environment. Thus, there is a gap for some comprehensive studies in these areas, especially from developing nations of Asia. These nations are suffering much in terms of ecological disorder, energy usage, and sustainable economic growth. This research attempts to cover these two different concepts. This study attempts to fill the literature gap regarding energy efficiency as the source of the EKC trajectory. Here, the combined effect of energy, economic, and environment underlined developing economies' analysis to understand the EKC trajectory and speed. The study's efficiency score indicates the current condition of energy efficiency, energy intensity, and environmental efficiency of the individual country based on the last twenty-three years of progress to depict the gap of EKC among underline nations. Thus, lower-middle-income and upper-middle-income countries have been selected to view their respective economic and environmental conditions with energy efficiency as per world bank classification. Therefore, the current study has novelty because of its sole combination of variables, two different angels of analysis, and the selection of nations from the Asian region concerning their income levels. This study can help policymakers, and business individuals decide the course of EKC occurrence and its trajectory for preferring the supportive sources of renewable energy with high efficiency and low intensity in their respective countries.

2. Materials and Methods

A twenty-three-year panel dataset of fifteen developing economies of Asia was taken from 1990 to 2013. The primary source of this dataset is "World Development Indicators." This dataset has been divided into two income categories classified by World bank 2021. Here, Nepal, Bangladesh, India, Mongolia, Pakistan, Philippine, Vietnam, and Sri Lanka sorted out as lower-middle-income economies. At the same time, China, Iran, Jordan, Malaysia, Thailand, Turkey, and Indonesia are upper-middle-income economies [33]. This study is based on two different methods of research. First, the ARDL method of econometrics utilizes the EKC theory of environment and economic growth. Secondly, the DEA method of operational research to assess the energy efficiency of underline countries.

Table 1 depicts the detail of the indicators used for this study. Here, ecological disorder (END) is the dependent variable, while renewable energy (ENC), economic growth (EGW), the square of economic growth (EGW^2), capital formation (FCF), and population growth (PG) are the independent variables.

Table 1. Variable description and source.

Description of Variables	Abbreviation	Unite	Source
Ecological disorder	END	Metric ton	WDI
Renewable energy	ENC	kg of oil equivalent	WDI
Economic Growth	EGW	GDP per capita	WDI
Square of Economic Growth	EGW^2	GDP-square per capita (real term)	WDI
Capital Formation	FCF	Annual growth rate	WDI
Population Growth	PG	Growth rate	WDI

2.1. Methodological Framework of ARDL

The model's technical specification is that the economic upswing and ecological disorder are positively associated in initial stages, whereas the square of GDP helps to reduce environmental depletion. The linear-quadratic equation confirms the presence of an inverted U-shaped EKC [34] and [35]. It can be written as:

$$END_{it} = \beta_0 + \beta_1 EN_{it} + \beta_2 EGW_{it} + \beta_3 EGW^2{}_{it} + \beta_4 FCF_{it} + \beta_5 PG_{it} + \mu_i \quad (1)$$

Equation (1) illustrates the linear quadratic equation to run the ARDL for the confirmation of EKC. Following the footprints of [36], Equation (2) establishes to assess the short-run ARDL results.

$$\begin{aligned} \Delta END_{it} = \beta_0 \; + \; & \sum_{i=1}^{k} \gamma_1 \Delta END_{i\,t-1} + \sum_{i=0}^{k} \alpha_1 \Delta END_{i\,t-1} + \sum_{i=0}^{k} \alpha_2 \Delta GW_{i\,t-1} \\ & + \sum_{i=0}^{k} \alpha_3 \Delta GW^2{}_{i\,t-1} + \sum_{i=0}^{k} \alpha_4 \Delta FCF_{i\,t-1} + \sum_{i=0}^{k} \alpha_5 \Delta PG_{i\,t-1} \\ & + \beta_1 ENC_{i\,t-1} + \beta_2 GW_{i\,t-1} + \beta_3 GW^2{}_{i\,t-1} + \beta_4 FCF_{i\,t-1} \\ & + \beta_5 PG_{i\,t-1} + \mu_{it} \end{aligned} \quad (2)$$

In the above equation, Δ represents the difference, whereas $t-1$ used for cross-section shows the model's previous years. The α and β are the coefficients of underline indicators. In the next step, the Error Correction Model (ECM) develops by formulating the following equation.

$$\begin{aligned} \Delta END_{it} = \; & \beta_0 + \textstyle\sum_{i=1}^{k} \gamma_1 \Delta END_{i\,t-1} + \sum_{i=1}^{k} \alpha_1 \Delta ENC_{i\,t-1} + \sum_{i=1}^{k} \alpha_2 \Delta GW_{i\,t-1} + \\ & \textstyle\sum_{i=1}^{k} \alpha_3 \Delta GW^2{}_{i\,t-1} + \sum_{i=1}^{k} \alpha_4 \Delta FCF_{i\,t-1} + \sum_{i=1}^{k} \alpha_5 \Delta PG_{i\,t-1} + \\ & \beta_1 ENC_{i\,t} + \beta_2 GW_{i\,t} + \beta_3 GW^2{}_{i\,t} + \beta_4 FCF_{i\,t} + \beta_5 PG_{i\,t} + \delta ECM_{i\,t} + \\ & u_{it} \end{aligned} \quad (3)$$

The coefficient of ECM' δ' demonstrates the speed of adjustment, and it should be with a negative sign to show the convergence towards the long run from the short run to achieve the equilibrium condition.

2.2. Hybrid Error Correction Model

The Error Correction Model might show an error correction of the first difference exclusively, which is as follows:

$$\Delta Y_t = Y_t - Y_{t-1} \quad (4)$$

Error Correction Model can use for quantitative computation, and it is necessary to point out that it is the base of the Auto Regressive Distributive Lag Model (ARDL). We have used the Error Correction Model to condition that, if the ARDL sum coefficient is equal to 1, by decreasing the constant terms. Consequently, the coefficient of error correction term long-run association can attain if and only if the transformation at term grows at the constant rate, N. Hence, the coefficient mathematical model Error Correction Model can be presented as:

$$Y_t = \beta_0 + \beta_1 Y_{t-1} + \beta_2 Z_t + \beta_3 Z_{t-1} + \vartheta_t \quad (5)$$

The proposed term Y_{t-1} is deducted from the ARDL both sides:

$$Y_t - Y_{t-1} = \beta_0 + \beta_1 Y_{t-1} + \beta_2 Z_t + \beta_3 Z_{t-1} - Y_{t-1} + \vartheta_t \quad (6)$$

$$\Delta Y_t = \beta_0 + \beta_1 Y_{t-1} + \beta_2 Z_t + \beta_3 Z_{t-1} - Y_{t-1} + \vartheta_t \quad (7)$$

Through addition and deducting $\beta_2 Z_{t-1}$ in the right-hand side of the mathematical model. The new equation is as follows:

$$\Delta Y_t = \beta_0 + \beta_1 Y_{t-1} + \beta_2 Z_t - \beta_2 Z_{t-1} + \beta_3 Z_{t-1} - Y_{t-1} + \beta_2 Z_{t-1} + \vartheta_t \quad (8)$$

or

$$\Delta Y_t = \beta_0 + (\beta_1 - 1) Y_{t-1} + \beta_2 Z_t + (\beta_2 + \beta_3) Z_{t-1} + \vartheta_t \quad (9)$$

To fulfill the condition of the Error Correction Model, the coefficient (Z_{t-1}) must be analogous to the deducted coefficient Y_{t-1}. So, the newly construed mathematical model is as follows:

$$\beta_1 - 1 = -(\beta_2 - \beta_3) \quad (10)$$

$$\beta_1 + \beta_2 + \beta_3 = 1 \quad (11)$$

Consequently, the term of error correction constant considers as:

$$\Delta Y_t = \beta_0 + \beta_2 Z_t - \tau(Y_{t-1} - Z_{t-1}) + \mu_t \quad (12)$$

$$\tau = -(\beta_1 - 1) = (\beta_2 + \beta_3) \quad (13)$$

If the variation in constant term increases at a continuous rate N, the association of the long-run phenomena is:

$$N = \beta_0 + \beta_2 N - \tau(y^* - Z^*) \quad (14)$$

$$\tau(y^* - Z^*) = \beta_0 + (\beta_2 - 1)N \quad (15)$$

$$y^* = \beta_0 + \frac{(\beta_2 - 1)N}{\tau} + Z^* \quad (16)$$

then the original order having and without having the value of log is considered as:

$$y_t^* = K Z_t^* \quad (17)$$

If we take the log from both sides, then it will be as:

$$Log y_t^* = Log K + log Z_t^* \quad (18)$$

Through using the anti-log of the new model, the long run will consider:

$$y^* = \exp\left[\frac{\beta_0 + (\beta_2 - 1)B}{\tau}\right] \quad (19)$$

where K represents the association between the variable Y and Z, the Error Correction Model is being used to measure the long-run relationship, and the Error Correction Model characterizes the previous imbalance in an existing factor. It can be:

$$\Delta N_t = \sum_{i=1}^{N} \tau_1 \Delta N_{t-1} + \tau_2 \Delta N_{t-1}\beta_2 N + YZ_t + \mu_t \tag{20}$$

2.3. Model Specification of DEA

DEA (Data Envelopment Analysis) is one of the many techniques for efficiency assessment. However, there are the following advantages to use the DEA method.

- Simultaneous analysis of outputs and inputs
- It is not necessary, a Priori, to define the frontier form
- Relative efficiency compared to the best observation
- Need no information on price

Let $V_k = (v_{ki}, \ldots, v_{kn})$ be the set of "n" environmental variables aimed at entity $k = 1, \ldots, K$. Environmental index (EVI) develops through underlying variables for each entity. Ranking the environmental performance of various entities is a general practice to develop an environmental index. It can differentiate with a choice of ordering ($\succcurlyeq$) defined on Rn. Therefore, the EVI can demonstrate through a mapping function such that $I:\ R^n \rightarrow\ R$, which satisfies

$$V_k \succcurlyeq V_l \Leftrightarrow I(V_k) \geq I(V_l)\forall k,\ l\ \in \{1, \ldots, K\} \tag{21}$$

The assessment of each fundamental variable, which can represent through the function of the transformation unit, may be improved $F = (f_1, \ldots,\ f_n)$ such that

$$F : (V_{k1}, \ldots, V_{kn}) \rightarrow (f_1(V_{k1}), \ldots,\ f_n(V_{kn})) \tag{22}$$

As pointed out by [37] and [38], an admissible transformation engages extension and translation in such a way that $f_i(vk_i) = \alpha_i v_{ki} + \beta_1,\ \alpha_i > 0$. In correspondence with EVI, the order of various underlying entities which are expected to be chosen as inconsistent and associated with any acceptable conversion and transformation of fundamental factors being assessed in construction of EVI.

$$V_k \succcurlyeq V_l \Leftrightarrow F(V_k) \geq F(V_l)\forall k,\ l\ \in \{1, \ldots, K\} \tag{23}$$

The geometric mean proved to choose with a useful index with strictly positive and ratio-scale variables—criteria for information loss in the direction of alternate combination techniques designed to develop indices [39]. The author of [40] used a non-compensatory approach of aggregation and discussed its usefulness. A nonparametric DEA methodology identifies a good frontier practice using a linear programming approach. It measures the comparative efficiency of underlying indicators based on outputs and inputs from comparable and measurable entities [41]. The DEA study by [42,43] used to measure the energy system performance, environmental performance, and productivity of different entities or decision-making units.

DEA's traditional use to measure environmental performance takes the difference between a good and a lousy output. For performance assessment, [42] introduced a fundamental academic foundation, which was the reason for the nonparametric DEA frontier practice's popularity to measure the wrong outputs. The vector $V_k = (v_{ki}, \ldots, v_{kn})$ is replaced by $X_k, \ldots, Y_k = X_{k1}, \ldots, X_{km}, Y_{k1}, \ldots, Y_{ks}$ (xk1,$\cdots$, xkm, yk1,$\cdots$, yks) to differentiate between inputs and outputs, where X_k and Y_k are input and output vectors, respectively. The input vector $X_k = (X_{k1},\ \ldots, X_{km})$ is used to produce the output vector

$Y_k. = (Y_{k1}, \ldots, Y_{ks})$. The inputs are $X \in R_+^p$ and the outputs are $Y \in R_+$. As a result, the production is the set of a potential combination of inputs and outputs:

$$
\begin{aligned}
S = \Big\{ (X,Y) : S = \sum_{k=1}^{K} x_{ik} z_k \le x_i,\ i = 1, \ldots, m \\
S = \sum_{k=1}^{K} y_{rk} z_k \le x_r,\ r = 1, \ldots, S \\
S = \sum_{k=1}^{K} z_k = 1\ i = 1, \ldots, m \\
z_k 0,\ k = 1, \ldots, K \}
\end{aligned}
\tag{24}
$$

In Equation (24) with constraint, the range-adjusted DEA model can be as follows:

$$
\begin{aligned}
\max \frac{1}{m+s} \left(\sum_{k=1}^{K} \frac{S_i^-}{R_i^-} + \sum_{k=1}^{K} \frac{S_r^+}{R_r^+} \right) \\
S = \sum_{k=1}^{K} x_{ik} z_k + S_i^- = x_{0i},\ i = 1, \ldots, m \\
S = \sum_{k=1}^{K} y_{rk} z_k - S_r^- = y_{0r},\ r = 1, \ldots, S \\
S = \sum_{k=1}^{K} z_k = 1\ i = 1, \ldots, m \\
z_k 0,\ S_i^- 0,\ S_r^- 0.
\end{aligned}
\tag{25}
$$

the x_{oi} is the i-th input and y_{or} is the r-th output for entity $o(o) \in \{1, \ldots, K\}; R_i^-$ and R_r^+ show the ranges for output r and input i, which can be defined as:

$$R_i^- = max\{x_{ki}, k = 1, \ldots, K\} - \min\{x_{ki}, k = 1, \ldots, K\}\mathrm{d}$$

$R_i^+ = max\{y_{ki}, k = 1, \ldots, K\} - \min\{y_{ki}, k = 1, \ldots, K\}$ The additive DEA model's objective function is the inefficiency measurement of the entity's slack-based values, which can use to measure energy efficiency. The constraints decide the maximum possible reduction form the maximum reduction and the recognized extension in inputs and outputs. A variable having zero to one shows that all the entities have zero value to exclude in EVI. It is necessary to separate the relevant constituent in the objective function (Equation (25) while an equivalent restraint is required. The final constraint $\sum_{k=1}^{K} z_k 1$ appears to be a convexity situation, ensuring that the ratio-scale measurement units do not vary in the objective function. Any permissible conversion for the original factors, $f_i(vk_i) = \alpha_i k_i + \beta_1$ major attention on slacks accommodate the shifting parameter and the scaling factor α_i can be controlled through the adjustment range. After getting the optimal solution form Equation (25), the environmental index can define as:

$$\mathrm{EI}(\mathrm{v_O}) = \mathrm{EI}(\mathrm{X_O}, \mathrm{Y_O}) = 1 - \frac{1}{m+s} \left(\sum_{k=1}^{K} \frac{\mathrm{S_i^{*-}}}{\mathrm{R_i^-}} + \sum_{k=1}^{K} \frac{\mathrm{S_r^{*+}}}{\mathrm{R_r^+}} \right) \tag{26}$$

* shows the variable of consistent optimum slack. The EI derived from Equation (26) satisfies the following properties [44]:

P1.0 ≤ EI ≥1;

P2.EI (V0) = 1 ⇔ Entity o is situated on the best practice frontier;

P3.EI (V0) is inconsistent with the measurement units of outputs and inputs;

P4.EI (V0) contains the properties of strongly monotonic;

P5.EI (V0) is a conversion invariant.

P1 represents that Equation (26) provides a standardized index between 0 and 1, while the higher values are associated with better performance.

P2 shows that the underlying entities are essential to developing the best frontier practice with index values less than 1. It can see from Equation (27) that the identification of underlying entities determines the frontier of best practice associated with non-zero Zk, which are determined by the employed model.

P3 shows that the EVI index values are invariant through the ratio of scale dimension variables.

P4 demonstrates that a decrease in any input or any output causes the highest index value.

P5 translates that the addition and subtraction of constants through any variables do not impact indexes' values, especially after the interval-scale factors encompassed in constructing EVI [45]. In Equation (27), the normalization of linear min-max is accepted, while the normalized weighted version of underlying factors for all the entities is not. Indeed, the mentioned practice makes Equation (27) feasible, resulting in the ease of assessing the environmental index. However, [46] highlighted that the weighted sum aggregation rule presupposed the full compensability among underlying variables that are fully replaceable with each other [47]. Since various dimensions of underlying variables are not entirely replaceable with each other, the assumption may not be appropriate for measuring an environmental efficiency index [48].

$$\begin{aligned} \mathrm{EVI}(\mathrm{V_k}) &= 1 - \tfrac{1}{m+s}\left[\sum_{i=1}^{m} \tfrac{x_{ki}-min_k\{x_{ki}\}}{R_i^-} + \sum_{r=1}^{S} \tfrac{min_k\{y_{kr}\}-y_{kr}}{R_{r+}}\right] \\ \mathrm{EVI}(\mathrm{V_k}) &= \sum_{i=1}^{m} \tfrac{1}{m+S}\left[\tfrac{\max\{x_{ki}\}-x_{ki}}{max_k\{x_{ki}-min_k\{x_{ki}\}}\right] + \tfrac{1}{m+S}\left[\sum_{r=1}^{S} \tfrac{y_{kr}-\min\{y_{ki}\}}{max_k\{y_{kr}\}-min_k\{x_{kr}\}}\right] \end{aligned} \tag{27}$$

For now, exact preference and equal weights do not provide insights and robust results. Due to the standard weights associated with each dimension, it is considerably hard to achieve a consensus. Therefore, this study provides an insight into the virtue of the nonparametric frontier method through an apprehensive environmental perspective. In contrast, commonly used inputs, such as capital and labor, may not be incorporated to construct the development of EVIs.

Generally, [49] excluded inputs to develop an environmental efficiency index since it generates per unit of lousy output. In this study, the wrong outputs are considered inputs as they both indicate the cost type, which implies that they follow the properties of "smaller the best." Considering the view, this study treats energy consumption as input and assumes that it would also follow the "smaller the best," considering the sustainable environment. Based on diversification indices that measure risk-free energy supplies by assuming that riskier energy supplies pose a more significant threat to energy security while at the same time reducing the energy security impact on energy efficiency and energy intensity. Therefore, risk-free energy supplies need to assess.

$$RIES - CR_i = HHI - CR_i \times DEP_i = D_i \sum_{i=1}^{N} W_{ij}{}^2 \times CR_j \tag{28}$$

$$RIES - CR_i = HHI - PE \times DEP_i = DEP_i \sum_{i=1}^{N} W_{ij}{}^2 \times \frac{1}{PE} \tag{29}$$

$$RIES - PE = HHI - PE \times DEP_i = DEP_i \sum_{i=1}^{N} W_{ij}{}^2 \times \frac{1}{PE} \times CR_i \tag{30}$$

The RIES represents the risk in energy supplies, *CR* is the country risk, *HHI* is the Herfindahl Hirschman Index, *DEP* is the energy dependency on energy suppliers, *PE* shows potential exports, $\mathrm{W_{IJ}} = \frac{\mathrm{X_{ij}}}{\sum \mathrm{X_{ij}}}\mathrm{X_{ij}}$ represents the contribution of energy suppliers in over-all energy imports of the economy. The fourth variable is a financial indicator, i.e., gross domestic product (GDP), which shows each country's capability to produce revenue against a specific amount of GHGs emissions. Out of these variables, GHG emissions and energy consumption are inputs, while total energy supplies and GDP are outputs. It is necessary to explain that Equations (6), (7), and (10) measure the energy efficiency, environmental index, and energy intensity, respectively, to analyze other countries or regions.

3. Results

This study aims to assess the existence of EKC theory in fifteen developing economies of Asia and attempts to probe the energy and environmental nexuses to view the trajectory of KEC.

3.1. Results of Panel ARDL

Table 2 represents the statistical summary of the dependent and independent variables of Equation (2). It shows that the values of minimum, maximum, kurtosis, standard deviation, average, and skewness of the indicator, as mentioned earlier, have reflected an improved understanding of the data and their distribution within the structure—the outcomes of the correlation matrix present in Table 3.

Table 2. Statistical summary.

	END	ENC	GW	GW^2	FCF	PG
Mean	2.47	923.66	3.66	27.86	12.48	1.62
Median	1.59	739.06	3.71	17.53	10.89	1.50
Maximum	14.54	3019.81	15.31	234.62	58.15	5.63
Minimum	0.03	115.70	−14.35	0.00	−48.21	0.14
Std. Dev.	2.30	665.90	3.80	33.26	15.95	0.79
Jarque-Bera	2.05	3.82	1.94	3.11	2.57	4.66
p-value	0.36	0.21	0.52	0.22	0.37	0.13

Source: authors own calculations.

Table 3. Correlation matrix.

	END	ENC	GW	GW^2	FCF	PG
END	1.00					
ENC	0.63	1.00				
GW	0.06	−0.01	1.00			
GW^2	−0.28	0.16	0.46	1.00		
FCF	0.00	−0.13	0.54	0.22	1.00	
PG	0.00	−0.06	0.35	0.36	0.14	1.00

Source: authors own calculations.

The correlation matrix results find that renewable energy usage (ENC) correlates with the ecological disorder (END). At the same time, all other concerned variables have a weak correlation with the ecological disorder. GDP-square variable (GW^2) shows the desired negative association with the ecological disorder. As the application of Panel ARDL apple concerning panel unit root, Table 4 represents the panel unit root results. According to the output, variables are stationary at different levels.

Table 4. Panel unit root.

Variables		Level I	Level I & T	1st difference I	1st difference I&T	Decision
END	LL & C	0.86	1.68	−1.67	−0.23	I(1)
		−0.8	−0.95	-0.04	−0.4	
		4.1	2.1	−5.73	−4.24	
	IPS	−1	−0.98	0	0	
ENC	LL & C	2.5	0.03	−3.19	−2.50	I(1)
		−0.99	−0.51	0	0	
		4.75	1.48	−5.22	−3.85	
	IPS	−1	−0.93	0	0	
GW	LL & C	−7.08	−6.58	–	–	I(0)
		0	0			
		−6.92	−5.77			
	IPS	0	0			
GW^2	LL & C	−6.06	−5.58	–	–	I(0)
		0	0			
		−6.86	−6.77			
	IPS	0	0			
FCF	LL & C	−4.21	−4.64	–	–	I(0)
		0	0			
		−6.40	−4.52			
	IPS	0	0			
PG	LL & C	0.34	−10.99	–	–	I(0)
		−0.63	0			
		10.75	−7.78			
	IPS	−1	0			

Source: authors own calculations. Note: Parentheses have Probability values.

To view the co-integration association between dependent and a set of independent variables of Equation (2), the study performs the bound test, depicted in Table 5, according to the bounds test results. F-statistics (estimated) is higher than the upper and lower critical value bunds. Thus, bounds test results accepted the co-integration of energy usage and environmental depletion alongside other demographic and economic indicators.

Table 5. Results of the bound test.

Equation (1)	Bound Test Value	Df	Conclusion
END/ENC, GW, GW^2, FCF, PG	F-statistics = 12.96 > 3.61 Probability = (0.00)	(6, 334) 6	Co-integration exists

Source: authors own calculations. Note: table CI cited the unrestricted intercept, and no critical trend values of Lower bound at 5% = 2.45 and unrestricted intercept and no trend critical values of Upper bound at 5% = 3.61.

Then, Panel ARDL applied to test the long-run influence of economic upswing on environmental depletion. The Panel ARDL projection for the long run prearranges in Table 6.

Table 6. Panel autoregressive-distributed lag (ARDL) (Long Run).

No. of Panels = 15 Dependent Variable = END				
Regressor	**Coefficients**	**Standard. Error**	**t-Statistics**	***p*-Value**
ENC	−0.26 ***	0.08	3.25	0.0012
GW	0.78 ***	0.33	2.36	0.0183
GW^2	−0.31 **	0.17	−1.82	0.0688
FCF	0.11 **	0.05	2.21	0.0271
PG	0.61	0.41	1.487	0.1370

Source: authors own calculations, Note: ** 5% and *** 1% show the statistical significance level.

The Panel ARDL findings confirmed EKC (inverted U-shape curve) for these selected developing Asian economies. Economic upswing (GW) also has a positive association in terms of ecological disorder and found that 0.78 units of CO_2 emissions are being generated by the economic upswing to pollute the environment. The results depict the positive influence of GDP growth on carbon dioxide emission in developing economies, evidenced by past research [50]. However, the confirmation of EKC proved by the coefficient value −0.31 of GDP-square (GW^2). The negative coefficient value of GW^2 represents the negative bond between GDP-square and carbon emission, which confirmed the reduction of carbon dioxide emission in the developing Asian economies. Thus, reducing carbon dioxide emissions due to improved economic upswing has shown the presence of inverted U-shaped EKC in developing economies and confirms the findings of [51,52].

It further finds that renewable energy usage participates negatively in carbon dioxide emissions in these developing economies. The renewable energy usage (ENC) coefficient is −0.26, which indicates that one percent of energy usage is a source of 0.26 carbon emission reduction emission. The coefficient of renewable energy usage is also significant at one percent. Hence, it proved that renewable energy usage growth helps reduce the pollution of these developing economies. The previous studies also establish the same affirmative influence of renewable energy usage on developing economies' carbon dioxide emissions [53,54].

Capital formation (FCF) has shown a role in terms of increasing CO_2 emissions. The results show that enhancement in capital formation has increased the environmental depletion in the developing economies. Results suggest that with ceteris paribus, a one percent increase in capital formation is a source of 0.11 percent of carbon dioxide emission, and [51,55] have shown the same evidence in their past study in which the improvement in capital formation was promoting the ecological disorder. The long-run panel ARDL does not significantly affect population growth (PG) on the ecological disorder. According to past research, population growth can be an essential indicator of any country's economy. However, it has no substantial evidence to affect the environmental conditions in the ecological disorder.

According to Table 7, the short-run results are insignificance for the economic upswing and environmental depletion in these selected developing economies. The short-run regressor consists of lag terms of the previous year. Almost all variables were found to be insignificant concerning the depletion of the environment in the preceding year. Thus, the short-run results confirmed that all economic (economic upswing, GDP-square, energy usage, fixed capital formation) and demographic indicators do not affect the carbon dioxide emission of these developing economies of Asia. This short-run analysis estimates through ECM. The coefficient value of ECMit-1 is 0.26, which shows convergence toward the long run from the short run to attain equilibrium condition. The significant negative value has proven the belongings of ECM to form the steadiness by dropping error. It estimates that the unsteadiness or errors are diminished by about 26 percent each year towards the long run from the short run, which helps attain equilibrium conditions among economic upswing, fixed capital formation, energy usage, and environmental depletion in these selected developing Asian economies. Regarding our other socio-economic variables' unemployment rate, we found it positive (as expected), but not statistically significant. Our findings are consistent with the results drawn by [51,56].

Table 7. Short-run panel ARDL with Error Correction Model.

No. of Panels = 15 Dependent Variable = END				
Variables	**Coefficients**	**Standard. Error**	**t-Statistics**	***p*-Value**
ENC	−0.26 ***	0.08	3.25	0.0012
GW	0.78 ***	0.33	2.36	0.0183
GW^2	−0.31 **	0.17	−1.82	0.0688
FCF	0.11 **	0.05	2.21	0.0271
PG	0.61	0.41	1.487	0.1370
dENC	−0.16 *	0.09	1.78	0.0751
dGW	0.37 *	0.21	1.75	0.0801
dGW^2	−0.14	0.11	1.27	0.2041
dFCF	0.08	0.05	1.60	0.1096
dPG	0.43	0.29	1.48	0.1389
ECM_{it-1}	−0.26 **	0.12	−2.17	0.0300

Source: authors own calculations, Note: * 10%, ** 5% and *** 1% show the statistical significance level.

Table 8 demonstrates the results of the augmented Dickey-Fuller (ADF) test. Generally, the average time level has no stationary series, even though all series are stationary having the first difference. If the ADF test did by using the first difference, generally, the insignificant supposition is rejected at the significance level of 1% or 5%. Consequently, this stated that the data are converted into the shape of stationary with the first difference.

Table 8. Unit root results of the augmented Dickey-Fuller test (ADF).

With Intercept				With Trends and Intercept		
Factor	**k Level**	**k 1st Difference**	**k Results**	**Factor**	**k 1st Difference**	**Result**
END	1.21	2.83	I(1)	1.53	2.32	I(1)
ENC	1.61	2.51	I(1)	2.53	3.52	I(1)
EGW	2.41	5.31	I(1)	2.39	5.43	I(1)
EGW^2	2.51	5.74	I(1)	1.63	6.32	I(1)
FCF	1.89	6.00	I(1)	1.39	5.97	I(1)
PG	1.19	4.92	I(1)	2.89	4.78	I(1)

Source: Author's own calculation by using E-Views 5. ENG stands for the ecological disorder, ENC shows the energy consumption, EGW is economic growth, EGW^2 is the square of economic growth, FCF is capital formation, PG shows population growth.

The results of the error correction model presented in Table 9.

Table 9. Error correction model results.

Variable	Ln(GDP)	St. Error	t-Statistics
END(−1)	k−0.89	(0.32)	[−2.45]
ENG(−1)	k−4.21	(1.96)	[−4.19]
EGW(−1)	k−42.51	(4.32)	[−5.36]
$EGW^2(-1)$	k−4.56	(1.54)	[−2.39]
FCF(−1)	k−5.39	(2.34)	[−1.42]
PG(−1)	−3.20	(2.730)	[−1.21]
C	327.16	–	k–
EC_{t-1}	−0.12 ***	(0.05)	[−2.69]

Source: Author's own calculation by using E-Views 5. Note: *** denotes 1% significance level. ENG stands for the ecological disorder, ENC shows the energy consumption, EGW is economic growth, EGW2 is the square of economic growth, FCF is capital formation, PG shows population growth.

The EC_{t-1} coefficient demonstrates the short-run adjustment rate to the long-run rate, whereas the adjustment rate was noted as 12%; this implies that the 12% imbalance is corrected per year.

The numerical coefficient of significance ensures the long-run causality among independent variables and dependent variables.

This study develops three indexes (environmental index (EVI), energy efficiency index (EE), and energy intensity index (EIN). The EVI index develops by using all indicators of Table 1. In contrast, the EE index utilizes energy efficiency and energy consumption as input indicators, and the EIN index has been developed by dividing energy consumption by GDP.

3.2. Results of DEA

Results of the DEA-based environmental index, energy index, energy intensity, and aggregated index are depicted as follows:

According to Figure 1, underline developing economies are passing through different environmental development phases compared to each other. Currently, a mixed condition is observed in this region in terms of environmental performance. Jorden has better conditions in this lineup, followed by Sri Lanka and Malaysia, while Philippine and Bangladesh are poor performers. The comprehensive set of indicators' choices are similar to the work done by [57–60].

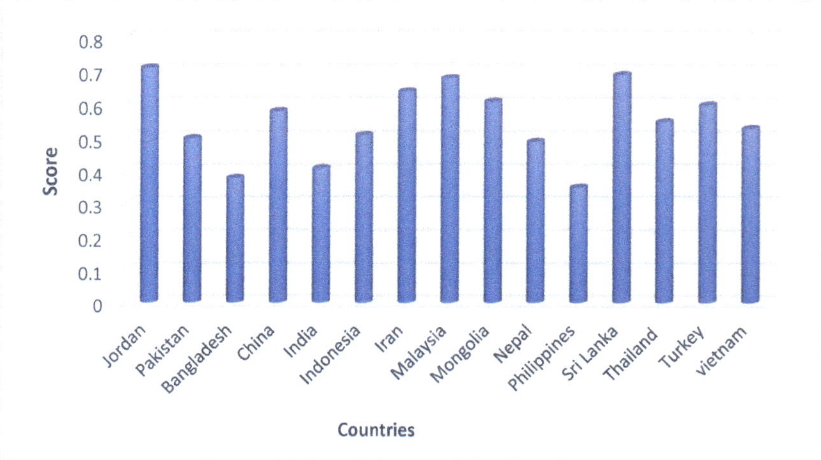

Figure 1. Environmental index. Source: authors' calculations.

Figure 2 shows the energy efficiency trend in these countries. China, India, and Turkey are energy efficient among these countries, while Nepal and Pakistan are the least energy-efficient countries. Thus, it may have happened due to more expenditure on China and India's renewable energy sources to meet their growing power demand for economic growth.

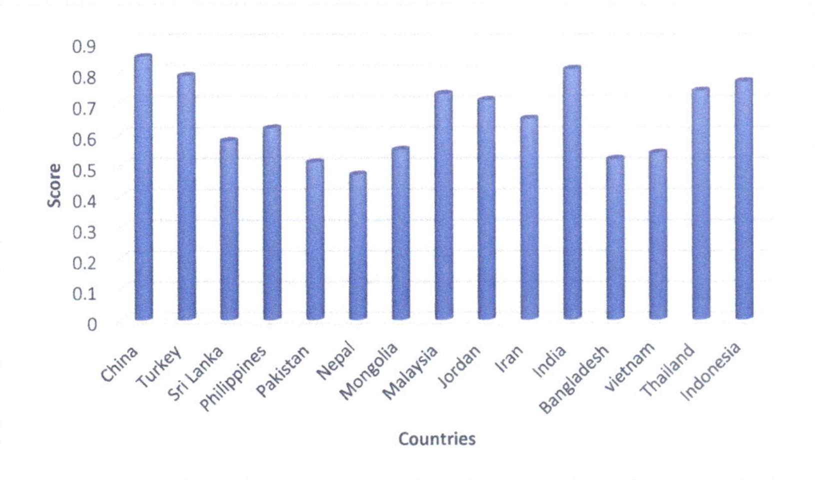

Figure 2. Energy efficiency index. Source: authors' calculations.

The greater energy intensity represents a higher price or cost transformed into real GDP. The level of energy intensity indicates that there is a decoupling of energy consumption and economic development. According to Figure 3, Iran is the most energy-intense economy among this dataset, followed by Nepal and Mongolia. The case of Iran's energy intensity can be valid as most energy-exporting countries find themselves in such conditions. In our results, Sri Lanka and Turkey are the least energy-intense economies. The same results are generated by [29] for the BRICS region.

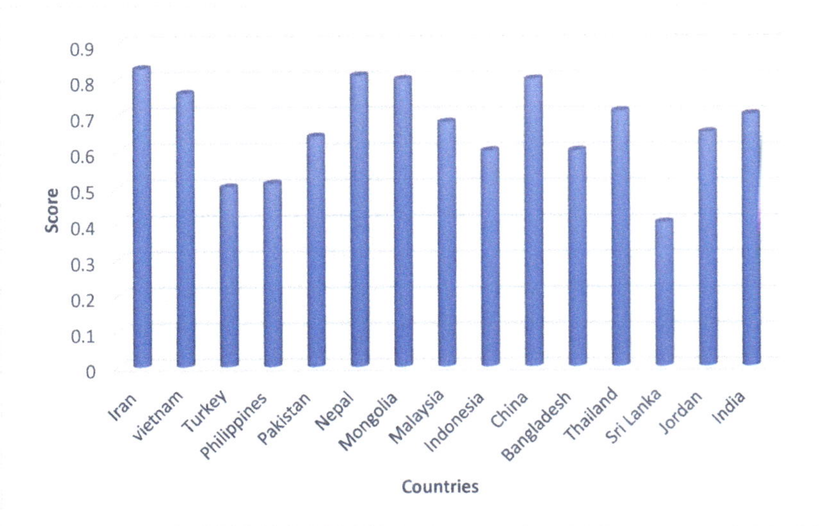

Figure 3. Energy intensity index. Source: authors' calculations.

Thus, all three indexes' distributions and frequencies confirm that these countries have no horizontal pattern regarding energy and environment, and there is an inconsistency between one country and the other. It may also confirm that individual initiatives matter for collective results.

Usually, decoupling expects to decrease environmental pressure from fossil-based energy production and consumption. The relationship between energy efficiency and economic factors show that energy efficiency improvements concentrate on decreasing fuel costs. However, its environmental effect relies on the nature of energy. Figure 4 shows a clearer picture of the aggregate performance of these countries.

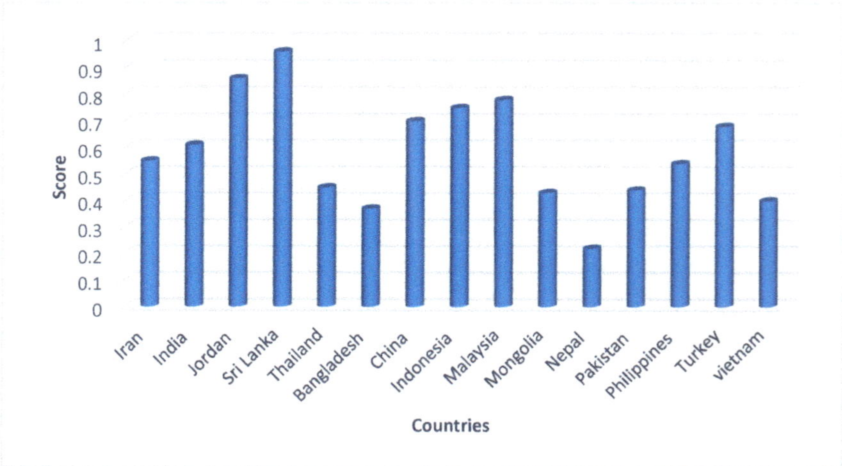

Figure 4. An aggregate score of all three indexes. Source: authors' calculation.

According to Figure 4., countries such as Nepal, Bangladesh, Vietnam, Pakistan, Magnolia, and Thailand are suffering to attain the best combination set of energy efficiency, energy intensity, and environmental protection. It shows that severe environmental issues (such as global warming) are linked with higher energy consumption due to rapid industrialization and urbanization [61]. Thus, renewable energy transformation can be the primary solution to enhance the energy efficiency, reducing energy intensity, and maintain sustainable environmental conditions, as also suggested by [30,62].

4. Discussion

This study aims to verify the existence of EKC and its trajectory with a fresh dataset of the fifteen developing economies of Asia. According to the results of panel ARDL, the theory of EKC exists in the underline developing economies. Here, the indicator of economic growth and its square term shows the positive and negative signs, respectively, EKC's confirmation statement. According to panel ARDL, the probability of EKC in these emerging economies exists at a 0.26 convergence rate in the long run. These results are in line with [27]. However, economic growth may not allow having happened as it requires energy sources, which are mostly carbon-based. Thus, the speed and trajectory of EKC heavily depend on energy sources. Therefore, energy may treat as an essential factor for EKC. The more the energy sources will be renewable and carbon-free, the more chance will be a smooth transaction towards EKC.

Most of the previous results confirm this notion for many developing economies, but these studies usually completed analysis. This study applied DEA to analyze the current condition of underline developing economies regarding energy efficiency, energy intensity, and environmental sustainability. Our DEA-based energy and environmental index results show that most of these developing economies are suffering to reduce energy intensity, increase energy efficiency, and maintain sustainable environmental conditions. The same results have been depicted by [31,32]. These results support the notion of EKC trajectory, which depended on the economic growth process and other favorable indicators, such as renewable energy usage and improvement of energy intensity with technological innovations.

The results of this study confirm that renewable energy consumption has a negative relation to carbon emission. Other than the square term of economic growth, some factors can help EKC's occurrence in these developing economies. However, these countries' energy efficiency is abysmal, which indicates that these economies depended on inefficient energy sources such as fossil fuel. Renewable energy sources could not only help in the occurrence of EKC, but they have the potential to enhance the trajectory rate of EKC as well. These findings are in line with previous studies [28,29,63].

The fixed capital formation results confirm that it also has significant features for carbon emission in these economies. It is due to the low usage of innovative technologies in these underline countries. Therefore, The EKC model shows that practical, efficient energy policies can reduce energy-based carbon dioxide emissions without damaging economic progress. For sustainable economic growth and to reduce greenhouse gas emissions, adopting clean and efficient energy sources is essential. Therefore, these developing economies' governments should focus on energy efficiency for long-term sustainable economic growth with less stress on environmental conditions. The same suggestions are made by [32,59].

Future research should ensure the results to be more general and broader. In this context, further criteria shall consider the selection of varying indicators. In applying a nonparametric frontier approach to measure the environmental vulnerability index, the study does not provide a strategy to include the decision-makers' preference weights. Therefore, further evaluation, such as on rank information and decision-makers' preferred weight, could also be included in the future.

5. Conclusions and Policy Recommendations

This study applied panel ARDL and DEA simultaneously to assess economic growth and energy consumption in ecological disorder. Empirical results of panel ARDL confirm the inverted U-shaped EKC for these underline emerging countries as the GDP square's coefficient is significant with a negative sign. It implies that underlined countries expect to follow the EKC theory. Renewable energy also shows the negative sign for an ecological disorder, which implies that more renewable energy use can help to mitigate carbon emission. The indexes of energy efficiency, energy intensity, and environment show that underline countries suffer from environmental conditions due to high energy intensity and low energy efficiency. As economic growth demands more and more energy supply, renewable energy is the only source to meet this demand without compromising the environment. Thus, the conversion and trajectory of EKC heavily depend on energy consumption and energy sources. According to the results mentioned above, although EKC exists for underline developing economies, the trajectory of EKC has a question surrounding what and how this will happen. Therefore, it needs to be considered the other fundamentals (such as renewable or zero-carbon energy sources) to enhance the probability of EKC theory and speedup of its trajectory.

In the wake of the COVID-19 pandemic and global economic recessions, which resulted in a drastic drop in oil and other fossil fuel prices, green energy and energy efficiency projects are losing their economic feasibility. It will endanger the achievement of the Paris agreement goals on climate change and several sustainable development goals. Based on this study's results, the policy recommendations for the developing countries are to adopt new supportive policies for the development of green energy and energy efficiency projects. The emerging economies should endorse the development and transformation of low-carbon concepts and adopt a sustainable energy system. One of the significant obstacles to developing renewable energy and energy efficiency projects is their difficulties accessing finance. These projects are considered risky projects; hence, many financiers are reluctant to finance these projects [51,52]. Therefore, in the current and post- COVID-19 era, the necessity of employing green finance tools is highlighted [63].

Other actions can help to mitigate air pollution. For example, identifying and monitoring air pollution sources from industrial energy consumption, fuel supplies of regulating petroleum, and boosting vehicle sectors to adopt green fuels can substitute petroleum products. On the other hand, the reliance on renewable energy may enhance developing economies' growth and reduce the usage of fossil fuels. Diversification of the energy basket and relying more on renewable energy resources can also enhance energy security [51,52]. Simultaneously, increasing energy efficiency will be considered a cost-effective way to decrease energy production's environmental influences. Therefore, this study suggests that sustainable renewable energy production can be one of the main factors for sustainable development goals.

Author Contributions: Conceptualization, done by Q.A. and J.Z.; methodology, software form by M.M.; validation and formal analysis did by F.T.-H.; investigation, resources, data curation, performed by K.J.; writing—original draft preparation is done by M.A.; writing—review and editing and policy recommendations by W.L.; visualization, and supervision by M.A. All authors contributed equally to this paper. This research paper is contributed by the authors mentioned above in the following way. All authors have read and agreed to the published version of the manuscript.

Funding: This research was funded by the Zhejiang province soft science project fund (2019C35019). Farhad Taghizadeh-Hesary acknowledges the financial support of the Japan Society for the Promotion of Science (JSPS) Kakenhi (2019–2020) Grant-in-Aid for Young Scientists (19K13742), and Grant-in-Aid for Excellent Young Researcher of the Ministry of Education of Japan (MEXT).

Conflicts of Interest: It is submitted that the manuscript mentioned above is initially written in all aspects and submitted for the possible publication in the journal *Sustainability*. This manuscript tries to fill the gap of literature for the other essential factors, mostly renewable energy, for the smooth and quick trajectory of EKC for developing economies. We declared no conflict of interest among all authors, and they unanimously agreed to possible publication in *Sustainability*.

References

1. Kelkar, U.; James, C.R.; Kumar, R. The Indian insurance industry and climate change: Exposure, opportunities, and strategies ahead. *Clim. Policy* **2006**, *6*, 658–671. [CrossRef]
2. IPCC. *Fifth Assessment Report of the Intergovernmental Panel on Climate Change*; Synthesis Report; IPCC: Paris, France, 2014.
3. Lyster, R. The idea of (Climate) justice, neoliberalism and the talanoa dialogue. *J. Hum. Rights Environ.* **2019**, *10*, 35–61. [CrossRef]
4. Hu, J.; Harmsen, R.; Crijns-Graus, W.; Worrell, E. Geographical optimization of variable renewable energy capacity in China using modern portfolio theory. *Appl. Energy* **2019**, *253*, 113614. [CrossRef]
5. Azizkhani, M.; Vakili, A.; Noorollahi, Y.; Naseri, F. Potential survey of photovoltaic power plants using Analytical Hierarchy Process (AHP) method in Iran. *Renew. Sustain. Energy Rev.* **2017**, *75*, 1198–1206. [CrossRef]
6. International Energy Agency World energy balances 2019. In *World Energy Balances: Overview*; International Energy Agency: Paris, France, 2019.
7. BP p.l.c. Energy Outlook 2019 edition. In *BP Energy Outlook 2019*; British Petroleum: London, UK, 2019.
8. United Nations National Accounts—Analysis of Main Aggregates (AMA). 1996. Available online: https://unstats.un.org/unsd/snaama (accessed on 3 October 2020).
9. World Energy Statistics. *World Energy Statistics 2018*; International Energy Agency: Paris, France, 2018. [CrossRef]
10. Ludwig, D.; Lanig, S.; Klärle, M. Location analysis for solar panels by LiDAR-Data with Geoprocessing-SUN-AREA. In Proceedings of the Environmental Informatics and Industrial Environmental Protection: Concepts, Methods and Tools, Berlin, Germany, 9–11 September 2009; pp. 83–89.

11. Haghi, E.; Raahemifar, K.; Fowler, M. Investigating the effect of renewable energy incentives and hydrogen storage on advantages of stakeholders in a microgrid. *Energy Policy* **2018**, *113*, 206–222. [CrossRef]
12. Liu, X.; Zhang, S.; Bae, J. The impact of renewable energy and agriculture on carbon dioxide emissions: Investigating the environmental Kuznets curve in four selected ASEAN countries. *J. Clean. Prod.* **2017**. [CrossRef]
13. Acar, Y.; Gürdal, T.; Ekeryllmaz, Ş. Environmental Kuznets Curve for CO_2 emissions: An analysis for developing, Middle East, OECD and OPEC countries. *Environ. Socio-Econ. Stud.* **2018**. [CrossRef]
14. Özokcu, S.; Özdemir, Ö. Economic growth, energy, and environmental Kuznets curve. *Renew. Sustain. Energy Rev.* **2017**, *72*, 639–647. [CrossRef]
15. Abbas, Q.; Nurunnabi, M.; Alfakhri, Y.; Khan, W.; Hussain, A.; Iqbal, W. The role of fixed capital formation, renewable and non-renewable energy in economic growth and carbon emission: A case study of Belt and Road Initiative project. *Environ. Sci. Pollut. Res.* **2020**. [CrossRef]
16. Lu, W.C. The impacts of information and communication technology, energy consumption, financial development, and economic growth on carbon dioxide emissions in 12 Asian countries. *Mitig. Adapt. Strateg. Glob. Chang.* **2018**. [CrossRef]
17. Agency, I.E. Global energy & CO_2 status report. In *Global Energy CO_2 Status Report*; International Energy Agency: Paris, France, 2017; pp. 1–15.
18. van den Bergh, J.; de Bruyn, S.M.; Heintz, R.J. The environmental kuznets curve hypothesis. In *Handbook of Environmental and Resource Economics*; Edward Elgar Pub.: Cheltenham, UK, 2013.
19. Uchiyama, K. *Environmental Kuznets Curve Hypothesis and Carbon Dioxide Emissions*; Development Bank of Japan: Tokyo, Japan, 2016; pp. 11–30. [CrossRef]
20. Galeotti, M.; Manera, M.; Lanza, A. On the robustness of robustness checks of the environmental kuznets curve hypothesis. *Environ. Resour. Econ.* **2009**, *42*, 551–574. [CrossRef]
21. Sun, H.; Kporsu, A.K.; Taghizadeh-Hesary, F.; Edziah, B.K. Estimating environmental efficiency and convergence: 1980 to 2016. *Energy* **2020**. [CrossRef]
22. Taghizadeh-Hesary, F.; Rasoulinezhad, E. Analyzing energy transition patterns in Asia: Evidence from countries with different income levels. *Front. Energy Res.* **2020**. [CrossRef]
23. Yoshino, N.; Taghizadeh-Hesary, F.; Otsuka, M. Covid-19 and optimal portfolio selection for investment in sustainable development goals. *Financ. Res. Lett.* **2020**. [CrossRef] [PubMed]
24. Sun, L.; Cao, X.; Alharthi, M.; Zhang, J.; Taghizadeh-Hesary, F.; Mohsin, M. Carbon emission transfer strategies in supply chain with lag time of emission reduction technologies and low-carbon preference of consumers. *J. Clean. Prod.* **2020**. [CrossRef]
25. Zafar, M.W.; Zaidi, S.A.H.; Sinha, A.; Gedikli, A.; Hou, F. The role of stock market and banking sector development, and renewable energy consumption in carbon emissions: Insights from G-7 and N-11 countries. *Resour. Policy* **2019**. [CrossRef]
26. Hanif, I. Impact of economic growth, nonrenewable and renewable energy consumption, and urbanization on carbon emissions in Sub-Saharan Africa. *Environ. Sci. Pollut. Res.* **2018**. [CrossRef]
27. Hanif, I.; Faraz Raza, S.M.; Gago-de-Santos, P.; Abbas, Q. Fossil fuels, foreign direct investment, and economic growth have triggered CO_2 emissions in emerging Asian economies: Some empirical evidence. *Energy* **2019**. [CrossRef]
28. Mohsin, M.; Abbas, Q.; Zhang, J.; Ikram, M.; Iqbal, N. Integrated effect of energy consumption, economic development, and population growth on CO_2 based environmental degradation: A case of transport sector. *Environ. Sci. Pollut. Res.* **2019**. [CrossRef]
29. Baloch, Z.A.; Tan, Q.; Iqbal, N.; Mohsin, M.; Abbas, Q.; Iqbal, W.; Chaudhry, I.S. Trilemma assessment of energy intensity, efficiency, and environmental index: Evidence from BRICS countries. *Environ. Sci. Pollut. Res.* **2020**. [CrossRef]
30. Mohsin, M.; Rasheed, A.K.; Sun, H.; Zhang, J.; Iram, R.; Iqbal, N.; Abbas, Q. Developing low carbon economies: An aggregated composite index based on carbon emissions. *Sustain. Energy Technol. Assess.* **2019**, *35*, 365–374. [CrossRef]
31. Asbahi, A.A.M.H.A.; Gang, F.Z.; Iqbal, W.; Abass, Q.; Mohsin, M.; Iram, R. Novel approach of Principal Component Analysis method to assess the national energy performance via Energy Trilemma Index. *Energy Rep.* **2019**. [CrossRef]

32. Turner, K.; Hanley, N. Energy efficiency, rebound effects and the environmental Kuznets Curve. *Energy Econ.* **2011**. [CrossRef]
33. Honma, S.; Hu, J.L. Total-factor energy efficiency of regions in Japan. *Energy Policy* **2008**. [CrossRef]
34. World Bank Group. *World Bank Country and Lending Groups*; World Bank Group: Washington, DC, USA, 2019.
35. Shahbaz, M.; Lean, H.H.; Shabbir, M.S. Environmental Kuznets Curve hypothesis in Pakistan: Cointegration and Granger causality. *Renew. Sustain. Energy Rev.* **2012**, *16*, 2947–2953. [CrossRef]
36. Shahbaz, M.; Sbia, R.; Hamdi, H.; Ozturk, I. Economic growth, electricity consumption, urbanization and environmental degradation relationship in United Arab Emirates. *Ecol. Indic.* **2014**, *45*, 622–631. [CrossRef]
37. Jayaraman, R.; Colapinto, C.; La Torre, D.; Malik, T. A Weighted Goal Programming model for planning sustainable development applied to Gulf Cooperation Council Countries. *Appl. Energy* **2017**, *185*, 1931–1939. [CrossRef]
38. Ebert, U.; Welsch, H. Meaningful environmental indices: A social choice approach. *J. Environ. Econ. Manag.* **2004**, *47*, 270–283. [CrossRef]
39. Zhou, P.; Delmas, M.A.; Kohli, A. Constructing meaningful environmental indices: A nonparametric frontier approach. *J. Environ. Econ. Manag.* **2017**, *85*, 21–34. [CrossRef]
40. Zhou, P.; Ang, B.W.; Poh, K.L. Comparing aggregating methods for constructing the composite environmental index: An objective measure. *Ecol. Econ.* **2006**, *59*, 305–311. [CrossRef]
41. Fusco, E. Enhancing non-compensatory composite indicators: A directional proposal. *Eur. J. Oper. Res.* **2015**, *242*, 620–630. [CrossRef]
42. Coelli, T.; Prasada Rao, D.S.; Battese, G.E. Additional topics on data envelopment analysis. In *An Introduction to Efficiency and Productivity Analysis*; Springer: New York, NY, USA, 2005; pp. 183–208. ISBN 978-0-387-25895-9.
43. Faere, R.; Grosskopf, S.; Lovell, C.A.K.; Pasurka, C. Multilateral productivity comparisons when some outputs are undesirable: A nonparametric approach. *Rev. Econ. Stat.* **1989**, *71*, 90–98. [CrossRef]
44. Tyteca, D. On the measurement of the environmental performance of firms—A literature review and a productive efficiency perspective. *J. Environ. Manag.* **1996**, *46*, 281–308. [CrossRef]
45. Cooper, W.W.; Park, K.S.; Yu, G. IDEA and AR-IDEA: Models for dealing with imprecise data in DEA. *Manag. Sci.* **1999**, *45*, 597–607. [CrossRef]
46. Aida, K.; Cooper, W.W.; Pastor, J.T.; Sueyoshi, T. Evaluating water supply services in Japan with RAM: A range-adjusted measure of inefficiency. *Omega* **1998**, *26*, 207–232. [CrossRef]
47. Bostian, M.; Fare, R.; Grosskopf, S.; Lundgren, T. Environmental investment and firm performance: A network approach. *Energy Econ.* **2016**, *57*, 243–255. [CrossRef]
48. Cooper, W.W.; Pastor, J.T.; Borras, F.; Aparicio, J.; Pastor, D. BAM: A bounded adjusted measure of efficiency for use with bounded additive models. *J. Product. Anal.* **2011**, *35*, 85–94. [CrossRef]
49. Ogundimu, E.O.; Collins, G.S. Predictive performance of penalized beta regression model for continuous bounded outcomes. *J. Appl. Stat.* **2018**, *45*, 1030–1040. [CrossRef]
50. Zambrano-Monserrate, M.A.; Carvajal-Lara, C.; Urgiles-Sanchez, R. Is there an inverted U-shaped curve? Empirical analysis of the Environmental Kuznets Curve in Singapore. *Asia-Pac. J. Account. Econ.* **2018**, *25*, 145–162. [CrossRef]
51. Dinda, S. Environmental Kuznets Curve hypothesis: A survey. *Ecol. Econ.* **2004**, *49*, 431–455. [CrossRef]
52. Munda, G.; Nardo, M. Noncompensatory/nonlinear composite indicators for ranking countries: A defensible setting. *Appl. Econ.* **2009**, *41*, 1513–1523. [CrossRef]
53. Kim, K.H. Inference of the environmental Kuznets curve. *Appl. Econ. Lett.* **2013**, *20*, 119–122. [CrossRef]
54. Sinha, A.; Bhatt, M.Y. Investigating the Environmental Kuznets Curve for India. *J. Environ.* **2016**, *3*, 1–6 [CrossRef]
55. Sica, E. Economic dualism and air quality in Italy: Testing the Environmental Kuznets Curve hypothesis. *Int. J. Environ. Stud.* **2014**, *71*, 463–480. [CrossRef]
56. Martchamadol, J.; Kumar, S. The Aggregated Energy Security Performance Indicator (AESPI) at national and provincial level. *Appl. Energy* **2014**, *127*, 219–238. [CrossRef]
57. Augutis, J.; Krikstolaitis, R.; Martisauskas, L.; Peciulyte, S. Energy security level assessment technology. *Appl. Energy* **2012**, *97*, 143–149. [CrossRef]
58. Iram, R.; Zhang, J.; Erdogan, S.; Abbas, Q.; Mohsin, M. Economics of energy and environmental efficiency: Evidence from OECD countries. *Environ. Sci. Pollut. Res.* **2019**. [CrossRef]

59. Xu, B.; Lin, B. Reducing carbon dioxide emissions in China's manufacturing industry: A dynamic vector autoregression approach. *J. Clean. Prod.* **2016**, *131*, 594–606. [CrossRef]
60. Shahbaz, M.; Sinha, A. Environmental Kuznets curve for CO2 emissions: A literature survey. *J. Econ. Stud.* **2019**, *46*, 106–168. [CrossRef]
61. He, W.; Abbas, Q.; Alharthi, M.; Mohsin, M.; Hanif, I.; Vinh Vo, X. Integration of renewable hydrogen in light-duty vehicle: Nexus between energy security and low carbon emission resources. *Int. J. Hydrogen Energy* **2020**. [CrossRef]
62. Hanif, I. Impact of fossil fuels energy consumption, energy policies, and urban sprawl on carbon emissions in East Asia and the Pacific: A panel investigation. *Energy Strateg. Rev.* **2018**. [CrossRef]
63. Färe, R.; Grosskopf, S. Directional distance functions and slacks-based measures of efficiency. *Eur. J. Oper. Res.* **2010**, *200*, 320–322. [CrossRef]

sustainability

Article

The Roles of Beijing-Tianjin-Hebei Coordinated Development Strategy in Industrial Energy and Related Pollutant Emission Intensities

Cong Hu [1], Biliang Hu [2,3], Xunpeng Shi [4] and Yan Wu [5,*]

1 School of Economics and Resource Management, Beijing Normal University, Beijing 100875, China; 201831410005@mail.bnu.edu.cn
2 Emerging Markets Institute, Beijing Normal University, Beijing 100875, China; hubiliang@bnu.edu.cn
3 The Belt and Road School, Beijing Normal University, Beijing 100875, China
4 Australia-China Relations Institute, University of Technology Sydney, Sydney Ultimo, NSW 2007, Australia; Xunpeng.Shi@uts.edu.au
5 School of Economics, Beijing Technology and Business University, Beijing 100048, China
* Correspondence: wuyan@btbu.edu.cn

Received: 28 July 2020; Accepted: 24 September 2020; Published: 25 September 2020

Abstract: This study investigates the different impacts of coordinated development in the Beijing–Tianjin–Hebei (BTH) region on industrial energy and pollution intensities based on the difference-in-difference (DID) method and the quantile DID method. The panel data cover industrial energy consumption and three wastes, which are industrial wastewater, sulfur dioxide, and dust emissions, from all 13 cities in the BTH region and 17 cities in Henan Province for the period 2007–2017. The study finds that China's BTH coordinated development strategy, on average, tends to restrain regional industrial energy intensity, especially in lower quantile level (0.1–0.4) cities. However, it tends to promote industrial energy intensity in higher quantile level (0.7–0.9) cities. The impacts on pollution intensities vary among industrial wastewater, sulfur dioxide, and dust emissions The results suggest that, in addition to paying attention to dust pollution caused by transportation integration in the BTH region, China should also pay more attention to green relocation of industries from Beijing to Hebei and strengthen coordinated environmental regulation while maintaining corporate interests.

Keywords: industrial energy intensity; pollution emission intensity; quantile DID method; Beijing–Tianjin–Hebei coordinated development; China

1. Introduction

Industrial pollution is one of the world's most serious environmental problems. In China, industries are the largest source of pollution, especially for air contamination. Any form of pollution that can be directly traced back to industrial practice is called industrial pollution. Most of the pollution on Earth can be traced back to some kind of industry [1]. Therefore, the reduction of industrial pollution has always been critical in dealing with environmental degradation around the world. Most developing countries facing the rapid growth of industrial pollution found it to be a serious problem that must be brought under control. However, major environmental disasters caused by industrial accidents still occur sometimes in developing countries.

As for China, environmental pollution mainly comes from traditional heavy industries, especially power plants, petrochemical industries, metal smelting, and machinery manufacturing, which produce a large volume of sulfur dioxide, dust, and wastewater. In the face of unprecedented economic and industrial growth levels, China rapidly developed its system of environmental governance [2,3]. It can be used to reform the highly inefficient and strictly regulated energy market [4] and to improve the regulation of industrial pollution [5]. The reasons may be the two-level pressures including domestic economic growth and international governance status [6]. In recent years, China's governing system resulted in a lot of environmental policy implementations from the central government [7], which have had a noticeable effect on improving the environment [8]. Although great achievements in some fields like solar photo voltaic, wind energy, and nuclear power have been made [9], China still faces many energy-related challenges including air pollution, urbanization, and climate change in the future [10].

The Beijing–Tianjin–Hebei (BTH) region is China's "capital economic circle", including Beijing, Tianjin, and 11 prefecture-level cities of Hebei province. It is one of the three major urban agglomerations and has the strongest industrial base in China, accounting for about 2% of the national land area, 10% of the national GDP, and 8% of the total population. However, there is a huge economic disparity in this region. In terms of per capita GDP in 2019, the per capita GDPs of Beijing, Tianjin, and Hebei were USD 23,600, USD 13,100, and USD 6,800, respectively. In terms of the industrial structure in 2019, Beijing's tertiary industry accounted for 83% and the secondary industry accounted for only 16%, while Tianjin and Hebei's secondary industries accounted for 36% and 41%, respectively. In terms of urbanization rate in 2019, the urbanization rates of Beijing, Tianjin, and Hebei were 86%, 83%, and 56% respectively.

The BTH region in which environmental degradation is prominent in China has been dominated by some energy-intensive industries like mining and steelmaking sectors, especially in Hebei province. The common problem of the three areas is the consumption of fossil energy and the three wastes it brings, including industrial wastewater, sulfur dioxide, and dust emissions, which are the main causes of environmental problems in this region. The three wastes are related to fossil energy because sulfur dioxide and dust emission are mainly produced by fossil energy consumption, while industrial wastewater, although not entirely produced by fossil energy, is overall strongly correlated with energy consumption.

In February 2014, Chinese President Xi Jinping proposed the coordinated development strategy of the BTH region, which is a national strategy in China. The strategy is to build a new capital economic circle and to promote innovation of regional development. First, it is to explore a mode of optimal development of this densely populated and economically intensive area. Second, it will take the lead in making breakthroughs in three key areas: ecological and environmental regulation, industrial upgrading and transfer, and the integration of the BTH transport sector. It is expected to ease the increasing pressure on resources and environment, to accelerate the transformation of economic development patterns, and to promote balanced development for the BTH region.

The Outline of the Coordinated Development Plan in the BTH region approved by the Chinese government in 2015 made a clear definition of the core functions of Beijing: the national political center, the cultural center, the international communication center, and the scientific and technological innovation center. The industries that do not accord with these certain orientations should be gradually relocated to Tianjin and Hebei province. Tianjin is identified as an important city but subordinate to the core city of Beijing, and thus, it is meant to take on complementary functions [11]. By 2019, nearly 10,000 enterprises had been relocated out of Beijing, most of which were high-end manufacturing and high-tech service industries. These industries may not fit Beijing's new orientation, but for Tianjin and Hebei, they can promote local industrial upgrade through the relocation. It is worth mentioning that the "transfer" is not transferring polluting industries. During the transfer process, enterprises should upgrade or switch to sustainable and environmentally friendly businesses. If they do not want to relocate, they can upgrade locally following the new and more stringent environmental protection standards. Therefore, the feature of this strategy is that it seeks to address environmental and social issues in the process of coordinated development while also laying the foundation for sustainable development through environmental constraints.

In recent years, some major projects involving the coordinated development of the BTH region have been launched. For example, Beijing and Hebei will jointly organize the 2022 Beijing Winter Olympics, Beijing and Tianjin are fully supporting the development of the Hebei Xiong'an New

Area, and the Chinese government has been supporting the orderly transfer and sharing of Beijing's scientific and technological innovation resources and high-quality public service resources in this region. More importantly, Beijing, Tianjin, and Hebei were together dedicated to deepening joint prevention and control of the increasing regional environmental degradation.

From the perspective of sustainable development, what impacts this strategy has had on fossil energy intensity and related pollution intensities in the BTH region are absent in the literature. To fill this gap, this paper is therefore intended to empirically study the roles of the BTH coordinated development strategy in industrial energy and pollution intensities based on the difference-in-difference (DID) method and the quantile DID method. The results of this study will help clarify the potential problems in the implementation of this strategy and provide some empirical evidence and policy implications for green and sustainable development in this region. The implementation of the BTH coordinated development strategy provides an effective quasi-natural experiment in assessing the impact of specific policy events on industrial energy intensity and related pollution intensities.

The study has made three contributions. First, it investigates the comprehensive impact of the BTH coordinate development strategy on industrial energy and pollution intensities, and the estimation results are robust based on a number of statistical tests. Second, this study applies the quantile DID method to evaluate the heterogeneous impact of the strategy on industrial energy intensity in the BTH region at different quantile levels of industrial energy intensity. Third, this study uses the data of industrial energy pollutions to estimate the different impacts of the strategy on the environment among different energy pollution sources. Such heterogeneous effects can help make policy implications based on different pollution sources under the context of the strategy.

There are three findings from this study. First, based on the DID method results, China's BTH coordinated development strategy, on average, tends to restrain regional industrial energy intensity in this region. Second, based on the quantile DID method results, the BTH coordinated development strategy tends to restrain the industrial energy intensity in lower quantile level cities but tends to increase the industrial energy intensity in higher quantile level cities. Third, the impacts of the BTH coordinated development strategy on the environment vary among three industrial energy-related pollution sources including wastewater, sulfur dioxide, and dust emissions.

In Section 2, we introduce the study areas and provide some basic statistics about this region. In Section 3, we conduct a literature review and analyze the mechanism between the BTH coordinated development strategy and industrial energy consumption. Section 4 estimates a regression model for industrial energy intensity based on the DID method. Section 5 assesses the impact of the BTH coordinated development strategy on industrial energy and pollution intensity empirically. Finally, Section 6 provides the conclusion, policy implications, and discussions.

2. The Study Areas

As shown in Figure 1, Beijing is located in the north of China and north of the BTH region. It borders Tianjin in the east and some cities of Hebei in the rest. It is the capital of the People's Republic of China, a municipality directly under China's central government, a national central city, and a megacity in the world. The total area of Beijing is 16,410 square kilometers, and it has 16 districts under its jurisdiction. By the end of 2019, there were 21.536 million permanent residents and 18.5 million urban residents, representing an urbanization rate of 86.6 percent. The GDP of Beijing in 2019 was RMB 3537.13 billion yuan, and the per capita GDP was RMB 164,000 yuan. The proportions of primary, secondary, and tertiary industries are 0.3%, 16.2%, and 83.5%, respectively. Besides, the energy consumption per unit of GDP in 2019 was about 0.25 tons of standard coal per RMB 10,000 yuan, which was the lowest in China.

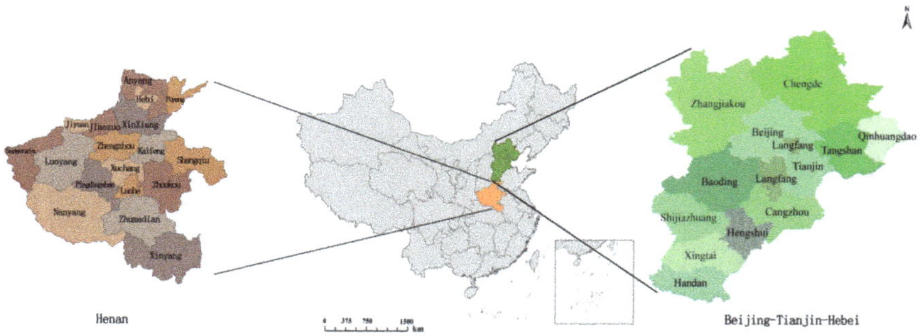

Figure 1. Locations of the Beijing–Tianjin–Hebei (BTH) region and Henan Province.

Tianjin, a municipality directly under the central government of China, is a national central city, a megacity, and the largest port city in north China. It borders Beijing in the west and some cities of Hebei in the rest. The total area of Tianjin is 11,966 square kilometers with 16 districts under its jurisdiction. By the end of 2019, there were 15.618 million permanent residents and 13.038 million urban residents, representing an urbanization rate of 83.5 percent. The GDP of Tianjin in 2019 was RMB 1410.43 billion yuan, and the per capita GDP was RMB 90,000 yuan. The proportions of primary, secondary, and tertiary industries were 1.3%, 35.2%, and 63.5%, respectively. Besides, the energy consumption per unit of GDP in 2019 was about 0.41 tons of standard coal per RMB 10,000 yuan. Now, the ecological civilization has also become an important element for the urban development of Tianjin under the coordinated development strategy of the BTH region [11].

Hebei province is located in the north of China, bordering the Bohai Sea in the east and Beijing and Tianjin in the inner ring. The total area of Hebei province is 188,800 km^2, and it has 11 prefecture-level cities under its jurisdiction, including Shijiazhuang, Tangshan, Qinhuangdao, Baoding, Handan, Xingtai, Zhangjiakou, Chengde, Cangzhou, Hengshui, and Langfang, with Shijiazhuang as its provincial capital. By the end of 2019, there were 75.920 million permanent residents and the permanent resident urbanization rate was 57.6 percent. The GDP of Hebei in 2019 was RMB 3510.45 billion yuan, and the per capita GDP was RMB 46,000 yuan. The proportions of primary, secondary, and tertiary industries were 10.3%, 39.7%, and 50.0%, respectively. Besides, the energy consumption per unit of GDP in 2019 was about 0.84 tons of standard coal per RMB 10,000 yuan, which was much higher than the national average of 0.51 tons of standard coal equivalent per RMB 10,000 yuan. Hebei province is rich in mineral resources because 156 kinds of mineral resources have been discovered, 39 of which are among the top 5 in China. It has formed a mining economic system with metallurgy, coal, building materials, and petrochemicals. In April 2017, the central government decided to establish Xiong'an New Area in Hebei province. In August 2019, the state council established a new China (Hebei) Pilot Free Trade Zone.

The control group, Henan Province, is located in central China, south of Hebei province, with a total area of 167,000 km^2. It has jurisdiction over 17 prefecture-level cities, including Zhengzhou, Kaifeng, Luoyang, Pingdingshan, Anyang, Hebi, Xinxiang, Jiaozuo, Puyang, Xuchang, Luohe, Sanmenxia, Shangqiu, Zhoukou, Zhumadian, Nanyang, and Xinyang, with Zhengzhou as the provincial capital and Jiyuan as a county-level city directly under the jurisdiction of Henan Province. By the end of 2019, there were 109.52 million permanent residents and the permanent resident urbanization rate was 53.2 percent. The GDP of Henan in 2019 was RMB 5425.92 billion yuan, and the per capita GDP was RMB 56,000 yuan. The proportions of primary, secondary, and tertiary industries were 8.5%, 43.5%, and 48.0%, respectively. Energy consumption per unit of GDP in 2019 was about 0.48 tons of standard coal per RMB 10,000 yuan. Henan province is also rich in mineral resources because 142 kinds of mineral resources have been discovered, 58 of which are among the top 5 in China.

In general, Beijing is dominated by the service industry, Tianjin is dominated by the processing and manufacturing industry and by the port service industry, and Hebei is dominated by the resource-intensive industry and agriculture because of the rich mineral resources and rural labor force. Therefore, we can also conclude the following socioeconomic characteristics in the BTH region. First, the issues of industrial pollution are very serious. The heavy industries in Hebei and Tianjin lead to great pressure on the environment, especially serious air pollution. Second, economic development is very unbalanced. Hebei is geographically advantaged with three port cities including Tangshan, Cangzhou, and Qinhuangdao, and most of the cities are very close to Beijing and Tianjin. However, there is a huge gap in economic development as the above statistics show. Third, as the capital city, Beijing has strong political, economic, cultural, scientific, and technological strength. However, there has always been a "siphon effect" of talents, funds, and other resources transferring from Hebei and even Tianjin to Beijing. Therefore, Beijing has a very weak radiation effect on the surrounding cities. Fourth, under the BTH coordinated development strategy, Tianjin has changed to a more sustainable urban development model. For example, the Sino-Singaporean Eco-city replaced Hanggu to become one subcenter of the Binhai New Area of Tianjin, as it is a national project that retains a high level of support from the central government for environment protection in the BTH region.

Based on the outline of the coordinated development plan, Beijing is the national center for political, cultural, and international exchanges and for scientific and technological innovation; Tianjin is the national advanced manufacturing research and development base, the northern international shipping core zone, and the financial innovation operation and reform pilot demonstration zone, and Hebei is an important base for modern trade and logistics in China, a pilot area for industrial transformation and upgrading, a demonstration area for new urbanization and urban-rural integration, and a supporting area for the BTH ecological environment. The overall orientation of the BTH region is "a world-class city cluster with the capital as the core, a leading area for coordinated regional development and reform, a new national engine for innovation-driven economic growth, and a demonstration area for ecological restoration and environmental improvement". The overall development plan can be concluded as "one core, two cities, three axes, four areas, and multiple nodes". "One core" means Beijing, and "two cities" refer to Beijing and Tianjin. They are the main engines for the coordinated development of the BTH region so the linkage between Beijing and Tianjin needs to be further strengthened. "Three axes" refer to the three industrial development and urban agglomeration axes, including Beijing–Tianjin, Beijing–Baoding–Shijiazhuang, and Beijing–Tangshan–Qinhuangdao. "Four areas" refer to the central core functional area, the eastern coastal development area, the southern functional extension area, and the northwest ecological conservation area. "Multiple nodes" include regional central cities such as Shijiazhuang, Tangshan, Baoding, and Handan and node cities such as Zhangjiakou, Chengde, Langfang, Qinhuangdao, Cangzhou, Xingtai, and Hengshui, with the focus on improving these cities' comprehensive carrying capacity and service capacity and on promoting industry and population aggregation in an orderly manner.

There are some reasons why the Chinese government made great effort to develop this region. First, the coordinated development of this region is conducive to solving Beijing's "big-city diseases" such as population expansion, traffic congestion, housing difficulties, environmental degradation, and resource shortage. Second, the BTH region lags behind China's Yangtze River Delta and Pearl River Delta but it has a huge development potential. Third, the economic capabilities are extremely uneven among cities. Beijing and Tianjin lead the country in urbanization rate and per capita GDP, while those of Hebei cities are below the national averages. For example, Beijing's pillar industries are finance, information technology, and science and technology research, but Tianjin and Hebei are dominated by medium- and low-end manufacturing industries with lots of pollution. The milestones are as follows. By 2017, remarkable progress was made in the orderly relocation of noncapital functions of Beijing and some breakthroughs occurred in key areas such as transportation integration, environmental protection, and industrial upgrading and relocation. By 2020, the population of Beijing should be controlled within 23 million, an integrated regional transport network will be shaped, the environment will effectively improved, major progress will be made in the coordinated development of industries, and the development gaps within the region should be narrowed. By 2030, the core functions of Beijing will be improved, the regional economic structure will become more reasonable, the environmental quality will improve, and the levels of public services should be balanced. In the end, the BTH region will become a region with strong international competitiveness and influence, thus playing a greater role in guiding and supporting China's economic and social sustainable development.

3. Literature Review and the Mechanism

As mentioned above, the BTH region is facing great environmental pressure in the process of rapid urbanization [12]. Pollution in this area is closely related to energy consumption [13], especially for energy consumption in industries and transportation [14]. Besides, though some of the cities in the BTH region have entered the postindustrial stage, none of them has crossed the turning point of the environmental Kuznets curve [15]. Because the BTH coordinated development strategy concerned in this paper focuses on environmental regulation, industrial upgrading, and transportation, a literature review is conducted accordingly.

First, for environmental regulation in the BTH region, recently, many studies made quantitative or qualitative analyses on its impact on energy consumption and gave some suggestions for further improvement. For example, by using a computable general equilibrium (CGE), Li et al. [16] found that, over the entire BTH area, the environmental policies could generate an average annual loss of 1.4% of gross regional product growth in the action plan scenario and of 2.3% in the enhanced action plan scenario. These results suggest that more joint measures are needed to promote energy conservation and emission reduction in the BTH region. For determining how policies and regulations support energy efficiency measure proliferation, Wang et al. [17] conducted a questionnaire survey of the enterprises in the BTH industrial transfer and found that the importance of awareness and investment priorities is great, so policy makers need to pay more attention to economic and legal tools and to appropriately increase supervision and punishment. By quantitatively analyzing the environmental policies of the BTH region, Zhang et al. [18] found that environmental regulations have direct and indirect spatial effects on industrial structures across regions and that environmental regulations have a long-term promotion effect on industrial structure upgrade and energy conservation.

Second, for industrial upgrades in the BTH region, the literature has made analyses about the effects of industrial upgrading on energy consumption for different industries or different regions. For example, by calculating the total factor energy efficiency (TFEE) of 27 industries in the BTH region, Li et al. [19] found that, because of the technological spillover effect from Beijing enterprise, Hebei has the highest total factor average energy efficiency in the production and supply of electric power and heat power industry, and Tianjin has the highest total factor average energy efficiency in the manufacture of raw chemical materials and chemical products and in the smelting and processing of ferrous metals. By measuring the energy rebound effect of industrial enterprises in the BTH region from 1996 to 2015, Li et al. [20] found that Hebei faces greater pressure to attain high energy conservation and emission reduction goals in the future than the other regions. By quantitatively analyzing the delinking indicators on industry growth and environmental pressures in the BTH region from 1996 to 2010, Wang and Yang [21] found that the carbon emissions in the BTH region were dominated by the secondary industry, which accounted for about 80% of total carbon emissions, and that the energy structure and energy intensity made significant contributions to the industrial decoupling progress. By quantitatively measuring the total factor carbon emission performance (TFCP) and the carbon emission mitigation potential (CMP) of 39 industrial sectors in the BTH region, Wang et al. [22] found that the manufacture of the nonmetallic mineral product sector and the production and distribution of electric power and heat power sector belong to the low TFCP–high CMP quadrant.

Third, for transportation in the BTH region, some studies analyzed the energy consumption and related pollutions brought by transportation development in the BTH region. For example, by using the panel data from 1995 to 2016, Guo and Meng [23] found that transportation energy intensity and the economic effect is the main factor increasing carbon dioxide emissions. They also found that the contributing factors to the carbon dioxide emission reduction in the transport sector are the energy structure effect, the freight turnover of unit industrial output effect, and the industrialization effect. By analyzing vehicular emission trends from road vehicles of the BTH region in the period 1999–2010, Lang et al. [24] found that, due to the rapid development of freight traffic, emissions of NO_X and PM_{10} kept increasing in Tianjin and Hebei. By analyzing the driving forces behind carbon emission of the BTH region from 2005 to 2013, Zhu and Li [25] found that the effect of energy intensity from the transportation sector always plays a negative role in Tianjin and Hebei but a positive one in Beijing.

How did the BTH coordinated development strategy affect industrial energy consumption in this region? The above studies focused on the environment and on sustainable development in this region. However, there is no direct literature on the comprehensive policy influence of the BTH coordinated development strategy on industrial energy intensity and related pollution emission intensities. As mentioned above, the strategy focuses on environmental regulation, industrial upgrade, and transportation; thus, we will analyze the relations between energy consumption and environmental regulations, transportation, and industrial upgrade combined with the existing literature.

The first one is environmental regulation. Environmental regulation will bring lower energy consumption for enterprises, and it can be divided into two situations based on the policy effects. One is that simple environmental regulations force a reduction of energy consumption, which may not be conducive to economic development. In the short term, it may harm the interests of enterprises, which will resist regulations, leading to weak policy effects. In the long term, it may force enterprises to upgrade their technology, but economic loss cannot be recovered [26]. The second one is to encourage industrial upgrading and technological progress while implementing environmental regulation, thus reducing energy consumption while maintaining corporate interests. Under the BTH coordinated development strategy, governments will break regional administrative restrictions; promote a revolution in energy production and consumption; promote green, circular, and low-carbon development; strengthen environmental protection and governance; and expand regional ecological space. The focus will be on the joint prevention and control of environmental pollution, the strengthening of environmental pollution control, the implementation of clean water action, the development of a circular economy and ecological protection, and plans to build some national parks and forest parks around the capital to actively tackle climate change. At present, Beijing, Tianjin, and Hebei have in-depth cooperation in various aspects such as improving coordination mechanisms, unified planning, unified legislation, unified standards, and joint law enforcement, and the results of collaborative environmental governance have been remarkable. For example, air quality in the three places has further improved, and the annual average concentration of PM2.5 has shown a downward trend. The PM2.5 average concentration in the BTH region has dropped by 46% compared with 2014, of which 85.9 $\mu g/m^3$ in Beijing dropped to 42 $\mu g/m^3$ in 2019, a decrease of 51%. Besides, in terms of ecological and environmental protection, China has supported ecological restoration in Zhangjiakou and Chengde and formulated an implementation plan for afforestation from 2015 to 2017 in the northwest ecological conservation area.

The second one is industrial transfer and upgrade. Under the BTH coordinated development strategy, the industries to be relocated and transferred from Beijing are mainly energy-intensive industries; tertiary industries like logistics bases and wholesale markets; public service sectors such as education, medical care, or training institutions; as well as some company headquarters. The transfer principle is a combination of the role of government and the role of market. In addition to Beijing and Tianjin, the central core function area of the "four areas" also includes Baoding and Langfang in Hebei province. These two cities will focus on the relocation and transfer of noncapital functions of Beijing and will take the lead in realizing interconnected development. It is worth noting that many regional cooperation projects have been completed, such as Beijing Automotive Industry Corporation Huanghua plant project, Beijing Hyundai Motor Cangzhou plant project, Sinopec Beijing Yanshan Branch Caofeidian ten-million-ton refining project, and Tianjin Binhai–Zhongguancun science park, etc. Industrial transfer and upgrade can have an impact on energy consumption through several channels. First, enterprises transferred from Beijing, such as high-end manufacturing and the internet industry, will promote industrial upgrading and have a strong technology spillover effect, thus reducing energy consumption in surrounding cities. Second, the transfer of industries such as higher education and science and technology R&D sectors, as well as preferential policies to attract talents in Xiong'an New Area, can improve the level of human capital and can reduce energy consumption. Third, the establishment of the Hebei Free Trade Zone in 2019, the first large-scale free trade zone in China covering the three provincial-level administrative regions of Beijing, Tianjin, and Hebei, will drive the transformation and upgrade of Hebei's manufacturing industry and will reduce energy consumption by participating in international competition. Fourth, taking advantage of the industrial advantages of Beijing and Tianjin, Hebei province accelerated the formation of many industrial clusters, such as Zhangjiakou renewable energy demonstration area, Beidaihe life health industry innovation demonstration zone, Shijiazhuang high-end biomedical industry base, and Baoding and Cangzhou auto equipment manufacturing industry. Until the end of 2019, Hebei province altogether accepted 9773 enterprises transferring from Beijing and Tianjin. Actually, quite a few studies have quantitatively measured the reduction of energy consumption and emissions caused by industrial transfer and upgrade. For example, Li et al. [27] found that rationalization and upgrade of manufacturing structures mitigate CO_2 emissions during the period of 2003–2014; Wang at al. [28] demonstrated that advancement of the industrial structure has increased carbon emission efficiency in China between 2003 and 2016; Zhu et al. [29] found that the increase in the proportion of secondary industries would increase energy-related smog pollutions in 73 key cities of China during 2013–2017; and Zhang et al. [30] found that the industrial structure can reduce energy-related haze pollution through the path of rationalization in China from 2006 to 2016, etc.

The third one is transportation. The BTH region will construct the Beijing–Tianjin–Hebei intercity transportation network based on rail transit and will build the world-class aviation hub as well as the port group of Tianjin–Hebei, thus improving regional integrated transportation. First, traffic construction and development in the short term will undoubtedly increase travel demand and energy consumption, but in the long term, the effects of transportation on economic growth are pronounced. Transportation integration can lead to talent flow and economic exchange, can promote industrial-technological progress and industrial upgrade, and then can significantly improve GDP, thus reducing energy consumption per unit GDP. Second, transportation integration can shorten logistics distance and can reduce energy consumption. For example, parcels before, especially air express parcels, generally arrived in Beijing first and then transferred to Tianjin and Hebei by truck. High-speed highway construction shortens logistics distance and reduces energy consumption. Third, the green transportation system in the BTH region, including promotion of new energy public transportation and encouragement of the purchase of new energy vehicles, will also reduce fossil energy consumption. Fourth, the integrated transportation planning of the BTH region is more efficient than decentralized planning and the energy efficiency of the transportation sector is also higher. Until 2020, the BTH region initially built a one-hour economic circle connected by high-speed rail and highways, which is about an hour from Beijing to surrounding cities. By 2030, an intercity railway network with four verticals, four horizontals, and one ring will be formed. The "four verticals" include Beijing–Baoding–Shijiazhuang–Xingtai–Handan, Beijing–New Airport–Hengshui, Cangzhou–Tianji–Chengde, and Qinhuangdao–Caofeidian–Binhai–Huanghua Port. The "four horizontals" include Beijing–Tianjin–Yujiabao, Beijing–Tongzhou–Tangshan–Caofeidian, Tianjin–Bazhou–Baoding, and Shijiazhuang–Cangzhou–Huanghua Port. One ring is the intercity transportation ring of Beijing.

Based on the above analyses, the BTH coordinated development strategy may have a big impact on industrial energy and pollution intensities, but the direction is uncertain and needs to be empirically tested. This paper is different from previous studies: first, this study investigates the comprehensive impact of the BTH coordinate development strategy on industrial energy and pollution intensities and the estimation results are robust based on a number of statistical tests; second, it applies the quantile DID method to evaluate the heterogeneous impact of the strategy on industrial energy intensity in the BTH region at different quantile levels of industrial energy intensity; and third, it uses the data of industrial energy pollutions to estimate the different impacts of the strategy on the environment among different energy pollution sources. Such heterogeneous effects can help to make policy implications based on different pollution sources under the context of the BTH coordinated development strategy.

4. Data, Variables, and DID Models

4.1. Data and Variables

This paper uses the DID method to investigate the impacts of the BTH coordinated development strategy on industrial energy intensity and related pollution emission intensities in this region by comparing the implementation of this strategy before and after between the treatment group and the control group [31]. Industrial energy intensity and the pollution emission intensity are defined as the ratios of industrial energy consumption and industrial pollution emissions to the value-added of the industry.

This paper uses panel data, containing 30 cities from 2007 to 2017, which includes all 13 cities in the BTH region and 17 cities covering all cities of Henan Province. Henan is chosen as the control group because it is adjacent to the BTH region and they are similar in industrial energy structure, the sources of industrial energy pollution, and their variation trends. Another reason is that, except for Henan, other provinces around the BTH region have only a few recent years of statistics on industrial energy consumption and do not cover the years before implementation of the BTH coordinated development strategy. Although some southern Chinese provinces have related data, the statistical indicators for industrial pollution emissions are inconsistent with the BTH region and Henan province and the structures, especially for the variation trends of industrial energy and pollution emissions, are obviously far from that of the BTH region. The data of industrial fossil energy consumption and industrial value-added come from each Provincial Statistic Yearbook [32–35]. The data of industrial energy-related pollution emissions and all control variables come from the China City Statistical Yearbooks [36].

According to previous studies, all the models control for the following variables, which are expected to have impacts on industrial energy intensity and related pollution intensities in the BTH region. Urbanization is defined as the proportion of urban population in the total population and is denoted as *UR* [37], the industrial structure is defined as the ratio of service industrial value-added over GDP and is denoted as *SER* [38], per capita GDP is the natural logarithm itself and is denoted as *PGDP* [39], foreign direct investment (FDI) is defined as the share of FDI stock over GDP and is denoted as *FDI/GDP* [40], and R&D is defined as the share of the R&D expenditure over GDP and is denoted as *RD* [41].

4.2. Empirical Models

In this paper, the DID method will be adopted and the implementation of the BTH coordinated development strategy will be put forward as a quasi-natural experiment. Therefore, the following DID models are constructed:

$$INDEN_{it} = \beta BTH_i \cdot Post_{it} + \gamma X_{it} + \alpha_i + \psi_t + \varepsilon_{it} \tag{1}$$

$$INDWA_{it} = \beta BTH_i \cdot Post_{it} + \gamma X_{it} + \alpha_i + \psi_t + \varepsilon_{it} \tag{2}$$

$$INDSD_{it} = \beta BTH_i \cdot Post_{it} + \gamma X_{it} + \alpha_i + \psi_t + \varepsilon_{it} \tag{3}$$

$$INDST_{it} = \beta BTH_i \cdot Post_{it} + \gamma X_{it} + \alpha_i + \psi_t + \varepsilon_{it} \tag{4}$$

Formulas (1)–(4) are the DID estimation models that take time and city as fixed effects into account. In all models, i stands for the city from 1 to 30 and t stands for the year from 2007 to 2017. The dependent variables $INDEN_{it}$, $INDWA_{it}$, $INDSD_{it}$, and $INDST_{it}$ respectively denote the industrial energy intensity, industrial wastewater emission intensity, industrial sulfur dioxide emission intensity, and industrial dust emission intensity. $Post_{it}$ is the dummy variable for the processing time effect of the BTH coordinated development strategy. For $Post_{it}$, the years after 2014 are set as 1 and the previous years are set as 0. BTH_i is the dummy variable for processing the treatment group, indicating whether the city is located in the BTH region. If it is a city in the BTH region, it is set as 1; otherwise, it is 0. $BTH_i \cdot Post_{it}$ is the interaction term between BTH_i and $Post_{it}$, which is the core variable concerning by the DID method. Besides, X_{it} is a series of control variables that may cause changes in industrial energy intensity and related pollution emission intensities, including urbanization rate, industrial structure, per capita GDP, FDI, and R&D. ψ_t is the year fixed effect, α_i is the city fixed effects, and ε_{it} is the random disturbance term.

The DID method focuses on the coefficient β of the variable $BTH_i \cdot Post_{it}$, for which the economic implication can be explained by the impact of the BTH coordinated development strategy on industrial energy and pollution intensities. If the coefficients of β are positive and statistically significant, the BTH coordinated development strategy tends to promote industrial energy intensity and related pollution emission intensities. On the contrary, if they are negative and statistically significant, the BTH coordinated development strategy tends to restrain industrial energy intensity and related pollution emission intensities.

5. Empirical Results and Analyses

5.1. BTH Coordinated Development Strategy and Industrial Energy Intensity

5.1.1. The DID Results

The city fixed effect, year fixed effect, and city time trend effect variables are gradually added in Model 1, and the standard errors are clustered at the city level. Compared with other regions, the industrial energy intensity may have some inherent change trend rather than the BTH coordinated development strategy effect. Ignoring the underlying trend change of dependent variables in the treatment group will produce bias of missing variables and make the estimated results unreliable. The interaction item between city and time trends is added in column (2) of Table 1. This interaction item control for the likely trends in the treatment group. Please note that, in this paper, all subsequent regressions are controlled for the city and year fixed effects and there may be no repeated reporting in the subsequent tables.

Table 1. Regression results of Model 1.

Dependent Variable	***Industrial Energy Intensity***	
	(1)	**(2)**
BTH·Post	−0.050 *** (0.000)	−0.043 *** (0.000)
PGDP	0.039 *** (0.000)	0.045 *** (0.000)
SERT	−0.003 *** (0.000)	−0.006 *** (0.000)
R&D	−76.671 *** (0.000)	−80.490 *** (0.000)
FDI/GDP	−0.424 (0.140)	−0.453 (0.118)
UR	0.087 *** (0.006)	0.096 *** (0.000)
R square	0.36	0.39
Year FE	Y	Y
City FE	Y	Y
BTH × Time trend		Y
Observations	330	330

Source: Authors' estimation. Notes: p-values are in brackets *** $p < 0.01$.

In Table 1, the BTH coordinated development strategy has a significantly negative impact on industrial energy intensity in this region. Listed in columns (1) and (2), the coefficients of the interaction terms are significantly negative at the 1% level. It shows that the BTH coordinated development strategy tends to restrain industrial energy intensity in this region. In addition, due to the extensive mode of rural ecological management, insufficient investment in environmental protection infrastructure, and technical limitations in Hebei Province, the rural ecological environment has been seriously polluted and destroyed, especially for the water source polluted by industrial wastewater and the soil polluted by solid waste. Therefore, attention should be paid to relevant environmental law enforcement efforts in the rural areas of this region.

As for the control variables, the regression results of FDI stock over GDP are negative but not statistically significant, indicating that FDI had no significant influence on industrial energy intensity. This is consistent with previous literature [42]. The regression results of the other control variables are consistent with theories and expectations.

5.1.2. The Quantile DID Results

In Table 2, the industrial energy intensities are divided into nine subpoints from low to high. Due to limited space, no other control variable results are shown. Based on the quantile DID method results, the BTH coordinated development strategy tends to restrain industrial energy intensity at the lower quantile levels from 0.1 to 0.4 but tends to promote industrial energy intensity at the higher quantile levels from 0.7 to 0.9. The results indicate that the BTH coordinated development strategy tends to restrain industrial energy intensity in the lower quantile level cities but tends to promote industrial energy intensity in the higher quantile level cities. The reason is perhaps that energy-intensive industries are mainly located in some cities of Hebei province and some of the industries transferred from Beijing are highly polluting, like steel companies, thus increasing the industrial energy intensity in some Hebei cities. However, cities with low energy intensity, such as Beijing and Tianjin, may optimize and upgrade their industries and may reduce their industrial energy intensity after the transfer.

Table 2. Regression results by quantile difference-in-difference (DID) method.

Quantile	0.1	0.2	0.3	0.4	0.5	0.6	0.7	0.8	0.9
BRI·Post	−0.095 ** (0.000)	−0.056 ** (0.000)	−0.068 ** (0.000)	−0.084 ** (0.007)	−0.154 (0.421)	0.216 (0.375)	0.066 ** (0.003)	0.062 * (0.032)	0.163 * (0.011)
Control variables	Y	Y	Y	Y	Y	Y	Y	Y	Y
R square	0.40	0.34	0.36	0.41	0.35	0.38	0.43	0.51	0.53

Source: Authors' estimation. Notes: p-values are in brackets * $p < 0.05$, ** $p < 0.01$.

Specifically, at the city level, Tangshan, Handan, Zhangjiakou, and Chengde of Hebei province rank high in industrial energy intensity and their industrial energy intensities rose in 2015. In the early stage, most of the transferring enterprises from Beijing to Hebei were high-energy-intensive, thus reducing air pollution in Beijing. Moreover, most of these enterprises were relocated to high-intensity industrial cities or nearby cities in Hebei province like the above ones. Therefore, how to improve the technical level of these enterprises transferred to Hebei and to eliminate backward production capacity among them must be of the utmost importance in the future. More environmental policies should be formulated to reduce energy consumption while maintaining corporate interests. For example, China can improve carbon emission trading markets in Beijing and Tianjin, can establish a new carbon emission trading market in Hebei province, and finally can realize carbon emission market integration in the BTH region.

5.2. Data Validity Analysis and Robustness Check

First, we tested whether the mean values of the dependent variable industrial energy intensity between the treatment group and the control group were equal after BTH coordinated development strategy. The null hypothesis of the mean test is that there is no significant difference. The results show that the *p*-value equals zero from the base period, which indicates that there is a significant difference in industrial energy intensity between the treatment group and the control group.

Second, this paper conducted a further parallel trend check before the BTH coordinated development strategy. The DID model does not require the mean values to be the same but hypothesizes that the trends between the control group and the treatment group must be the same before policy implementation. To support this assumption, we set up a year dummy variable representing different years, and the cross term *BTH·Year* represents the possibly different variation trend of energy intensity in the treatment group compared with the control group. Figure 2 indicates that the trends of these two groups do not have significant differences in all three years before implementation of the BTH coordinated development strategy. The coefficients of the dummy variables are not statistically significant in the three years before implementation of the BTH coordinated development strategy. Furthermore, from the year of implementation of the BTH coordinated development strategy, the impact of the policy on industrial energy intensity was significantly negative and gradually decreasing with the passage of time. In the three years after implementation of the BTH coordinated development strategy, the industrial energy intensity decreased by 4.25% in the year of implementation and decreased by 6.37% three years after implementation.

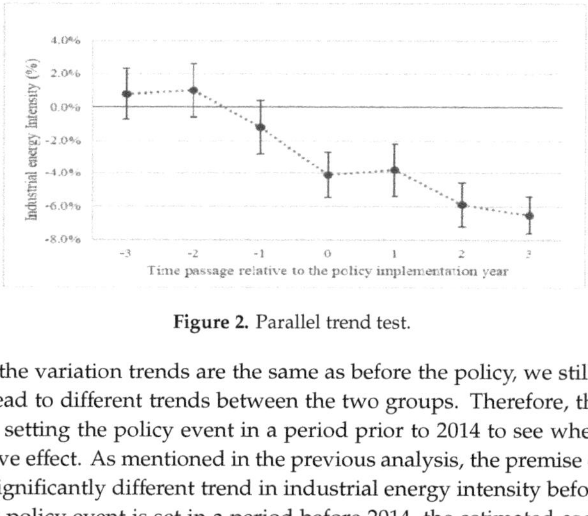

Figure 2. Parallel trend test.

Third, even if the variation trends are the same as before the policy, we still need to test other policies that may lead to different trends between the two groups. Therefore, this paper performs the placebo test by setting the policy event in a period prior to 2014 to see whether there is still a significantly negative effect. As mentioned in the previous analysis, the premise of the DID method is that there is no significantly different trend in industrial energy intensity before implementation of the policy. If the policy event is set in a period before 2014, the estimated coefficient of the core variable may be not significant. If the results are contrary to our expectations, such as significantly negative, it means that there are some potential unobserved policy factors other than the BTH coordinated development strategy that could affect industrial energy intensity in this region. To ensure the robustness of empirical results, the policy impacting years are set as 2008, 2009, 2010, 2011, 2012, and 2013 and the years after them. The corresponding estimated results are reported in Table 3. It shows that the estimated coefficients of variable *BTH·Post* are all insignificant, which proves that the DID results are robust again.

Table 3. Placebo test for robustness check.

Policy Year	2008	2009	2010	2011	2012	2013
BRI·Post	−0.033 (0.232)	0.062 (0.344)	0.025 (0.345)	−0.042 (0.563)	−0.047 (0.423)	−0.083 (0.272)
Control variables	Y	Y	Y	Y	Y	Y
R square	0.25	0.34	0.33	0.32	0.26	0.29

Source: Authors' estimation. Notes: *p*-values are in brackets.

5.3. The BTH Coordinated Development Strategy and Industrial Energy Pollution Intensities

The effects of the BTH coordinated development strategy on different sources of energy-related pollutions may vary. For example, Beijing started replacing coal with electricity for heating in 2013, followed by Tianjin and some cities of Hebei province after implementation of the coordinated development strategy, which greatly reduced industrial sulfur dioxide and dust emissions in the BTH region [43]. In this part, this paper will examine the impacts of the BTH coordinated development strategy on different sources of industrial energy-related pollution including the three wastes, which are industrial wastewater, sulfur dioxide, and dust emissions, based on the DID method.

First, in Table 4, the variable *BTH·POST* is significantly negative at the 10 percent level in Model 2, significantly negative at the 5 percent level in Model 3, and positive but insignificant in Model 4. The results indicate that the BTH coordinated development strategy tends to decrease industrial wastewater and sulfur dioxide emission intensities but may have no significant impact on dust emission intensity though the coefficient is positive. The main reason may be that, while environmental regulation and industrial upgrades have reduced the three waste emissions, some transportation infrastructure construction and operation, such as high-speed rails and highways for regional transportation development, have caused much dust pollution. Specifically, at the city level, Zhangjiakou, Baoding, Chengde, Tangshan, and other cities which are constructing expressways and high-speed railways on a large scale to build the one-hour economic circle in the BTH region may encounter more serious dust pollution issues.

Table 4. Regression results of Models 2–4.

	Model 2	Model 3	Model 4
Dependent Variable	*Wastewater*	*Sulfur Dioxide*	*Dust Emissions*
BTH·Post	−0.003 * (0.077)	−0.034 ** (0.019)	0.323 (0.282)
PGDP	0.035 *** (0.001)	0.227 *** (0.000)	0.265 ** (0.034)
SERT	−0.001 * (0.056)	−0.003 ** (0.012)	−0.014 ** (0.027)
R&D	−1.239 (0.149)	−1.781 *** (0.008)	−1.842 (0.162)
FDI/GDP	−0.113 ** (0.020)	−1.052 (0.601)	−3.852 (0.435)
UR	0.033 *** (0.001)	0.210 ** (0.048)	0.415 ** (0.025)
R square	0.35	0.33	0.22
City FE	Y	Y	Y
Year FE	Y	Y	Y
BTH * Time trend	Y	Y	Y
Observations	330	330	330

Source: Authors' estimation. Notes: p-values are in brackets * $p < 0.1$, ** $p < 0.05$, and *** $p < 0.01$.

As to the countermeasures for the industrial three wastes, the most direct way to mitigate these industrial pollutions is to use clean technologies or facilities. There are some available methods for the BTH region to mitigate industrial pollutions. For example, end-of-pipe treatment technologies for the water- and energy-intensive coal-fired power industry with high emissions in the BTH region could reduce SO_2, NO_X, and dust emissions by 89%, 90%, and 88%, respectively, while consuming an average of 2% less energy and 8% more water as tradeoffs [44]. As the most polluted region in China caused by the coal-based heating system, the integration of large-scale heat pumps can potentially result in at least 9.5% energy savings and 9.28% reduced CO_2 emissions compared to the baseline of 2015 for the whole BTH region by 2030 while ensuring economic feasibility [45]. Besides, based on a slacked-based data envelopment analysis (DEA) model by cluster benchmarking of 861 wastewater treatment plants (WWTPs) in China, the technology gap ratio confirmed that large WWTPs operated more efficiently than small ones [46].

It is worth mentioning that this paper does not make an empirical analysis of the industrial solid waste pollution, which is also a crucial environmental issue in China, because some city data are not available. From the latest available data, the solid waste utilization rates of Beijing, Tianjin, and Hebei in 2019 were 80%, 98%, and 75%, respectively, which are all higher than the average of 65% in China. However, there is still potential for further improvement, particularly in Beijing and Hebei. In the process of coordinated development, corresponding environmental regulations on industrial solid waste should be strengthened to improve the utilization rate.

For the control variables, the impact of FDI stock over GDP on industrial wastewater emission intensity is significantly negative but the impacts on sulfur dioxide and dust emission intensities are insignificant. In the BTH region, foreign enterprises invest very little in energy-intensive industries but more in labor-intensive light industries which may produce mainly wastewater. As a result, the technology spillover effect can be generated in these industries and relevant enterprises will reduce industrial wastewater discharge. The impact of R&D expenditure over GDP on industrial sulfur dioxide intensity is significantly negative but the impacts on industrial wastewater discharge and dust emission intensities are insignificant. The results suggest that research spending should be focused more on reducing industrial wastewater and dust emissions for further regional sustainable development. The regression results of other control variables are consistent with the theories and expectations.

6. Conclusions, Policy Implications, and Discussions

This study investigates the different impacts of the coordinated development of the BTH region on industrial energy and pollution intensities based on the DID method and theh quantile DID method. The panel data cover industrial energy consumption and three wastes, which are industrial wastewater, sulfur dioxide, and dust emissions, from all 13 cities in the BTH region and 17 cities in Henan Province for the period 2007–2017.

The study finds that, first, based on the DID method results, the dummy variable is significantly negative at the one percent level, indicating that China's BTH coordinated development strategy on average tends to restrain regional industrial energy intensity in the BTH region. Second, based on the quantile DID method results, the BTH coordinated development strategy tends to restrain industrial energy intensity in lower quantile level (0.1–0.4) cities; still, it tends to promote industrial energy intensity in higher quantile level (0.7–0.9) cities. Third, the impacts of BTH coordinated development strategy on the environment vary among industrial wastewater, sulfur dioxide, and dust emissions. The BTH coordinated development strategy tends to decrease industrial wastewater and sulfur dioxide emission intensities because the coefficients are statistically significant at the 10 percent level and the 5 percent level but have no significant impact on dust emission intensity though the coefficient is positive.

The findings of this study can generate some policy implications. First, China should pay more attention to the green transfer and clean energy use of industries from Beijing instead of transferring outdated production capacities that may increase industrial fossil energy consumption and related pollution emissions. More attention should be given to cities that have a high energy intensity. Second, it is necessary to control dust emission during the integrated development of transportation by enforcing strict standards and by establishing an ecological traffic system, especially in the construction of expressways and high-speed railways. Third, on the premise of respecting the rules of market economy and the rules of fairness in world trade, relevant subsidy policies can be formulated with the goal of promoting environmental protection talents and of upgrading energy conservation and emission reduction technologies in Hebei, so as to change the serious unsustainability and environmental pollution problems. Fourth, standards for wastewater and sulfur dioxide emissions from some industries such as oil refining, steel making, and metallurgy can be tightened more, and governments can continue making efforts to replace coal with electricity or gas for heating and can extend the action to a wider region to reduce industrial sulfur dioxide and dust emissions, like some small towns and rural areas in Hebei province. Fifth, except for stimulating upgrade of regional industrial structures to service industry domination, the governments of some Hebei cities, like Tangshan, Handan, and Zhangjiakou, should provide more funds for energy conservation and emission reduction-related research. Sixth, the government should strengthen Beijing's role in driving industrial and technological upgrade in neighboring cities. For example, the new Free Trade Zone in Beijing established in 2020 will focus on service trade and will develop high-end industries such as the digital economy; Beijing ranked first in scientific research level on the Nature Index for five consecutive years. All of these advantages can be put to good use in the coordinated development of the BTH region. Seventh, relying on some major projects like the Xiong'an New District, Hebei free trade area, and the Beijing–Zhangjiakou 2022 Olympic Winter Games, Hebei cities should accurately position their regional comparative advantages, such as existing resources, talents, and industrial bases, to undertake related high-end industrial relocation from Beijing; to absorb the radiation effect to accelerate the upgrade of local industry and technology, thus forming the ability to retain talent and capital; and then further to own the ability to attract talent and capital, fundamentally changing the Beijing siphon phenomenon. However, it is worth noting that the central government's decision to build the Xiong'an New Area in Hebei has downgraded the national importance of Tianjin's Binhai New Area, which is the second state-level pilot zone in China after the Pudong district of Shanghai [11]. Relevant policies should be formulated to balance the relationship between the two state-level new areas so that they can complement each other. Eighth, we should specifically develop the four regional central cities, including Shijiazhuang, Tangshan, Baoding, and Handan, which will be models of environmental protection, industrial upgrade, and green transportation for neighboring node cities. Ninth, strengthening the integration of customs clearance in the BTH region can shorten customs clearance time and can reduce transportation cost of goods imported from Beijing via air transportation and those imported from Tianjin via sea transportation. At last, as the current environmental policy loses the interest of enterprises, the participation enthusiasm of industrial transfer and green upgrade is not high. The coordinated environmental regulation should be strengthened to reduce energy consumption while maintaining corporate interests, like establishing a carbon emission trading market in Hebei Province.

Although the above research gives some findings and policy implications on this topic, it should be acknowledged that, due to the limitation of the latest data, for which the time series is till 2017, whether the effects of the BTH coordinated development strategy on industrial energy and related pollutant emission intensities will continue, to this day, is unknown. Except for the DID method, other empirical techniques and data if available can be used to test some of the major aspects of the BTH coordinated development strategy, such as the impact of investment in the Xiong'an new area on energy consumption and related pollutions. If more relevant data can be obtained to form a larger control group, the DID results based on propensity score matching (PSM) can be also used to test the robustness.

Author Contributions: Conceptualization, C.H. and Y.W.; methodology, X.S.; software, Y.W. and X.S.; validation, B.H. and X.S.; formal analysis, Y.W. and C.H.; data curation, C.H.; writing—original draft preparation, C.H. and Y.W.; writing—review and editing, C.H. and X.S.; supervision, B.H. and X.S.; project administration, B.H.; funding acquisition, B.H. All authors have read and agreed to the published version of the manuscript.

Funding: We are grateful for the financial support provided by the National Social Science Foundation Major Project of China (No. 19ZDA100), by the National Development and Reform Commission of China (No. 221100004; 240200004), by International Poverty Reduction Center in China (No. 240200003), by the Natural Science Foundation of China (No. 71828401), by the Social Science Fund of Beijing Education Commission (No. SM201910011011; PXM2019_014213_000007), and by the Beijing Social Science Foundation (No. 17JRC012).

Conflicts of Interest: The authors declare no conflict of interest.

References

1. Freeman, H.; Harten, T.; Springer, J.; Randall, P.M.; Curran, M.A.; Stone, K. Industrial Pollution Prevention! A Critical Review. *J. Air Waste Manag. Assoc.* **1992**, *42*, 618–656. [CrossRef]
2. Shi, H.; Zhang, L. China's environmental governance of rapid industrialisation. *Environ. Politics* **2006**, *15*, 271–292. [CrossRef]

3. Mol, A.P.; Carter, N.T. China's environmental governance in transition. *Environ. Politics* **2006**, *15*, 149–170. [CrossRef]
4. Deng, H.; Farah, P.D. China's energy policies and strategies for climate change and energy security. *J. World Energy Law Bus.* **2020**, *13*, 141–156. [CrossRef]
5. Chow, G.C. *China's Environmental Policy: A Critical Survey*; Working Paper 206; Princeton University, Department of Economics, Center for Economic Policy Studies: Princeton, NJ, USA, 2010; Available online: http://www.princeton.edu/~{}ceps/workingpapers/206chow.pdf (accessed on 26 August 2020).
6. Wu, F. *Energy and Climate Policies in China and India: A Two-Level Comparative Study*; University Press: Cambridge, UK, 2018.
7. Lieberthal, K. China's governing system and its impact on environmental policy implementation. *China Environ. Ser.* **1997**, *1*, 3–8. Available online: https://www.cleanairnet.org/asia/1412/articles-36867_ces1a.pdf (accessed on 26 August 2020).
8. Pan, X.; Pan, X.; Li, C.; Song, J.; Zhang, J. Effects of China's environmental policy on carbon emission efficiency. *Int. J. Clim. Chang. Strat. Manag.* **2019**, *11*, 326–340. [CrossRef]
9. Reins, L.; Van Eynde, S.; Van Hende, K.; Gao, A.M.Z. China's Climate Strategy and Evolving Energy Mix: Policies, Strategies and Challenges. *Carbon Clim. Law Rev.* **2015**, *9*, 256–269.
10. Grumbine, R.E.; Xu, J. Recalibrating China's environmental policy: The next 10 years. *Biol. Conserv.* **2013**, *166*, 287–292. [CrossRef]
11. Wang, W.; Wang, Y.P.; Kintrea, K. The (Re)Making of Polycentricity in China's Planning Discourse: The Case of Tianjin. *Int. J. Urban Reg. Res.* **2020**, *44*, 857–875. [CrossRef]
12. Wang, S.; Ma, H.; Zhao, Y. Exploring the relationship between urbanization and the eco-environment—A case study of Beijing–Tianjin–Hebei region. *Ecol. Indic.* **2014**, *45*, 171–183. [CrossRef]
13. Yue, W.; Jingyou, W.; Mei, Z.; Shi, L. Spatial Correlation Analysis of Energy Consumption and Air Pollution in Beijing-Tianjin-Hebei Region. *Energy Procedia* **2019**, *158*, 4280–4285. [CrossRef]
14. Li, M.; Mao, C. Spatial Effect of Industrial Energy Consumption Structure and Transportation on Haze Pollution in Beijing-Tianjin-Hebei Region. *Int. J. Environ. Res. Public Health* **2020**, *17*, 5610. [CrossRef] [PubMed]
15. Ding, Y.; Zhang, M.; Chen, S.; Wang, W.; Nie, R. The environmental Kuznets curve for PM 2.5 pollution in Beijing-Tianjin-Hebei region of China: A spatial panel data approach. *J. Clean. Prod.* **2019**, *220*, 984–994. [CrossRef]
16. Li, N.; Zhang, X.; Shi, M.; Hewings, G.J.D. Does China's air pollution abatement policy matter? An assessment of the Beijing-Tianjin-Hebei region based on a multi-regional CGE model. *Energy Policy* **2019**, *127*, 213–227. [CrossRef]
17. Wang, J.; Yang, F.; Zhang, X.; Zhou, Q. Barriers and drivers for enterprise energy efficiency: An exploratory study for industrial transfer in the Beijing-Tianjin-Hebei region. *J. Clean. Prod.* **2018**, *200*, 866–879. [CrossRef]
18. Zhang, G.; Zhang, P.; Zhang, Z.G.; Li, J. Impact of environmental regulations on industrial structure upgrading: An empirical study on Beijing-Tianjin-Hebei region in China. *J. Clean. Prod.* **2019**, *238*, 117848. [CrossRef]
19. Li, J.; Xiang, Y.; Jia, H.; Chen, L. Analysis of Total Factor Energy Efficiency and Its Influencing Factors on Key Energy-Intensive Industries in the Beijing-Tianjin-Hebei Region. *Sustainability* **2018**, *10*, 111. [CrossRef]
20. Li, G.; Sun, J.; Wang, Z. Exploring the energy consumption rebound effect of industrial enterprises in the Beijing–Tianjin–Hebei region. *Energy Effic.* **2019**, *12*, 1007–1026. [CrossRef]
21. Wang, Z.; Yang, L. Delinking indicators on regional industry development and carbon emissions: Beijing-Tianjin-Hebei economic band case. *Ecol. Indic.* **2015**, *48*, 41–48. [CrossRef]
22. Wang, C.; Zhan, J.; Bai, Y.; Chu, X.; Zhang, F. Measuring carbon emission performance of industrial sectors in the Beijing–Tianjin–Hebei region, China: A stochastic frontier approach. *Sci. Total Environ.* **2019**, *685*, 786–794. [CrossRef]
23. Guo, M.; Meng, J. Exploring the driving factors of carbon dioxide emission from transport sector in Beijing-Tianjin-Hebei region. *J. Clean. Prod.* **2019**, *226*, 692–705. [CrossRef]
24. Lang, J.; Cheng, S.; Wei, W.; Zhou, Y.; Wei, X.; Chen, D. A study on the trends of vehicular emissions in the Beijing-Tianjin-Hebei (BTH) region, China. *Atmos. Environ.* **2012**, *62*, 605–614. [CrossRef]
25. Zhu, X.; Li, R. An Analysis of Decoupling and Influencing Factors of Carbon Emissions from the Transportation Sector in the Beijing-Tianjin-Hebei Area, China. *Sustainability* **2017**, *9*, 722. [CrossRef]

26. Simpson, R.D.; Bradford, R.L. Taxing variable cost: Environmental regulation as industrial policy. *J. Environ. Econ. Manag.* **1996**, *30*, 282–300. [CrossRef]
27. Li, Z.; Shao, S.; Shi, X.; Sun, Y.; Zhang, X. Structural transformation of manufacturing, natural resource dependence, and carbon emissions reduction: Evidence of a threshold effect from China. *J. Clean. Prod.* **2019**, *206*, 920–927. [CrossRef]
28. Wang, K.; Wu, M.; Sun, Y.; Shi, X.; Sun, A.; Zhang, P. Resource abundance, industrial structure, and regional carbon emissions efficiency in China. *Resour. Policy* **2019**, *60*, 203–214. [CrossRef]
29. Zhu, L.; Hao, Y.; Lu, Z.N.; Wu, H.; Ran, Q. Do economic activities cause air pollution? Evidence from China's major cities. *Sustain. Cities Soc.* **2019**, *49*, 101593. [CrossRef]
30. Zhang, M.; Sun, X.; Wang, W. Study on the effect of environmental regulations and industrial structure on haze pollution in China from the dual perspective of independence and linkage. *J. Clean. Prod.* **2020**, *256*, 120748. [CrossRef]
31. Zhang, H.; Wu, K.; Qiu, Y.; Chan, G.; Wang, S.; Zhou, D.; Ren, X. Solar photovoltaic interventions have reduced rural poverty in China. *Nat. Commun.* **2020**, *11*, 1–10. [CrossRef]
32. Beijing Bureau of Statistics. *Beijing Statistical Yearbook*; China Statistics Press: Beijing, China, 2008.
33. Tianjin Bureau of Statistics. *Tianjin Statistical Yearbook*; China Statistics Press: Beijing, China, 2008.
34. Hebei Bureau of Statistics. *Hebei Statistical Yearbook*; China Statistics Press: Beijing, China, 2008.
35. Henan Bureau of Statistics. *Henan Statistical Yearbook*; China Statistics Press: Beijing, China, 2008.
36. National Bureau of Statistics. *China City Statistical Yearbook*; China Statistics Press: Beijing, China, 2008.
37. Wang, Q.; Lin, J.; Zhou, K.; Fan, J.; Kwan, M.-P. Does urbanization lead to less residential energy consumption? A comparative study of 136 countries. *Energy* **2020**, *202*, 117765. [CrossRef]
38. Shi, T.; Zhang, W.; Zhou, Q.; Wang, K. Industrial structure, urban governance and haze pollution: Spatiotemporal evidence from China. *Sci. Total Environ.* **2020**, *742*, 139228. [CrossRef]
39. Wang, Y.; Wang, C.; Wang, H. Research on the quantitative relationship between the generation of industrial solid waste and Per capita GDP of Henan Province. *Energy Procedia* **2011**, *5*, 593–597. [CrossRef]
40. Mielnik, O.; Goldemberg, J. Foreign direct investment and decoupling between energy and gross domestic product in developing countries. *Energy Policy* **2002**, *30*, 87–89. [CrossRef]
41. Petrović, P.; Lobanov, M.M. The impact of R&D expenditures on CO_2 emissions: Evidence from sixteen OECD countries. *J. Clean. Prod.* **2020**, *248*, 119187. [CrossRef]
42. Liao, X.C.; Shi, X.P. Public appeal, environmental regulation and green investment: Evidence from China. *Energy Policy* **2018**, *119*, 554–562. [CrossRef]
43. Cheong, T.S.; Li, V.J.; Shi, X. Regional disparity and convergence of electricity consumption in China: A distribution dynamics approach. *China Econ. Rev.* **2019**, *58*, 101154. [CrossRef]
44. Wang, C.; Li, Y.; Liu, Y. Investigation of water-energy-emission nexus of air pollution control of the coal-fired power industry: A case study of Beijing-Tianjin-Hebei region, China. *Energy Policy* **2018**, *115*, 291–301. [CrossRef]
45. Yuan, M.; Zinck Thellufsen, J.; Lund, H.; Liang, Y. The first feasible step towards clean heating transition in urban agglomeration: A case study of Beijing-Tianjin-Hebei region. *Energy Convers. Manag.* **2020**, *223*, 113282. [CrossRef]
46. Jiang, H.; Hua, M.; Zhang, J.; Cheng, P.; Ye, Z.; Huang, M.; Jin, Q. Sustainability efficiency assessment of wastewater treatment plants in China: A data envelopment analysis based on cluster benchmarking. *J. Clean. Prod.* **2020**, *244*, 118729. [CrossRef]

MDPI
St. Alban-Anlage 66
4052 Basel
Switzerland
Tel. +41 61 683 77 34
Fax +41 61 302 89 18
www.mdpi.com

Sustainability Editorial Office
E-mail: sustainability@mdpi.com
www.mdpi.com/journal/sustainability

www.ingramcontent.com/pod-product-compliance
Lightning Source LLC
Chambersburg PA
CBHW061604220926
51850CB00005B/418
* 9 7 8 3 0 3 6 5 5 7 1 3 7 *